AF838597

RADIOISOTOPES – APPLICATIONS IN PHYSICAL SCIENCES

Edited by **Nirmal Singh**

Radioisotopes – Applications in Physical Sciences
Edited by Nirmal Singh

Published by InTech
Janeza Trdine 9, 51000 Rijeka, Croatia

Copyright © 2011 InTech
All chapters are distributed under the Creative Commons Attribution 3.0 license, which allows users to download, copy and build upon published articles even for commercial purposes, as long as the author and publisher are properly credited, which ensures maximum dissemination and a wider impact of our publications. After this work has been published by InTech, authors have the right to republish it, in whole or part, in any publication of which they are the author, and to make other personal use of the work. Any republication, referencing or personal use of the work must explicitly identify the original source.

As for readers, this license allows users to download, copy and build upon published chapters even for commercial purposes, as long as the author and publisher are properly credited, which ensures maximum dissemination and a wider impact of our publications.

Notice
Statements and opinions expressed in the chapters are these of the individual contributors and not necessarily those of the editors or publisher. No responsibility is accepted for the accuracy of information contained in the published chapters. The publisher assumes no responsibility for any damage or injury to persons or property arising out of the use of any materials, instructions, methods or ideas contained in the book.

Publishing Process Manager Masa Vidovic

Technical Editor Teodora Smiljanic

Cover Designer InTech Design Team

Additional hard copies can be obtained from orders@intechopen.com

Radioisotopes – Applications in Physical Sciences, Edited by Nirmal Singh
p. cm.
978-953-307-510-5

We are IntechOpen, the world's leading publisher of Open Access books

Built by scientists, for scientists

4,000+
Open access books available

116,000+
International authors and editors

120M+
Downloads

151
Countries delivered to

Our authors are among the
Top 1%
most cited scientists

12.2%
Contributors from top 500 universities

WEB OF SCIENCE™

Selection of our books indexed in the Book Citation Index in Web of Science™ Core Collection (BKCI)

Interested in publishing with us?
Contact book.department@intechopen.com

Numbers displayed above are based on latest data collected.
For more information visit www.intechopen.com

Meet the editor

Dr. Nirmal Singh received his Bsc (Honors School), MSc (Honors School) and PhD from the Panjab University, Chandigarh. He was Assistant Professor, Associate Professor, Professor and finally Emeritus Professor at Panjab University, Chandigarh. He was Chairman of the Physics Department from 2004-2007 and also Coordinator of the Medical Physics/Nuclear Medicine Msc Course at the University (2006-09). He received an Emeritus Fellowship by the University Grant Commission, New Delhi and the Emeritus Scientist award by the Council of Scientific and Industrial Research, Government of India. His major fields of research are experimental nuclear physics and radiation physics and has supervised about two dozen PhD students. He has visited, delivered lectures and worked in many accelerator laboratories in India, USA and Europe. Dr. Singh has been reviewing research papers for many international journals of repute. He is a life member of the Indian Physics Association, Indian Association of Physics Teachers, Panjab Academy of Sciences And Radiation Society of India.

Contents

Preface

Isotopes are atoms of the same element having different atomic mass because they have a different number of neutrons in their nuclei with the identical number of protons. Some isotopes are unstable due to a specific combination of neutrons and protons which occurs naturally or can be artificially produced by bombarding the target atoms with neutrons in a nuclear reactor or with charged particles in an accelerator. These unstable nuclei attain their stability by emitting nuclear radiations such as neutrons, alpha particles, beta (positron/electron) or gamma rays and are termed as radioisotopes, while the term radioactivity is used to describe these spontaneous, energy emitting, atomic transitions that involve changes in the state of the nucleus of an atom. Radioactivity was discovered by A. H. Becquerel in 1896 when he was investigating the fluorescence of a double sulphate of uranium and potassium, using a photographic plate. Marie Curie coined the word radioactivitè, investigated this property in a number of minerals containing uranium, which she found to be more active and subsequently polonium was discovered. Readers are referred to read more in NCRP Report No.58, A Handbook of Radioactivity Measurements and Procedures.

Radiation from radioisotopes plays a very important role in all fields, useful in our life such as soil and earth science, radioactive dating, environment, archeology, agriculture, biochemical analysis, radiotherapy and cancer treatment, medical diagnosis, nuclear medicine, biological sciences, sterilization of medical products, non-destructive elemental analysis and testing of materials, oceanography, pharmaceutical, radioisotope power systems for space applications and many more.

The basic aim of this book is to present a very comprehensive review. Applications of radioisotopes have been reviewed and compiled in chapters contributed by world known authors in their respective specialized fields of research. The book contains three sections; Radioisotopes and some physical aspects, Radioisotopes in environment and Radioisotopes in power system for space applications.

In Tech Open Access Publisher has put in a lot of effort to complete the task of publishing this book for the immense benefit of the scientific and technical community, to discharge their social obligation and serve mankind all over the world. With this background I acknowledge the contributions put in these volumes by my expert colleagues and I endorse my special thanks to In Tech Open Access Publisher to assign me the job as an editor and for giving me an opportunity to review the all chapters published in the book.

Nirmal Singh
Emeritus Professor,
Department of Physics,
Panjab University, Chandigarh,
India

Part 1

Radioisotopes and Some Physical Aspects

1

Natural Occurring Radionuclide Materials

Raad Obid Hussain and Hayder Hamza Hussain
College of Science/Kufa University
Iraq

1. Introduction

The stellar material, from which the earth was formed, about 4.5 billion years ago, contained many unstable nuclides (Scholten and Timmermans, 1996). Some of the original primordial nuclides, whose half-lives are about as long as the earth's age, are still present. Radiation comes from outer space (cosmic), the ground terrestrial, and even from within over bodies. It present in the air we breathe, the food we eat, the water we drink and in the construction materials used to build our houses. So, radiation is all around us, it is naturally in our environment and it has been since the birth of our planet (Maher and Raed, 2007). Radioactivity of soil environment is one of the major sources of exposure to human (Abusini, 2007).

The ^{235}U, ^{232}Th series and natural ^{40}K are the main source of natural radioactivity in soil (Yasir et al., 2007; Vosniakos et al., 2002). Since these natural occurring radio nuclides materials, (NORMs) such as ^{238}U, ^{232}Th, ^{235}U, and ^{40}K have very long half- lives (up to 10^{10} years), their presence of soils and rocks can simply be considered as permanent. The geological and geographical conditions are the major factors effects, the natural environmental radioactivity and the associated external exposure due to gamma radiation. Thus these radiation levels appear at different levels in the soil of each region in the world (UNSCEAR, 2000).

Issue in terms of radiological protection exposure to natural source of radiation becomes an important. In 1992 the national radiological protection Board (NRPB), estimated that radon accounts for approximately 50% of annual dose of radiation from all sources in the most of the world (Ibrahim, 1999).

An average person receives a radiation dose of about 300 millirem per year from natural sources compared to a dose of about 50 millirem from produced material source of radioactive materials such as medical x-ray (UNSCEAR, 1988). Exposure of public to radiation from any sources is unlikely. The European committee has issued a draft proposal for revision of the basic safety standards for the protection of workers and the general against the dangers of ionizing radiation (Marcelo and Pedro, 2007).

The United Nations Scientific Committee on the Effects of Atomic Radiation established that the world mean dose from natural radiation sources of normal area is estimated to be 2.4 $mSv.y^{-1}$ while for all man- made sources including exposure, is about 0.8 $mSv.y^{-1}$ (UNSCEAR, 1993; Valter at el., 2008). Thus 75% of the radiation dose received by humanity is come from natural radiation source. It is clear that the assessment of gamma radiation dose from natural source is of particular importance as natural radiation is the largest contributor to the external dose of the world population (UNSCEAR, 1988).

Since predominate part of the environmental radiation is found in the upper soil layer, this knowledge ensures radiological control. The ^{238}U, ^{232}Th and ^{40}K have a non-negligible radioactivity (WHO, 1993; José at el., 2005). The high radioactivity of ^{226}Ra and ^{228}Ra and their pressure in soil require particular attention. It is known that even a small amount of a radiation substance may produce a damaging biological effects and that ingested and inhaled radiation can be a serious health risk (Rowland, 1993).

The radiological impact of natural radio nuclides is due to the gamma ray exposure of the body and irradiation of lung tissue from inhalation of radon and its daughters. In general exposure to ionizing radiation often comes from medical diagnosis and therapy application in just food and air and environmental sources. The radiation from the last one cannot be switched of thus the environmental radioactivity surveillance becomes, therefore a necessity. Generally, in Iraq and especially in the middle region area, there is a lack of the scientific information on radioactivity contents of naturally occurring radioactive materials in soil especially in terms of environmental radiological studies.

Based on these facts, one can certify that the knowledge of natural occurring radionuclide materials (MORMs), such as ^{238}U, ^{232}Th and ^{40}K, is an important pre-requisite for evaluation of the rate of exposure absorbed dose by the population in order to estimate their radiological impacts and to establish a data base which will be used as reference to radiation observer in the studied area (NCRP, 1987).

2. Background

According to the source of radiation, radioactivity in the environment may be classified into two general categories; artificial and natural. Natural radioactivity comes from naturally-occurring, Uranium, Thorium and Actinium radioactive series as well as from radioisotopes like Rubidium-87, Indium-115, Lanthanum-138, Neobynium-144, Samarian-147, Luteium-176, Hafinium-174, Vanadium-150, Gadolinium-152, Platinum-190 and 192, Rhenium-187 and Potassium-40. Except for K-40 these non-series radioisotopes occur scarcely. In contrast, K-40 is ubiquitous (Vosnikos et al., 2003). In the other hand artificial activity arises mainly from discarded sources, radioactive wastes, and radioactive fallout in the nature.

There are four distinct natural series: Uranium, Actinium, thorium, and Neptunium as listed in (Table 1). Only uranium, actinium, and thorium series are found in natural. Since the isotope ^{237}Np has a half-life much shorter than the age of the earth (about 5 billion of years), virtually all neptunium decayed within the first 50 millions of years after the earth formed.

Only uranium, actinium, and thorium series are found in natural. Since the isotope ^{237}Np has a half-life much shorter than the age of the earth (about 5 billion of years), virtually all neptunium decayed within the first 50 millions of years after the earth formed.

Series	First Isotope	Half-life(years)	Last Isotope
Uranium	^{238}U	4.5×10^{9}	^{206}Pb
Actinium	^{235}U	7.10×10^{8}	^{207}Pb
Thorium	^{232}Th	1.39×10^{10}	^{208}Pb
Neptunium	^{237}Np	2.14×10^{6}	^{209}Bi

Table 1. Natural series of uranium, actinium, thorium and neptunium

2.1 The NORM decay series

Uranium and thorium are not stable; they decay mainly by alpha-particle emission to nuclides that themselves are radioactive. Natural uranium is composed of three long lived isotopes, ^{238}U, a smaller proportion of ^{235}U and an even smaller proportion of ^{234}U, the decay-series daughter of ^{238}U. Natural thorium has one single isotope, ^{232}Th. Each of these nuclides decays to an unstable daughter leading, in turn, to a whole series of nuclides that terminate in one or other of the stable isotopes of lead. Under normal circumstances, in a natural material, the $^{235}U/^{238}U$ ratio will be fixed and all nuclides in each of the series will be in equilibrium.
Gamma spectrometry of materials containing these nuclides can only be effectively done with a detailed understanding of the decay chains of the nuclides involved.

2.1.1 Uranium series

The products of the decay are called radioactivity series. This series starts with the Uranium-238 isotope, which has a half-life 4.5×10^{10} year as shown in Figure 1 (Henery and John, 1972; Littlefield and Thorley, 1974). Since nuclides have very long half-life, this chain is still present today. The radionuclide ^{238}U decays into ^{234}Th emitting an alpha-particle, the newly formed nuclide is also unstable and decay further (Figure 1). Finally, after total of 14 such steps, emitting 8 alpha particles and 6 Beta particles, accompanied by gamma radiation, stable lead is formed. This series is said to be in secular equilibrium because all their daughters following ^{238}U have shorter half-life than the parent nuclide ^{238}U (Benenson, 2002).
This decay series includes the ^{226}Ra which has half-lives of 1600 year and chemical properties clearly different from those of uranium. ^{226}Ra decay into ^{222}Rn which is an inert noble gas that not form any chemical bonds and can escape into the atmosphere and attacks rapidly to aerosols and dust particles in the air deposited. The radiation emitted at the decay of these products, can cause damage to the deep lungs.

2.1.2 Actinium series

It is also known as Uranium-235 series and starts with ^{235}U and by successive transformations and up in a stable lead ^{207}Pb. It comprises 0.72% of natural uranium. Although only a small proportion of the element, its shorter half-life means that, in terms of radiations emitted, its spectrometric significance is comparable to ^{238}U. The decay series, shown in (Figure 2), involves 12 nuclides in 11 decay stages and the emission of 7 alpha particles (ignoring a number of minor decay branches). Since its abundance is very small, it dose not taken into account in the measurements (Harb, 2004).
Within this series, only ^{235}U itself can readily be measured, although ^{227}Th, ^{223}Ra and ^{219}Rn can be measured with more difficulty. Even though the uncertainties may be high, measurement of the daughter nuclides can provide useful support information confirming the direct ^{235}U measurement or giving insight into the disruption of the decay series.

2.1.3 Thorium series

Natural thorium is 100% ^{232}Th. The decay series is shown in (Figure 3). Six alpha particles are emitted during ten decay stages. Four nuclides can be measured easily by gamma spectrometry: ^{228}Ac, ^{212}Pb, ^{212}Bi and ^{208}Tl. The decay of ^{212}Bi is branched - only 35.94% of decays produce ^{208}Tl by alpha decay. The beta decay branch produces ^{212}Po that cannot be measured by gamma spectrometry. If a ^{208}Tl measurement is to be used to estimate the thorium activity, it must be divided by 0.3594 to correct for the branching (Harb, 2004).

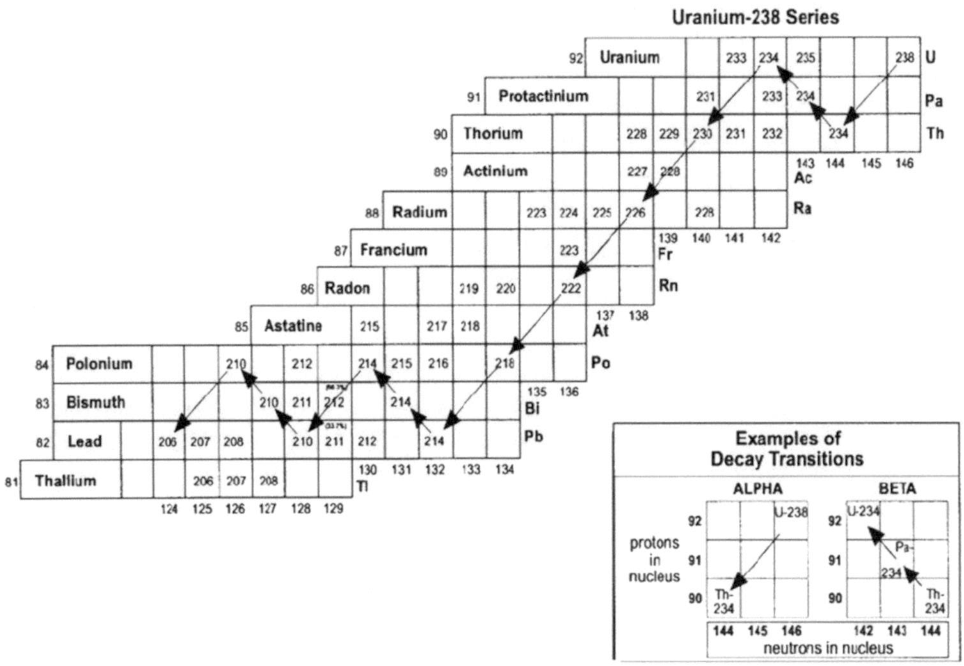

Fig. 1. Schematic diagram of the Uranium-238 series.

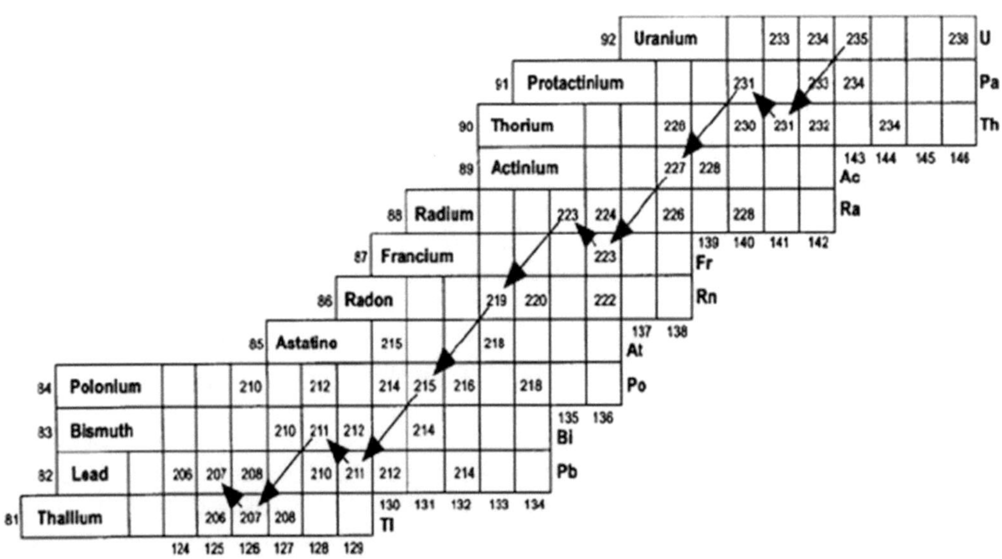

Fig. 2. A schematic diagram of Uranium-235 series (actinium).

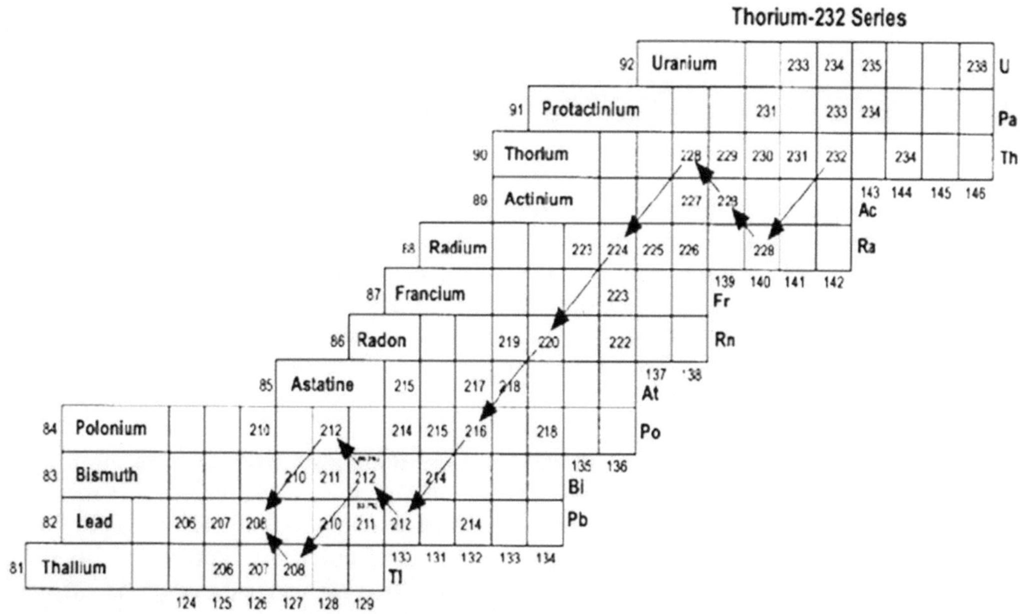

Fig. 3. A schematic diagram of the Thorium-232 series.

2.1.4 Potassium radionuclide

In 1905, J.J. Thompson discovered the radioactivity in ^{40}K is what makes everybody radioactive, it is present in body tissue. This radionuclide can be decayed by three general modes:

a. Positron emission.
b. K- electron capture.
c. Beta emission.

In first mode, ^{40}K radionuclide disintegrates directly into the ground state of ^{40}Ca by the emission of Beta- particle of energy 1321 keV in probability of 88.8% of the decays and no gamma emission is associated with this type of formation (Podgorsak, 2005).
Through the second mode, ^{40}K nuclide can be transformed into stable state (ground state) of ^{40}Ar by two ways, in the first one, ^{40}K disintegrates directly with one jump into ground state of ^{40}Ar with sixteen hundredths of the decays go by electron capture. In the second way, ^{40}K nuclide can be decayed indirectly into the ground state of ^{40}Ar by two stages. firstly, ^{40}K decay into the first excited state of ^{40}Ar. Secondly, the excited nuclide ^{40}Ar, decayed into ground state, accompanied by gamma radiation of 1460 keV energy in probability of 11% of the ^{40}K atoms undergo this change. In the last one (beta emission), a proton will be decayed into positron and ^{40}K changed into ^{40}Ar by probability of 0.0011%.

2.2 Biological effects of radiation

The study of the biological effects of radiation is a very complex and difficult task for two main reasons.

1. The human body is a very complicated entity with many organs of different sizes, functions, and sensitivities.
2. Pertinent experiments are practically impossible with humans.

The existing human data on the biological effects of radiation come from accidents, through extrapolation from animal studies, and from experiments in vitro. How and why does radiation produce damage to biological material? To answer the question, one should consider the constituents and the metabolism of the human body. In terms of compounds, about 61 percent of the human body is water. Other compounds are proteins, nucleic acids, fats, and enzymes. In terms of chemical elemental composition, the human body is, by weight, about 10 percent H, 18 percent C, 3 percent N, 65 percent 0, 1.5 percent Ca, 1 percent P, and other elements that contribute less than 1 percent each. To understand the basics of the metabolism, one needs to consider how the basic unit of every organism, which is the cell, functions.
The understandings of natural radiation concepts are essential for radiation protection purpose. The presences of radionuclides in soil affect the common people immensely. Since, the natural radionuclides form 10% of the average annual dose to the human body from all other types of radiation (UNSCEAR, 1993) and exposure to ionizing radiation, in generally considered undesirable at all levels.
Researchers drew attention to the low level exposure; there are three ways, through which the radio nuclides enter the human body: (1) direct inhalation of air born particulates, (2) ingestion through the mouth and (3) entry through the skin (Dipak et al., 2008). Direct exposure to skin is also responsible for radioactive contamination. Some of radionuclide which inters the lung by inhalation affects the blood. Their effectiveness, depend primarily upon two factors:

1. Kind of the radionuclide.
2. Physiological of the exposed person

The effects of radioactive in take depend upon the physical and chemical form and the root through which the radionuclide inter the body. These effects may cause damage to genetic organs, and eye defects and skin smear and destroy the circulatory system and lung cancer.
Exposure to low radiation ray lead to somatic infirmities like cancer and genetic defects such as mutation and chromosome aberrations. Gene modifications may result such conditions and diseases as asthma, diabetes, anemia. Genetic changes are passed on from one generation to another (Gerrado, 1974).
When people are exposed to certain levels of ^{238}U, ^{232}Th and ^{40}K for a long period of time cancer of the bone and hazard cavity may result (Nour, 2004). When radium inters the body by ingestions and inhalation, its metabolic behavior in similar to that of calcium faction of it will be deposited in bone where the remaining fraction being distributed uniformly in the soft tissues, thus the most radiotoxic and most important, among the several radionuclide in the radioactive decay chain the two natural series of uranium and thorium are ^{226}Ra and ^{228}Ra. The biological radiation effectiveness can be dividing into two types:

2.2.1 Body effectiveness

Cell is the basic unit of living tissue. Cells are complex structures enclosed by a surface membrane. DNA (deoxyribonucleic acid) is existed in the central of nucleus and considered as code of the structure, function, and replication of the cell. The famous "double helix" of the DNA molecule has a diameter of about (2nm). The induction of cancer or of hereditary disease by low levels of ionizing radiation is believed to be related to damage of the DNA molecules. This can be happen direct by ionization of the molecule, or indirectly through ionization of the water molecules in the cell (Cottingh and Greenwood, 2001). A single broken start and of DNA is rapidly repaired by cellular enzyme system, the unbroken strand of the DNA acting as template.

The water represents nearly 80% of the human body. When a body exposed to an ionized radiation, the large effect will be happened on the water molecules. The break up of water molecule may Produce $(.OH^{-}.)$ on that is highly reactive chemically and may attack the DNA molecule.

If, at the same time, there is adjacent damage to the other cell that may be errors in the repair process. The cell may die or damage to cell which cause later uncontrolled cell division. The incident radiation on the body, the water molecule will be ionized the water molecule and free electron will be liberated. The mechanism of this process can be summarized in the following equations.

$$H_2O \longrightarrow H_2O^{+} + {}^{-}e$$

(-e) is called dried electron, its energy proportional with the incident radiation energy. This dried electron will be losses its energy through its path in the body and caused some ionizing and excitation of other atoms and nuclides.

At low energy some water molecules shall be capture these dried electrons produced water electron which denoted by e- symbol. The velocity of water electron e- in the human cells is less than of dried electrons by factor of 105 times (Podgorsak, 2005).

The water positive ion (H_2O^{+}) will be interacts with the free Hydrogen according to the following equation:

$$H_2O^{+} + H_2O \longrightarrow {}_{\bullet}OH + H_3O^{+}$$

The water electron e- will be interacting with H_3O^{+} reduced water and free hydrogen root ($H_{\bullet}$):

$${}^{-}e + H_3O^{+} \longrightarrow H_2O + H_{\bullet}$$

It is found that the effectiveness of free hydroxyl root five times of that of free hydrogen root. This free root will be interacting with the organic molecules and other constituent of the cell causing change in their chemical properties which leads to distortion in their functions which may be lead to death. The effect degree of ionizing radiation depends on the type, energy and intensity of radiation and exposure time. The body effective classified into two types:

2.2.2 The early effectiveness

The germ cells in human are the most sensitive in the body if they exposed to radiation, in a dolts these cells presents in double number as compared with children in infancy this increase related to cell division and generation of completely similar new cells.

During division there are spindle-like particulars cells as chromosomes that reveal a hung number of granulated particles with a special arrangement called as genes the later are the responsible for the individually inherited character. the destruction or change of the geneses or chromosomes many reveal a character that not previously present in parents this change called as genetic mutation .

Radiation will increase the probability of occurring of genetic mutation and childhood abnormalities or infection with certain genetic discuses. If the divided cell exposed to radiation, these cells may undergo abnormal cell division, in the same time these cells may have the capacity proof reading of genetic mistakes that faced with hence may repair any

Biological disturbance. The radiation in this harm, but in accordance, they have new characters that may transform to embryos ending with exposed obvious genetic mutations. These genetic mutations not always harmful, in contrast, may result in favorable characters like the gain a good quality fruits in shape or size in the same time, animal's cell if exposed to genetic mutation may result in improved characters.

This type take place when the whole body expressed to radiation for high dose through short time some results can be appear in a few days or weeks, likes, reducing in weigh change in the blood cells hair loss and redness of the skin, and sometime the death is probable.

3. Experimental study

3.1 Gamma spectrometer

In this study we used gamma spectroscopy to determine the ^{238}U, ^{232}Th and ^{40}K in surface soil layer around the uranium mine at Najaf city.

Gamma spectroscopy is one of many famous techniques are used to measure the NORMs contents in the different environmental elements. It has many advantages such as high accuracy, measure wide energy range and different type samples and not need a chemical method in sample preparation. Beside these advantages gamma spectrometry of NORM is still difficult for a number of reasons. First, the activity levels are low and, if statistically significant results are to be obtained, need long count periods, ideally on a gamma spectrometer whose construction and location are optimized for low activity measurements. The second difficulty is the matter of spectrometer background (i.e. a large number of peaks that one might see in background spectra). Many of these are due to the NORM nuclides in the surroundings of the detector. Any activity in the sample itself must be detected on top of all that background activity. In many cases, it will be necessary to make a peaked-background correction in addition to the normal peak background continuum subtraction. All of those difficulties are then compounded by the fact that there are a large number of mutual spectral interferences between the many nuclides in the decay series of uranium and thorium.

The gamma rays levels were measured by integral counting using a spectrometer consist of a scintillation detector NaI(Tl) of ($2'' \times 2''$) crystal dimension with resolution value of 6.48% for line energy of 662 keV, scalar, shielding and specially designed sample container that allowed the sample to surround the scintillation detector at the top and on the sides. This system was computer controlled. The detector was connected to the amplifier through preamplifier unit; an analog to digital converter (ADC) of 4096 channels was assembled to the system. The spectroscopic measurements and analysis were performed via the CASSAY software into the PC of the laboratory.

In order to reduce the background radiation due to different radiation hazard, the detector was maintained in vertical position and shield by a cubic chamber of two layers starting with copper of 2mm thick followed by lead of 10 cm thick. The cosmic rays, photons and electrons, are reduced to a very low level by the 10 cm of lead shielding. This interaction will produced x-ray with low energy which can be suppressed by the copper layer (Aziz, 1981). The x-rays can be also come from radioactive impurities like antimony in the lead.

The spectrometer was calibrated for energy by acquiring a spectrum from radioactive standard sources of known energies like ^{60}Co (1332 keV, 1773 keV) and ^{137}Cs (662 keV). To measure the counting efficiency of the system, ^{22}Na, ^{57}Co, ^{60}Co, ^{109}Cd, ^{133}Ba and ^{137}Cs standard sources of gamma rays were used (Table 2). The relative intensities of the photo-peaks corresponding to their gamma rays lines have been measured.

Isotope	E (keV)	I %	Isotope	E (keV)	I %
^{22}Na	1274.5	99.95	^{133}B	80.99	34
^{57}Co	122.1	85.6		276.39	7.16
	136.4	10.88		302.85	18.3
^{60}Co	1173.2	99.97		356	62
	1332.5	99.98		383.85	8.9
^{109}Cd	88.03	3.6	^{137}Cs	661.6	85.1

Table 2. Energies and transition probabilities of standard sources (Heath, 1997)

3.2 Study area and sampling

Twenty five soil samples were collected in area of approximately 40000 m^2, located around the uranium mine at Najaf governorate. The latitude and longitude of this area are 31° 52′ 254″ N and 22° 26′ 221″ E. We used systematic grid sampling system involves subdividing the area of concern by using a square and collecting samples from the nodes (intersections of the grid lines). The origin and direction for placement of the grid is done, where the mine was centred in grid. From that point, a coordinate axis and grid is constructed over the whole site. The distance between sampling locations was 50 m (Figure 4). Systematic grid sampling is often used to delineate the extent of contamination and to define contaminant concentration gradients (IAEA, 2004).

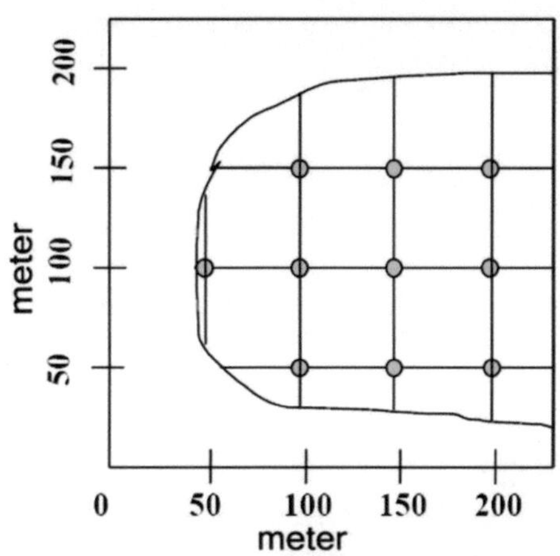

Fig. 4. Systematic grid sampling method

In order to measure the NORMs in soil surface, 25 soils samples were collected, one sample average from each point, was taken by digging a hole at a depth of 35 cm before the ground surface. The soil texture for all samples was very similar. The collected samples were transferred to labeled closed polyethylene bags and taken to the laboratory of radiation detection and measurement in the physics department, collage of science, university of Kufa. In this work a 1.4 litter polyethylene marinelli beaker was used as a sampling and measuring container. Before use, the containers were washed with dilute hydrochloric acid

and rinsed with distilled water. The soil samples were prepared for analysis by drying, sieving and kept moisture free by keeping 24 hours in the oven at 100C°. They were mechanically crushed and sieved through of 0.8 mm pore size diameter sieved to get homogeneity (R2).
To remove completely the air from sample, the later was pressed on by light cap of the marinelli beaker. The respective net weights were measured and record with a high sensitive digital weighting balance with a percent of ±0.01%. After that about 1 kg of each sample was then packed in a standard marinelli beaker that was hermetically sealed and dry weighted. The sample was placed in face to face geometry over the detector for along time measurement.

3.3 Specific activity

The definition of activity refers to the number of transformations per unite time. Since the fundamental unite time is the second the quantity activity is measured in disintegrations per second or (dps). In 1950, the international joint commission on standards Unit and constants of radioactivity define the curie by accepting 37 billion dps as curie of radioactivity regardless of its source or characteristics. The SI derived unit of activity is the Becquerel (Bq) and is that quantity of radioactive material in which one atom is formed per second or under goes one disintegration per second (1 dps).
The activity of ^{238}U was estimated from the 1765 keV gamma transition energy of ^{214}Bi (17% possibility). Also the activity of ^{232}Th was measured from the 2614 keV gamma transition energy of ^{208}Tl (100% possibility) whereas ^{40}K activity was determined using the 1460 keV gamma ray line (10.7% possibility) (Vosniakos, 2003).
The specific activity is defined as activity per unite mass of radioactive substance and the reported in units such as Curie per gram or Becquerel per kilogram (Bq/kg). The specific activity of each radionuclide was calculated using the following equation (UNSCEAR, 2000).

$$A = \frac{A'}{\varepsilon \cdot I_{\gamma} \cdot m \cdot t}$$

Where A the specific activity of the radionuclide in Bq/kg, A' the liquid count, ε the counting efficiency, I_{γ} the percentage of gamma emission probability of the radionuclide under study, t the counting time in second and m the mass of the sample in kg.

3.4 Radium equivalent activity

To represent the activity levels of ^{238}U, ^{232}Th and ^{40}K which take into account the radiological hazards associated with them, a common radiological index has been introduced. This index is called radium equivalent activity (Ra_{eq}) and is mathematically defined by (UNSCEAR, 2000).

$$Ra_{eq}(Bq/kg) = A_U + 1.43A_{Th} + 0.077A_K$$

Where A_U, A_{Th} and A_K are the specific activities of Uranium, Thorium, and potassium respectively. This equation is based on the estimation that 10 Bq/kg of ^{238}U equal 7 Bq/kg of ^{232}Th and 130 Bq/kg of ^{40}K produced equal gamma dose. The maximum value of Ra_{eq} must be less than 370 Bq/kg. Also the Ra_{eq} value of 370 Bq/kg is equivalent to the annual dose equivalent of 1.5 mSv/y, which we assumed to be the maximum permissible dose to human from their exposure to natural radiation from soil in one year.

3.5 Absorbed does rate in air

Absorbed dose rate defined as the ratio of an incremental dose (dD) in a time interval (dt).

$$AD = \frac{dD}{dt}$$

Gamma dose rate in air, one meter above the ground, is used for the description of terrestrial radiation, and is usually expressed in nGy/h or pGy/h. the absorbed dose rate due to gamma radiation of naturally occurring radionuclide (^{238}U, ^{232}Th, and ^{40}K), were calculated on guidelines provided by (UNSCEAR, 2000).

$$AD(nG/h) = 0.462A_U + 0.621A_{Th} + 0.0417A_K$$

Where 0.462, 0.621 and 0.0417 are the conversion factors for ^{238}U, ^{232}Th and ^{40}K assuming that the contribution natural occurring radionuclide can be neglected as they contribute very little to total dose from environmental background.

3.6 Annual effective doses

To estimate annual effective doses, account must be taken of (a) the conversion coefficient from absorbed dose in air to effective dose and (b) the indoor occupancy factor. The average numerical values of those parameters vary with the age of the population and the climate at the location considered. In the UNSCEAR 1993 Report, the Committee used 0.7 Sv.Gy/y for the conversion coefficient from absorbed dose in air to effective dose received by adults and 0.8 for the indoor occupancy factor, i.e. the fraction of time spent indoors and outdoors is 0.8 and 0.2, respectively. These values are retained in the present analysis. From the data summarized in this Chapter, the components of the annual effective dose are determined as follows: (UNSCEAR, 1993).

$$\text{Indoor (nSv)} = \text{absorbed dose nGy/h} \times 8760\ \text{h} \times 0.8 \times 0.7\ \text{SvG/y}$$

$$\text{Outdoor (nSv)} = \text{absorbed dose nGy/h} \times 8760\ \text{h} \times 0.2 \times 0.7\ \text{SvG/y}$$

The resulting worldwide average of the annual effective dose is 0.48 mSv, with the results for individual countries being generally within the (0.3 - 0.6) mSv range. For children and infants, the values are about 10% and 30% higher, in direct proportion to an increase in the value of the conversion coefficient from absorbed dose in air to effective dose.

3.7 Hazard index

To reflect the external exposure, a widely used hazard index, called the external hazard index (H_{ex}), which is defined as following:

$$H_{ex} = \frac{A_U}{370} + \frac{A_{Th}}{259} + \frac{A_K}{4810}$$

There is another hazard index called internal hazard index (H_{in}), which is given by equation.

$$H_{in} = \frac{A_U}{185} + \frac{A_{Th}}{259} + \frac{A_K}{4810}$$

The values of the index must be less than the unity in order to keep the radiation hazard to be insignificant unity corresponds to the upper limit of radiation equivalent activity (370 Bq/kg).

4. Results and discussion

The spectra of twenty five surface soil samples surrounded the abandoned Uranium mine hole have been analyzed. The specific activity of ^{238}U, ^{232}Th, ^{40}K and Radium equivalent activity (Ra_{eq}) are given in Table 3. The specific activity (Bq/kg) varied from 37.31 to 1112.47 (mean = 268.16), 0.28 to 18.57 (mean = 6.68) and 132.25 to 678.33 (mean = 277.49) for ^{238}U, ^{232}Th and ^{40}K respectively.

The obtained results are comparable to the worldwide average recommended by UNSCEAR which are 30, 35 and 400 Bq/kg for ^{238}U, ^{232}Th and ^{40}K respectively (UNSCEAR, 2000). It was found that all values of ^{238}U specific activities are higher than the worldwide average whereas those of ^{232}Th are less than it. For ^{40}K, it is clear that the specific activities, with the exception of five samples, are found to be less than worldwide average.

Obviously demonstrate that the minimum and maximum specific activity values of ^{238}U are least by factor of 4 and 37 higher than the corresponding values obtained worldwide average. The large variation between the specific activities obtained for ^{238}U and other two radionuclides can be easily ascribed to the high content of uranium in the neglected waste of drilling and exploration operations on the surface soil surrounding the mine. The contour maps (radiological maps) of the activity distribution of ^{238}U, ^{232}Th and ^{40}K in the study area are shown in Figures 5, 6 and 7. From Figure 5, we can observe three regions with a highest specific activity values of ^{238}U situated at northeast, east and south-west portions of the hole mine. In contrast, Figure 7 indicates that high concentrations of ^{40}K occupies the same positions of ^{238}U while for ^{232}Th there are no placements have activities require attention as shown in Figure 6.

The calculated Ra_{eq} values for all samples were also presented in Table 3. It may be seen that Ra_{eq} oscillates between 52.727 and 1189.845 with an average of 299.09 Bq/kg. It is observed that the values of Raeq in twenty one samples were less than the acceptable safe limit of 370 Bq/kg (OECD, 1979; UNSCEAR, 1982; UNSCEAR, 1988). As shown in Table 3 there are four values greater than worldwide average. As a rule, the matter whose Ra_{eq} exceeds 370 Bq/kg is discouraged (Beretka and Mathew, 1985). Figure 8 demonstrates the distribution of Ra_{eq} and it appears three positions have highest values.

The calculated absorbed dose rate of samples was listed in Table 3. As shown in Table 3, the values ranged from 25.02 to 553.01 with an average value of 139.61 nG/h which is nine fold higher than the world average of 15 nG/h recommended by UNSCEAR (UNSCEAR, 2000). It can be seen that all values were much higher than the world average.

Sample code	specific activity (Bq/kg)			Ra_{eq} (Bq/kg)	AD nG/h
	^{238}U	^{232}Th	^{40}K		
S11	72.17±4.43	3.38±0.69	253.19±8.79	96.49	46.00
S12	59.37±4.02	0.28±0.20	184.47±7.51	73.97	35.29
S13	213.24±7.62	5.63±0.89	238.84±8.54	239.68	111.97
S14	39.76±3.29	0.28±0.20	174.70±7.30	53.61	25.82
S15	480.12±11.43	8.02±1.06	415.06±11.26	523.54	244.10
S21	138.07±6.13	2.81±0.63	176.23±7.34	155.65	72.88
S22	645.42±13.26	10.13±1.19	521.04±12.61	700.02	326.20
S23	312.64±9.23	6.47±0.95	409.87±11.19	353.45	165.54
S24	249.18±8.24	4.08±0.76	283.43±9.30	276.83	129.47
S25	37.31±3.19	3.66±0.72	132.25±6.36	52.72	25.02
S31	122.82±5.78	2.39±0.58	153.63±6.85	138.06	64.63
S32	285.95±8.82	12.94±1.35	219.60±8.19	321.36	149.30
S33	122.82±5.78	5.35±0.87	153.52±6.85	142.29	66.46
S34	273.97±8.64	6.47±0.95	211.96±8.05	299.54	139.43
S35	78.43±4.62	4.08±0.76	232.12±8.42	102.13	48.44
S41	142.43±6.23	4.92±0.83	205.85±7.93	165.31	77.44
S42	324.07±9.39	7.03±0.99	264.49±8.99	354.48	165.11
S43	860.29±15.31	13.65±1.39	678.33±14.39	932.04	434.21
S44	175.38±6.91	3.80±0.73	229.67±8.38	198.49	92.96
S45	1112.47±17.41	18.57±1.62	459.96±11.85	1189.84	553.01
S51	167.48±6.75	7.32±1.01	302.06±9.60	201.20	94.51
S52	276.69±8.68	10.41±1.21	156.98±6.92	303.66	140.84
S53	139.16±6.16	7.32±1.01	158.21±6.95	161.81	75.43
S54	85.78±4.83	7.88±1.05	201.27±7.84	112.54	52.91
S55	288.94±8.87	10.27±1.20	320.38±9.89	328.29	153.22
Min.	37.31	0.28	132.25	52.72	25.26
Max.	1112.47	18.57	678.33	1189.84	553.01
mean	268.16	6.68	277.49	299.09	139.61

Table 3. Specific activity, Radium equivalent activity and absorbed dose rate of soil samples.

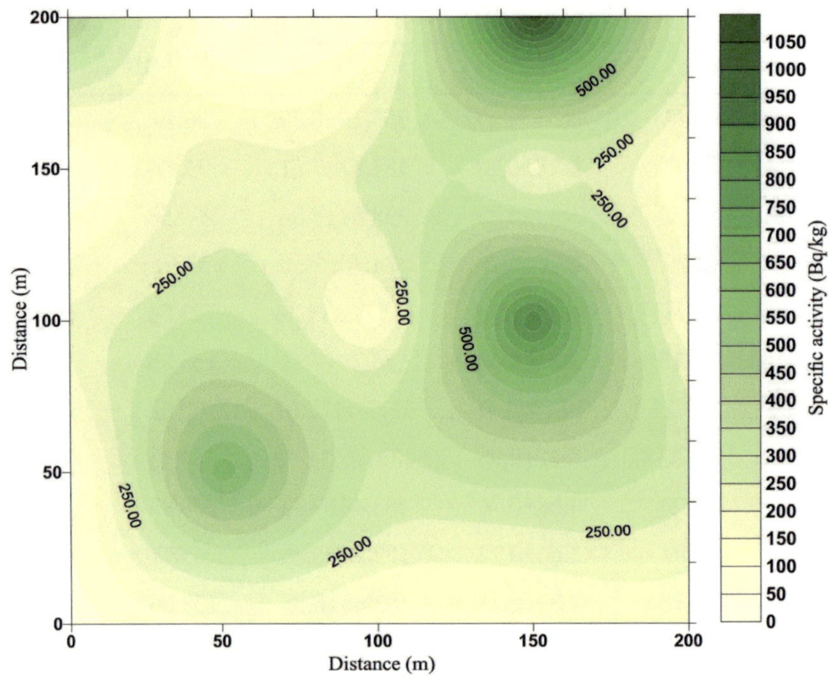

Fig. 5. Specific activity distribution of ^{238} U.

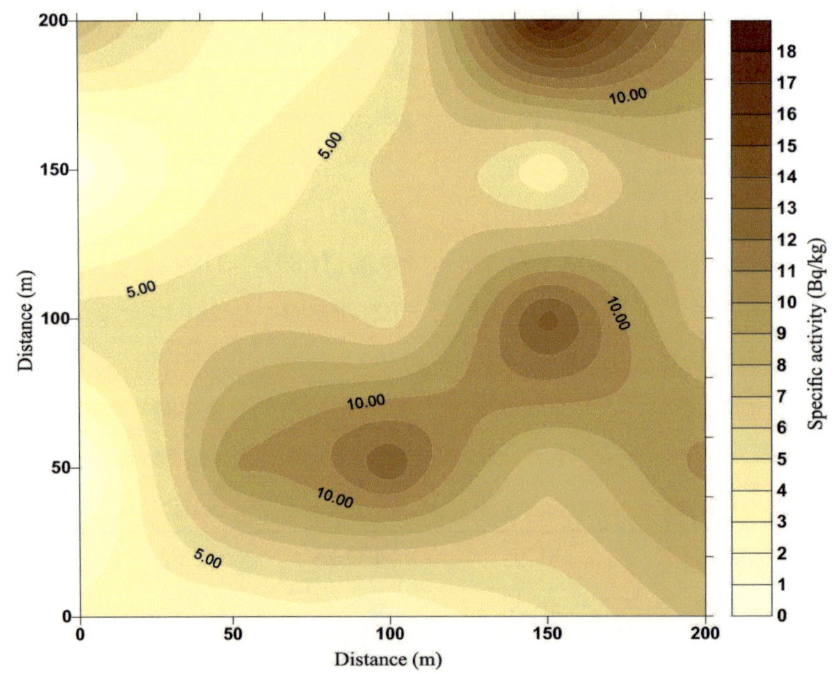

Fig. 6. Specific activity distribution of ^{232}Th.

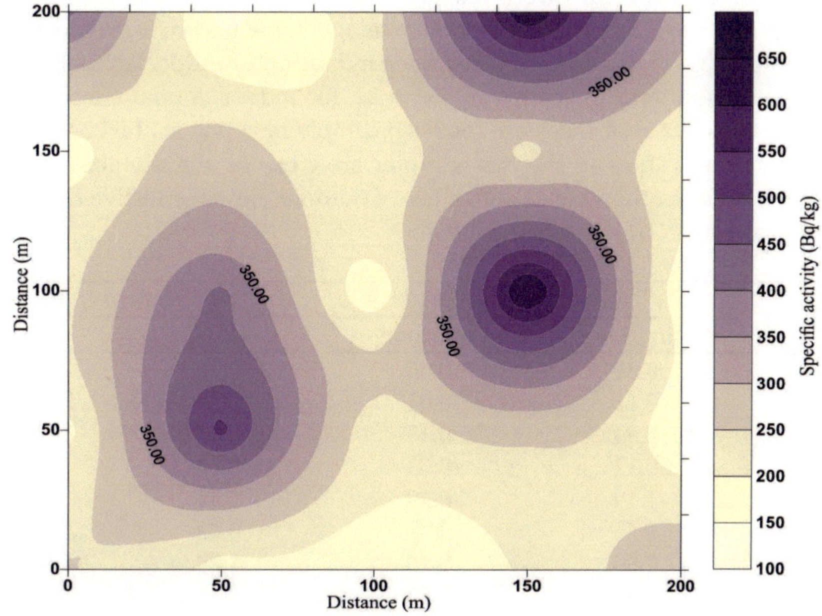

Fig. 7. Specific activity distribution of ^{40}K.

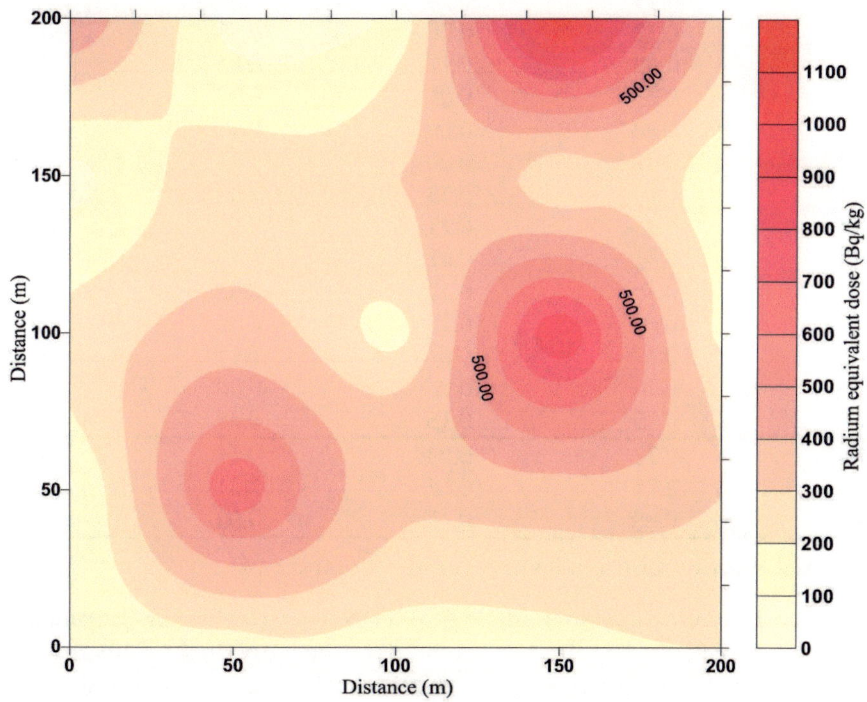

Fig. 8. Distribution of Radium equivalent in surface soil around mine.

The annual effective dose values were calculated and listed in Table 4. They were found to be in the range 0.123 to 2.713 mSv/y with an average value 0.68 mSv/y and from 0.031 to 0.6780 with an average value of 0.17 mSv/y for indoor and outdoor annual effective dose respectively. In general and as shown in Table 4, for indoor annual effective dose, It is important here to notice that there are fourteen sample have values higher than the word average whereas, the values of the rest samples are close or slightly above of the world average value of soil. In other words, all values of outdoor annual effective dose were below the worldwide average.

Sample code	Annual dose (mSv)		Hazard index	
	indoor	outdoor	H_{ex}	H_{in}
S11	0.22	0.05	0.26	0.45
S12	0.17	0.04	0.20	0.36
S13	0.54	0.13	0.64	1.22
S14	0.12	0.03	0.14	0.25
S15	1.19	0.29	1.41	2.71
S21	0.35	0.08	0.42	0.79
S22	1.60	0.40	1.89	3.63
S23	0.81	0.20	0.95	1.80
S24	0.63	0.15	0.74	1.42
S25	0.12	0.03	0.14	0.24
S31	0.31	0.07	0.37	0.70
S32	0.73	0.18	0.86	1.64
S33	0.32	0.08	0.38	0.71
S34	0.68	0.17	0.81	1.55
S35	0.23	0.05	0.27	0.48
S41	0.38	0.09	0.44	0.83
S42	0.81	0.20	0.95	1.83
S43	2.13	0.53	2.51	4.84
S44	0.45	0.11	0.53	1.01
S45	2.71	0.67	3.21	6.22
S51	0.46	0.11	0.54	0.99
S52	0.69	0.17	0.82	1.56
S53	0.37	0.09	0.43	0.81
S54	0.26	0.06	0.30	0.53
S55	0.75	0.18	0.88	1.66
Min.	0.12	0.03	0.14	0.24
Max.	2.71	0.67	3.21	6.22
Mean	0.68	0.17	0.80	1.53

Table 4. Annual effective dose and hazard indexes of soil samples.

The international commission on Radiological Protection (ICRP) has recommended the annual effective dose equivalent limit of 1 mSv/y for the individual members of the public and 20 mSv/y for the radiation workers (ICRP, 1993). The worldwide average annual effective dose is approximately 0.5 mSv and the results for individual countries being generally within the 0.3 to 0.6 mSv range (UNSCEAR, 2000).

In addition, the calculated values of hazard index for the soil samples were ranged from 0.142 to 3.216 with an average value of 0.808 and from 0.243 to 6.222 with an average value of 1.533 for external (H_{ex}) and internal (H_{in}) respectively as mentioned in Table 2.

Out of 25 positions, 4 for H_{ex} and 13 for H_{in}, have values very higher than unity. Since these values are dispersed randomly within a limited area around the min hole, therefore, according to the report of European Commission in Radiation Protection, the area study is not safe and posing significant radiological threat to the population (European Commission, 1999).

5. Conclusion

The surface soil layer around the uranium mine hole has uranium activities greater than worldwide average; this can mainly due to the waste of drilling and exploration left on the surface layer of soil surrounding the mine.

The thorium activities were within normal level in the studied area. Generally, potassium radionuclide in soil samples was in the range of worldwide average.

The absorbed dose rates of studied area are higher than the criterion limit of gamma radiation dose rate with an average of nine times.

Finally, from the radiation protection point of view the studied area is considered to be not safe inhabitants because the values of both internal and external hazard indexes associated with the samples are higher than unity. Thus, the human inside the area are supposed to acquire radiological complication.

6. References

Abusini M. (2007). Determination of Uranium, Thorium and Potassium Activity Concentrations in Soil Cores in Araba valley, Jordan, *Radiation Protection Dosimetry*, 128: 213-216.

Benenson W. (2002). *Hand book of physics*. Fourth Edition. Spriger –Velarg, New York, INC.

Beretka J. and Mathew P. J. (1985). Natural Radioactivity of Australian Building Materials, Industrial Wastes and by Products, *Health Physics*, 48: 87-95.

Cottingh W. N. and Greenwood D. A. (2001). *An introduction to Nuclear physics*, second edition, University of Cambridge, United kingdom.

Dipak G., Argha D., Sukumar B., Rosalima S., Kanchan K. P. (2008). Measurement of Natural Radioactivity in Chemical Fertilizer and Agriculture Soil: Evidence of High Alpha Activity, *Environmental Geochemistry and Health*, 30: 79-86.

European Commission (1999). *Radiological Protection Principle concerning the Natural Radioactivity of Building Materials.* Radiation Protection 112, European Commission ,Brussels.

Gerrado C. Maxino (1974). Radioactive Potassium, *The nucleous*, 12: 4-8.

Harb S. (2004). *On the Human Radiation Exposure as Derived From the Analysis of Natural and Man-Made Radionuclides in Soil*, Ph.D. Thesis, Hannover University.

Henery S. and John R. A. (1972). *Introduction to Atomic and Nuclear Physics*, Fifth edition, Holt, Rineart and Winston INC.

IAEA (2004). *Soil Sampling for Environmental Contaminants*, International Atomic Energy Agency, TECDOC-1415, Vienna, Austria.

Ibrahim N. (1999). Natural Activities of ^{238}U, ^{232}Th and ^{40}K in Building Materials, *J. Environ. Radioact.*, 43: 255-258.

ICRP (1993). International Commission on Radiological Protection, publication 65,Annals of the ICRP 23(2), Pergamon press, Oxford.

José A. dos Santos, Jorge J. R., Cleomacio M. da Silva, Suêldo V. S. and Romilton dos Santos A. (2005), Analysis of the ^{40}K Levels in Soil using Gamma Spectrometry, *Brazilian Archives of Biology and Technology*, 48: 221-228.

Littlefield T.A. and Thorley N. (1974). *Atomic and Nuclear Physics*, Third edition. Van Nostrand Reinhold company, London, UK.

Maher O. El-Ghossain and Raed M. Abu Saleh (2007). Radiation Measurements in Soil in the Middle of Gaza-Strip Using Different Type of Detectors, *The Islamic University Journal (Series of Natural Studies and Engineering)*, 1 5: 23-37.

Marcelo F. M. and Pedro M. J. (2007). Gamma Spectroscopy in the Determination of Radionuclides Comprised in Radioactive Series, *International Nuclear Atomic Conference-INAC*, Santos, SP, Brazil.

NCRP (1987), National Council on Radiation Protection and Measurements, *Exposure of the Population in the United States and Canada from natural background radiation*. No.94, USA.

Nour K. A. (2004), Natural Radioactivity of Ground and Drinking Water in Some Areas of Upper Egypt, *Turkish J. Eng. Env. Sci.*, 28: 345-354.

OECD (1979). *Exposure to Radiation From the Natural Radioactivity in Building Materials*, Report by a Group of Experts of the OECD Nuclear Energy Agency, Organization for Economic Cooperation and Development, Paris, France.

Podgorsak E .B. (2005). *Radiation Physics for Medical Physicist*, Springer Berlin Heidelberg, New York, USA.

Rowland R. E. (1993). Low-Level Radium Retention by the Human Body; A Modification of the ICRP Publication 20 Retention Equation, *Health Phys*, 65: 507-513.

Scholten L. C. and Timmermans. C. W. M. (1996). Natural Radioactivity in Phosphate Fertilizers, *Fertilizer Resrch.*, 43: 103-107.

UNSCEAR (1982). *Sources effects and risks of ionizing radiation*, Report to the General Assembly, With Annexes, United Nations Scientific Committee on the Effects of Atomic Radiation, New York, United Nations.

UNSCEAR (1988). *Sources effects and risks of ionizing radiation*, Report to the General Assembly, With Annexes, United Nations Scientific Committee on the Effects of Atomic Radiation, New York, United Nations.

UNSCEAR (1993). *Sources effects and risks of ionizing radiation*, Report to the General Assembly, With Annexes, United Nations Scientific Committee on the Effects of Atomic Radiation, New York, United Nations.

UNSCEAR (2000). *Sources effects and risks of ionizing radiation*, Report to the General Assembly, With Annexes, United Nations Scientific Committee on the Effects of Atomic Radiation, New York, United Nations.

Valter A. B., Francisco J. F. F., William C. P. (2008). Concentration of Radioactive Elements (U, Th and K) Derived From Phosphatic Fertilizers in Cultivated Soils, *Brazilian Archives of Biology and Technology*, 51: 1255-1266.

Vosniakos F., Zavalaris K., Papaligas T. (2003). Indoor Concentration of Natural Radioactivity and The Impact to Human Health, *Journal of Environmental Protection and Ecology*, 4: 733-737.

Vosniakos F., Zavalaris K., Papaligas T., Aladjadjiyan A. and Ivanova D. (2002). Measurements of Natural Radioactivity Concentration of Building Material in Greece, *Journal of Environmental protection and Ecology*, 3: 24-29.

WHO (1993). *Guideliens for Drinking –Water Quality "Recommendations*, World Health Organization, Geneva.

Yasir M.S., Majid A. Ab., Yahaya R. (2007). Study of Natural Radionuclides and Its Radiation Hazard Index in Malaysia Building Material, *Radioanal. Nucl. Chem.*, 273: 539-541.

2

Research Reactor Fuel Fabrication to Produce Radioisotopes

A. M. Saliba-Silva[1], E. F. Urano de Carvalho[1], H. G. Riella[2] and M. Durazzo[1]
[1]Nuclear Fuel Center of Nuclear and Energy Research Institute Brazilian Commission of Nuclear Energy, São Paulo,
[2]Chemical Engineering Department of University of Santa Catarina, Florianopólis, Brazil

1. Introduction

This chapter describes the manufacturing technology of fuel used in research reactors that produce radioisotopes. Besides this production, the research reactors are also used for materials testing. The most common type of research reactors is called "MTR" - Materials Testing Reactor. The MTR fuel elements use fuel plates, which are quite common around the world. There was a historic development in that fuel type over the years to reach the current state-of-art in this technology.

The basic MTR fuel element is an assembled set of aluminum fuel plates. It consists of regularly spaced plates forming a fuel assembly. These spaces allow a stream flow of water that serves as coolant and also as moderator to nuclear reaction. The fuel plates have a meat containing the fissile material, which is entirely covered with aluminum. They are manufactured by adopting the traditional assembling technique of dispersion fuel briquette inserted in a frame covered by aluminum plates, which are welded with subsequent rolling. This technique is known internationally under the name "picture-frame technique". Powder metallurgy techniques are used in the manufacture of the fuel plate meats, making briquettes using ceramic or metallic composites. The briquette is made with powdered nuclear material and pure aluminum powder, which is the structural material matrix of the briquette.

Using UF_6 in the chemical plant, it is able to produce several intermediate compounds of uranium. One of these compounds is UF_4, which is the main raw material to produce metallic uranium. It could be made by several routes. The production of metallic uranium uses the UF_4 reduction through calcio- and magnesiothermic reaction. The metallic uranium is alloyed with Al, Si or Mo. Previously, stable uranium oxides were used as MTR fuels, but they had very small densities to accomplish a good operational performance of the reactors. The fuel material candidate mostly prone to be used in nuclear research reactors is based on alloys carrying more U-density toward the fuel meat. In present state, the U-Mo alloys are good candidates, but it would not be subject of the present chapter since it on its path to be certified to future use in research reactors. Currently, the most used material is U_3Si_2 LEU, which is low enriched uranium enriched up to 20% of ^{235}U isotope, which is the nuclear fissile material.

The production procedures of U_3Si_2 fuel fabrication will be discussed in this chapter, starting from U_3Si_2 fabrication and powder manufacture. This powder is mixed with aluminum powder and pressed, resulting in a solid briquette with good mechanical strength. After quality inspection, the briquette becomes the fuel plate meat.

The fuel plate manufacturing procedures will be described according to the picture frame technique. This technique includes the assembling of the briquette inside the frame sandwiched with cover plates. The assembly is welded and hot and cold rolled to get the fuel plate, where the fuel meat is completely sealed inside aluminum. All the process and quality control during fabrication will be commented ahead.

Once the plates having been fabricated, the fuel assembly is finally made by fixing the fuel plates and the other mechanical components, such as nozzle, handle and screws. This finishing process to produce the element is also commented in this chapter.

The characteristics of the fuel plates must meet specifications of each particular research reactor characteristics. Inspections and qualifications are carried out in various stages of fuel plate manufacture.

As this chapter describes nuclear fuel manufacturing for research reactors, the sub-items of fabrication process can be divided into the following topics: evolution of nuclear fuel materials for MTR fuel; production of uranium hexafluoride (UF_6); production of uranium tetrafluoride (UF_4); production of metallic uranium; U_3Si_2 production; production of fuel cores from U_3Si_2 powder and aluminum; production of fuel plates with U_3Si_2-Al dispersion briquettes; assembling of fuel elements; recovery of uranium; effluent treatment; quality control.

In this chapter, the experience of IPEN/CNEN-SP (Energy and Nuclear Research Institute of Brazilian Commission of Nuclear Energy, São Paulo, Brazil) will be given as a productive route to produce MTR nuclear fuels for research reactors, since this is the main expertise of all the authors of this chapter.

2. Evolution of nuclear materials for research reactors fuel

The use of radioisotopes in medicine is certainly one of the most important social uses of nuclear energy. Radiopharmaceuticals are radioactive substances that help doctors to make important decisions for treatments in oncology, cardiology, neurology, among other areas. For patients, the diagnoses represent safety and pain relief, as in the case of samarium-153 use, which is employed to relieve bone pain caused by metastatic tumors.

Nuclear medicine is a medical specialty that uses radioactive material for diagnostic tests and therapeutic purposes. Although it is often confused with radiotherapy, the last application has a lot of different procedures and applications. The main distinction between the two specialties is the way both use the radioactive material. While radiotherapy (radiation therapy) uses sealed sources (or closed), which emit radiation outside the patient, nuclear medicine uses open sources of radiation, administered in vivo (oral or intravenous). If, in radiotherapy, radiation is directed toward the point to be discussed, in nuclear medicine is the body own metabolism of the patient who is in charge of carrying radioactive material into the organ to be examined or treated.

The success of nuclear medicine in diagnosis is due to its ability to show the functioning of various body organs, avoiding the use of invasive techniques such as biopsy and catheterization. The use of ultrapure iodine-123 to examine thyroid function is one example. By scintigraphy, a diagnostic imaging technique, which has several medical applications, made possible to measure the uptake of iodine by the thyroid and thereby assess the functioning of the gland. Another radioactive element widely used to study the various functions of the human organism is technetium-99m. This isotope can be chemically combined with various organic complexes, which evaluate liver disorders, bone and brain, among others. In bone scintigraphy, the radioactivity of technetium reveals the existence of tumors from six to eight months before they have reached sufficient size to be picked up by X-ray examinations. With this, it is possible to start treatment much earlier with greater cure perspective.

Nuclear reactors that produce radioisotopes are called research reactors. This type of reactor is also used to perform tests on materials and nuclear fuels in the development phase. The modern research reactors are designed with both purposes, radioisotope production and testing of materials, and for this reason are called Multipurpose Reactors.

Unlike power reactors, which are well known and are intended to generate heat for electricity generation, the research reactors or the modern multipurpose reactors aim to generate neutrons used for radioisotopes production or for testing materials in terms of verify their performance under irradiation. Unlike power reactors, research reactors operating with much higher power density, which is necessary to get high neutron fluxes. For this reason its fuel is usually in the form of a metal plate, usually covered by aluminum. They are very different from the fuel rod with ceramic pellets (UO_2) as used in the fuel for power reactors.

The research reactors moderated and cooled with light water and using plate-type fuel elements has been named MTR type reactors (Materials Testing Reactor). After the construction of the first MTR, a joint venture of ORNL (Oak Ridge National Laboratory) and ANL (Argonne National Laboratory) operated it since March 31, 1952. Many research reactors around the world uses MTR type fuel elements, which are formed by assembling fuel plates fabricated by a well-known and established technique of assembling a core, commonly named fuel meat, which incorporates the fissile material, a frame plate and two cladding plates, with subsequent deformation by hot and cold-rolling (picture frame technique) (1) (2).

Initially, the fuel plates usually used as the core material an uranium-aluminum alloy (U-Al) containing 18 wt% of highly enriched uranium (93 wt% ^{235}U) (1) (3). Even in the 50's, with the concern about nuclear weapons non-proliferation, the research reactors began to use fuels containing low-enriched uranium (20 wt% ^{235}U) (4). With enrichment lowering, in order to maintain the reactivity and lifetime of the reactor cores, it became necessary to increase the amount of uranium in each fuel plate. In the U-Al alloy, the uranium concentration had to be increased to 45 wt% to compensate the decrease in the enrichment level.

Fuel plates containing the meat based on the U-Al alloy with 18 wt% of highly enriched uranium were easily fabricated. However, difficulties arise in fabricating fuel plates with meats of U-Al alloy containing 45 wt% of low-enriched uranium, because of the fragility and propensity for segregation of this alloy (4) (5) (6). An alternative to overcome this problem was the use of cores manufactured by powder metallurgy, which used dispersions of uranium compounds in aluminum and could incorporate quantities of low-enriched uranium significantly greater. For instance, the Argonauta reactor (10 MW), in Rio de Janeiro, Brazil, started its operation in 1956 and was developed by the Argonne National Laboratory, USA. This pioneer Brazilian research reactor used fuel plates with the meat based on an U_3O_8-Al dispersion containing 39 wt% of U_3O_8 with low enrichment (7).

Efforts were made to increase the concentration of uranium in this type of dispersion fuel, getting 65 wt% of U_3O_8 in the fuel fabricated for the Puerto Rico Research Reactor of the Puerto Rico Nuclear Center to the end of the 70's (8).
Aiming at obtaining more and more high neutron fluxes, the development of research reactors with higher power required a continuous production of fuels, which used highly enriched uranium (93 wt% ^{235}U), yielding higher specific reactivity and economics, since these fuels could stay longer in the reactor core (long life). The 100 MW HFIR (High Flux Isotope Reactor) used dispersion U_3O_8-Al with 40 wt% U_3O_8 (9) and the ATR (Advanced Test Reactor), with 250 MW, used the same type of dispersion with 34 wt% highly enriched U_3O_8 (10). In addition to the U_3O_8-Al dispersions, UAl_x-Al dispersions were commonly used (x is approximately 3), all these fuels systems still using highly enriched uranium. At this time, in late 70's, the highest uranium density obtained inside the fuel was 1.7 gU/cm^3, which was quite well qualified.
Since highly enriched uranium was easily obtainable in the 70's, the commercial reactors that were using the low-enriched uranium started gradually to convert their cores to highly enriched fuel. Thus, it reached a total of approximately 156 research reactors in 34 countries using highly enriched uranium, resulting in an annual circulation of approximately 5000 kg of this material (11). In 1977, arose again the concern about the proliferation risk associated with loss of fuel during manufacture, transport and storage, leading to restriction by the U.S. government's sale of uranium with high enrichment (above 90 wt% ^{235}U) and producing an impact on the availability and use of the highly enriched fuel for research reactors.
From 1978 programs, it was established for the enrichment reduction, aimed at developing the technology base for replacement of highly enriched uranium by low-enriched uranium (less than 20 wt% ^{235}U) in research reactors. The main program, still active today, is the *RERTR Program* (Reduced Enrichment for Research and Test Reactors), which aims to develop the technology necessary to convert the reactors that use highly enriched uranium (= or > 20% ^{235}U) by low-enriched uranium (less than 20% ^{235}U). During the existence of this program more than 40 research reactors have been converted. At this time, the decrease of enrichment has demanded an effort bigger that previously, because, in most high power research reactors, which are designed to operate in extremes, this substitution involved the development and qualification of new fuels with maximum possible concentration of uranium, which limits are imposed for manufacturability and performance under severe and prolonged irradiation.
In this context, the developments were based initially on increasing the concentration of uranium in the fuel currently used at the beginning of RERTR program, until the practical limit of 2.3 gU/cm^3 in the case of UAl_x-Al and 3.2 gU/cm^3 in the case of U_3O_8-Al. Also, an effort was made in developing new fuels that would allow obtaining uranium densities of 6-7 gU/cm^3, well above the density that can be achieved with the UAl_x-Al and U_3O_8-Al fuel. The development of new fuels would allow the conversion to low enriched from virtually all existing research reactors.
High density of uranium in the dispersion can only be achieved by using the dispersion of fissile compounds with high uranium content. Figure 1 shows the potential of various uranium compounds. The technological limit for the use of dispersions is 45% by volume of fissile material dispersed, since it must be kept a solid aluminum matrix as dispersant. The uranium silicides and U_6Fe compounds were initially considered promising.

The problem encountered in using these intermetallics with high concentrations of uranium as fissile material in the form of dispersions in aluminum is related to its dimensional stability during operation, leading to swelling of the fuel plates and therefore the problems that compromise the thermohydraulic security of the reactor. In mid-1988, based on results from irradiation tests (12) (13), the U_3Si_2-Al based dispersion fuel was qualified by the U.S. Nuclear Regulatory Commission and released for sale with uranium densities up to 4.8 gU/cm^3, with a swelling consistent with the commonly used dispersions (14).

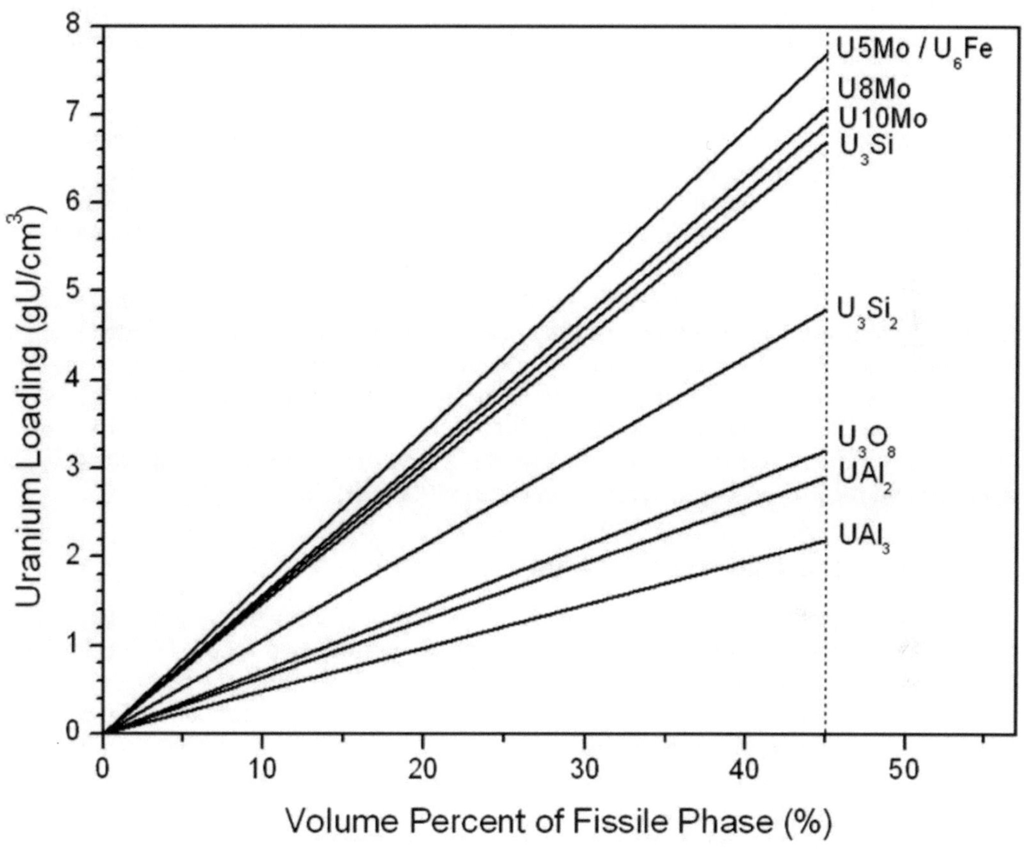

Fig. 1. Density of uranium in terms of concentration of dispersed phase for different fissile uranium compounds.

Research continued aiming at the use of intermetallic with even higher concentrations of uranium, such as U_3Si, U_3SiAl and U_6Fe as fissile material in the form of dispersions in aluminum. However, results of irradiation tests showed an unacceptable dimensional stability of these new fuels. Due to its high concentration of uranium (96 wt%) the U_6Fe was mainly considered (15), and the research was virtually abandoned in 1986 due to high swelling observed in irradiation tests, coupled with the promising results obtained with the U_3Si_2-Al dispersion, being considered a viable alternative (16).
Only through the use of U_3Si_2 as fissile material in the dispersion with aluminum was not possible to convert all research reactors. Many research reactors are awaiting a high-

performance technology solution finale, needing a uranium density of 6-9 gU/cm^3. In an effort to convert these reactors, other high density fuels has been studied, including dispersions based on U-Mo, U_3SiCu, $U_3Si_{1.5}$, $U_3Si_{1.6}$, $U_{75}Ga_{15}Ge_{10}$, $U_{75}Ga_{10}Si_{15}$ and uranium nitrides. Still, innovative manufacturing techniques has been investigated, which are based on hot isostatic compaction (HIP - Hot Isostatic Pressing) or increasing the volume fraction of U_3Si_2 beyond 50% (the limit currently accepted for this technology is 45%) or using wires of U_3Si and/or $U_{75}Ga_{10}Si_{15}$ and/or $U_{75}Ga_{15}Ge_{10}$ metallurgically bonded with aluminum in a geometry such that result in plates with a density close to 9 gU/cm^3 in the fuel core. The nearest alternative to be commercially deployed is the UMo alloy dispersion in aluminum, which enables to achieve the density near to 8 gU/cm^3. The performance under irradiation of this type of fuel is being tested with promising results. However, it is not a commercial fuel yet.

Thus, currently the most advanced manufacturing technology commercially available for the MTR type fuel plates is based on the U_3Si_2-Al dispersion, with a concentration of U_3Si_2 resulting in a uranium density into the fuel meat of 4.8 gU/cm^3. The next commercially available technology will probably use a dispersion of UMo alloy with 7-10 wt% Mo, resulting in a uranium density of between 6 and 8 gU/cm^3.

Each type of MTR fuel element is produced in accordance with a manufacturing specification and a set of manufacturing drawings agreed between the fabricator and the reactor operator or his representative. The specification sets down the scope and general conditions, the requirements of manufacturing method, together with the inspection requirements and acceptance criteria. In addition to the specification, an inspection schedule is normally produced which includes all of the supporting documentation such as the inspection and record sheet and certification (17; 18; 19; 20; 21).

3. Production of uranium tetrafluoride

The UF_4 has a specific role in nuclear fuel technology. It is an important intermediate product, being the basic substance to produce either uranium as metal (U^o) or uranium hexafluoride (UF_6) (17).

Uranium tetrafluoride (UF_4) is a green crystalline solid that melts at about 96°C and has an insignificant vapor pressure. It is slightly soluble in water. UF_4 is less stable than uranium oxides and produces hydrofluoric acid in reaction with water; thus it is a less favorable form for long-term disposal. The bulk density of UF_4 varies from about 2.0 g/cm^3 to about 4.5 g/cm^3 depending on the production process and the properties of the starting uranium compounds. Uranium tetrafluoride (UF_4) reacts slowly with moisture at room temperature, forming UO_2 and HF, which are very corrosive.

In principle, several other compounds may also be used for the production of metal and hexafluoride uranium, however, the use of UF_4 is prescribed by technological and economic considerations. It is considerably easier to obtain metallic uranium from UF_4, due to the reactivity of UF_4 mixture with reducing agent (mainly Ca and Mg) with large thermal outcome, which makes easy the production of uranium ingot.

According to the production process, the UF_4 must have certain specifications in regard to its purity. The content of uranium oxides and uranyl fluoride (UO_2F_2) may vary and also its density and its granulometric composition. The major technical requirement for tetrafluoride is observed during metallic uranium fabrication. It must contain at least 96% of tetrafluoride, virtually free of impurities. It should be anhydrous and having sufficiently high density.

When the reduction process to produce metallic uranium is performed at higher pressures and lower temperatures, normal tolerances up to 4% of UO_2 + UO_2F_2 should be reduced. It is recommended that the tapped density of loose UF_4 should be greater than 1 g.cm^{-3}.The good quality of the metallic uranium to be produced should have the UF_4specification as displayed in Table 1.

Elements	Al	B	Cd	C	Co	Cr	Cu	Fe	Mn	Ni	Si
% in mass	70	0,2	0,1	150	5	25	40	75	15	40	30

Table 1. Specification of the limits for main impurities in UF_4

In the case the oxide content is high, there would be larger losses of metal with the slag. As the reaction develops high amount of heat, it should be avoided evolution of tetrafluoride volatile components such as water and ammonia. During smelting, the metal is slightly contaminated with impurities from reducing agent and crucible. For this reason, UF_4 should be pure enough to allow slight contamination degree during this process. It must also be sufficiently dense. The load consists of blending of UF_4 powder and chips of calcium or magnesium tetrafluoride. The higher the density of UF_4, the greater the density of the load, and the greater the amount of heat involved per unit volume of the furnace (17).
The production of uranium tetrafluoride can be made by several processes which are divided into two groups, namely dry (fluorination of uranium oxide or hexafluoride reduction) and aqueous (preparation of UF_4 from U^{+6} salt) pathway (18) (19) (20) (21) (22).
The first task of obtaining UF_4 were carried through water (22) (23) by the end of the 19th century, and from an industrial standpoint that prevailed till the beginning of 20th century. The process essentially comprises the steps of reducing the uranium contained in uranyl fluoride solutions, uranyl chloride or uranyl sulfate up to its tetravalent state, followed by UF_4 precipitation by adding hydrofluoric acid.
With the development of dry processes, the aqueous processes were abandoned because they had difficulties in filtration, washing and drying, in spite of their simplicity and safety. Nowadays, the production via aqueous route is only used in plants to produce UF_4 for small quantities, which is the present experience of IPEN in producing LEU UF_4. Nevertheless, IPEN also developed the Brazilian technology for dry route.

3.1 Procedures for obtaining UF_4 via wet process

3.1.1 Preparation of UF_4 from salts U^{+6}

The UF_4 preparation methods through water have been developed mostly by the British and its modifications were based on work done by Bolton in 1866 (24; 25)
Essentially, the process consists in reducing the uranium, contained in solutions of uranyl fluoride, uranyl chloride and uranyl sulfate to the tetravalent state and the precipitation of uranium tetrafluoride by adding hydrofluoric acid. Several compounds of uranium have been used as starting materials and various reducing agents have been used. An overview of the process can be obtained from the reaction of uranyl fluoride with stannous chloride and sodium hyposulphide.

$$UO_2F_2 + SnCl_2 + 4HF \leftrightarrow UF_4 + 2H_2O + SnCl_2F_2 \quad (1)$$

$$UO_2F_2 + Na_2S_2O_4 + UF_4 + 2HF \leftrightarrow Na_2SO_3 + SO_2 + H_2O \quad (2)$$

An alternative to this process is the replacement of the electrolytic reduction by reducing agents that prevents the possible contamination with the reducing agent. This process has been adopted in countries like USA, Spain, Australia, Japan, Canada, England, South Africa and India (26; 27; 28; 29; 30; 31; 32). For the production of UF_4 with nuclear purity from UO_2F_2 acids solutions, some fundamental stages are required such as obtaining the solution, reduction to uranium valence and precipitation of the formed U^{+4}. These stages are shown in Figure 2, as schematized operations.

3.1.2 Obtaining UO_2F_2 solutions

Uranium hexafluoride is a crystalline substance at normal pressure and temperature conditions. At the temperature of 900C under a pressure of 3kgf/cm2, UF_6 becomes gas and when it is injected into water, it hydrolyzes immediately according to the following:

$$UF_6 + H_2O \rightarrow UO_2F_2 + 4HF \tag{3}$$

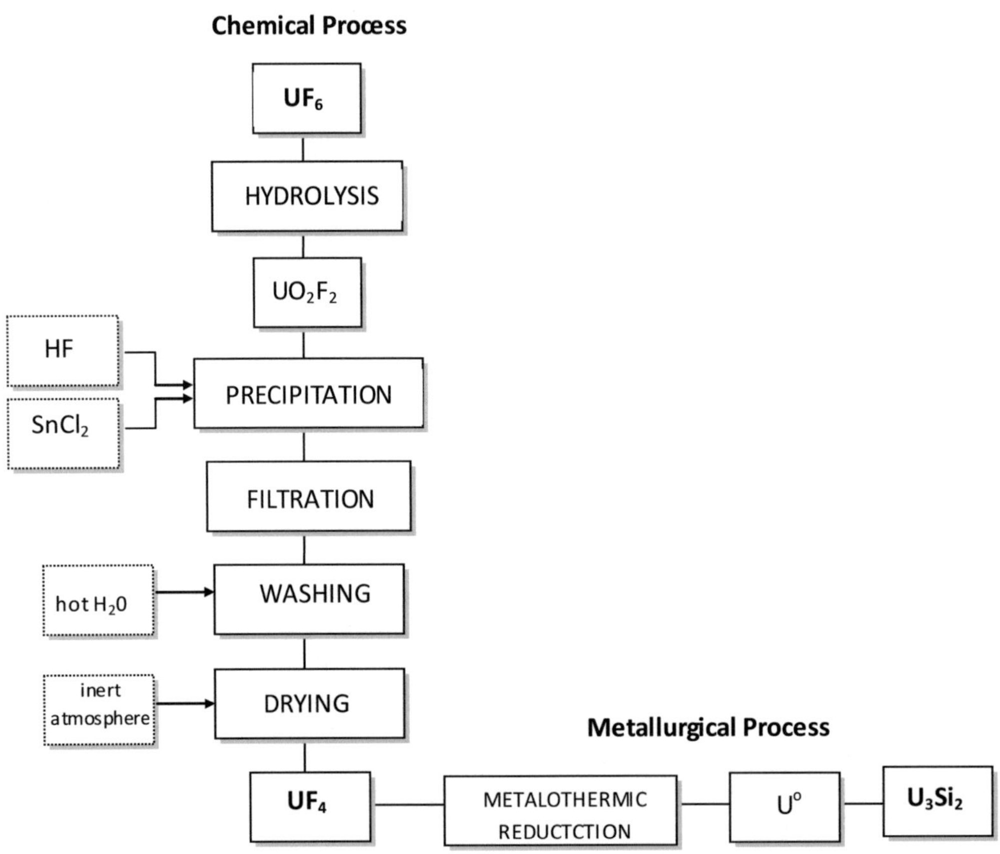

Fig. 2. Wet Process to produce UF_4

Table 2 shows the chemical characteristics of UO_2F_2 solution obtained from UF_6 hydrolysis.

Uranium (g/L) 60									
Fluoride (g/L) 17									
Metallic impurities (g/mL)									
Cd	B	P	Fe	Cr	Ni	Mo	Zn	Si	Al
<0.1	0.2	<100	1500	100	40	<2	100	300	40
Mn	Mg	Pb	Sn	Bi	V	Cu	Ba	Co	
10	15	<2	<2	<2	<3	3	1	<10	

Table 2. Chemical characteristics of UO_2F_2 solution

3.1.3 Chemical reduction of UF_6 to UF_4

Uranium in its tetravalent state is very important in different technological processes. Essentially, the preparation process (aqueous way) from solutions containing uranyl ion (hexavalent) involves the reduction towards tetravalent state, and later precipitation as UF_4 using HF solution. In aqueous solutions, these reductions can be carried out by chemical, electrochemical or photochemical methods.

All the trials for the preparation of UF_4 using chemical reduction have been carried out using UO_2F_2 solution inside a stainless steel reactor, coated with Teflon. The solution has been heated under continuous stirring to reach a temperature set, and the reducing agent has been added. Next, the precipitating agent solution is slowly added to UO_2F_2 in solution with hydrofluoric acid (HF). Tests have been carried out using some reducing agents, such as $SnCl_2$, CuCl, $FeCl_2$, $Na_2S_2O_4$.

$$UO_2F_2 + SnCl_2 + 4HF \rightarrow UF_4 + SnClF_2 + 2H_2O \quad (4)$$

$$UO_2F_2 + 4HF + Fe \rightarrow UF_4 + FeF_2 + 2H_2O \quad (5)$$

$$UO_2F_2 + CuCl + 4HF \rightarrow UF_4 + CuClF_2 + 2H_2O \quad (6)$$

$$UO_2F_2 + Na_2S_2O_4 + 2HF \rightarrow UF_4 + Na_2SO_3 + H_2O \quad (7)$$

Upon UF_4 precipitation the suspension is left in rest up to reaching room temperature. After over 12 hours, it was performed the solid/liquid separation by vacuum filtration, washing and drying in a muffle kiln. The salts obtained were all identified as being uranium tetrafluoride. According to the results shown in Figure 3, it is evident that, from all used reducing agents, only $SnCl_2$ and $FeCl_2$ have shown significant results in regards of getting UF_4. Nevertheless, $SnCl_2$ is more consistent reducing agent at higher temperature of process.

The influence of the temperature upon UO_2F_2 and UO_2 contents in obtained UF_4 is shown in Figure 4. It was employed $SnCl_2$ as the reducing agent in this study to precipitate UO_2F_2 solution. The residual moisture is dried at 130°C. The tin content in all obtained UF_4 has shown to be in the range of 0.15 - 0.15%.

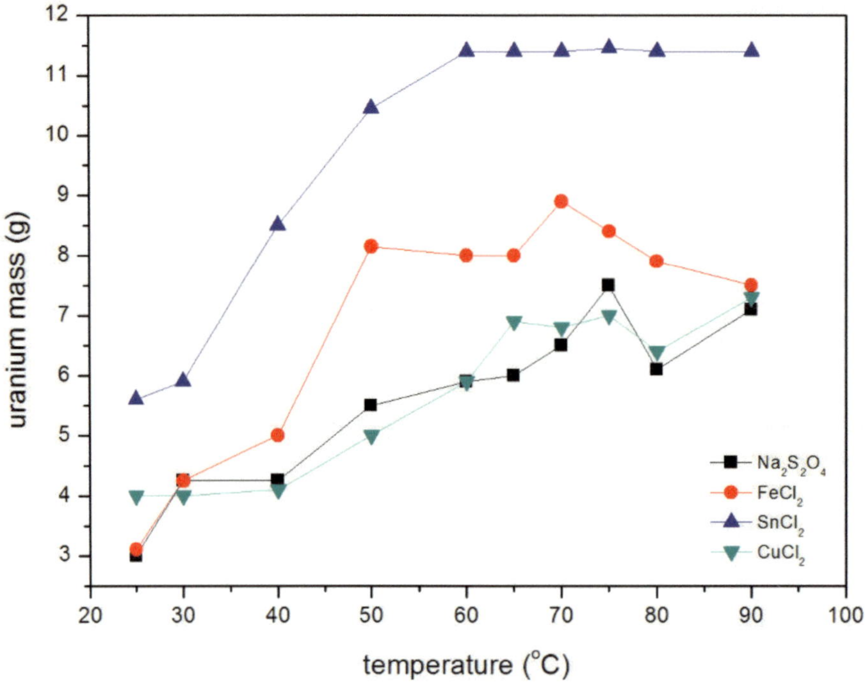

Fig. 3. Influence of reducing agent as a function of obtaining UF_4

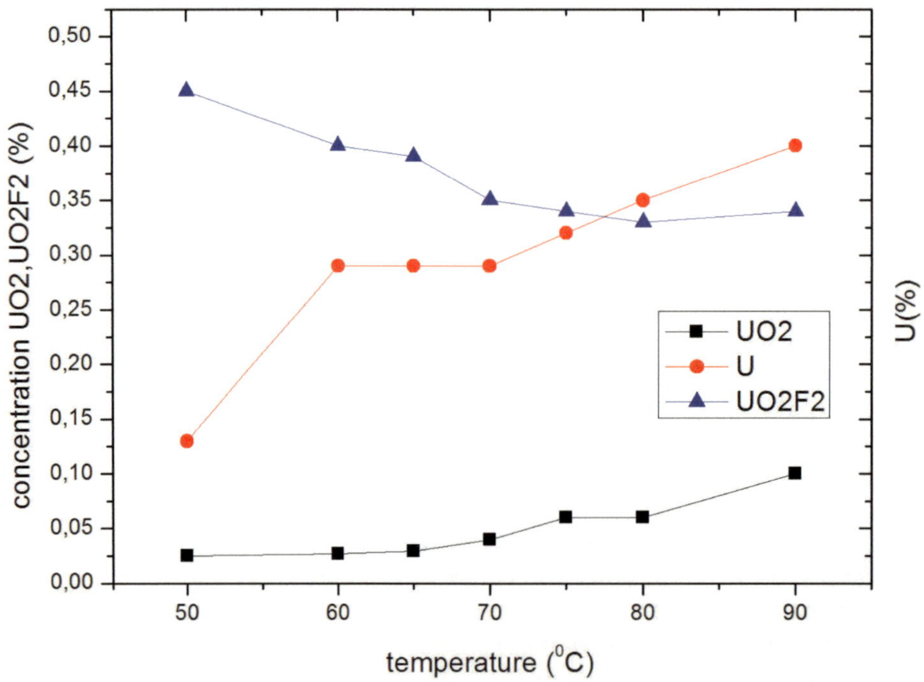

Fig. 4. Influence of the temperature as a function of the contents of UO_2 F_2 and UO_2 in UF_4

3.1.4 Obtaining UF4

As shown previously, the process for obtaining UF_4 by reduction precipitation using $SnCl_2$ had the best results and achieved an yield of 98% of UF_4 precipitation. The precipitation with HF solution is relatively slow and tends to accelerate as the temperature rises (17; 18). This is important, since it avoids excessive precipitate hydration and facilitates the sedimentation, filtration and drying operations. The full reaction is represented by:

$$UO_2F_2 + SnCl_2 + 4HF \rightarrow UF_{4\,pp} + SnCl_2F_2 + 2H_2O \qquad (8)$$

During the uranium processing stages, the goal is to achieve an end product with high purity and showing physical and chemical characteristics appropriate for the preparation of nuclear fuel.
Table 3 lists the suitable chemical and physical characteristics of UF_4 for a later reduction to obtain metallic uranium.

	at 130°C			inert atmosphere at 400°C				
Uranium (%)	74.20			75.0				
Fluoride (%)	24.60			27.90				
UF_4 (%)	97.50			99.85				
UO_2F_2 (%)	0.29			0.34				
UO_2 (%)	0.06			0.29				
HF(%)	0.23			0.12				
Moisture (%)	0.33			<0.03				
Crystallization H_2O	4.50			<100				
Met. Impurities	Fe	Cr	Ni	Mo	Al	Mn	Cu	Sn
(µg/g)	<20	<10	<10	<5	<10	<5	<5	0,1
Density (g/cm^3)	6.70							
Granulometry (m)	15.0							
Specific Surface (m^2/g)	0.21							

Table 3. Chemical and Physical Properties of UF_4 produced by an aqueous route

3.1.5 Preparation of UF_4 from UO_2

The UF_4 obtained by reaction with UO_2 with hydrofluoric acid is easily made. The reaction can be summarized as follows:

$$UO_2\,(s) + 4HF\,(aq) \leftrightarrow UF_4\,(s) + 2H_2O \qquad (9)$$

This process has some advantages over the other processes. Since the reaction occurs at low temperatures, the reactor can be constructed using materials as polyethylene, polypropylene or carbon steel with plastic coating, while other processes require equipment built with metal (monel, inconel, nickel) which increases the cost of a plant.
In Figure 5, the x-ray diffractogram spectra are presented for UF_4 produced by the method via NH_4HF_2 (bifluoride route) and by aqueous route. Typical SEM image of precipitated UF_4 is presented in Figure 6. It displays a granular structure with relevant amount of porosity.

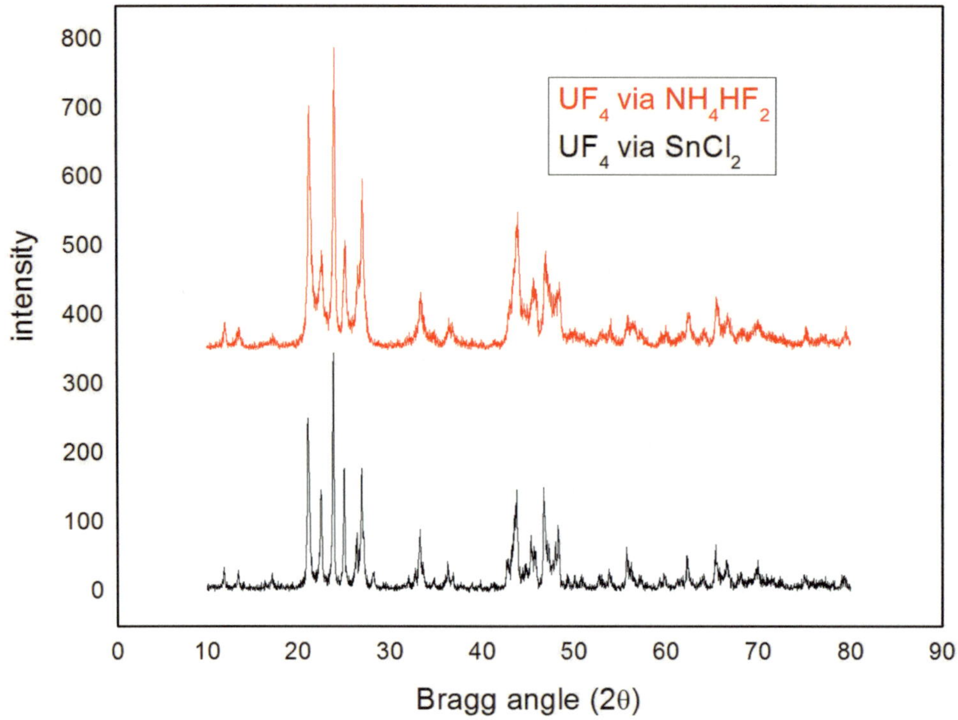

Fig. 5. X-ray diffraction pattern of UF_4 produced by the bifluoride route and from the aqueous route.

The UF_4 fabrication using fluorination media with ammonium bifluoride is perfectly feasible. The ammonium bifluoride is a by-product effluent generated during the UF_6 conversion to AUC[1]. UF_4 obtained by this route has the same crystalline structure presented by the aqueous process, as demonstrated by the x-ray spectrum. Besides, it has the correct chemical and physical characteristics for metallothermic production of metallic uranium. Even presenting a lower relative tapped density; this property will not be a problem, because this is an alternative process that has as main goals the recovery of uranium, ammonium and the fluorides of the liquid effluents generated in the process of UF_6 reconversion. This UF_4 will be lately diluted in the UF_4 charges produced by the aqueous route. The development of this process (bifluoride route) not only provides an efficient process for uranium recovery from secondary sources, as also eliminates the environmental pollution by discarding the bifluoride. It also provides a chemical compound with chemical and physical characteristics very similar to the aqueous route ($SnCl_2$).

[1] Ammonium uranyl carbonate ($UO_2CO_3 \cdot 2(NH_4)_2CO_3$) is known in the uranium processing industry as AUC and is also called uranyl ammonium carbonate. Ammonium uranyl carbonate is one of the many forms called yellowcake in this case it is the product obtained by the heap leach process. This compound is important as a component in the conversion process of uranium hexafluoride (UF_6) to uranium dioxide (UO_2). In aqueous process uranyl nitrate is treated with ammonium bicarbonate to form ammonium uranyl carbonate as a solid precipitate and ammonium bifluoride as by-product (41).

Fig. 6. SEM image of some UF_4 particles, produced by the bifluoride(a) route e via SnCl2 (b).

3.2 Procedures for obtaining UF_4 by dry process

3.2.1 Preparation of UF_4 by fluorination of UO_2

The achievement of UF_4 by this process was adopted in Canada, France, the former Czechoslovakia, South Africa, United States, Portugal, Brazil, Germany and Sweden (17; 21; 23).

The sequence of operations is to reduce UO_3 by hydrogen, followed by treatment with HF resulting UO_2 anhydrous at atmospheric pressure.

$$UO_3 (s) + H_2 (g) \leftrightarrow UO_2 (s) + H_2O (v) \quad (10)$$

$$UO_2 (s) + 4HF (g) \leftrightarrow F_4 (s) + 2H_2O (v) \quad (11)$$

The reduction of UO_2 is performed at temperatures of 500-700°C. Another alternative is the reduction of U_3O_8 recommended when you have storage problems UO_3, being extremely hygroscopic.

$$U_3O_8\,(s) + 2H_2\,(g) \leftrightarrow 3UO_2\,(s) + 2H_2O\,(v) \tag{12}$$

In such a process is commonly used the moving bed or fluidized bed reactor type.
Preparation of UF_4 by reaction of the UO_3 with NH_3 and HF gaseous
The process consists of only one step to produce UF_4. The mixture consisting of NH_3 and HF is treated with UO_3 at 500-700°C. This reaction is fast and produces high purity UF_4:

$$3UO_3 + 2NH_3 + 12\,HF \leftrightarrow 9H_2O + N_2 + 3UF_4 \tag{13}$$

The UF_4 fabrication by the reaction of uranium oxides with fluorinated hydrocarbons (freon) is as follows:

$$2CF_2Cl_2 + UO_3 \leftrightarrow UF_4 + CO_2 + Cl_2 + COCl_2 \tag{14}$$

The literature shows results of reactions of different freons with uranium oxides UO_2, U_3O_8 and UO_3 (27; 29; 33). The reactors used in this process cannot be constructed using nickel, copper, platinum and stainless steel, since they undergo chemical attack of reagents, besides this reaction promotes pyrolysis under carbon presence. The reactors are constructed with graphite or calcium fluoride, which may cause contamination to the obtained UF_4. The advantages of this method are equipment simplicity and the possibility of applying this reaction to all the uranium oxides.

3.2.2 Preparation of UF_4 from metallic uranium or uranium hydride (UH_3)

By fluoridation at high temperatures uranium metal can be quickly converted into uranium tetrafluoride by the reaction below:

$$U + 3/2H_2 \overset{250^\circ C}{\leftrightarrow} UH_3 \tag{15}$$

$$UH_3 + 4HF \overset{200^\circ C}{\leftrightarrow} UF_4 + 7/2H_2 \tag{16}$$

Uranium metal is industrially manufactured from UF_4. In the absence of advantage in obtaining first elemental uranium and transform it into UH_3, then get to UF_4.

3.2.3 Procedures for obtaining UF_4 by dry ammonium bifluoride with (NH_4HF_2)

The fluorination of UO_2 is made with NH_4HF_2, a white solid; it has low vapor pressure and can be operated freely since it is non-toxic. Initially, UO_2 is mixed with bifluoride, 20% above the stoichiometric amount. The bifluoride crystal is easily crushed and the mixture of $UO_2 + NH_4HF_2$ is made in a monel 400 container to prevent contamination.
The conversion of bifluoride at room temperature occurs after approximately 24 hours, although under such conditions the water formed in the reduction may be retained in the precipitate. The elimination of NH_3 and water is facilitated by the reaction of UO_2 and NH_4HF_2 at 150°C:

$$2UO_2 + 5NH_4HF_2 \leftrightarrow 3NH_3 + 4H_2O + 2NH_4UF_5 \tag{17}$$

At this temperature, only 8 hours are necessary to promote the fluorination. The material is loaded into an aluminum container with calcium fluoride and heated inside a furnace. The furnace is fitted with a condensing tube with a relief valve, which releases the water and ammonia from the fluoridation reaction to a reservoir and retains the excess of sublimed bifluoride.

During the fluorination and/or decomposition, the formation of UO_2F_2 probably occurs. This is a significant happening, since it may reduce the efficiency of reduction in the next step.

In a second step of the process, under vacuum distillation, NH_4UF_5 is decomposed in UF_4 with the NH_4F by this reaction:

$$NH_4\,UF_5 \leftrightarrow UF_4 + NH_4F \tag{18}$$

3.2.4. Preparation of UF_4 by the reaction of ammonium bifluoride with UO_3

The UF_4 can be prepared by reaction of ammonium fluoride or bifluoride with UO_3 according to the equation:

$$3UO_3\;6NH_4HF_2 + 9H_2O \leftrightarrow 3UF_4 + 4NH_3 + N_2 \tag{19}$$

Although the United States have been among the first to study the process (34) Canada is the country that developed this process (35; 36)

4. Production of metallic uranium

There are several possibilities to produce metallic uranium (41; 26; 42). Magnesiothermic reduction of UF_4 is one of them and it is a known process since early 1940's (7; 8). The IPEN technology uses this route in 1970-80's for production 100kg ingots of natural uranium. For LEU U-production, it is necessary to handle safe mass (less than 2.2 kg U), to avoid possible criticality hazards. IPEN presently produces around of 1000g LEU ingots via magnesiothermic process and in future may produce 2000g or more. This range of uranium weight is rather small if compared to big productions of natural uranium. Metallic uranium is reported (9) to be produced with 94% metallic yield when producing bigger quantities. The magnesiothermic process downscaling to produce LEU has small possibilities to achieve this higher metallic yield. This is due to the design of crucibles, with relatively high proportion of surrounding area, which is more prone to withdraw evolved heat from the exothermic reaction during uranium reduction. Normally, calciothermic reduction of UF_4 is preferred worldwide, since the exothermic heat is higher (-109.7 kcal/mol) compared to smaller amount of -49.85 kcal/mol using magnesium as the reducer (10). Nevertheless, IPEN chose magnesiothermic because it is easier to be done, avoiding no handling of toxic and pyrophoric calcium. Moreover, the magnesiothermic process is cheaper, so, it brings economical compensation for its worse metallic yield than calcium reduction process. In addition, the recycling of slag and operational rejects is highly efficient and there are virtually insignificant LEU uranium is lost (23).

The magnesiothermic reaction is given by:

$$UF_4 + 2Mg = U + 2MgF_2\;\Delta H = -\,49.85\text{ kcal/mol (at 640°C)} \tag{20}$$

As magnesium thermodynamics is less prompt to ignite than calcium, the batch reactor is heated up to the temperature around 640°C. The routine shows that this ignition normally happens some degrees bellow this temperature (9). Nevertheless, several reactions may occur during heating of the UF_4+Mg load. Moisture is normally present in the charge, either caught during UF_4 handling after drying or during crucible charging. During heating, as the temperature crosses the water boiling point (>100°C), all moisture becomes water vapor. This vapor not only bores its passage through the load but easily oxidize the reactants in this pathway by the following reactions (30):

$$UF_4 + 2H_2O \rightarrow UO_2 + 4HF \quad (21)$$

$$2UF_4 + 2H_2O \rightarrow 2UO_2F_2 + 4HF \text{ (via } UF_3(OH) \text{ and } UOF_2 \text{ steps)} \quad (22)$$

As the loading of the charge is not fully sealed to avoid atmosphere contact, some O_2 is entrapped in the system, leading also to reactants oxidation by:

$$2UF_4 + O_2 \rightarrow UF_6 + UO_2F_2 \quad (23)$$

Producing some UF_6 that transforms into UO_2F_2 by the following reaction:

$$UF_6 + 2H_2O \rightarrow UO_2F_2 + 4HF \quad (24)$$

and also occurring magnesium oxidation (very fast above 620°C) by:

$$2Mg + O_2 \rightarrow 2MgO \quad (25)$$

The presence of the UO_2 and UO_2F_2 in the produced UF_4 accumulates with previous oxidized ones during the dehydration. All these compounds formation worsens the metallic yield of uranium production.

In this work, it is discussed the effect of LEU UF_4 precipitated via hydrolyzed UF_6 and its potential variability in reactivity. The chemical UO_2F_2 residual content in dried UF_4 is also analyzed for its potential relevance in the uranium production. The tapped density of dehydrated and loaded UF_4 is also commented as affecting the reactivity process of uranium production. The magnesiothermic ignition is also analyzed since the heating time of the charge may affect the reactivity of the load. The reaction sequence after ignition is theoretically proposed as a possible sequence of chemical and physical events. The evidences in the slag solidification on crucible wall, during the reaction process to reduce UF_4 towards U°, is very enlightening to guide towards the interpretation of the reaction blast.

The IPEN's magnesiothermic reduction process of UF_4 to metallic uranium (in the range of 1000g) could be synthesized as:

1. In preparation for the mass reduction of a single batch, it is used with a standard charge of reactants of 1815 ± 5g of the mixture Mg + UF_4 (1540 ± 1g LEU UF_4) containing 15% excess of stoichiometric Mg content. For purpose of homogenization, the charge of UF_4 + Mg is divided into 10 layers, which are tapped one by one inside the crucible. All this operation is carried out inside a glovebox to prevent nuclear contamination. This sequence is illustrated in Figure 7.
2. After placing the reactants inside the graphite crucible, a variable amount of CaF_2 is tapped over the UF_4+Mg load in the crucible to fully complete the reaction volume. This amount is dependent on tapped density and UF_4+Mg blending, which varies in function to UF_4 fabrication. The crucible is made of fully machined graphite volume with enough resistance to produce safe nuclear uranium amount around 1000g. This crucible was designed to withstand the blast impact of metallothermic reaction, as well as thermal cycles of heating and cooling without excessive wear in order to be used in several batches.

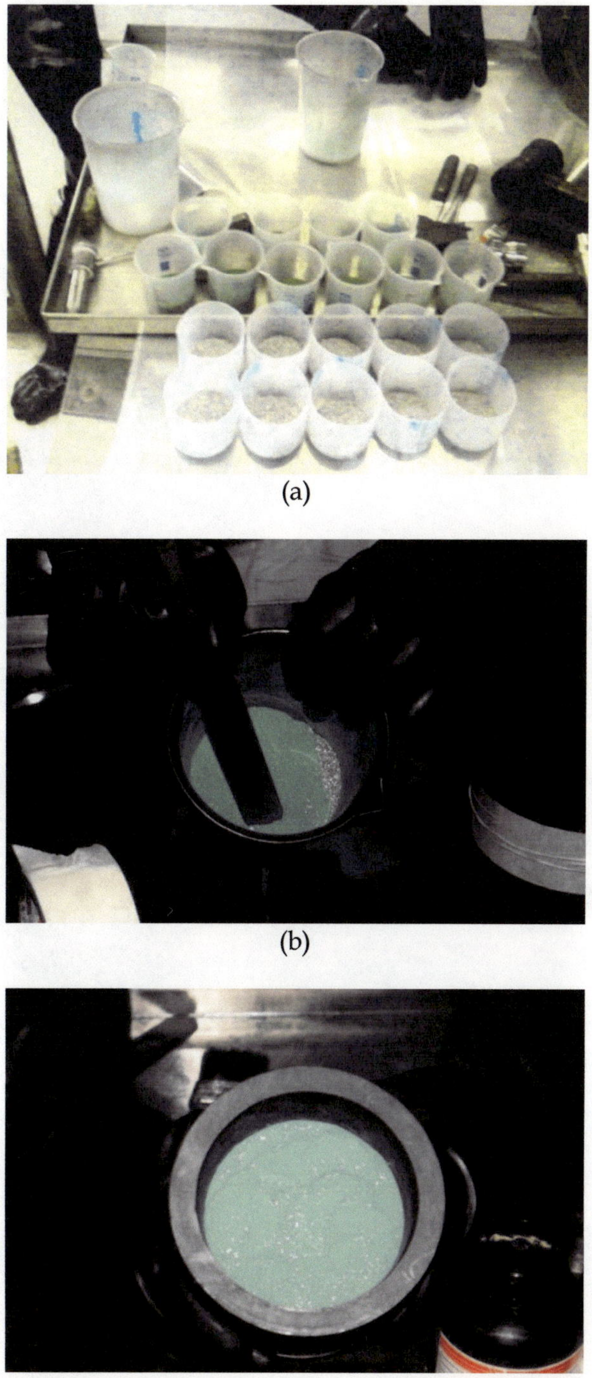

(a)

(b)

(c)

Fig. 7. Sequence of UF4+Mg charging in IPEN's magnesiothermic method to produce metallic uranium. (a) 10 layer preparation of UF_4 (green) and Mg (metallic bright); (b) blending of material; (c) full charge after tapping the 10 layers.

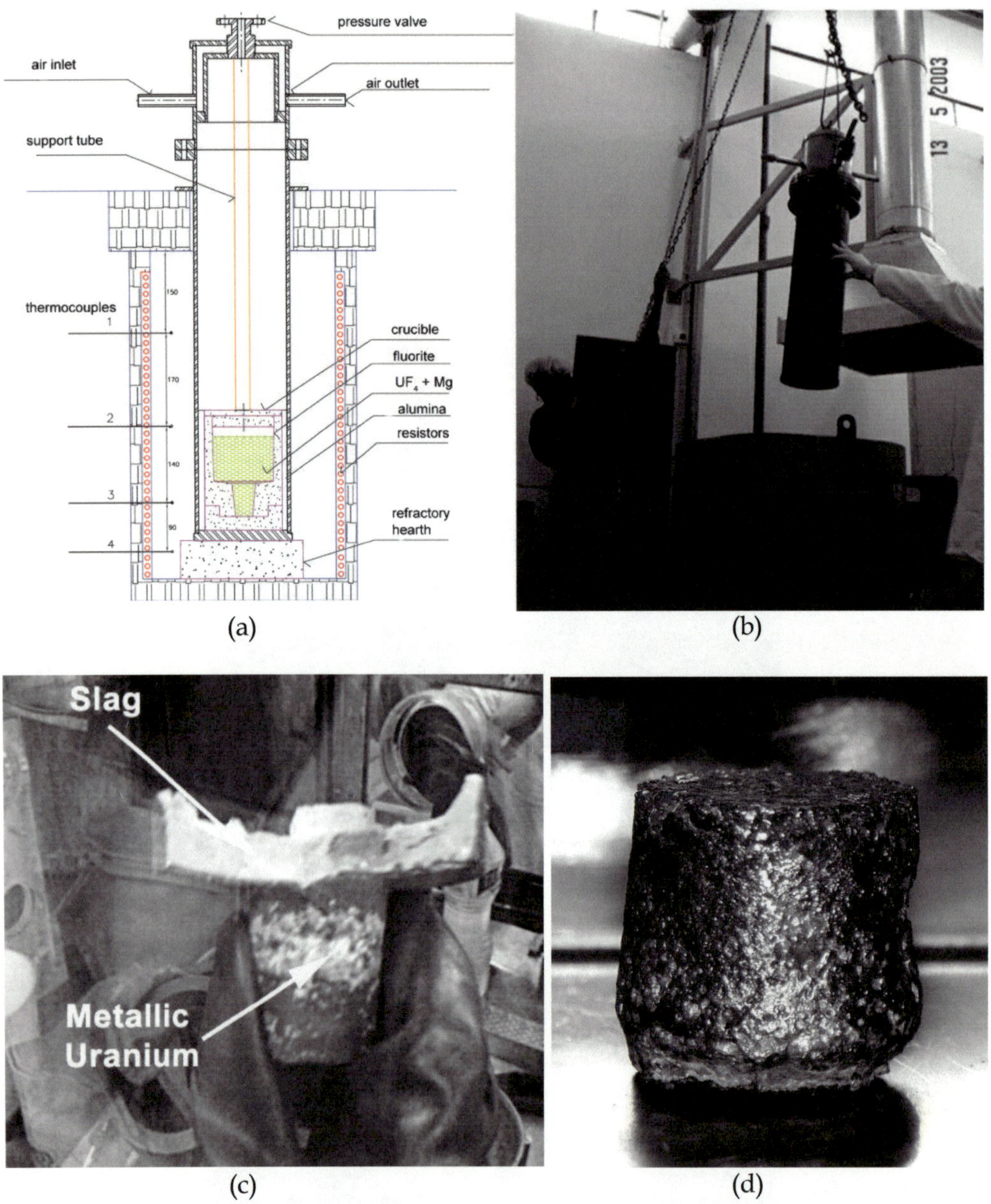

Fig. 8. (a) Schematic drawing of pit furnace, reactor vessel and crucible; (b) Charging of the reactor vessel inside the pit furnace; (c) Raw metallic uranium and upper deposited slag after removing from the crucible; (d) Metallic uranium after cleaning.

3. After closed with the top cover, the crucible is inserted inside a stainless steel cylindrical reactor vessel, made of ANSI 310, which allows argon fluxing during batch processing (1 L/min with 2 kgf/cm^2 of pressure). As shown in Fig. 8 (a-b), the whole crucible + reactor are placed in resistor pit furnace with four programmable zones having the possibility of raising the temperature up to 1200°C.
4. The reaction vessel is set to heat up to 620°C. At this level, the reaction ignition is expected. The total heating time and waiting for ignition is about 180 minutes from heat time to temperature setting point.
5. The reaction of UF_4 with Mg produces an intense exothermic heat release inside the crucible. It is considered as an adiabatic reaction. It produces metallic uranium and MgF_2 slag in liquid form. Both products deposit in the crucible bottom are easily taken apart after opening the crucible. Some products project over the crucible wall and freeze there.
6. This full reaction happens in a noticeable time between 800 and 1200ms from ignition to final deposit. This control is measured by sound waves, using an accelerometer.
7. After the reaction, 10 minutes is awaited for full solidification of reaction products inside the furnace. Then the furnace is turned off and the reactor vessel is lifted out of the furnace. There is a 16 hours for cooling before its opening. This avoids firing of metallic uranium in contact with atmosphere.
8. The disassembling of reduction set is performed inside a glove box. The top and bottom covers of the crucible are removed. By means of rubber soft hammering, it is able to withdraw the uranium ingot. The MgF_2 slag is removed by mechanical cleaning. The metallic uranium is pickled in nitric acid 65%vol and the final mass of metallic uranium is measured and its density evaluated by Archimedes' method.

5. Production of uranium silicide

The intermetallic U_3Si_2 is produced from metallic uranium (47). This alloy is produced from a uranium ingot and hyperstoichiometric silicon addition (7.9% Si). The induction furnace (15 kW) should be submitted to 2.10^{-3} mbar vacuum and flushed with argon-atmosphere. Then the melting is carried out. The blend is molten inside an induction furnace using zirconia crucible reaching more than 1750°C, as this intermetallic requests this level of temperature to be properly homogenized before solidification. No other crucibles, than a zirconia one could bear the aggressive environment created by uranium attack on linings. The load arrangement of uranium and silicon, as shown in Fig. 9, is then charged inside the crucible. It was planned to help the sequence of melting during the several stages that passes the alloy formation until reaching the final intermetallic composition. The quality of this intermetallic produced in this way normally meets the requirements as nuclear material. The X-ray diffractogram (Fig. 9) confirms the necessary proportion of phases presents in the produced powder of this alloy, which should be more than 80wt% of crystalline phases. As rule of thumb, the chemical amounts of boron, cadmium, cobalt, lithium should be less than 10μg/g individually. The other may reach hundreds of μg/g up 1000 μg/g. Carbon could reach up to 2000 μg/g. Isotopic concentration of ^{235}U is 19.75±0.20wt%. The required density is 11.7g/cm^3.

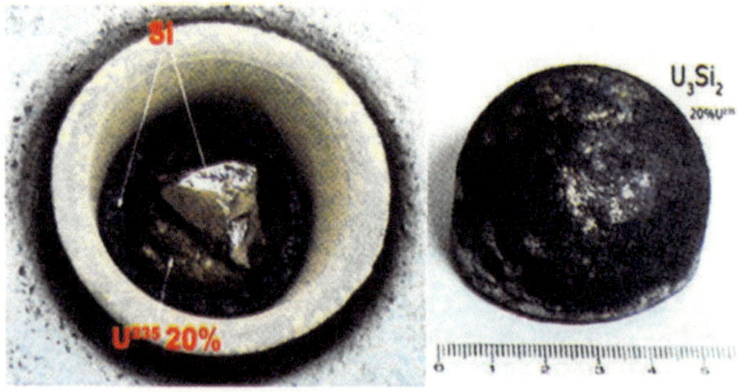

Assembling of U_3Si_2 crucible load

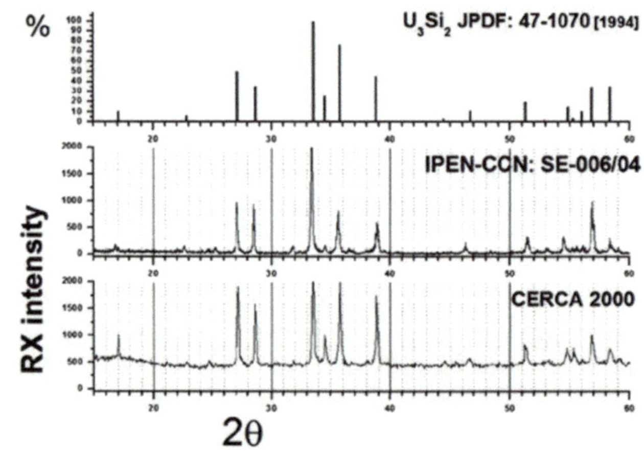

X-ray diffractogram of U_3Si_2 produced in Ipen compared

Fig. 9. Crucible arrangement of before melting to produce the intermetallic. U_3Si_2 product and its x-ray diffractogram results compared to CERCA product and JPDF 47-1070 for pure U_3Si_2.

6. Production of MTR nuclear fuel

The reference industrial process to produce plate-type fuel involves roll-milling together the fissile core, or fuel meat (a blend of an uranium compound and aluminum powders), and the cladding (aluminum alloy plates). This process can draw on considerable feedback from experience, since nearly all research reactors use this type of fuel. The process has seen large-scale implementation with NUKEM, in Germany, UKAEA, in the United Kingdom, CERCA, in France, and Babcock, in the United States.

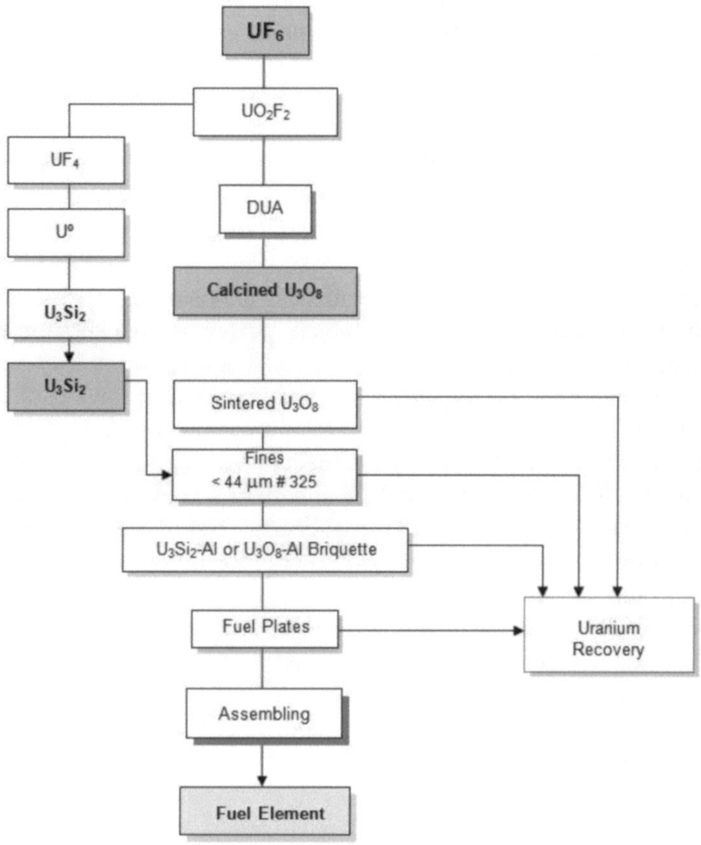

Fig. 10. Fabrication process of silicide fuel elements.

In general, the MTR type fuel element fabrication process using silicide (U_3Si_2) can be divided into the following main steps: hydrolysis of UF_6 through its reaction with water; production of uranium tetrafluoride (UF_4); production of metallic uranium; U_3Si_2 powder production from uranium metal; production of fuel cores from U_3Si_2 and aluminum powders; production of fuel plates with U_3Si_2-Al dispersion; assembling of fuel elements; recovery of uranium; effluent treatment; quality control.

The simplified block diagram of the fabrication process for silicide fuel elements is shown in Figure 10. The manufacturing process of the fuel begins with the UF_6 processing. The UF_6 is enriched to 19,75 wt% ^{235}U, a enrichment level that categorize the fuel as LEU (low enriched uranium). Bellow the main stages of manufacture of such fuel are discussed.

6.1 Fuel cores production from U_3Si_2 and aluminum powders

The U_3Si_2 ingot produced in the previous step is transferred to a glove box with inert atmosphere of argon, since the U_3Si_2 is pyrophoric. Inside the glove box, the ingot is subjected to a preliminary grinding, resulting in granules less than 4 mm in size with the smallest fraction of fines (< 44 μm) possible. This operation is performed with the aid of a manual crusher. After doing the preliminary grinding, the material is placed directly on a set of sieves, and then sieved by hand. The sieve set comprises a coarse sieve with 4 mm opening, a fine sieve with 150 μm opening and a background compartment. The granules with a diameter greater than 4 mm are crushed again. The granules with size between 4 mm and 150 μm are collected for final grinding and particles smaller than 150 μm are collected separately for particle size classification.

The U_3Si_2 obtained after the preliminary grinding is manually milled again. The material collected during the preliminary grinding (between 4 mm and 150 μm) is processed in this step. The grind is done carefully, with intermediate sieving, to classify the powder in the range from 150 to 44 μm. The specification allows 20 wt% fines fraction (below 44 μm) as maximum. The fraction above the specification (150 μm) is sent back to the final grinding system. The fraction inside the specified range (between 150 μm and 44 μm) is collected and stored. The fraction of fines (< 44 μm) is collected and stored separately. The glove box contains a vibrating screening machine, which performs the separation of three size fractions of silicide powder, above 150 μm, between 150 and 44 μm and bellow 44 μm. The batch U_3Si_2 powder composition is adjusted to have maximum fines content in the level of 20 wt%, as specified.

The next process step is the fabrication of the fuel cores, which will form the core of the fuel plates, or fuel meats. The core of the fuel plate contains U_3Si_2 as the fissile material. This core is fabricated by means of powder metallurgy techniques and is normally called briquette or fuel compact. Initially, the mass and composition of the briquette are calculated based on the analyzed values of total uranium and isotope enrichment of the U_3Si_2 powder. The criterion for calculating the briquette mass is the amount of the isotope ^{235}U specified for the fuel and the dimensions of the briquette. Based on the calculated mass of the briquette, the silicide Al powders mass are determined separately and mixed together to ensure that the specified ^{235}U amount is uniformly distributed. These charges are cold pressed to form the fuel compacts, and the briquettes are measured and weighed. The final dimensions of the fuel meat in the finished fuel plate are set by specification and the volume of the briquette is calculated from these data by their values of thickness, width and length. The thickness of the briquette is obtained by multiplying the specified thickness of the fuel meat by the deformation dimension resulted after rolling operation, assuming zero enlargement. The core content of voids depends only on the volume fraction of fuel powder content. To optimize the final geometry of the rolled core, the briquette has rounded corners, and the volume of the corners is included in the calculation of volume.

The difference between the volume of the briquette, obtained as described above, and the volume of the fuel powder, as determined by the division between the mass of the powder and its density, determines the amount of aluminum powder to be added to the mass of the briquette. As the theoretical density of the system cannot be achieved during the compaction of the briquette, the volume of aluminum is reduced by the amount of pores that remain after pressing. The total mass of the briquette is given by the calculated mass of fissile material powder added to the calculated mass of aluminum powder.

According to the calculation for the masses of U_3Si_2 and aluminum powders, the charges for pressing are weighed separately. The weighing is carried out in glass bottles specially designed for installation in a homogenizer. Once the powders are weighed, the charge is mixed inside a glove box with inert atmosphere. This blending ensures that the specified amount of ^{235}U is homogeneously distributed throughout the briquette to be pressed. The weighing operation is performed carefully and, after homogenization, the cautious handling of the charge is critical to avoid segregation.

The homogenization operation is performed using a special homogenizer with a capacity for simultaneous mixing of eight charges. The duration of homogenization is 120 minutes under rotation of 36 rpm and angle of 45°. To prepare the briquettes, the homogenized charges are pressed at room temperature using a hydraulic press with capacity for 700 tons, which is placed in a glove box. The pressing pressure is adjusted to get the desired thickness, keeping the residual porosity from 5 to 7% by volume. The bottle containing the homogenized charge is transferred from the glove box used for homogenization to the glove box used for pressing. Within this glove box, the charge of a briquette is emptied into the die cavity with the inferior puncture initially raised. The powder is placed in layers with the aid of a special smoother to prevent segregation and to minimize the variation of the thickness of the briquette, lowering the punch inferior gradually until all the charge is loaded, when the punch is fully lowered to its position during the pressing. Then, the superior punch is inserted and pressure is applied and maintained for 15 seconds. The entire array is then opened to eject the briquette and the punch superior, which is manually removed. The thickness of the briquette is defined based on final specifications valid for the fuel meat. This thickness is theoretically calculated and then adjusted through manufacturing tests.

Immediately prior to the transfer of the briquettes to be used in the manufacture of fuel plates, they are vacuum degassed at 2 x 10^{-3} torr in a retort. The temperature is 250 °C kept for 1 hour. After remaining inside the degassing retort for the time and temperature specified, the briquette is removed for cooling, keeping the vacuum system working until the room temperature is reached. Thus, the briquettes that will compose the cores of the fuel plates are used in the new phase of processing, or assembling the sets for rolling. Figure 11 illustrates the process for preparing the briquettes and the set.

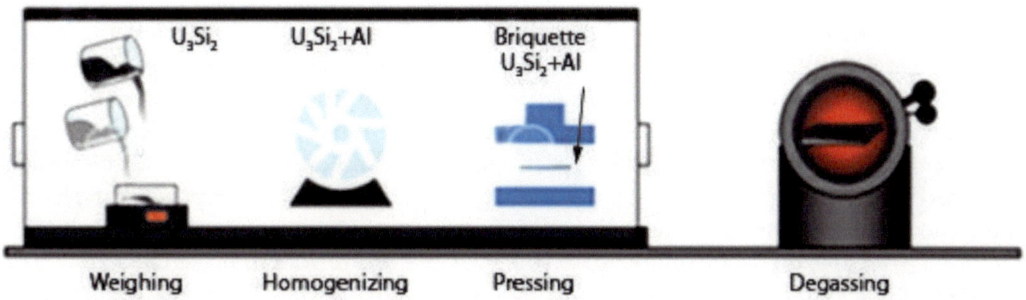

Fig. 11. Process for briquettes preparation and degassing.

6.2 Production of fuel plates with U_3Si_2 – Al dispersion

The technology of fuel plates manufacture adopts assembling and rolling of a set composed by the fuel meat (briquette), a frame plate and two cladding plates. In this way, after the rolling operation, it is fabricated a fuel plate containing inside the fuel meat totally isolated from the environment, which is done through the perfect metallurgical bonding between the core and frame with the claddings. The frame and cladding plates are made from commercial aluminum Al 6061 alloy (48).

In order to prepare the rolling assemblies, the frame plate is heated in a furnace at 440 °C. The cold briquette is then assembled inside the frame plate. Once cooled the frame, the briquette should be perfectly housed and fixed in the frame cavity by mechanical interference. The other cladding plates are placed above and below the frame plate with the core, completing then the assembling to be rolled. This assembly set is then fixed in a rotating welding bench and welded at its edges. The welding is TIG type protected with argon. A continuous welding bead is done on the four corners of the assembly, leaving the ends free in order to allow air to be exhausted in the first rolling pass. Figure 12 illustrates the procedure of preparing the assemblies for rolling.

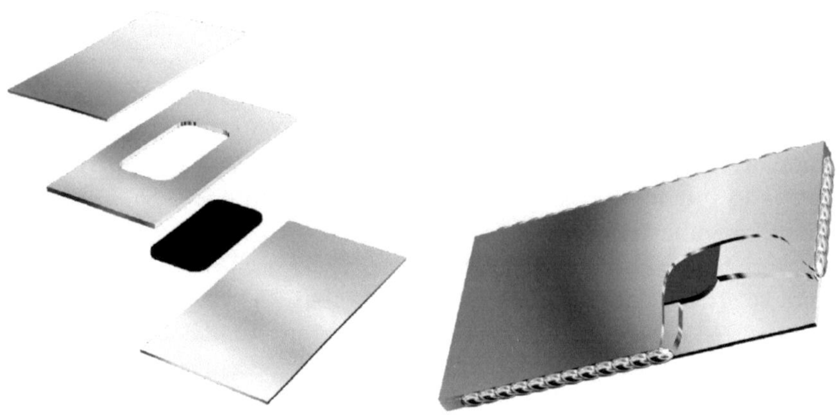

Fig. 12. Diagram illustrating the assembling of the set core-frame-claddings.

The welded assemblies are properly identified and inserted in a furnace for 60 minutes at a temperature of 440 °C. The hot rolling is performed in several passes following a well-established rolling schedule. The rolling schedule defines thickness reduction per pass in order to control the end defects and the final dimensions of the fuel meat. The rolling schedule is determined by theoretical calculations and empirical data from manufacturing tests and must guarantee the metallurgical bonding and the control and reproducibility of the fuel meat deformation. The rolling mill usually has an accuracy of 0.025 mm and is equipped with rolling cylinders coated with a chrome layer. It is important the perfect lubrication of the rolling cylinders. Between each pass, the assemblies are reheated for 15 minutes. After the final hot-rolling pass, the fuel plates are identified again in the same position of the initial identification in a region outside the fuel meat, using mechanical marker.

After hot rolling, a blister test is performed to test the metallurgical quality of bonding between meat-frame-claddings. The hot rolled plates are heated at 440 °C for 1 hour. After removal from the furnace, the fuel plates are visually inspected for observation and recording of bubbles (47; 48). Fuel plates that present bubbles are registered as reject and forwarded for chemically recover of uranium.

The cold rolling operation is performed in the same rolling mill used in the hot rolling. In this operation the specified thickness is achieved with precision. The total cold reduction is approximately 10% in thickness and is applied in one or two passes. During cold rolling, the length of the fuel meat is checked, ensuring the fulfillment of the specification for the minimum core length and for the thickness of the fuel plate.

After cold rolling, the fuel plates are pre-cut for facilitate handling during the subsequent fabrication operations, as flattening, radiography and final cut. The fuel plates obtained in cold rolling have their surfaces still undulating, requiring a flattening operation. This operation is performed using a roll-flattener, which is basically consisted with a group of flattener cylinders controlled by a position adjustment system to keep the cylinders in a flat position. Only one pass is enough to flatten the fuel plates.

The next step is the final cut of the fuel plate to reach the specified dimensions. This cut is made using a guillotine cutter machine and is oriented by x-ray radiography. This radiography is obtained by using an industrial system set, where the fuel meat can be perfectly positioned inside the fuel plate and, then, the plate receives line tracing to guide the final cut. Next, the fuel plates are degreased in acetone and pickled in a solution of NaOH 10wt% for 1 minute at 60 °C. Then, they are washed in water for 1 minute, neutralized in cold 40wt% HNO_3 for 1 minute, rinsed again in running demineralized water for 5 minutes (spray), washed by immersion in hot demineralized water and dried manually with the aid of hot air blast. Figure 13 shows a drawing of the fuel plate, illustrating its fuel meat. Figure 14 shows the sequence of operations performed to manufacture the fuel plates.

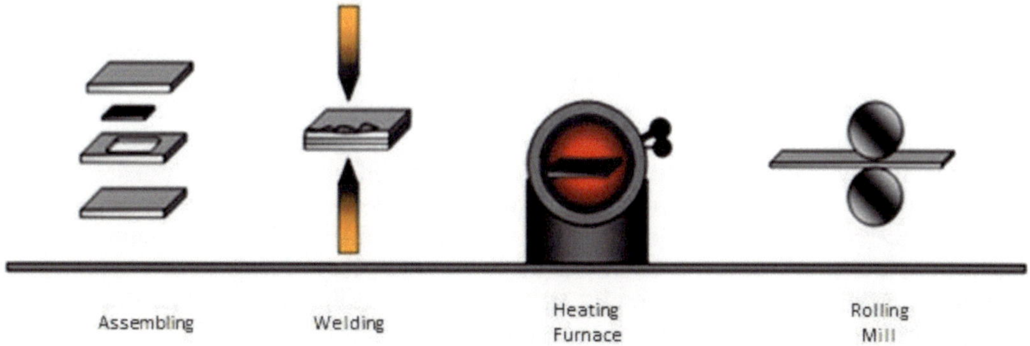

Fig. 13. Illustration of the process for preparing the assemblies and rolling

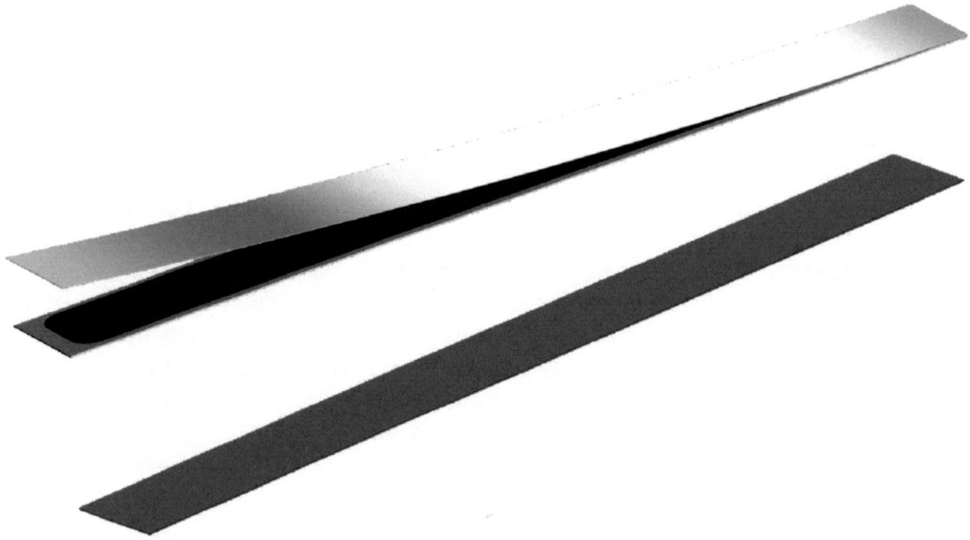

Fig. 14. The schematic illustration of the finished fuel plate (after rolling).

The finished fuel plates are characterized dimensionally, measuring in its length, width and thickness. Fuel plates that do not meet the dimensional specifications are rejected and sent for uranium recovery.

After the final cut, two new radiographs are obtained. The first one aims at checking the position of the fuel meat inside the fuel plate, as well as to verify its dimension, length and width. The second radiography aims to check the uranium distribution homogeneity in the fuel meat and also its integrity, as well as the possible presence of "white spots" and fissile particles outside the fuel meat zone.

To check the reproducibility and stability of the manufacturing process of fuel plates, the residual porosity of the fuel meat of all fuel plates produced are determined using the Archimedes principle.

Every 24 fuel plates produced, one fuel plate is separated to characterize the end defects in the fuel meat, which are basically the cladding thickness reduction in the area of the "dog-boning", inspection of the "diffuse zone" (end of the fuel meat) for studying the "fish tail" defect and to do the final geometry inspection of the fuel meat. This analysis is performed destructively to allow metallographic image analysis. In the case of fuel plate production routine, the quality analysis samples is randomly made (1:20) over all produced plates to check possible defects that do not meet specifications. In case, the sample is rejected then a second fuel plate is randomly taken from the batch and is destroyed to be examined. If this second sampled plate also proves defective then the entire batch is rejected. This metallographic analysis is performed using standard metallographic techniques and specific equipment for this purpose. All fuel plates rejected are forwarded for uranium chemical recovery.

The metallurgical bonding quality of the assembled plates set, after rolling, is checked by means of bending tests. This test is performed at two occasions, after pre-cutting and after the final cut. This test is performed in the leftover material from the cutting operations. The material is extensively bent in an angle of 180° and in reverse. In case of bonding failure, which is easily detected by visual inspection, the fuel plate is rejected and sent to uranium chemical recovery.

6.3 Assembling of fuel elements

In IPEN, two types of fuel elements are manufactured. The standard fuel element consists of 18 fuel plates, 2 side plates (right and left), a nozzle, a handling pin and 8 screws. The control fuel element is composed of 12 fuel plates, two side plates (right and left), two guide plates, a nozzle, a dashpot and 12 screws. The dimensional characteristics of the fuel elements are specified. All structural components of the fuel element are manufactured according to designs that are part of the specifications.

The process begins with the assembling of fuel plates to form a case that is the structural body of the fuel element. The plates are fixed to the side plates (left and right) by mechanical clamping. Subsequently, the nozzle is fixed. For the standard fuel element, the handling pin is fixed on the side opposite to the nozzle. In the case of the control fuel element, the dashpot is fixed on the side opposite to the nozzle. After cleaning and inspection, the fuel element is packed and stored until transportation to the reactor. Figure 15 illustrates the steps for the fuel elements assembling process.

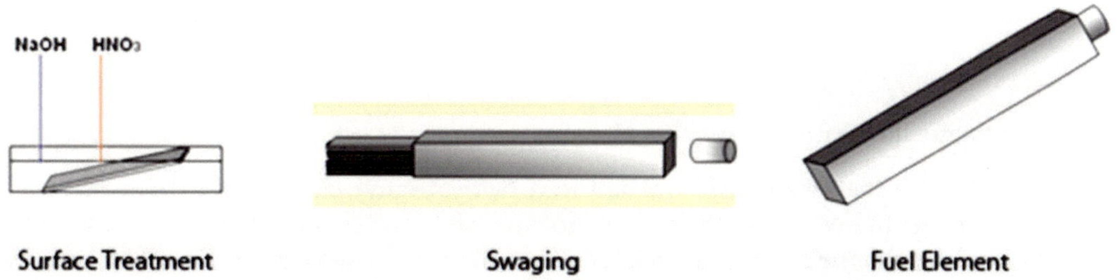

Fig. 15. The process of assembling the fuel elements.

After fixing the fuel plates in the side plates to form the main case, the next component to be installed is the nozzle. The nozzle is used to fix the fuel elements in the reactor core. It is fixed by screws at the lower end of the main case. The nozzle is aligned with the case of fuel element through an adjustment operation by using precision measuring instruments. The holes in the nozzle that are used to fix the side plates are already machined. The holes to hold the external fuel plates at the nozzle are machined with the nozzle already fixed in the side plates, with the aid of a milling machine. The screws used are made with aluminum and are already qualified and properly cleaned before use. The final tightening is done after a previous dimensional characterization, once verified the alignment of the nozzle in the main case. If alignment does not meet the specification, it is adjusted. In the case of the control fuel element, the procedure for fixing the nozzle and dashpot is the same as described above.

The handling pin is used to handle the standard fuel element inside the reactor pool. It is installed at the upper end of the main case, which contains two holes where the handling pin is fixed by clinching. In this operation, the ends of the handling pin, which have cavities, are deformed by pressure with the aid of a drilling machine. In the case of the control fuel element, this pin is replaced by the dashpot, which is aimed at damping the control or security bars that operates within this type of fuel elements. Figure 16 illustrates the standard fuel element and its components.

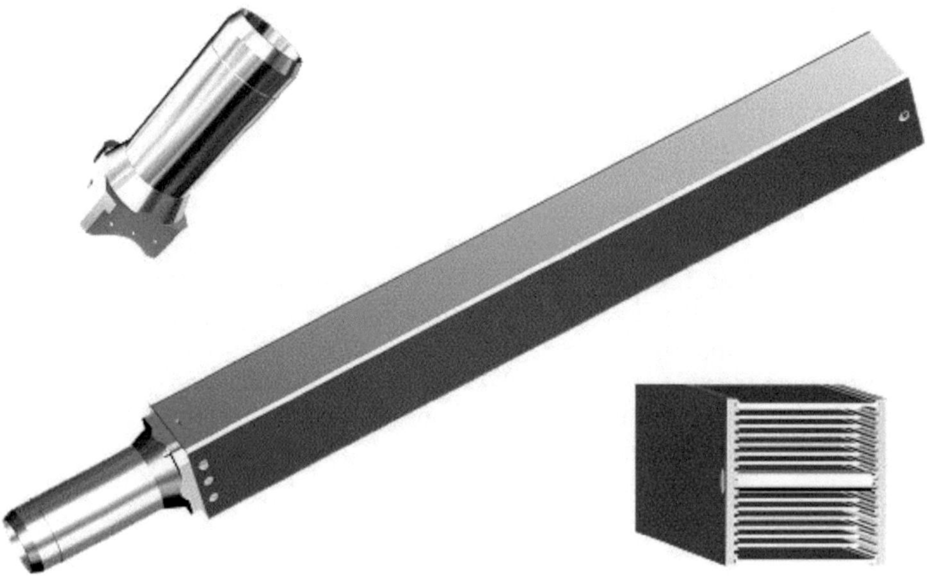

Fig. 16. Schematic illustration of the fuel element produced at IPEN.

Once qualified, the fuel element is washed in a bath of ethyl alcohol and dried manually with the aid of a jet of hot air. After this cleaning, a visual inspection is conducted, especially inside the cooling channels (the channels between the fuel plates), trying to detect possible obstructions caused by chips or foreign material. After washing and inspection, the fuel element is transferred to the reactor.

7. Uranium recovery and effluent treatment

A great variety of uranium residues must be recovered by chemicals means. A major source of such residues is uranium remaining in crucibles after melting and pouring. The recovery of solid or liquid uranium residues is vital because quantities are generated in every step of the process and this is a valuable material that must be recovered for reuse. Figure 17 displays a schematic diagram of the process showing the flow of products and residues.

The first step of the chemical recovery process is usually acid leaching to solubilize the uranium content. Any of several purification steps may then be employed to separate impurities such as iron, chromium, nickel, silicon, boron, etc. The end product of chemical recovery process is UF_4 which can be reduced to metal and then recycled. A typical sequence of chemical processing steps to recover uranium compounds from leach liquor is solvent extraction with tributyl phosphate, dinitration of purified uranyl nitrate solution to produce uranium trioxide (UO_3), and hydrogen reduction and hydrofluorination of UO_2 to UF_4. The technology of these operations is similar to that used in processing normal uranium.

Since chemical recovery will usually involve aqueous mixtures of uranium compounds, nuclear safety limits the critical dimensions of process equipment and imposes bath quantities within safe limits. If these factors are properly provided for chemical recovery unit design, the process operating costs will not be substantially raised by nuclear safety requirements.

The aggregate amount of scrap recycled via chemical recovery may reach 10% or more of finished fuel material weight. Chemical recovery is naturally more costly than direct recycle of metallic scrap to remelt. These considerations justify various expedients to by-pass chemical recovery by recycling metallic scrap. However, particular emphasis is given to the recovery of all residues solids and liquids because of the higher intrinsic value of the enriched material.

As an example, the IPEN process to produce U_3Si_2 involves metallic uranium as an intermediate product, through magnesiothermic reduction which produces slags containing uranium. The recovery process consists on slag lixivium of calcined by-products from metallic uranium reduction. The results from researching this process confirmed that this method could be integrated in treatment and recovery routines of uranium. The chemical route avoids dealing with metallic uranium since this material is unstable, pyroforic and extremely reactive. On the other hand, U_3O_8 is a stable oxide with low chemical reactivity, and it justifies the slags calcination of metallic uranium reduction by-products. This calcination occurs under oxidizing atmosphere and transforms the metallic uranium into U_3O_8. Some experiments have been carried out using diferente nitric molar concentrations, acid excess contents and temperature control of the lixivium process. The nitric lixivium main chemical reaction for calcined metallic uranium slags is represented by the equation:

$$U_3O_8\ (s) + 8\ HNO_3\ (I) \rightarrow 3\ UO_2(NO_3)_2\ (I) + 2\ N0_2(g) + 4H_2O(l) \quad (26)$$

The adopted process has the following parameters:

- Temperature and time: calcination of metallic uranium slag at 600°C during 3h;
- Granulometric control: sieving and segmentation of calcined slag in the range of 100-200 mesh;
- Concentration: lixivium adjustment of HNO_3 at 1 molar; HNO_3 excess (120%);
- Lixivium temperature: 40 - 50°C;
- Agitation: 300 rpm, turbine stem type (45° inclination).

As results, the full lixivium took 9 hours; the fluoride concentration in lixivium was 0,002g/L. Lixivium made at lower temperatures and lower nitric concentrations reduced both the magnesium and calcium fluorides solubility and the corrosion effect caused bifluoride ions was not prominent. This ensured a stable and secure lixivium from the operational point of view. The nitric dissolution of metallic uranium slags produced uranyl nitrate solution, which has been reused as a feed-in compound for uranium purification system made by solvent extraction method, using diluted n-tributhylphospate. The purified uranium product was then precipitated as ammonium diuranate (ADU) at 60°C, by injecting ammonium gas diluted with air. Aiming at returning the recovered product to the fuel fabrication cycle with nuclear quality level, the purified ADU was converted into uranium tetrafluoride (UF_4) by U_3O_8 route. The final yield in U content was 94%, proving the viability of IPEN´s slag recovering from uranium magnesiothermic reduction.

8. Acknowledgements

Thanks are due to IPEN for providing generously the technology of Nuclear Fuel Center, fully exemplified in this chapter, providing so nuclear know-how to a more peaceful, safer and healthy world. We are especially thankful to our colleagues who provided lots of information shown here, mainly Mr. Davilson Gomes da Silva who made many of the illustrations to qualify better this text.

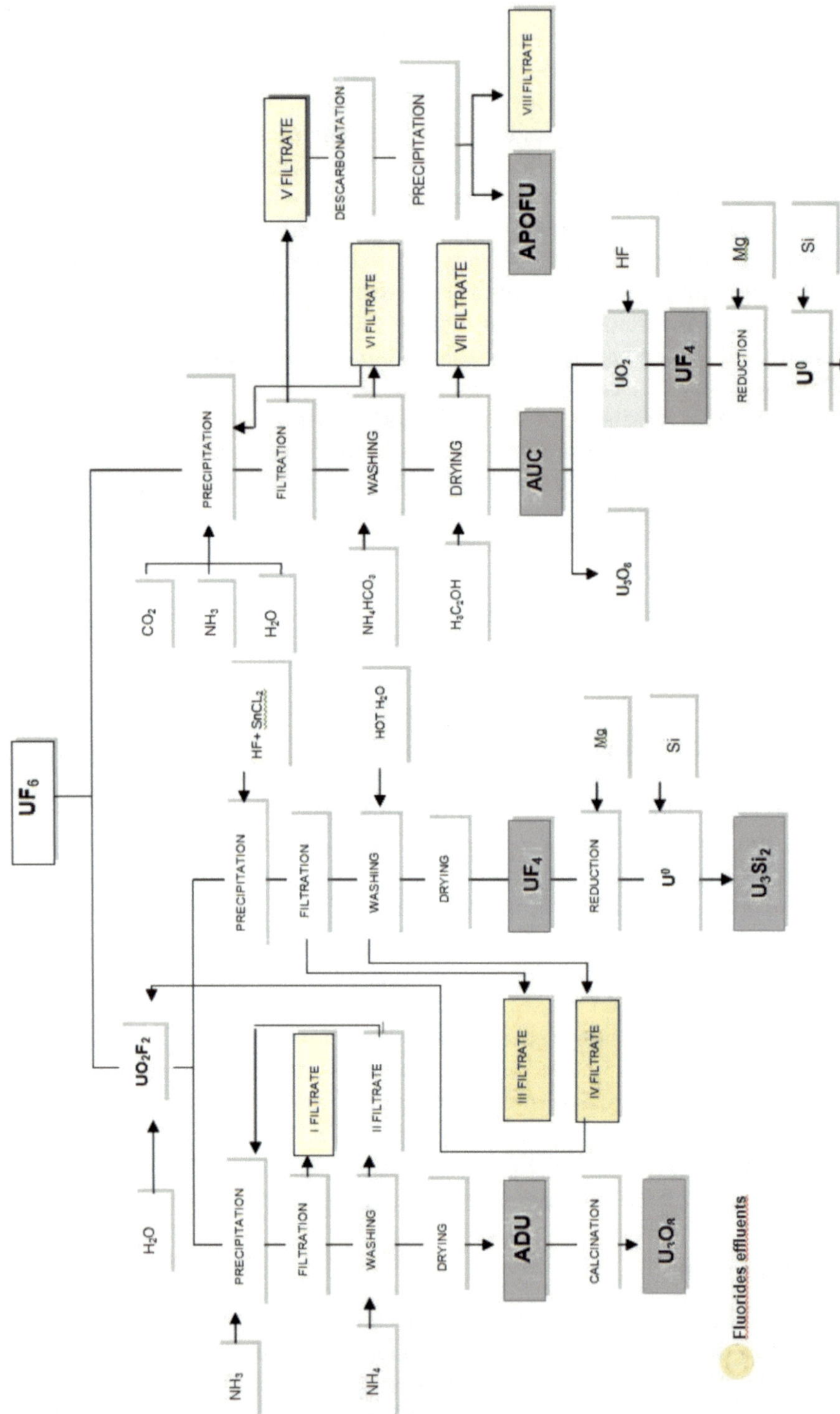

Fig. 17. Flowsheet MTR fuel processing (products and residues solid, liquids) (55; 54)

9. Conclusions

This chapter gave a general idea of the MTR fuel elements production for multipurpose and researching reactors that are producing radioisotopes throughout the world. Nowadays, the level of uranium enrichment is envisaged to be 20% (LEU), according to ruling requests of RERTR program. The given example of this production derived from IPEN/CNEN-São Paulo-Brazil, which produces through a well stablished routine to fabricate its own MTR fuel elements. Nevertheless, the technique to produce such elements has many variants, which are applied diversely from plant to plant.

As a final consideration, the future of fuel elements material, based on RERTR request, should also supply many high performance research reactors needing higher core densities of 6 to 9 gU/cm^3. This demand is not possible with U_3Si_2 elements, since its operational upper limit is less than 5 gU/cm^3. So, the presently envisaged product to reach this request is based on U-Mo alloy. Nevertheless, this product is not ready yet. Future prognosis are very confident that alloys U + 7 to 10wt%Mo should meet up this ability. This alloy production is still in experimental-pilot level, by this moment (2011), but with very consistent and pertinent results. For those willing to follow the development of this research, we indicate the transaction pages of RERTR and RRFM[2], where all papers and results are displayed freely.

10. References

[1] Cunningham, J. E. and Boyle, E. J. MTR-Type fuel elements. International Conference on Peaceful uses atomic energy. [ed.] United Nations. 1955, Vol. 9, pp. 203-7.

[2] Kaufman, A.R. *Nuclear reactor fuel elements, metallurgy and fabrication.* New York, NY, USA : Interscience, 1962.

[3] Holden, A.N. *Dispersions Fuel Elements.* New York, USA : Gordon & Breach, 1967.

[4] Cunningham, J.E., et al. Fuel Dispersions in Aluminium-Base Elements for research reactors. [ed.] Atomic Energy Comission. 1958, Vol. 1, pp. 269-97.

[5] Saller, H.A. Reaction Technology and Chemical Processing - Preparation, Properties and Cladding of Aluminum-Uranium Alloys. [ed.] United Nations. 1956, Vol. 9, pp. 214-20.

[6] Thurber, W.C. And Beaver, R.J. Segregation in Uranium-Aluminum Alloys and its Effect on the Fuel Loading of Aluminum-Base Fuel Element. [ed.] USAEC. 1958, pp. 9-29.

[7] Lennox, D.H. And Kelber, C.N. *Summary Report on the Hazards of the Argonaut Reactor.* Lemont, Mi : s.n., 1956. ANL – 5647.

[8] Kucera, W.J., Leitten, C.F. And Beaver, R.J. Specifications and Procedures Used in Manufacturing U3O8-Aluminium Dispersion Fuel Elements for Core I of the Puerto Rico Research Reactor. *Oak Ridge National Lab.* 1963.

[9] Knight, R.W., Binns, J. And Adamson Jr, G.M. Fabrication Procedures for Manufacturing High Flux Isotope Reactor Fuel Elements. *Oak Ridge National Lab.* Jun 1968.

[10] R.I., Beaver, Adamson Jr, G.M. And Patriarca, P. Procedures for Fabricating Aluminium-Base ATR Fuel Elements. *Oak Ridge National Lab.* Oak Ridge, Tenn.,, June, 1964.

[2]RERTR – Reduced Enrichment for Research and Test Reactors Program]: http://www.rertr.anl.gov/
RRFM – http://www.euronuclear.org/meetings/rrfm2011/transactions/RRFM2011-transactions.pdf

[11] Travelli, A. Current Status of the RERTR Program. *Development Fabrication and Application of Reduced-Enriched Fuels for Research and Test Reactor: Proceedings Held in Argonne.* 12-14 Nov 1980.

[12] J.L., snelgrove, Et Al. The Use of U3Si2 Dispersed in Aluminum in Plate-Type Fuel Elements for Research and Test Reactors. Oct 1987.

[13] Copeland, G.L., et al. Performance of Low-Enriched U3Si2-Al Dispersion Fuel Elements in the Oak Ridge Research Reactor. *RERTR.* Oct 1987.

[14] U.S.Regulatory Commission. Safety Evaluation Report related to the Evaluation of Low-Enriched Uranium Silicide-Aluminum Dispersion Fuel for Use in Non-Power Reactors. Jul 1988.

[15] Nazaré, S. New Low Enrichment Dispersion Fuels for Research Reactors Prepared by PM-Techniques. *J. Nucl. Mat.* 1984, Vol. 124, p. 14.

[16] —. Low Enrichment Dispersion Fuels for Research and Test Reactors. *Powder Met. Intern.* 1986, Vol. 18, 3, p. 150.

[17] Galkin, N.P., Et Al. *Technology of uranium.* Jerusalem : Israel Program for Scientific Translations, 1966. Cap.11.

[18] Snelgrove, J. L., Et Al. The use of U3Si2 dispersed in aluminum in plate-type fuel elements for research and test reactors. [ed.] October Argonne National Lab. Oct 1987.

[19] Van Wiencek, J. Uranium and fabrication. *Chemical Engineering Progress.* May 1954, p. 230.

[20] Seneda, J.A., Et Al. Recovery of uranium from the filtrate of ammonium diuranate prepared from uranium hexafluoride. *Journal of Alloys and Compounds.* 2001, Vols. 323-324, pp. 838-841.

[21] Instituto De Pesquisas Energéticas E Nucleares. *Implantação de um Centro de Processamento de Combustíveis no IPEN – Plano Diretor.* Ciclo do Combustível Nuclear. São Paulo : s.n., 1997. Relatório Interno MC.PT.0001.97.0.

[22] Cussiól, A.F. *Tecnologia para a preparação de tetrafluoreto de urânio. Fluoretação de UO2 obtido a partir de diuranato de amônio.* São Paulo, SP, Brasil : Escola Politécnica, Universidade de São Paulo, 1974. Dissertação de mestrado.

[23] Gisrgis, B.S. And Rofail, N.H. Reactivity of various UO3 modifications in the fluorination to UF4 by Freon 12. *J. of Nuclear Materials.* 1992, Vol. 195, p. 126.

[24] Gmelin, L. *Gmelins Handbuch der Anorganischen Chemie.* 8. Berlin : Springer-Verlag, 1980. C-8.

[25] Harrington, C.D. And Ruehle, E. *Uranium production technology.* . New Jersey : Van Nostrand, 1969.

[26] Katz, J. J. And Rabinowitch, E. *The chemistry of uranium, part 1 - The elements, its binary and related compounds.* New York : McGraw-Hill, 1951.

[27] Mellor, J.W. *A comprehensive treatise on inorganic and theorical chemistry.* London : Longmans, 1932. Vol. 12.

[28] Bolton, H.C. *Bull.Soc.Chim.* 1866, Vol. 6, 2, p. 450. apud Kat,J.J. & Rabinowitch, Z. The chemistry of uranium. Part I. The element, its binary and related compounds. New York, McGraw-Hill, 1951. p.355.

[29] —. *Z. Chem.* 1866, Vol. 2, 2, p. 353. apud KAT,J.J. & RABINOWITCH, Z. The chemistry of uranium. Part I. The element, its binary and related compounds. New York, McGraw-Hill, 1951. p.355..

[30] Alfredson, P.G. Australian experiments in the production of yellow cake and uranium fluorides. [ed.] International Atomic Energy Agency. *Production of yellow cake and uranium fluorides: proceedings of an Advisory Group meeting.* Jun 5-8, 1979, pp. 149-78.

[31] Allen, R.J., Petrow, H.G. And Magno, P.J. Precipitation of uranium tetrafluoride from aqueos solution by catalytic reduction. *Ind. Eng. Chem., 50(12).* 1958, Vol. 50, 12, pp. 1748-9.

[32] —. Preparation of sense, metal grade uranium tetrafluoride from uraniferous ores. [ed.] UNITED NATIONS. *Peaceful uses of atomic energy:proceedins of the 2nd international conference.* Sept. 1-13, 1958, Vol. 4.

[33] Gispert Benach, M., Et Al. Obtención de UF4 por reducción eletrolítica. I Estudios a escala de laboratorio. *Energ. Nucl.* 1972, Vol. 16, 80, pp. 623-36.

[34] Esteban Duque, A., Et Al. Producción de UF4 por reducción eletrolítica. II planta Piloto. *Energ. Nucl.* 1973, Vol. 17, 82, pp. 113-21.

[35] Scott, C.D., Adams, J.B. And Bresee, J.C. Fluorox process: production of UF6 in fluidized bed reactor. [ed.] Oak Ridge National Lab. 1960.

[36] Vogel, G.L., Et Al. Fluidized-bed techniques in producing uranium hexafluoride from ores concentrates. *Ind. Eng, Chem.* 1958, Vol. 50, 12, pp. 1744-7.

[37] Zidan, W. Elseaidy, I., , 26-29 October 1999, B. C.De Bariloche, Patagonia, Argentina. General Description and Production Lines Of The Egyptian Fuel Manufacturing Pilot Plant. *Proc.7th the Meeting of the International Group on Research ractors.* May 26-29, 1999.

[38] Opie, J.V. The preparation of pure uranium tetrafluoride by a wet process. . [ed.] Mallinckrodt Chemical Works. Apr 1, 1946.

[39] Ashbrook, A.W. And Smart, B.C. A review and update of refining pratice in Canada. In: International Atomic Agency. Production of yellow cake and uranium fluorides: proceedings od an Advisory Group meeting. Held in Paris, June 5-8, 1079. Vienna, 1980. p.261-8.

[40] Lenahan, K.J. Eldorado wet way process. In: Uranium 82: 12th annual hydrometallurgical meeting, held in Toronto, Aug. 29-Sept.1.1982.

[41] Harper, J. And Williams, A. E. *Factors influencing the magnesium reduction of uranium tetrafluoride.* In: Extraction and Refining of Rarer Metals, London: The Institute of Mining and Metallurgy, 1957, p. 143-162.

[42] Yemelyanov, V. S. E Yevstyukhin, A. I. *The metallurgy of nuclear fuel. London.* London : Pergamon Press, 1969.

[43] Huet, H. And Lorrain, C. Le procédé de magnésiothermie pour la préparation de l'uranium métallique. *Énergie Nucléaire.* 1967, Vol. 9, 3, pp. 181-188.

[44] Kubaschewski, O. High temperature reaction calorimetry - potencialities, limitations and application of results. *Thermochimica Acta.* 1978, Vol. 22, pp. 199-209.

[45] Beltran, A.D., Rivas Diaz, M. And Sanchez, A. F. Fabricacion de uranio metal. [ed.] Publicacion Bimestral de la Junta de Energia Nuclear. *Energia Nuclear.* mayo-junio 1972, pp. 295-320.

[46] Rand, M.H. And Kubaschewski, O. *The Thermochemical Properties of Uranium Compounds.* London : Oliver e Boyd, 1963.

[47] F., Fornarollo, Et Al. Recuperação de urânio em escórias geradas na produção de urânio metálico por magnesiotermia. *Revista Brasileira de Pesquisa e Desenvolvimento.* Novembro 2008, Vol. 10, 3, pp. 158-164.

[48] Saliba-Silva, A. M., Et Al. Fabrication of U3Si2 powder for fuels used in IEA-R1 nuclear research reactor. *Materials Science Forum.* IPEN, 2008, Vols. 591-93, pp. 194-199.

[49] Wiencek, T., Prokofiev, I. And Mcgann, D. Development and compatibility of Magnesium- Matrix Fuel Plates Clad With 6061 Aluminum Alloy", Proc. The International Meeting on Reduced Enrichment For research And Test Reactors. [ed.] ANL. *RERTR 1998.* Oct 18-23, 1998.

[50] Copeland, G. And Martin, M. Development of High - Uranium Loaded U3O8 - Al Fuel Plates. *Nuclear Technology.* 1982, Vol. 56.

[51] Tzou, G., Et Al. Analytical approach to the cold- and- hot bond rolling of sandwich sheet with outer hard and inner soft layers". *J. of Mater. Processing Technology.* Vols. 125-126, pp. 664-669.

[52] Frajndlich, E. U., Saliba-Silva, A. And Zorzetto, M. Xxi Rertr Meeting. Alternative Route For UF6 Conversion Towards UF4 to Produce metallic Uranium. *RERTR/DOE, 19º.* [Online] 1998. http://www.rertr.anl.gov/Fuels98/Elita.pdf.

[53] International Bio-Analytical Industries, Inc. Ammonium Uranyl Carbonate MSDS. [Online] [Cited: 05 16, 2011.] http://www.ibilabs.com/Ammonium%20Urany%20Carbonate%20MSDS.htm.

[54] Silva Neto, J. *Dry uranium tetrafluoride process preparation using the uranium hexafluoride reconversion process effluents.* São Paulo : CPG-IPEN/USP , 2008. MSc. Dissertation of University of São Paulo (in Portuguese).

[55] Frajndlich, E.U.C. *Chemical treatment study of an ammonium fluoride solution at the uranium reconversion plant.* São Paulo : CPG-IPEN/USP , 1992. MSc.Dissertation, University of São Paulo (in Portuguese).

3

Application of Enriched Stable Isotopes in Element Uptake and Translocation in Plant

Shinsuke Mori[1], Akira Kawasaki[2], Satoru Ishikawa[2] and Tomohito Arao[2]
[1]*NARO Western Region Agricultural Research Center*
[2]*National Institute for Agro-Environmental Sciences*
Japan

1. Introduction

Isotope technique including radioisotopes and stable isotopes is useful and potent tool for various scientific areas. Especially, enriched stable isotopes are indispensable tools for researchers in biological systems (Stürup et al. 2008).

Stable isotope ratios are usually used in examining the biogeochemical cycling of light elements such as carbon(C), oxygen (O), nitrogen (N) and sulphur (S) in the environment. Thermal ionization mass spectrometry (TIMS) for the isotope analysis has been the most standard technique for many years. However, for TIMS analysis, time for sample preparation is needed because sample need to ensure efficient ionization. On the other hand, ICP-MS analysis has some advantages that sample preparation is simple and high sample throughput for isotope experiments where a large amount of samples need to be analyzed (Stürup et al. 2008). The disadvantage to resolve in isotope analysis using ICP-MS is spectroscopic interferences in the process of analysis. It is therefore needed to be resolved these interferences.

When plant physiologists investigate mineral absorption mechanisms in roots of plant, evaluation of symplastic mineral absorption capacity in roots cell in kinetics and time course experiments is very important because mineral translocation in shoots is mainly contributed to capacity of symplastic absorption in roots. In these experiments, radioisotopes methods are mainly used for element uptake in plants. Radioisotopes in solute were the most useful markers used in nutrient uptake and translocation in plants because they are chemically similar to the solute and can be distinguished from non-labeled solutes already contained in the roots (Davenport 2007). However, there are limitations to this method, including radioisotope administrative restriction and the restricted half-life of the radioisotope. Isotope tracer experiments, using a stable isotope, are very similar to those using a radioisotope on element to analyse plant mechanisms (Stürup et al. 2008). Accurate and precise determination of mineral isotope ratios is required for analysis of enriched stable isotopes. Inductive coupled plasma mass spectrometry (ICP-MS) has now become the effective and potent technique for enriched stable isotope tracer experiments due to increased availability. Therefore, the application of enriched stable isotopes in various biological systems increased rapidly.

There are so many research using enriched stable isotopes used as tracers aquatic and terrestrial ecosystems, animals and humans (See review of Stürup et al. 2008). However, there are a few researches using enriched stable isotopes element in plants. Recently, Stürup et al. (2008) reviewed that application of enriched stable isotopes as tracers in biological systems including aquatic ecosystem, terrestrial ecosystem, animals and humans in detail. Therefore, we did not focus on aquatic ecosystem animal and human in this chapter.

In this chapter, we therefore provide a review of some example using isotope technique. Especially, we focus on the application of enriched stable isotopes element uptake and translocation in plants. Our new method for evaluation of symplastic absorption of roots introduced in Section 4 has some merits, compared to radioisotopes techniques. Application of stable isotopes will become a new tool to evaluate element behavior in plants.

2. Application of stable isotopes in plants

The biochemical cycling of light element such as carbon(C), oxygen(O), nitrogen(N) and sulphur(S) have been studying using stable isotopes. The mechanisms of photosynthesis and of element uptake and translocation in plants was clarified by these studies using stable isotopes ratios such as C,O,N and S. Recently, the application of enriched isotopes of such as Mg, Cu, Ca, K and Cd behavior in plants rapidly increased with the development of ICP-MS analysis techniques. There are several studies on element uptake and translocation in plant using enriched stable isotopes (Table1).

Isotopes	Aim of study and method	Reference
^{10}B, ^{11}B	Characterization of boron uptake and translocation in sunflower plant. After preculture under nutrient solution containing ^{11}B,a short time experiment were conducted under nutrient solution containing low or high ^{11}B.	Dannel et al. (2000)
^{10}B	After preculture grown in nutrient solution containing boron, uptake experiment was conducted in solution containing enriched stable isotopes of ^{10}B.	Takano et al. (2002)
^{113}Cd	Intact leaves and cell sap of Cd accumulator plant were subjected to ^{113}Cd-NMR and H-NMR analysis for identification of the form of Cd in leaves.	Ueno et al. (2005)
^{113}Cd and ^{114}Cd	To examine Cd uptake in roots of *solanum* species with different Cd accumulation in shoot, uptake experiments were conducted using ^{113}Cd and ^{114}Cd.	Mori et al. (2009b)
^{113}Cd	Cd accumulation stage in soybean seed was examined in hydroponic solution using enriched isotope of ^{113}Cd.	Yada et al. (2004), Oda et al. (2004)
^{113}Cd	Cd uptake mechanisms in soybean was examined using ^{113}Cd isotopes in pot and field experiment	Kawasaki et al. (2004,2005)

Table 1. Element uptake and translocation in plant using enriched stable isotopes

Dannel et al. (2000) characterized the boron uptake and translocation from roots to shoots in sunflower using the stable isotopes ^{10} B and ^{11}B. In the report, after sunflower plant was precultured with high(100μM) or low (1μM) ^{11}B supply, plants were treated under differential ^{10}B supply condition. The results suggested that B uptakes are mediated by two transport mechanisms. First mechanism is passive diffusion which is indicated by the linear components. Second mechanism is energy dependent process which is indicated by the saturated components. Kawasaki et al. (2004, 2005) conducted that an isotope tracer technique with ^{113}Cd has been used in pot and field experiments. They examined that the most critical stages of soybean in which Cd absorbed via roots was transferred into the seeds. Cd absorbed before the beginning seed stage causes an increase of Cd concentration in seeds. Yada et al.(2004) reported that soybean plants were grown in hydroponic solution and supplied ^{113}Cd via roots for 48 h at early growth stage to investigate Cd accumulation pathway in soybean seed using enriched isotope of ^{113}Cd. Cd accumulated in leaves was translocated to seeds at seed beginning maturity stage. Oda et al. (2004) also indicated that the Cd absorbed from full pod to full seed was the most contributive to raise the Cd amount of seeds. Ueno et al. (2005) reported that *Thlaspi caerulescens* which is Cd hyperaccumulator plants have been grown hydroponically with a highly enriched ^{113}Cd isotope to investigate the form of Cd in the leaves using ^{113}Cd nuclear magnetic resonance (NMR) spectroscopy. They identified that cadmium binds with malate in the leaves. Several enriched isotopes such as ^{111}Cd, ^{113}Cd and ^{114}Cd will become a new tool to evaluate Cd behavior in plants. Several studies stated above suggest that enriched isotope is a very potent technique for tracking the distribution, uptake, translocation and recycling in biological system. Now, many enriched element stable isotopes except B and Cd are able to purchase in chemical forms such as metallic or oxide. In the future, the benefit of enriched stable isotopes techniques would be paid much attention in plant and environmental science areas.

3. Several methods for evaluating symplastic element uptake in plants

Intensive studies on the absorption mechanisms of various elements by plant roots have been conducted. There are evidence on mineral uptake and translocation in plants. It is well known that ion absorption in plant roots shows a saturated curve in kinetics experiments, indicating that a type of proteinaceous transporter mediates ion absorption (Epstein and Hagen 1952). Plant physiologists examining ion absorption in plant roots have given much attention to ion transport via the symplast across the plasma membrane (Epstein 1973). However, when ion absorption experiments were conducted, it was found that the apoplastically absorbed ions needed to be washed out of the apoplast to determine the symplastically absorbed ions across the plasma membrane or the determination of absorption is overestimated (Glass 2007). Therefore, it is necessary to eliminate the apoplastically bound ions to evaluate the symplastically absorbed ion content in the roots. To evaluate symplastic cadmium(Cd) and other elements absorption in roots, several methods have generally been used in the past: (1) expose the plant material to Cd radioisotopes and subsequent desorption using unlabelled Cd in the root apoplast (Hart et al. 1998, 2002, 2006), (2) plant material is exposed to Cd radioisotopes under conditions at 2°C and 22°C (Zhao et al. 2002, Uraguchi et al. 2009), (3) metabolic inhibitors such as DNP or CCCP (Cataldo et al.1983, Ueno et al. 2009), (4) centrifuge method (Yu et al. 1999, Mitani and Ma 2005, Ma et al. 2004, Ueno et al. 2008), (5)estimation of desorption from roots with time(Lasat et al. 1998)

Regarding evaluation for symplastic element uptake in roots using radioisotopes, this method is used for symplastic element uptake in roots. Hart et al. (1998, 2002, 2006) reported that Cd uptake experiment was conducted in nutrient solution containing ^{109}Cd-labeled $CdSO_4$ and apoplastic ^{109}Cd were desorbed using excessive nonlabelled Cd. As other method, Nakanishi et al. (2006) evaluated that apoplastic Cd in the roots was washed in 0.5 mmol L^{-1} ethylenediaminefetraacetic acid (EDTA) for 1 min. Lasat et al. (1996) evaluated that symplastic Zn uptake in roots of Zn hyperaccumulator and nonaccumulator *Thlaspi* species apoplastic ^{65}Zn in roots desorbed by excessive unlabelled $ZnCl_2$ solution after Zn uptake experiment was conducted using ^{65}Zn raidoisotopes. There is merit that this method is able to detect radioisotope element with high sensitivity. However, there are limitations to this method, including radioisotope administrative restriction and restricted half of the radioisotope. Additionally, the radioisotope technique has toxicological concern. It is required for handling its isotopes to be careful.

Regarding evaluation of symplastic element uptake in roots using differences in the amounts of Cd absorbed at 2°C and 25°C. Uptake of element at 2°Cwas assumed to represent mainly apoplastic binding in the roots whereas the difference in uptake between 22°C and 2°C represented metabolically dependent influx. Zhao et al. (2002) reported that apoplastic and symplastic uptake in two *Thlaspi* species from Cd and Zn depletion in solution using radioisotope tracer. Uraguchi et al.(2009) reported that genotypic variation in cadmium accumulation in rice and evaluated that symplastic Cd uptake in roots of rice using the method of subtraction the Cd content in the roots at 2°C from the Cd content in the roots at 25°C. This method using unlabeled Cd is easy to handle because there is no administrative limitation not using radioisotope elements. However, this method needs double seedlings for evaluation. Additionally, this method cannot be evaluated using same seedling. This method is not easy for dicotyledonous plant such as *Solanum melongena* to handle.

As for methods using metabolic inhibitors, Cataldo et al. (1983) reported that Cd uptake dependent on energy in roots is suppressed by dinitrophenol as metabolic inhibitor. In this study, using dinitrophenol as a metabolic inhibitor, the 'metabolically absorbed' fraction was shown to represent 75 to 80% of the total absorbed fraction at concentration less than 0.5 μmol, and decreased to 55% at 5 μmol.

Regarding centrifuge method, tap roots of plants were harvested and 2 cm root tips were excised. Then, cut ends were washed in distilled water and blotted dry. For each sample, 30 roots were used. The cut ends were washed in distilled water quickly and blotted dry. The tips were placed in a 0.45 mM filter unit with the cut ends facing down and centrifuged at 2,000g for 15 min at 4°C to obtain the apoplastic solution. After centrifugation, root segments were frozen at -80°C for 2 h and then thawed at room temperature. The symplastic solution was prepared from frozen-thawed tissues by centrifugation at 2,000g for 15 min at 4°C. Ma et al. (2004) evaluated that symplastic Si uptake of wild type rice and mutant rice using this centrifuge method. Additionally, Mitani and Ma (2005) also evaluated that symplastic Silicon uptake in rice, tomato and cucumber which differ from Si accumulation capacity using this method. Ueno et al.(2009) reported that symplastic Cd uptake is estimated by cell sap obtained from centrifuge method. To check the purity of apoplastic solution, the activity of malic dehydrogenase in apoplastic and symplastic solution was determined. The activity of malic dehydrogenase in apoplastic solution was below one-twentieth and approximately one-fortieth of symplastic solution. This method is valuable for evaluation of symplastic Cd concentration in roots because Cd concentration in roots cell cap was directly determined. However, evaluation using root tips possibly is not representative of most root tissues. Rain et al.(2006) pointed out that there are the difference of K_m value in kinetics experiment between whole roots and root tips.

As other evaluation method of roots fraction, Lasat et al.(1998) evaluated that each fraction of cell wall, cytoplasm and vacuole by each efflux fraction from roots. They investigated that difference of Zn fraction in roots such as cell wall, cytoplasm and vacuole using this method.

4. Application of enriched stable isotopes in element uptake and translocation in plant

In this section, we introduce that our new method for evaluation of symplastic ion absorption, especially cadmium (Mori et al. 2009a). Several methods stated above is evaluation that apoplastically bound element is desorbed by some elements after element absorption experiment. Our method is that symplastic Cd absorption capacity is evaluated by difference of enriched isotope of ^{113}Cd and ^{114}Cd. Cadmium (Cd) is a hazardous heavy metal with regards to human health and is dispersed in natural and agricultural environments principally through human activities (Wanger, 1993). Arable land contains, to some extent, Cd, reportedly in the range, 0.04–0.32M, even in non-polluted soil (Keller, 1995; Wanger, 1993). This results in Cd accumulation in the edible parts of crops. Recently, the Codex Alimentarius Commission (2005) adopted a maximum concentration of 0.05 mg Cd kg^{-1} (fresh weight) recommended for fruiting vegetables. Approximately 7% of 381 samples of eggplant (*Solanum melongena*), 22% of 165 samples of okra (*Abelmoschus esculentus*), and 10% of 302 samples of taro (*Colocasia esculenta*) contained Cd concentrations above this limit in a field and market-basket study during 1998–2001 in Japan (Ministry of Agriculture Forestry and Fisheries of Japan, 2002); despite the fact that these crops were cultivated in non-polluted fields. Under these circumstances, new technologies for reducing the Cd level in crops are urgently required in Japan. Therefore, it is important to elucidate the mechanisms mediating Cd absorption, accumulation, and translocation in these crops. The crop conditions were represented by low Cd concentration experimental mediums.

4.1 Validity of our method for evaluation of symplastic Cd uptake in roots using enriched isotopes of ^{113}Cd and ^{114}Cd

When ion absorption experiments were conducted, it was found that the apoplastically absorbed ions needed to be washed out of the apoplast to determine the symplastically absorbed ions across the plasma membrane or the determination of absorption is overestimated(Glass 2007). Therefore, it is necessary to eliminate the apoplastically bound ions to evaluate the symplastically absorbed ion content in the roots. There are several methods to eliminate apoplastic ions as stated above. In this section, we introduced our new method for symplastic Cd absorption in roots of *Solanum melongena* using enriched isotopes of ^{113}Cd and ^{114}Cd.

The enriched isotopes of ^{113}Cd (^{106}Cd, 0.16%; ^{108}Cd, 0.135%; ^{110}Cd, 0.81%; ^{111}Cd, 2.53%; ^{112}Cd, 2.61%; ^{113}Cd, 93.29%; ^{114}Cd, 0.46%; ^{116}Cd, 0.01%) and ^{114}Cd (^{106}Cd, 0.05%; ^{108}Cd, 0.05%; ^{110}Cd, 0.05%; ^{111}Cd,0.05%; ^{112}Cd, 0.05%; ^{113}Cd, 5.6%; ^{114}Cd, 93.6%; ^{116}Cd,0.8%) used in the present study were purchased from Isoflex (San Francisco, CA,USA) in metallic form and dissolved in diluted HNO_3. The enriched isotopes of ^{114}Cd contained the 5.6 % of ^{113}Cd.

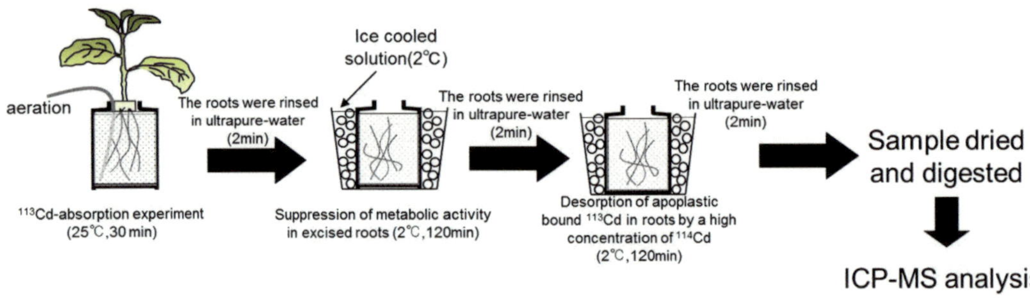

(modified from Mori et al. 2009a)

Fig. 1. Absorption experiment procedure for evaluating symplastic ^{113}Cd absorption in roots

The procedure for evaluating symplastic Cd absorption in the roots, using enriched isotopes ^{113}Cd and ^{114}Cd, is illustrated in Fig. 1. The roots of intact seedlings were rinsed in ultrapure water for 2 min and then exposed to a 500 mL ^{113}Cd solution containing 0.5 mmol L^{-1} $CaCl_2$ and 2 mmol L^{-1} 2-morpholinoethanesulfonic acid monohydrate Tris (hydroxymethyl) aminomethane (MES–Tris) (pH 6.0) at 25°C for 30 min (Fig. 1). The levels of ^{113}Cd were 40 nmol or 400 nmol in the ^{113}Cd treatment. A-B shown in Fig.2 indicates that ^{113}Cd absorbed in roots consists of apoplastic ^{113}Cd and symplastic ^{113}Cd (Fig.2 A, B). To suppress metabolically dependent symplastic absorption from the apoplast, the roots were excised from each seedling and immersed in a cold Cd-free buffer solution (2 mmol L^{-1} MES–Tris [pH 6.0], 0.5 mmol L^{-1} $CaCl_2$) at 2°C for 120 min (Fig. 1, Fig.2 C). The apoplastic-bound ^{113}Cd in the roots from 40 or 400 nmol ^{113}Cd treatment was then desorbed by immersing the roots in the same cold buffer solution at 2°C containing a 50-fold concentration of ^{114}Cd (2 or 20 μmol) for 120 min (Fig.1, Fig.2 D, E, F). The excised roots were then rinsed in ultrapure water for 2 min. Harvested samples were dried in an oven at 75°C for 3 days until dry. After digestion of dried sample, we then determined ^{113}Cd and ^{114}Cd contents in roots by ICP-MS analysis. To confirm the validity of this method, we compared our Cd absorption results with the Cd absorption results obtained at 25°C and 2°C using unlabeled $CdCl_2$ reagent. The experimental procedure was as follows. The Cd-absorption experiments were conducted for 30 min using 500 mL solutions containing 2 mmol L^{-1} MES–Tris (pH 6.0), 0.5 mmol L^{-1} $CaCl_2$ and different concentrations of Cd (40 or 400 nmol) at 25°C. After the absorption experiment, the excised roots from each seedling were rinsed with ultrapure water for 2 min. For the Cd-absorption experiment at 2°C, plants were transferred to an ice-cold pretreatment solution containing 2 mmol L^{-1} MES–Tris (pH 6.0) and 0.5 mmol L^{-1} $CaCl_2$ for 120 min. The Cd-absorption experiment at 2°C was conducted for 30 min. In the unlabeled Cd-absorption experiment at different temperatures, the amount of Cd reportedly absorbed into roots at 2°C was estimated to be apoplastically bound Cd on the assumption that metabolically dependent absorption would be suppressed at low temperature. Therefore, the difference in the amount of Cd absorbed at 2°C and at 25°C represents symplastic Cd absorption depending on metabolic energy. All absorption experiments were replicated three times. Each procedure illustrated in Figure1 signifies a schematic representation shown in Fig. 2.

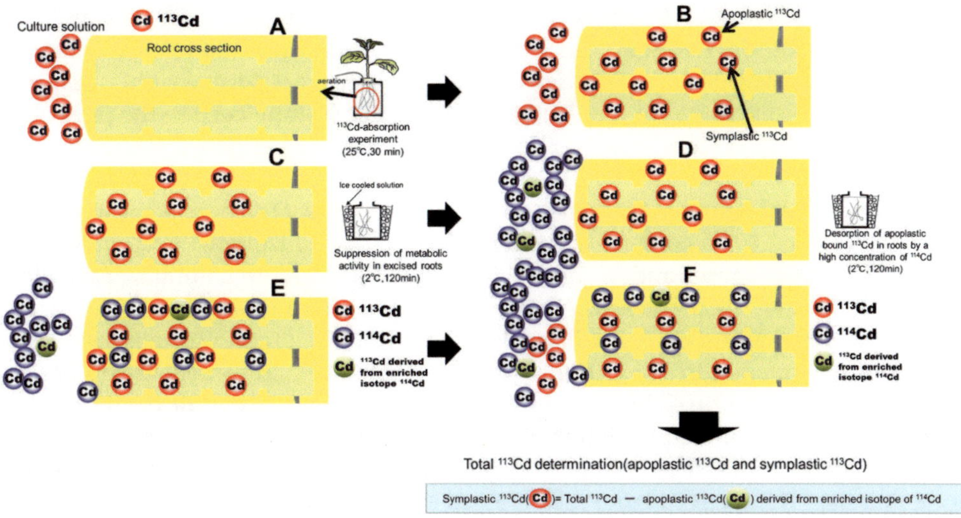

Fig. 2. A schematic representation of Cd absorption and desorption in roots using different enriched isotopes

4.2 Determination of ^{113}Cd, ^{114}Cd and the Cd contents in the roots

Approximately 0.05–0.1 g of dried roots was transferred and digested in a 10 mL Teflon tube containing 3 mL HNO_3. After digestion, the digested solution was diluted and 10 ng mL^{-1} of indium (In) was added to each diluted solution as an internal standard for ^{114}Cd determination. For ^{113}Cd determination, 10 ng mL^{-1} of tellurium (Te) was added as an internal standard. The concentrations of ^{113}Cd and ^{114}Cd in the digested solutions were determined by ICP-MS (ELAN DRC-e; Perkin Elmer SCIEX, Concord, ON, Canada). The concentrations of Cd in the digested solutions from the Cd-absorption experiment using unlabeled $CdCl_2$ reagent were determined by ICP atomic emission spectroscopy (VISTA-PRO; Varian, Palo Alto, CA, USA). It is well known that MoO interferes spectroscopically in determining the concentration of Cd in ICP-MS analysis (Kimura et al. 2003; May and Wiedmeyer 1998). In addition, it has been shown that it is necessary to remove Mo from the digested solution to avoid spectroscopic interference by molybdenum oxides (Oda et al. 2004; Yada et al. 2004). Therefore, for the ^{113}Cd and ^{114}Cd count intensities, we monitored the spectroscopic interference of the molybdenum oxides (^{97}Mo^{16}O and ^{98}Mo^{16}O) detected in the 10 ng mL^{-1} Mo standard solution. The contribution rate of spectroscopic interference of the putative ^{97}Mo^{16}O and ^{98}Mo^{16}O for ^{113}Cd and ^{114}Cd contents was negligibly small in both treatments (40 and 400 nmol). Therefore, we considered that we could ignore spectroscopic interference of oxidative molybdenum in determining the ^{113}Cd and ^{114}Cd contents in the ICP-MS analysis.

As shown in Fig. 3, after desorption of apoplastic ^{113}Cd by excessive ^{114}Cd, distribution of ^{113}Cd and ^{114}Cd in roots is as follow. (1) apoplastic bound ^{114}Cd is derived from desorption solution of excessive ^{114}Cd. (2) apoplastic bound ^{113}Cd is derived from desorption solution of excessive ^{114}Cd. (3) symplastic ^{113}Cd is derived from ^{113}Cd- uptake experiment. Therefore, ^{113}Cd content in roots is the sum of (1) and (2). Symplastic ^{113}Cd is the subtraction between total ^{113}Cd and ^{113}Cd derived from an enriched stable of ^{114}Cd. As shown in Fig. 1, the total ^{113}Cd contents in the roots signifies the ^{113}Cd contents in the roots after the desorption experiment (Fig. 1). The total ^{113}Cd content in the roots at 40 and 400 nmol Cd was 23.0 ± 4.3 and 87.7 ± 5.6 mg kg^{-1} (dry weight), respectively (Table2). In contrast, the ^{114}Cd content at 40 and 400 nmol Cd was 117.3 ± 9.4 and 644.5 ± 33.7 mg kg^{-1} (dry weight), respectively (Table2). The purification rate of the ^{114}Cd-enriched stable isotope used in the present study was 93.60%; whereas, the composition rate of ^{113}Cd in the ^{114}Cd-enriched stable isotope was 5.6%. The total ^{114}Cd content in the roots after desorption of 20 µmol ^{114}Cd was approximately 5.5-fold higher than that using 2 µmol ^{114}Cd (Table 2),suggesting that the apoplastically bound ^{113}Cd content, derived from the enriched isotope ^{114}Cd, increased with an increase in the concentration of ^{114}Cd in the desorption solution. Actually, the apoplastically bound ^{113}Cd contents, derived from the enriched isotope ^{114}Cd (2 and 20 µmol) were 6.6 ± 0.5 and 36.6 ± 1.8 mg kg^{-1}, respectively (Table 2); these values were calculated using equation in Fig.3. The contribution rate of ^{113}Cd content derived from the enriched stable isotope of ^{114}Cd for total ^{113}Cd in the roots was 28.6% for the 40 nmol ^{113}Cd treatment. In contrast, the contribution rate of ^{113}Cd content derived from ^{114}Cd for total ^{113}Cd content in the roots was 41.8% for the 400 nmol ^{113}Cd treatment (Table 2). These results indicate that the ^{113}Cd derived from the enriched stable isotope of ^{114}Cd must be subtracted from the total ^{113}Cd content in the roots to evaluate the symplastic ^{113}Cd in the roots. The symplastic ^{113}Cd contents for the 40 and 400 nmol treatments, calculated using equation in Fig.3, were 16.4 ± 3.7 and 51.0 ± 3.8 mg kg^{-1}, respectively (Table 2). In the present study, we disregarded the contribution of ^{114}Cd derived from the enriched isotope of ^{113}Cd because the composition rate of ^{114}Cd in the enriched isotope of ^{113}Cd was considerably lower than that of ^{113}Cd in the enriched isotope of ^{114}Cd.

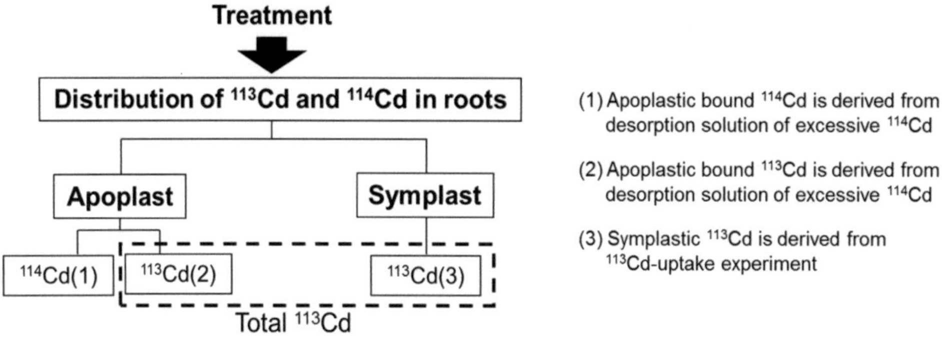

Symplastic ^{113}Cd = Total ^{113}Cd − ^{113}Cd derived from an enriched stable isotope of ^{114}Cd

^{113}Cd derived from an enriched stable isotope of ^{114}Cd = Total $^{114}Cd \times 0.056^{*}$

* Composition rate of ^{113}Cd in enriched stable isotope of ^{114}Cd

Fig. 3. Calculation of symplastic ^{113}Cd content in roots.

4.3 Comparison of the symplastic Cd contents in the roots between the two methods

To examine the validity of the new method for evaluating the symplastic Cd content in roots using ^{113}Cd and ^{114}Cd enriched isotopes, we compared the symplastic Cd content in roots using differences in the amounts of Cd absorbed at 2°C and 25°C with unlabeled Cd with the results obtained in the present study using the new method. In conventional Cd-absorption experiments, the Cd contents in roots at 40 and 400 nmol Cd in a 25°C treatment were 19.2 ± 1.6 and 84.4 ± 3.4 mg kg^{-1} (dry weight), respectively (Table 3). In contrast, the Cd contents in roots at 40 and 400 nmol in the 2°C treatment were 4.1 ± 0.3 and 28.1 ± 0.73 mg kg^{-1} (dry weight), respectively.

The symplastic Cd contents at 40 and 400 nmol were estimated to be 15.1 ± 1.3 and 56.4 ± 2.7 mg kg^{-1}, respectively, which was evaluated using the difference in the amount of Cd absorbed at 2°C and at 25°C.

In the ^{113}Cd-absorption experiment, the symplastic ^{113}Cd contents in the roots at the 40 and 400 nmol ^{113}Cd treatments were 16.4 ± 3.7 and 51.0 ± 3.8 mg kg^{-1}, respectively (Table 2, 3). Therefore, the symplastic ^{113}Cd content after using the enriched isotopes was similar to the symplastic Cd content evaluated from the difference between the amount of Cd absorbed at 2°C and at 25°C. These results indicate that it is possible to evaluate the contents of symplastic Cd in roots using ^{113}Cd and ^{114}Cd enriched isotopes using the method proposed in the present study.

There have been many reports on Cd absorption in roots eliminating apoplastic bound Cd in Durum wheat, soybean and hyperaccumulator plants, such as *Thlaspi caerulescens* (Cataldo et al. 1983; Hart et al. 1998, 2002, 2006; Zhao et al. 2002). In these studies, the symplastic Cd content in the roots was determined by subtracting the Cd content in the roots at 2°C from the Cd content in the roots at 25°C; the Cd content was determined using a radioisotope of ^{109}Cd or a metabolic inhibitor. These methods have frequently been used to evaluate nutrient element absorption in roots. Radioisotopes in solute were the most useful markers used in these studies because they are chemically similar to the solute and can be distinguished from non-labeled solutes already contained in the roots (Davenport 2007). However, there are limitations to this method, including radioisotope administrative restriction and the restricted halflife of the radioisotope. Although the method involving a temperature difference between 2 and 25°C that was used in the present study is easy to handle because there is no radioisotope administrative restriction, there is, however, a limitation to this method: the symplastic Cd content in the roots cannot be evaluated using the same seedlings. This method has the advantage of no radioisotope administrative restriction and no restrictive radioisotope half-lives. In addition, this method uses half the number of seedlings that are required for the method using the temperature difference between 2 and 25°C because the symplastically absorped Cd in the roots can be evaluated using roots from the same seedlings. In addition, the method proposed in the present study is applicable to other plants, not only *S. melongena*. We indicated that it is possible to evaluate symplastic Cd in roots using ^{113}Cd and ^{114}Cd enriched isotopes. The proposed method will contribute to research on symplastic ion absorption in plant roots stated below.

40nM			
Total ^{114}Cd	**Total ^{113}Cd**	**^{113}Cd derived from enriched ^{114}Cd**	**Symplastic ^{113}Cd**
117.3±9.3	23.0±4.3	6.6±0.53	16.4±3.7
400nM			
Total ^{114}Cd	**Total ^{113}Cd**	**^{113}Cd derived from enriched ^{114}Cd**	**Symplastic ^{113}Cd**
644.5±33.7	87.7±5.6	36.6±1.8	51.0±3.8

Table 2. ^{114}Cd and ^{113}Cd content in roots (modified from Mori et al. 2009a)

40nM			
Symplastic ^{113}Cd	Cd(25°C-2°C)	Cd(25°C)	Cd(2°C)
16.4±3.7	15.1±1.3	19.2±1.6	4.1±0.3
400nM			
Symplastic ^{113}Cd	Cd(25°C-2°C)	Cd(25°C)	Cd(2°C)
51.0±3.8	56.4±2.7	84.4±3.4	28.1±0.7

Table 3. Comparison of the symplastic Cd content in roots (modified from Mori et al. 2009a)

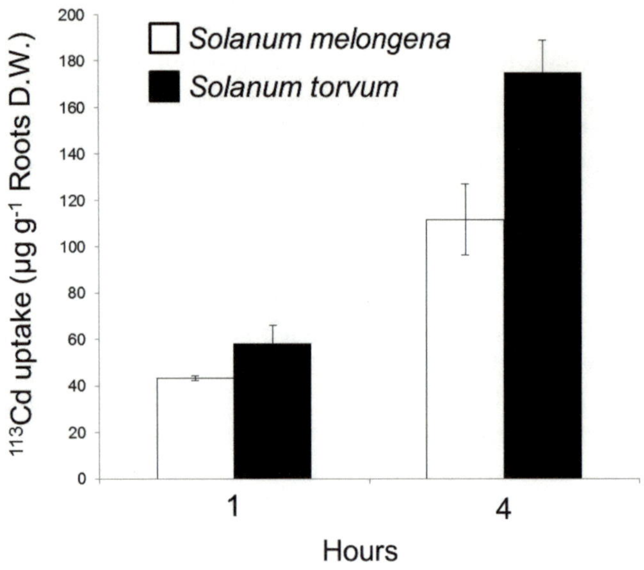

Fig. 4. Symplastic Cd absorption in roots of *Solanum melongena* and *Solanum torvum* with time. Experiment method is followed by the procedure illustrated in Fig.1 (modified from Mori et al.2009b)

We used the new method using enriched stable isotopes for evaluation of symplastic Cd absorption in roots of *solanaceous* plants (*Solanum melongena* and *Solanum torvum*) with contrasting root-to-shoot Cd translocation efficiencies (Mori et al. 2009a,b).

It is well known that efficiency of Cd translocation from roots to shoots is significantly higher in *S. melongena* than *S. torvum* (Arao et al. 2008, Mori et al. 2009a,b, Yamaguchi et al. 2011). Takeda et al.(2007) found that the Cd concentration in eggplant fruits could be reduced by grafting with *Solanum torvum* rootstock. Additionally, Arao et al.(2008) reported that although the Cd accumulation in shoots of *S. torvum* was lower than that found in *S. melongena*, there was no difference in the Cd content in roots of both plants when grown in culture solution. This result suggests that *S. torvum* develops noteworthy physiological mechanisms to suppress Cd translocation from roots to shoots, corresponding to the results observed in previous reports (Arao et al., 2008). Arao et al. (2008) suggested that symplastic Cd absorption and xylem loading capacity might be ascribed to the difference of Cd concentration in the shoots of *S. melongena* and *S. torvum*. We evaluated the symplastic Cd absorption rate in roots using enriched isotopes ^{113}Cd and ^{114}Cd. In time course-dependent experiments, the symplastic ^{113}Cd absorption rate for both plants increased with time (Fig. 4). In addition, the symplastic ^{113}Cd absorption rate of *S. melongena* was slightly higher than that of *S. torvum* at 4 h (Fig. 4). We examined kinetics analysis by similar method using enriched stable isotopes of ^{113}Cd and ^{114}Cd (Mori et al. 2009b). A kinetic study revealed that the symplastic Cd concentrations in the roots increased with increasing external Cd concentrations, but saturated at a higher concentration. The saturated curve obtained in this study suggests that absorption in both cultivars is mediated by a transporter that exhibits a similar affinity for Cd.. Moreover, the symplastic Cd concentrations slightly differed between the roots of *S. melongena* and *S. torvum*. Based on the reaction curves obtained, the K_m value was estimated to be 380 and 352 nmol L^{-1} for *S. melongena* and *S. torvum*, respectively. The corresponding V_{max} values were 152 and 101.5 μg root dw^{-1} 0.5 h^{-1}. The V_{max} value of *S. melongena* was approximately 1.5-fold higher than that of *S. torvum*, which suggests that the density of the Cd transporter in the root cell membranes of *S. melongena* is higher than in *S. torvum*. In this experiments, If the symplastic Cd absorption in roots is estimated by the conventional method using the difference of temperature at 2 and 25°C, it is required time consuming and double seedlings for experiment preparation.

5. Conclusion

For biological system analysis, the application of ICP-MS in enriched stable isotope tracer experiments has increased because ICP-MS has now become the preferred technique. An enriched stable isotope technique would be potent and useful tool for biological system experiments including element uptake, distribution and chemical form in plants. In this chapter, we introduced our one example of element uptake system using enriched isotope of ^{113}Cd and ^{114}Cd. This method has several merits compared to conventional methods if ICP-MS instrument is able to use. Application of enriched isotopes such as ^{113}Cd and ^{114}Cd would attain a new insight for plant biological system and will become a new tool to evaluate element behavior in plants.

6. Acknowledgment

This work was partly supported by the Program for the Promotion of Basic Research Activities for Innovative Biosciences (PROBRAIN).

7. References

Arao, T.; Takeda, H. & Nishihara, E. (2008). Reduction of cadmium translocation from roots to shoots in eggplant (Solanum melongena) by grafting onto Solanum torvum rootstock. *Soil Science and Plant Nutrition*, Vol. 54, pp.555–559.

Cataldo, D.A.; Garland, T.R. & Wildung, R.E. (1983). Cadmium uptake kinetics in intact soybean plants. *Plant Physiology*, Vol.73, pp.844–848.

Codex Alimentarius Commission 2005: Joint FAO/WHO Food Standards Programme. Twenty-eighth session, 4–9 July 2005, Rome, Italy. Report of the 37th session of the Codex Committee on Food Additives and Contaminants, 25–29 April 2005. Para. 175, Appendix XXVI. The Hague, the Netherlands. Available at URL: http://www.codexalimentarius. net/web/reports.jsp]ALINORM 05/28/12

Dannel, F.; Pfeffer, H. & RÖmheld, V. (2000). Characterization of root boron pools, boron uptake and boron translocation in sunflower using the stable isotopes ^{10}B and ^{11}B. *Australian Journal of Plant Physiology*, Vol.27, pp.397-405.
Davenport, R.J. (2007). Ion uptake by plant roots. In Plant Solute Transport. Eds AR Yeo and TJ Flowers, pp. 193–213. Blackwell Publishing, Oxford.
Epstein, E. (1973). Mechanisms of ion transport through plant cell membranes. *International Review of Cytology*, Vol.34, pp.123–168.
Epstein, E. & Hagen, C.E. (1952). A kinetics study of the absorption of alkali cations by barley roots. *Plant Physiology*,Vol.27, pp.457–474.
Glass, A.D.M. (2007). The apoplast: a kinetic perspective. In The Apoplast of Higher Plants: Compartment of Storage,Transport, and Reactions. The Significance of the Apoplast for the Mineral Nutrition of Higher Plants. Eds B Sattelmacher and WJ Horst, pp. 87–96. Springer-Verlag, Dordrecht.
Hart, J.J.; Welch, R.M.; Norvell, W.A. & Kochian, L.V. (2002). Transport interactions between cadmium and zinc in roots of bread and durum wheat seedlings. *Physiologia Plantarum*, Vol.116, pp.73–78.
Hart, J.J.; Welch, R.M.; Norvell, W.A. & Kochian, L.V. (2006). Characterization of cadmium uptake, translocation and storage in near-isogenic lines of durum wheat that differ in grain cadmium concentration. *New Phytologist*, Vol.172, pp.261–271.
Hart, J.J.; Welch, R.M.; Norvell, W.A.; Sullivan, L.A. & Kochian, L.V. (1998). Characterization of cadmium binding, uptake, and translocation in intact seedlings of bread and durum wheat cultivars. *Plant Physiology*, Vol.116, pp.1413–1420.
Kawasaki, A. & Oda, H. (2005). Application of ^{113}Cd as a tracer in evaluation of cadmium uptake by soybean under field conditions. Japanese *Journal of Soil Science and Plant Nutrition*, Vol.76, pp.261–267 (in Japanese with English summary).
Kawasaki, A.; Oda, H. & Yamada, M. (2004). Application of enriched ^{113}Cd-tracer to the soil pot experiment of soybean plant. *Japanese Journal of Soil Science and Plant Nutrition*, Vol.75, pp.667–672 (in Japanese with English summary).
Kimura, K.; Yoshida, K.; Sugito, T. & Yamasaki, S. (2003). Correction of interference by Mo oxide on Cd measurement by ICPMS. *Japanese Journal of Soil Science Plant Nutrition*, Vol.74, pp.493–497 (in Japanese with English summary).
Keller, C. (1995). Application of centrifuging to heavy metal studies in soil solutions. *Communication Soil Science and Plant Analysis*, Vol.26, pp.1621–1636.
Lasat, M.M.; Baker,A.J.M. & Kochian L.V. (1996). Physiological Characterization of Root Zn^{2+} Absorption and Translocation to Shoots in Zn Hyperaccumulator and Nonaccumulator Species of *Thlaspi*. *Plant Physiology*, Vol.112, pp.1715-1722.
Lasat, M.M.; Baker, A.J.M. & Kochian, L.V. (1998). Altered Zn compartmentation in the root symplasm and stimulated Zn absorption into the leaf as mechanisms involved in Zn hyperaccumulation in *Thlaspi caerulescens*. *Plant Physiology*, Vol.118, pp.875–883.
Ma, J.F.; Mitani, N.; Nagao, S.; Konishi, S.; Tamai, K.; Iwashita, T. & Yano, M. (2004). Characterization of the Silicon Uptake System and Molecular Mapping of the Silicon Transporter Gene in Rice. *Plant Physiology*, Vol. 136, pp. 3284–3289
May, T.W. & Wiedmeryer, R.H. (1998). A table of polyatomic interferences in ICP-MS. *Atomic Spectroscopy*, Vol.19, pp.150–154.
Ministry of Agriculture Forestry and Fisheries of Japan 2002:Survey of the cadmium contained in domestic vegetables. Available at URL:http://www.maff.go.jp/cd/PDF/C12. pdf,p34, p38, p25 (in Japanese)

Mitani, N. & Ma, J.F. (2005). Uptake system of silicon in different plant species. *Journal of Experimental Botany*, Vol. 56, pp. 1255–1261.

Mori, S.; Kawasaki, A.; Ishikawa, S. & Arao, T. (2009a). A new method for evaluating symplastic cadmium absorption in the roots of Solanum melongena using enriched isotopes ^{113}Cd and ^{114}Cd. *Soil Science and Plant Nutrition*, Vol.55, pp. 294–299.

Mori, S.; Uraguchi, S.; Ishikawa, S. & Arao, T. (2009b). Xylem loading process is a critical factor for determining Cd accumulation in the shoots of *Solanum melongena* and *Solanum torvum*. *Environmental and Experimental Botany*, Vol.67, pp.127-132.

Nakanishi, H.; Ogawa, I.; Ishimaru, Y.; Mori, S. & Nishizawa, K.N. (2006). Iron deficiency enhances cadmium uptake and translocation mediated by the Fe^{2+} transporters OsIRT1 and OsIRT2 in rice. *Soil Science and Plant Nutrition*, Vol.52, pp.464–469

Oda, H.; Yada, S.; Kawasaki, A. (2004). Uptake and transport of Cd supplied at different growth stages in hydroponically cultured soybean plants. *Biomedical Research on Trace Element*, Vol.15, pp.289–291 (in Japanese with English summary).

Rains, D.W.; Epstein, E.; Zasoski, R.J. & Aslam, M. (2006). Active silicon uptake by wheat. *Plant and Soil*, Vol.280, pp.223–228.

Stürup, S.; Hansen, H.R. & Gammelgaard, B. (2008). Application of enriched stable isotopes as tracers in biological systems: a critical review. *Analytical and Bioanalytical and Chemistry*, Vol.390, pp.541–554.

Takano, J.; Noguchi, K.; Yasumori, M.; Kabayashi, M.; Gajdos, Z.; Miwa, K.; Hayashi, H.; Yoneyama, T. & Fujiwara, T. (2002). Arabidopsis boron transporter for xylem loading. *Nature*, Vol.420, pp.337-340.

Takeda, H.; Sato, A.; Nishihara, E. & Arao, T. (2007). Reduction of cadmium concentration in eggplant (*Solanum melongena*) fruits by grafting with solanum torvum rootstock. *Japanese Journal of Soil Science and Plant Nutrition*, Vol.78, pp.581–586 (in Japanese with English summary).

Ueno, D.; Iwashita, T.; Zao, F.J. & Ma, J.F. (2008). Characterization of Cd translocation and identification of the Cd form in xylem sap of the Cd-hyperaccumulator *Arabidopsis halleri*. *Plant and Cell Physiology*, Vol.49, pp.540–548.

Ueno, D.; Koyama, E.; Kono, I.; Ando, T.; Yano, M. & Ma, J.F. (2009). Identification of a Novel Major Quantitative Trait Locus Controlling Distribution of Cd Between Roots and Shoots in Rice. *Plant and Cell Physiology*, Vol. 50, pp.2223-2233

Ueno, D.; Ma, J.F.; Iwashita, T., Zhao, F.J. & McGrath, S.P. (2005). Identification of the form of Cd in the leaves of a superior Cd-accumulating ecotype of *Thlaspi caerulescens* using ^{113}Cd-NMR. *Planta*, Vol.221, pp.928–936.

Uraguchi, S.; Mori, S.; Kuramata, M.; Kawasaki, A.; Arao, T. & Ishikawa, S. (2009). Root-to-shoot Cd translocation via the xylem is the major process determining shoot and grain cadmium accumulation in rice. *Journal of Experimental Botany*, Vol.60, pp.2677-2688

Wanger, G.J. (1993). Accumulation of cadmium in crop plants and its consequences to human health. *Advanced Agronomy*, Vol.51, pp.173–212.

Yada, S.; Oda, H. & Kawasaki, A. (2004). Uptake and transport of Cd supplied at early growth stage in hydroponically cultured soybean plants. *Biomedical Research on Trace Elements*, Vol.15, pp.292–294 (in Japanese with English summary).

Yamaguchi, N.; Mori, S.; Baba, K.; Yada, S.; Arao, T., Kitajima, N.; Hokura, A. & Terada, Y. (2011). Cadmium distribution in the root tissues of *solanaceous* plants with contrasting root-to-shoot Cd translocation efficiencies. *Environmental and Experimental Botany*, Vol.71, pp.198–206

Yu, Q.; Tang, C.; Chen, Z. & Kuo, J. (1999). Extraction of apoplastic sap from plant roots by centrifugation. *New phytologist*, Vol.143, pp.299-304.

Zhao, F.J.; Hamon, R.E.; Lombi, E.; McLaughlin, M.J. & McGrath, S.P. (2002). Characteristics of cadmium uptake in two contrasting ecotypes of the hyperaccumulator *Thlaspi caerulescens*. *Journal of Experimental Botany*, Vol.53, pp.535–543.

4

Diffusion Experiment in Lithium Ionic Conductors with the Radiotracer of ^{8}Li

Sun-Chan Jeong
Institute of Particle and Nuclear Studies (IPNS)
High Energy Accelerator Research Organization (KEK) 1-1 Oho
Japan

1. Introduction

Radioactive nuclides have been used in materials science for many decades. Besides their classical application as tracers for diffusion studies, nuclear techniques (i.e. Mössbauer Spectroscopy, Perturbed Angular Correlation, β-Nuclear Magnetic Resonance, Emission Channeling, etc.) are now being routinely used to gain microscopic information on the structural and dynamical properties of the bulk of materials via hyperfine interactions or emitted particles themselves (Wichert & Diecher, 2001). These nuclear techniques were primarily developed in nuclear physics for detecting particles or γ-radiations emitted during the decay of the radioactive nuclides. More recently these techniques have also been applied to study complex bio-molecules, surfaces, and interfaces (Prandolini, 2006). With the advent of most versatile 'radioactive isotope beam (RIB) factory' represented by the on-line isotope separator (ISOL)–based RIB facility (see Fig. 1), the possibilities for such investigations have been greatly expanded during the last decade (Cornell, 2003).
At the tandem accelerator facility of Japan Atomic Energy Agency (JAEA)-Tokai, a RIB facility, TRIAC (Watanabe et al., 2007)-Tokai Radioactive Ion Accelerator Complex- is operating since 2005. In the facility, short-lived radioactive nuclei produced by proton or heavy ion induced nuclear reactions can be accelerated up to the energy necessary for experiments. The energy is variable in the range from 0.1 to 1.1 MeV/nucleon, which is especially efficient for studies of the bulk of materials by using the RIBs as tracers. It allows us to implant (incorporate) the RIBs into specimens at a proper depth, avoiding the difficulties caused by the surface (e.g. diffusion barrier like oxide layers that often hampers the incorporation of those radioactive isotope probes into the materials of interest). In the facility, the separation and the implantation of radioactive probes are integrated into one device, as shown in Fig.1. Although the main concerns of the facility are nuclear physics experiments, as an effort to effectively use the available radioactive isotope beams at the TRIAC for materials studies, we have developed a diffusion tracing method by using the short-lived radioactive nuclei of ^{8}Li as diffusion tracers. The method has been successfully applied to measure diffusion coefficients in a typical defect-mediated lithium ionic conductor (refer to Chandra, 1981 for ionic conductors). We found that the present method is very efficient for the micro-diffusion, where the diffusion length is about 1μm per second.

In the following, the experimental method using ^{8}Li as a diffusion tracer and its application for measuring diffusion coefficients in inter-metallic lithium compounds will be reviewed, and then we will discuss possible extensions of the present method to study lithium diffusion with higher sensitivity such as the diffusion across interface in micrometer scale and the diffusion in nano-scale.

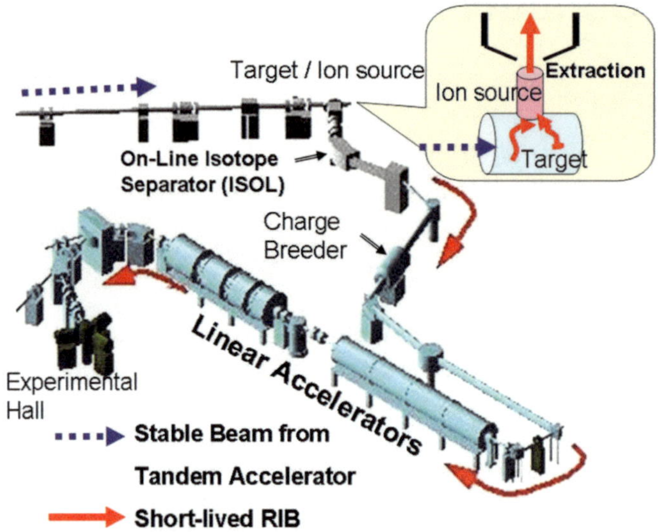

Fig. 1. Layout for ISOL-based RIB production at the TRIAC: Radioactive nuclei (e.g. uranium fission fragments) are produced by nuclear reactions induced in a thick target by a stable beam (e.g. UC_2 target by the irradiation of 30-MeV proton from the JAEA tandem accelerator which is not shown here). The target kept at a high temperature (~2000K) permits the fast diffusion of the reaction products into the ion source where they are ionized by plasma impact or surface ionization, as schematically shown in the inset. The singly charged ions (usually positively charged, 1+) are then extracted, mass-separated in a magnetic dipole field of the ISOL, and further, after being boosted to higher charge states by a charge breeder, accelerated to the energies necessary for experiments. For producing the radiotracer of ^{8}Li, the heavy ion beam of ^{7}Li on a ^{13}C-enriched graphite target was used (see the subsection 2-2).

2. Non-destructive on-line diffusion experiment

Over more than half a century, diffusion studies in solids using radioactive tracers (Tujin, 1997) have played an important role in understanding the underlying mechanism of atomic transport in solids, which is of great importance in a number of branches of materials science and engineering. Conventional diffusion studies by means of the radiotracer method in conjunction with a serial sectioning technique have been performed as follows (Wenwer at al., 1996): A small amount of a suitable radioactive isotope of the diffusing element is deposited onto the sample surface of interest. After a diffusion annealing at a temperature of T for a time of t, the sample is sectioned in parallel to the initial surface. An appropriate counting device measures the specific tracer activity in each section, which is proportional to the concentration of the diffusing species. The concentration-depth profile of the tracer is then compared with the solution of Fick's second law under the experimental boundary conditions, yielding the tracer diffusion coefficient at the temperature T in the sample. The choice of an appropriate serial sectioning technique depends on the average diffusion length related to the annealing time and temperature, often by $2(Dt)^{1/2}$, where D is the tracer diffusion coefficient in the sample. The method is consequently destructive.

Although the conventional radiotracer method for diffusion studies has yielded the most accurate diffusion coefficients, the method has not yet been applied for some elements because of no-availability of radiotracers with adequate lifetimes (a rather long lifer time is needed for the process of annealing and sectioning). Among them, ^{8}Li, the radioactive isotope of Li with a half-life of 0.84 s is of special interest for practical issues; how well Li ions move in the secondary Li ion batteries. Fast Li diffusion is desirable in battery materials, i.e., Li ionic conductors for materials of electrodes and solid electrolyte. For studies on the macroscopic diffusivity of Li in Li ionic conductors, various electro-chemical methods (Sato et al., 1997) have been usually adopted up to now. However, the diffusion coefficients are scattered over several orders of magnitude, strongly depending on the method used for the measurement. Therefore, the diffusion coefficients measured in different ways, e.g., by using the radiotracer of Li, are highly required to settle down such disagreements. Such an experimental knowledge on the Li diffusion in as-developed materials for the battery is also of importance in the recent general efforts to design the battery by simulations based on the first principle.

2.1 Principle of the measurement of Li diffusion coefficients with the radiotracer of ^{8}Li

The radiotracer ^{8}Li decays through β-emission to ^{8}Be with a half lifetime of 0.84 s, which immediately breaks up into two α-particles with energies continuously distributed around 1.6 MeV with a full width at half maximum (FWHM) of 0.6 MeV (Bonner et al., 1948).

As for a diffusion tracer, special attention has been paid on the energy loss of the α-particles in the sample of interest, which is sensitive to the diffusion length of about 1 μm. In an ideal case when the radiotracer emits monochromatic α-particles, the amount of incidental energy loss of the α-particles on their passage to the surface of the solid of interest depends on the position of the decaying emitter; the measured energies of the α-particles passed through the solid are closely related to the decaying positions of the tracer. The time evolution of the energy spectra is therefore supposed to be a measure of the diffusivity of the tracer in the solid. The energy spectra are broadening with increasing diffusion time; the tracer diffusion coefficients could be simply obtained by the time-dependent widths of the measured energy spectra if the inherent energy of the emitted charged particles is well defined. In the present case, however, the inherent energy distribution of the α-particles is continuous and broad (Bonner et al., 1948). Although the correspondence between the emitted position and measured energy of the charged particles is not as simple as above, it was shown in the simulation that the tracer diffusion coefficient could be obtained from the time-dependent yields of α-particles emitted by diffusing ^{8}Li with the help of the simulation (Jeong et al., 2003).

Figure 2 shows schematically the principle of the measurement: Implanting the beam of ^{8}Li with a properly adjusted energy into a depth, which is deeper than the average range of α-particles, we can make a situation where most α-particles stop in the sample. After implantation, since the primary implantation profile is broadening by diffusion, the α-particles emitted by ^{8}Li diffused toward the surface can survive and come out of sample with measurable energies. Then a charged particle detector located close to the sample surface could selectively detect α-particles from ^{8}Li diffusing toward the sample surface, since the implantation-depth is deeper than the average range of α-particles in the present case. Therefore, the temporal evolution of α-particle yields that come out of the sample would be a measure of the diffusivity of Li.

It should be noted that the present diffusion time is different from that of the conventional radiotracer method for diffusion studies (Wenwer at al., 1996) because the tracer in the present method diffuses all the time of the measurement. This is the reason why we call the present method as a non-destructive on-line measurement of diffusion.

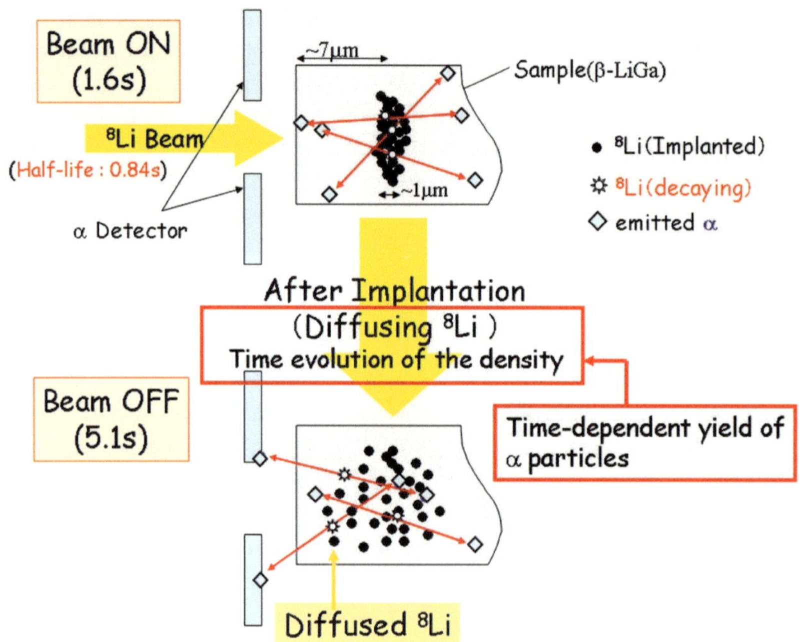

Fig. 2. Schematic view of the principle for measuring diffusion coefficients in β-LiGa (a typical sample of the inter-metallic Li compounds presently used) when the short-lived α-emitting ^{8}Li was used as the diffusion tracer. The yield of α particles measured at a time is a measure of the diffused distribution of ^{8}Li primarily implanted in the sample with a depth of about 7 μm.

2.2 Experimental set-up

Along the idea described in the previous subsection, we made an experimental set-up as shown in Fig. 3. All components were installed in a chamber evacuated to 1×10^{-4} Pa.

For producing ^{8}Li ($T_{1/2}$ = 838ms), we have used a neutron transfer reaction of ^{13}C (^{7}Li, ^{8}Li), using a sintered target of 99%-enriched ^{13}C. The 99%-enriched ^{13}C graphite disk (~10mm in diameter with a thickness less than 1mm) was mounted to the catcher position of the surface ionization type ion source with a beam window of 3-μm thick tungsten (Ichikawa et al., 2003). The target was bombarded with a 67-MeV $^{7}Li^{3+}$ beam with an intensity of about 100 pnA (particle nano-Ampere). The produced ^{8}Li was ionized, mass-separated as a radiotracer beam by the ISOL, and then injected to the post accelerators of the TRIAC, as shown in Fig.1.

Provide by the TRIAC, the radiotracer beam ^{8}Li of about 4 MeV with an intensity of about 10^4 particles/s was periodically implanted to a sample of β-LiGa with a following time sequence; 1.6 s for implantation (beam-on) and 5.1 s for subsequent diffusion (beam-off). With the incident energy, the ^{8}Li radiotracer can be implanted into the implantation-depth of about 7 μm from the front surface of the sample of β-LiGa. The α particles coming out of the sample were measured as a function of time by an annular solid-state detector (SSD) installed close to the front surface of the sample as shown in Fig. 3. The sequence was repeated to obtain good statistics, where the time-zero was always at the beginning of the implantation. Before starting the measurement, the sample was set at a temperature where the diffusion coefficient would to be measured.

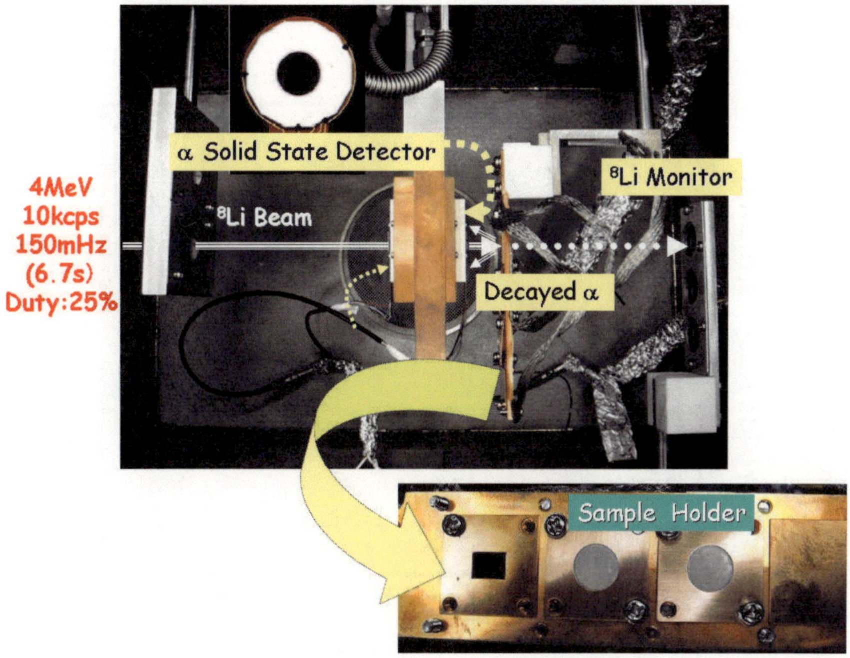

Fig. 3. Experimental set-up. The energetic and pulsed tracer beam of ^{8}Li provided by the TRIAC are implanted into the sample and the decayed α particles are measured as a function of time by the solid state detector located in front of the sample. The condition of the tracer beam is given, which includes energy, intensity, repetition frequency of the pulsed beam and its duty factor for the beam-on time.

2.3 Data analysis for diffusion coefficients

Using the experimental set-up in Fig. 3, a test experiment has been performed to measure the diffusion coefficients in the sample of LiAl compound (Jeong et al., 2005a, 2005b). Indeed, the diffusion coefficient of Li has been successfully obtained with an accuracy of better than 25% as a result of the comparison between the experimental and simulated time-dependent yields of α-particles. In the following, we present how diffusion coefficients are extracted in our present method, by applying to the measurement of the Li diffusion coefficients in LiGa, which is an inter-metallic Li compound and known as a good Li ionic conductor.

Figure 4 shows a normalized time spectrum of the yield of α-particles measured at room temperature for Li_xGa_{1-x}, x=0.54 in atomic ratio. The spectrum is presented by the ratios (i.e. time-dependent yields of α-particles divided by the α radioactivity of ^{8}Li at the time of interest). In this way was excluded the trivial time-dependency in the yield of α-particles just governed by the lifetime of ^{8}Li. The values of the ratios, therefore, should be constant over time if ^{8}Li does not diffuse at all. However, the experimental values as shown in Fig. 4 gradually increase with time and then fall off. Based on the relative time-dependency of α-particle yields (i.e. time dependent ratios), a diffusion coefficient was extracted by comparing with a Monte Carlo simulation where one-dimensional Fickian (Gaussian) diffusion was assumed. In Fig. 4 is also presented the time spectrum simulated with the diffusion coefficient best reproducing the experimental data as a result of comparisons to be described in the following.

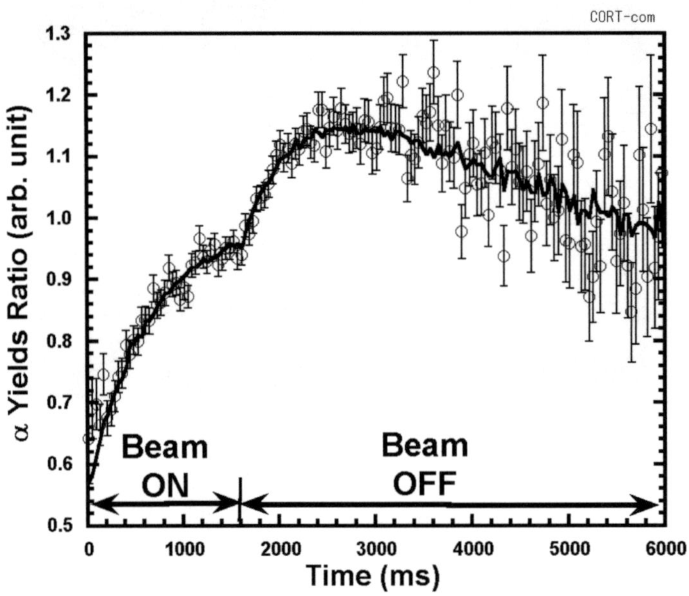

Fig. 4. Time spectrum of the α particle yields normalized by the α radioactivity of ^{8}Li implanted during the time of "Beam ON". The time spectrum, best simulated with the diffusion coefficient of $1.05 \times 10^{-7} cm^2/s$, is shown by solid line for comparison.

In the simulation, as described in detail in our previous publication (Jeong et al., 2003), we first defined the incident energy and energy spread of the ^{8}Li beam from the energy spectrum measured before implantation. Using the incident condition of the beam, we simulated the concentration-depth profile of ^{8}Li implanted in the sample and the time-evolution of the profile when a certain diffusion coefficient of Li in the sample was assumed. And then were simulated the energies of α-particles emitted from the time-dependent (diffusing) profiles of ^{8}Li, by taking into account the energy loss and straggling on their passage from the emitted position to the sample surface. Finally, integrated over the energies larger than 400 keV, the time-dependent α-particle yields associated with the diffusion coefficients assumed in the simulation were obtained and then compared with the experimental time spectrum, after being normalized in the same way as preformed for the experimental data. It should be noted that the resultant is the macroscopic diffusion coefficients, since the simulation neglects the isotopic characters of diffusing elements. The parameters (mean and FWHM) describing the concentration-depth profile, and the energy loss and straggling of α-particles were estimated by using the SRIM-2003 code (Ziegler, 1985), which is widely used in this kind of application with high reliability.

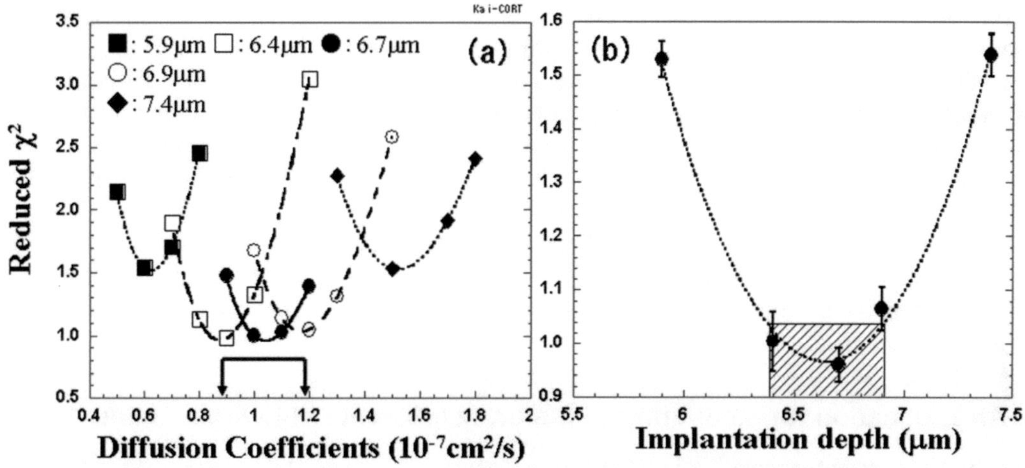

Fig. 5. (a) Reduced χ^2 values are compared for different implantation depths and correspondingly varied diffusion coefficients. Each of them was calculated with the spectrum simulated with a pair of implantation depth and diffusion coefficient. For each value of the implantation depth, the minimum value of the reduced χ^2 was extracted by a quadratic curve fitting as indicated by lines in (a), and then was plotted as a function of the corresponding depth in (b). By fitting the χ^2 values with a quadratic function of the depth [dotted line in (b)], the hatched region was estimated, especially as a constraint on the uncertain range of the implantation depth in the simulation. The allowed range of the depth was finally transformed into that of the diffusion coefficients [arrowed region in (a)], considering as a systematic error in the determination of the diffusion coefficient in the present case. (Fig. 2 from Jeong et al., 2008)

For comparison, we performed the χ^2 test, the likelihood test between the experimental and simulated time spectra. In the case of maximum likelihood, the reduced χ^2, the value of χ^2 simply divided by the number of data points in the present case, should be minimum and approximately close to 1. During the test of likelihood, two input parameters in the simulation were examined: diffusion coefficient and implantation depth. In the present simulation, the depth was explicitly considered as a variable input in order to take into account the roughness of the sample surface (less than 1μm) and range-uncertainty in the SRIM code. These two parameters were found to be most sensitive to the result of the simulation, i.e. the time-dependent structure of α-particle yields, but strongly inter-correlated in such a way that, e.g., simulations with somewhat higher (lower) diffusion coefficient and deeper (shallower) implantation could yield similar results as compared in Fig. 5(a). For different implantation depths, the χ^2 tests were explicitly performed with various diffusion coefficients. An example of the test is shown in Fig. 5(a), where the values of the reduced χ^2 were compared for different combinations of implantation depths and diffusion coefficients assumed in the simulation. For a value of the implantation depth, a diffusion coefficient giving rise to maximum likelihood (minimum value of the χ^2) can be identified. The minimum values of the χ^2 and the corresponding implantation depths are further compared in Fig. 5(b). Then a pair of the implantation depth and the diffusion coefficient corresponding to the most minimum value of the reduced χ^2 in the comparison is finally obtained; the diffusion coefficient of $(1.05\pm0.15)\times10^{-7}cm^2/s$ with the implantation depth of 6.64 ($\pm$0.25) μm in the present case. As indicated in Fig. 5 for the diffusion coefficients [arrowed in Fig. 5(a)] and implantation depth [hatched in Fig. 5(b)], the uncertainty in the determination of the coefficient comes mainly from the uncertain implantation depth. Here, we have considered the uncertain implantation depth as the main cause of systematic error inherent in the present method; otherwise the diffusion coefficient could be determined more accurately. It should be noted that the present results are not sensitive to the width of the primary concentration profiles of the tracer (Jeong et al., 2003).

As the reference spectrum for normalization, an experimental time spectrum of the α-radioactivity of ^{8}Li implanted in pure Cu was used. It allows us to avoid the systematic errors caused by the beam on/off operations, since no significant diffusion effects were observed in the case.

3. An application for measuring diffusion coefficients in Li ionic conductors

The β-phase of inter-metallic Li compounds, such as β-LiAl, LiGa and LiIn, has been considered as possible electrode materials in Li ionic batteries because of their high diffusion coefficients at room temperature for Li ions (Wen & Huggins, 1981). They are common in lattice structure; NaTl structure (Ehrenberg et al., 2002) composed of two interpenetrating sublattices, each forming a diamond lattice with a homogeneity range of around stoichiometric atomic ratio of Li (48~56 at. % Li for LiAl, 44~54% at. % Li for LiGa, 44~54 at. % Li for LiIn). In order to understand the motion of Li in the β phase as an ionic conductor, the defect structure in the Li compounds, closely related to the fast ionic motion, has been intensively studied via the measurement of electrical resistivity and density with help of the standard x-ray diffraction analysis (Sugai et al., 1995; Kuriyama et al., 1996). There exist three kinds of defects (see Fig. 6); vacancies on Li sites (V_{Li}), defects on anti-sites that replaced by Li (Li_A, A=Al, Ga, In) and complex defects (V_{Li} - Li_A). By forming the complex defects, the ionic motion of Li is suppressed or assisted depending on the kinds of anti-site atoms; Li diffusion rather slows down in β-LiAl while becomes rather faster in β-LiIn, although Li diffusivity almost linearly depends on the constitutional vacancy concentration on the Li sublattice (V_{Li}) (Tarczon et al., 1988). The high diffusion coefficient in the Li compounds is associated with the constitutional vacancy concentration on the Li sublattice, which is relatively large as compared to the usual metal alloy. The thermodynamic behavior of the Li vacancy has also been inferred from the anomalous electrical resistivity ("100K" anomaly) observed at around 95K near the critical composition corresponding to the Li-deficient region of β-LiAl (Kuriyama et al., 1980), which is considered as an order-disorder transition of vacancies on the Li sublattices (Brun et al., 1983).

The macroscopic ionic motion of Li has been so far inferred from the analysis of the electrical response to the applied voltage (electro-chemical method) (Sato et al., 1997). The values of diffusion coefficients, essentially obtained in such indirect ways, are often scattered over several orders of magnitude, strongly depending on the method of data analysis for finally extracting the diffusion coefficients.

Under such general situation, we have applied our method for measuring diffusion coefficients of Li, especially in β-LiGa where the diffusion coefficients has not been well measured although Li diffusion is known to be fastest among the Li inter-metallic compounds. The high diffusivity of Li in β-LiGa is associated with an especially large, constitutional vacancy concentration on the Li sublattice, almost three times lager than in β-LiAl. It would be also very interesting to observe, directly in terms of diffusion coefficient, the order-disordering transition of the vacancies on Li sites as well as the effect on the Li diffusion associated with the formation of the complex defects.

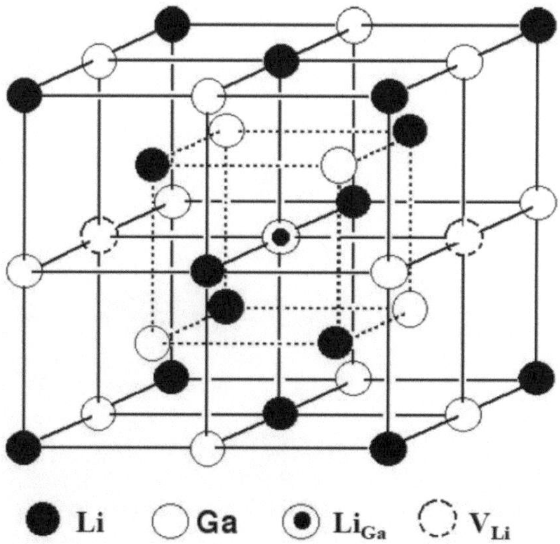

Fig. 6. Crystal structure of NaTl-type inter-metallic compound LiGa contained defects (Li: Li atoms at the Li-sites, Ga: Ga atoms at the Ga-sites, Li_{Ga}: Li atoms at the Ga-sites (i.e. anti-site defects, where Ga-sites are replaced by Li atoms), V_{Li}: vacancies at the Li-sites). The most Li-poor β-LiGa (44 at. % Li) can have two V_{Li} in unit cell, although the concentration of Li_{Ga} is close to zero (see Fig. 7).

3.1 Materials

The samples of β-LiGa with the Li composition of 43~54 at. % were prepared by direct reaction of desired amounts of lithium (99.9%) and gallium (99.999%) in a tantalum crucible. The crystallization was performed by the Tammam-stöber method as reported in Yahagi, 1980. And the crystal was found to be polycrystalline by X-ray diffraction analysis.

The composition of Li was determined by the electrical resistivity measurements, relying on the systematic correlation between them (Kuriyama et al., 1996) as shown in Fig. 7. The concentrations of the point defects, $[V_{Li}]$ and $[Li_{Ga}]$, strongly depends on Li compositions; with increasing Li composition from 43 to 54 at. %, $[V_{Li}]$ decreases from 11.4 to 2.8%, while $[Li_{Ga}]$ increases from 0 to 5.1%. V_{Li} is the dominant defect for Li-poor compositions, Li_{Ga} is the dominant defect for the Li-rich ones, and mixing of the two defects extends throughout the entire phase region. The coexistence of V_{Li} and Li_{Ga} is expected to form V_{Li}-Li_{Ga} complex defects (Kuriyama et al., 1996) as reported for the defect structure of β-LiAl (Sugai et al., 1995), which would play an important role in reducing the strain energy caused by the point defects in the lattice matrix. Especially, almost the same amounts of V_{Li} and Li_{Ga} exist for the composition of about 51 at. % Li.

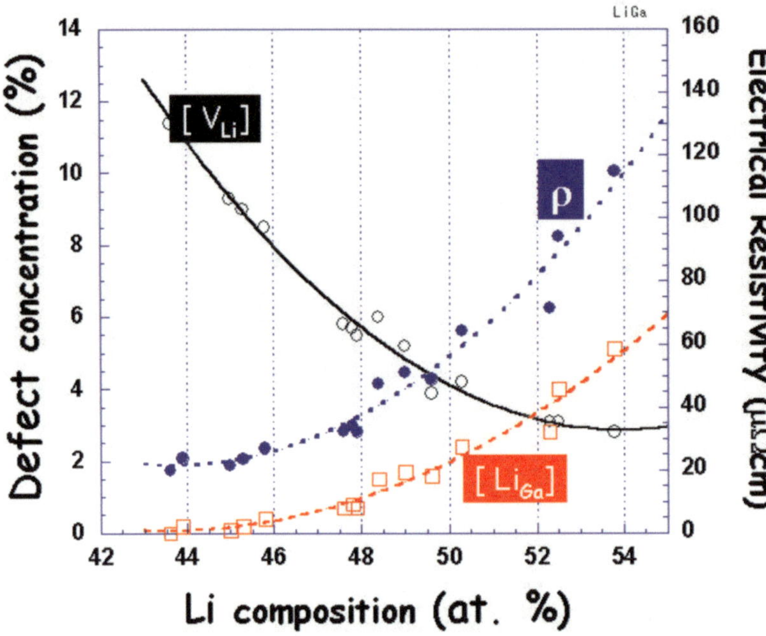

Fig. 7. Defect concentrations ([V_{Li}], [Li_{Ga}]), electrical resistivity (ρ) vs. Li composition in atomic % for LiGa. The data points were taken from Kuriyama et al. 1996, and plotted for comparison.

The sample sliced in a form of disk with a diameter of 10mm and a thickness of 1mm was installed on the sample holder in the experimental chamber shown in Fig. 3. The surface of the sample was polished to the roughness less than 1 μm before set-up.

3.2 Li composition dependence of Li diffusion in Li inter-metallic compounds

Since the concentration of the defects is characterized by the Li composition under control in synthesis (Yahagi, 1980), as demonstrated in Fig. 7, the Li inter-metallic compounds have attracted much attention as a typical Li ionic conductor for investigating the high diffusivity of Li ions in a well-defined environment of defects. The detailed study on the diffusivity of Li in the Li inter-metallic compounds is further of interest, since the compounds have been considered as possible negative electrodes for Li ion secondary batteries more efficient than the commercially available (Wen & Huggins, 1981; Saint et al., 2005).

For β-LiAl and β-LiIn with the composition ranging between about 48 and 53 at. % Li, the diffusion coefficients of Li have been measured by a pulsed field gradient nuclear magnetic resonance (PFG-NMR) method in Tarczon et al., 1988, where was shown a strong correlation between the defect structure and the Li diffusion coefficient through their respective Li composition-dependency. The diffusion coefficient at room temperature was observed to become higher monotonously with decreasing Li composition - in other words, increasing concentration of V_{Li}. It was pointed out, furthermore, that the monotonous dependence of Li diffusivity on the Li composition was found to be slightly modified, depending on the species of the anti-site atoms.

While the monotonous composition-dependence of Li diffusivity in both β-LiAl and β-LiIn strongly suggested that the Li atoms diffuse via a vacancy mechanism (i.e. by a vacancy-atom exchange process through nearest neighbored paths on the Li sublattice), the slight modification in the correlation was partly understood by the coexistence of two types of defects, namely vacancies in the Li sublattice and Li anti-site atoms in the aluminum or indium sublattice forming compound defects (V_{Li}-Li_{Al}, or V_{Li}-Li_{In}) (Kishio & Britain, 1979; Tarczon et al., 1988). Owing to the elastic relaxation of atoms against the strain induced in the vicinity of the point defects, the interaction between the two types of defects once forming a compound defect can be attractive or repulsive according to the relative size of the ions on the sublattice replaced by Li (atomic size effect). The Li anti-site atom Li_{Al} in β-LiAl produces compressional strain (expanded lattice), since the radius (0.68Å) for the Li ion in a closed shell configuration (Kittel, 2005) is larger than that (0.50 Å) for the aluminum ion, while the anti-site atom Li_{In} in β-LiIn induces dilatational strain (contracted lattice) because of ionic radius (0.8 Å) for indium larger than that of the substitutional Li ion. On the other hand, the vacancy V_{Li} always produces dilatational strain. Indeed, as supposed by the atomic size effect, the diffusion in β-LiIn was observed to be enhanced by the presence of the repulsive interaction between different types of defects, more than expected by the single vacancy diffusion mechanism (Tarczon et al., 1988).

Based on the atomic size effect, the interaction between V_{Li} and Li_{Ga} in β-LiGa is supposed to be attractive as in β-LiAl, because the radius (0.62 Å) of gallium ion is slightly smaller than that of Li ion. The strength of the interaction in β-LiGa is expected to be weaker than observed in β-LiAl and β-LiIn, since the radii of the constituent ions are quite close to each other. In addition to the specific interaction - attractive and weak, LiGa has stable β-phase over the Li composition range wider than investigated so far in β-LiAl and β-LiIn, consequently allowing us to address the Li diffusion in the concentration range of V_{Li} almost three times wider than before (Kuriyama et al., 1996).

The diffusion coefficients of Li in β-LiGa have not yet been measured in detail, but measured only by the electro-chemical method (Wen & Huggins, 1981), where the Li composition was not well defined. Using the on-line diffusion tracing method introduced in the previous section 2, we have measured the diffusion coefficients of Li in β-LiGa with the composition in the range of about 43 to 54 at. % Li. Of special interest is how the Li diffusion in β-LiGa depends on the Li composition, consequently the concentration of the V_{Li} defect.

3.2.1 Abnormal Li diffusion: Enhanced or suppressed by the formation of defect complex

In Fig. 8, the time-dependent normalized α-particle yields are compared for different Li compositions, i.e. 43.6, 50.0 and 53.2 at. % Li. Referring to the time dependence of the α-particle yields, the stoichiometric β-LiGa has the highest diffusivity of Li among three, most quickly rising and falling down. This observation is quite different from those observed for β-LiAl and β-LiIn (Tarczon et al., 1988) which are iso-structural with the β-LiGa.

The abnormal behavior of Li diffusivity in β-LiGa is well identified in Fig. 9, where the diffusion coefficients in β-LiGa at room temperature are presented as a function of Li composition together with the data for β-LiAl and β-LiIn reported in Tarczon et al., 1988. As shown in Fig. 9, the Li diffusion coefficients in β-LiAl and β-LiIn decrease monotonously with increasing Li composition, but the correlation is changing due to the coexistence of vacant Li sites V_{Li} and anti-site Li atoms on the aluminum (or indium) sites Li_{Al} (or Li_{In}), as discussed earlier in terms of the size effect of constituent atoms. Assuming the single vacancy diffusion in LiGa, the diffusion coefficients are supposed to keep increasing in the Li-poor composition in similar ways as observed in both β-LiAl and β-LiIn. However, what observed in the present measurement goes the other way; the diffusion becomes faster around stoichiometric atomic ratio, i.e. showing a maximum around the Li composition of 48~50 at. %.

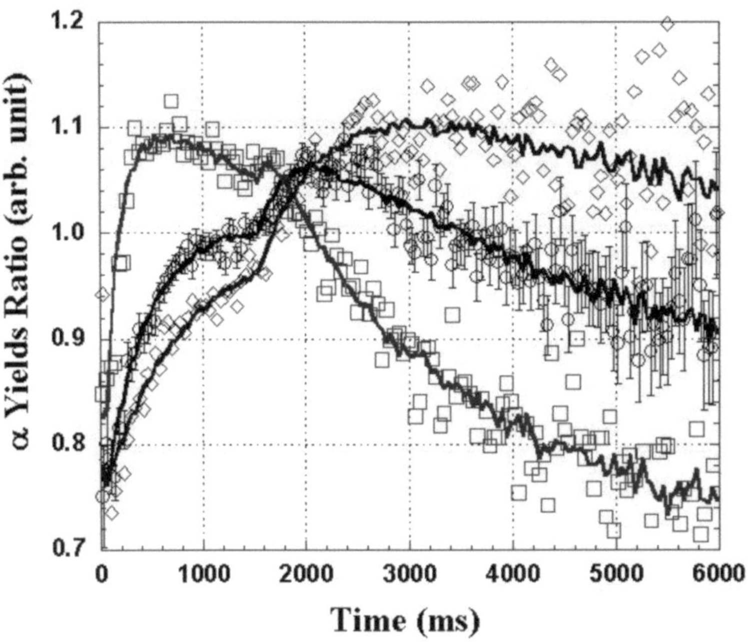

Fig. 8. Normalized time spectra of α yields obtained at room temperature for β-LiGa with Li atomic compositions of 43.6(○), 50.0(□) and 53.2(◇) at. %. The results from the simulation are also shown as solid lines. In the simulation, best reproducing the data as a result of χ^2 tests, the diffusion coefficients of $2.4x10^{-7}cm^2/s$, $8.5x10^{-7}cm^2/s$ and $1.2x10^{-7}cm^2/s$ were assumed for 43.6, 50.0 and 53.2 at. % Li, respectively.

Although the data for β-LiAl and β-LiIn were available only in a limited range of Li composition from 48.3 to 53 at. % Li, it would be interesting to note that the diffusion coefficients in β-LiGa varied in a similar way (increase with decreasing Li content) between those values in β-LiAl and β-LiIn in the corresponding range of Li composition. This would be intuitively understandable, since the iso-structural β-LiGa with a comparable size of the constituent atoms should have an interaction between V_{Li} and Li_{Ga} with intermediate strength as compared to those discussed in the case of β-LiAl and β-LiIn. Therefore, the composition-dependence of Li diffusivity in β-LiAl and β-LiIn could be considered as the lower and upper limits of the Li diffusivity in β-LiGa over the Li composition range of interest, respectively; two complementary explanations might be possible by referring to the respective composition-dependency observed in β-LiAl and β-LiIn.

Referring to the tendency in β-LiIn, the diffusion of Li in very Li-poor β-LiGa seems to be suppressed. This could happen, for example, by assuming the formation of defect complex such as V_{Li}-V_{Li} and/or V_{Li}-Li_{Ga}-V_{Li}, since the concentration of V_{Li} become much (almost three times at maximum) larger (Kuriyama et al., 1996) in the more Li-poor composition than investigated in β-LiIn (Tarczon et al., 1988). It should be noted that the number of vacant Li sites in a unit cell volume (8 for Li and 8 for Ga) is about two for the most Li-poor β-LiGa (43.6 at. % Li), whereas there exist about one vacant Li site in every two-unit cells for the most Li-poor β-LiAl (48.3 at. % Li) and β-LiIn (48.4 at. % Li) (refer to Fig. 6). Here, we assumed the random distribution of the vacancies over the available sites.

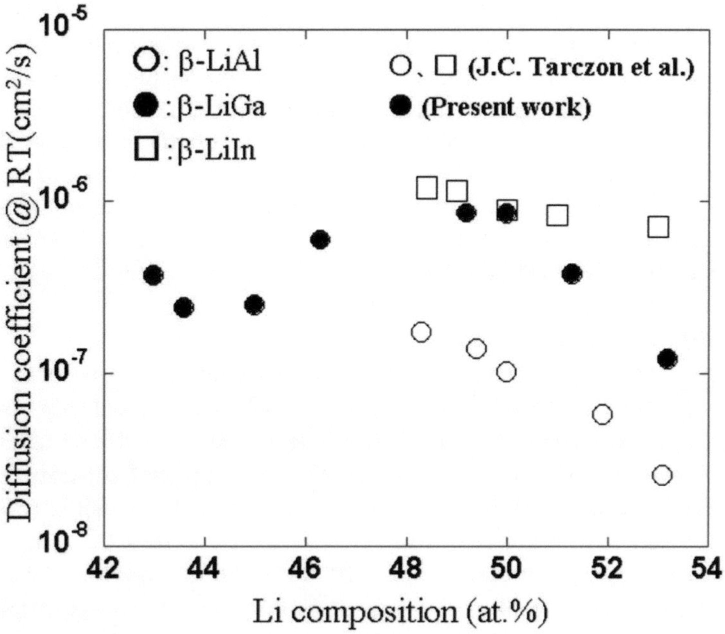

Fig. 9. Li composition-dependence of diffusion coefficients of Li at room temperature. The diffusion coefficients measured at room temperature are given as function of Li composition in atomic % for β-LiGa (present work), β-LiIn and β-LiAl (from Tarczon et al., 1988), respectively. The most Li-poor composition of LiGa shown in the figure is out of the β-phase (Fig.5 from Jeong at al., 2009).

Alternatively, assuming the diffusivity of Li in the most Li-poor β-LiGa as a good extension of the diffusivity of Li in the Li-poor β-LiAl (48.3 at. % Li) because of nearly zero concentrations of Li_{Al} and Li_{Ga} in the corresponding Li composition, the diffusivity of Li observed around the stoichiometric β-LiGa could be considered to be enhanced by the coexisting defects of V_{Li} and Li_{Ga}. Under the coexistence of V_{Li} and Li_{Ga}, the motion of the vacancies on the Li site, supposedly the carriers of Li atom, seems to be strongly promoted. This suggests that the interaction between V_{Li} and Li_{Ga} would be stronger and rather repulsive than expected by the atomic size effect, but not as strong as observed in β-LiIn. In addition, as a specific characteristic of the interaction in β-LiGa, we found strong composition-dependence of the interaction that appears to become stronger when a comparable amount of two types of point defects exists.

In the present measurement, although we found abnormal Li diffusion in very Li-poor composition of β-LiGa, the characteristic of the anomaly in diffusion, i.e. whether the diffusion is suppressed or enhanced, is not conclusive. A detailed theoretical consideration is highly required for quantitative discussion. From the experimental point of view, however, it would be interesting to extend the measurement to the more Li-poor composition of LiIn recently confirmed to have the β-phase in the same composition (Asano, 2000) as investigated for β-LiGa in the present work.

So far, we have compared our data for LiGa with those obtained by a pulsed field gradient nuclear magnetic resonance (PFG-NMR) method in Tarczon et al, 1988. The chemical diffusion coefficients of Li in LiGa obtained by an electrochemical method at 415°C were reported earlier (Wen & Huggins, 1981), where the self-diffusion coefficients of Li were also calculated by considering that the diffusivity of Ga was appreciably lower that that of Li. The self-diffusion coefficients at 415°C were found to be constant 1.1×10^{-6} and 5.0×10^{-7} cm^2/s on the Li poor and rich sides of LiGa, respectively, with a sudden change between 48 and 47.6 at. % Li. Those values were almost one order of magnitude smaller than those extrapolated to the temperature of 415°C from our present data. It should be noted that the similar amount of discrepancies in the values of diffusion coefficients were also found in the case of LiAl, by comparing the data obtained by PFG-NMR (Tarczon et al, 1988) and the electrochemical method (Wen et al, 1979).

3.3 Order-disordering of Li vacancies

Figure 10 shows time spectra of the yield of α-particles (represented by the ratios normalized to 1 at the end of beam-on) measured at different temperature for the stoichiometric LiIn. With decreasing temperature, the more α-particles are observed at later time, which means that the diffusion becomes slower at lower temperature. Interestingly, there observed a big change in the time structure of α-particle yields (normalized) at a certain temperature, where the temperature was varied by almost the same amount for the neighboring measurements compared in Fig. 10. Based on this relative time-dependency of α-particle yields (time dependent ratio) as demonstrated in Fig. 10, diffusion coefficients at various temperatures could be estimated by the way discussed previously.

In order to observe such a sudden change for β-LiGa, we have also performed a detailed measurement, where the temperature was varied by a fine step (about 5-degree difference in temperature between neighboring measurements), especially below room temperature.

The diffusion coefficients of Li in β-LiGa and β-LiIn with a near stoichiometeric composition of Li are displayed in Fig. 11 as a function of inverse temperature. For both samples, the diffusion coefficients suddenly change at a certain temperature, as observed time-dependent ratio spectra in Fig. 10, and follow Arrhenius behavior in the region of higher temperature. The sudden change in the value of the diffusion coefficient for β-LiGa of 44 at. % Li occurs at 234 (±2) K. The anomalous electrical resistivity (i.e. sudden change in the value of resistivity) is also observed at the same temperature, as shown in Fig. 11. The resistivity measurements were carried out using a van der Pauw method as used for β-LiAl (Sugai et al., 1995). This observation is closely related to the thermal properties of the structural defects, already observed as the anomalies in heat capacity (Hamanaka et al., 1998; Kuriyama et al., 1986) and nuclear-spin lattice relaxation (Nakamura et al., 2007) at 233 K near the critical composition of the Li-poor β-LiGa. It has been suggested that these phenomena are related to order-disorder transformation of the Li vacancies in the compounds. By the neutron diffraction measurements for β-LiAl (Brun at al., 1983), a sudden change around 100K in the electrical resistance ("100K" anomaly as mentioned previously) has been understood by an ordering of the vacancies on the Li sublattice below the transition temperature; the structure below 100K is body centered tetragonal; the unit cell contains 10 Li and 20 Al positions, with the vacancies located at 2*a* positions.

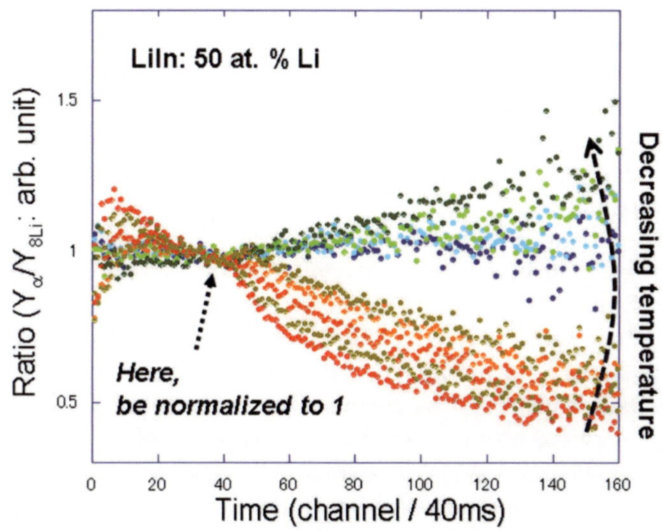

Fig. 10. Time spectra of α-particle yields measured at various temperatures for LiIn. The spectra were corrected for removing trivial time dependence and further normalized properly for easy comparison.

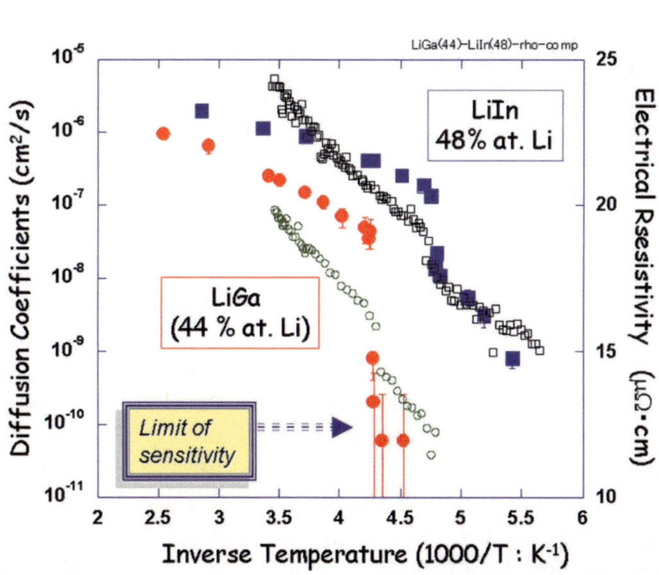

Fig. 11. Temperature-dependence of diffusion coefficients (closed symbols) and electrical resistivity (open symbols) for β-LiGa with 44 at. % Li, and β-LiIn with 48 % at. Li.

The ordering of the vacancies would produce a sharp drop in the Li diffusion coefficients at the ordering temperature, since the vacancies are supposed to be carriers of Li atom (i.e. ordered vacancies suppress the random exchange of Li atoms via vacancies). The observed amount of change, more than two orders of magnitude in the value of diffusion coefficients in the case of β-LiGa, is quite impressive as compared to those observed in the measurement of electrical resistivities where just a small change (at most 1/10) can be seen at the transformation temperature. The transition temperature is known to shift to lower temperature, and the change in the electrical resistivity becomes invisibly smaller with increasing Li composition, where a large fraction of the Li vacancies forms complex defects with anti-site Li atoms on Ga sublattices (Kuriyama et al, 1986). Therefore, with higher sensitivity, the present method could be applied for better investigating the characteristics of the transformation that are supposed to be correlated with the concentrations of structural defects (i.e. Li compositions).

Around the transition temperature, the time structure of the normalized α-particle yields could not be explained in terms of one component of diffusion, i.e. one diffusion coefficient, implying that the diffusion coefficients would not be singly determined in the transition region. The temperature window, in which the transition appears to occur, is about 10K.

At the lower temperature followed by a sudden change around 234 (±2) K for β-LiGa, the diffusion coefficients are observed as a constant, which is the lower limit of diffusion coefficients accessible by the present method; for diffusion coefficients less than about 10^{-10} cm^2/s, any significant effect in the yields of α-particles due to the diffusing ^8Li could not be observed because of the short life-time of the radiotracer.

4. Aiming at lithium micro- and nano- scope

As an extension of the present method, we consider the possibility to trace the Li macroscopic behavior across the interface in hetero-structural Li ionic conductors in micrometer scale, to be termed ^8Li microscope.

The time spectra shown in Fig. 10 represent dynamical movement of Li in the sample between as-implanted position and surface during a cycle for measurement. For a single layer (diffusion in homogeneous sample) as discussed in case of Fig.2, the present method can trace, most efficiently, the Li movement within one-dimensional distance of about 7μm for about 7s. For slow diffusion, Li is still moving toward the surface for the time of measurement, the α-particle yields are monotonically increasing with time. For moderate diffusivity, a maximum is observed when Li is reflected by the surface. After the maximum, the yield is simply decreasing with time since Li is diffusing into the bulk. For double layers (hetero-structural sample) whose interface exist in between, i.e. introducing an interface between the as-implanted position and the surface, we could observe time structure different from the case of single layer. In other words, we could observe how Li interacts with the interface, e.g. if Li is precipitated, perfect- or half-reflected on the interface. This idea could be applied to take a dynamical picture of Li in Li ion micro-batteries consisting of thin films of several μm in thickness. Therefore, the present method could be called ^8Li microscope by analogy with the neutron transmission image of Li in a secondary Li ion battery, where the picture of Li, actually ^6Li, in the battery was taken by the neutron radiography with a resolution of mm (Takai et al., 2005).

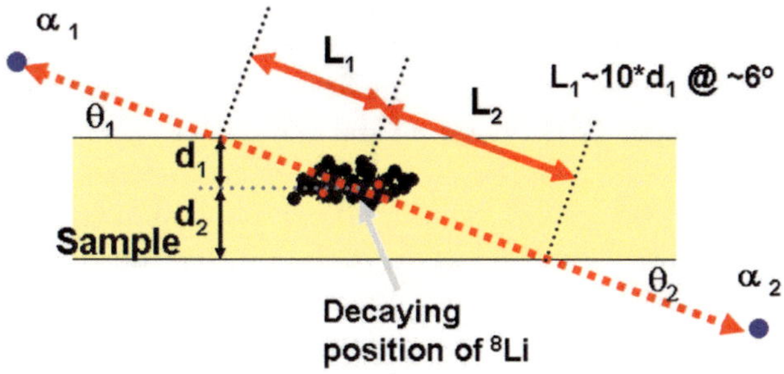

Fig. 12. Schematic layout for a coincident measurement of two α-particles emitted with angles of $\theta_1=\theta_2$ relative to the surface of sample. The actual path lengths (L_1 and L_2) in the sample are enhanced 10 times as compared to those considered in Fig. 2 when applying limitation to the emission angle of 6°.

For higher sensitivity, from micro-scale to nano-scale, we consider a coincident measurement of two α-particles emitted from position of β-decaying ^8Li. The micrometer sensitivity of the method discussed so far is partly coming from the broad energy distribution of α-particles on decaying. Therefore, the coincident measurement is supposed to dramatically improve the sensitivity since the coincident α-particles have the same energy at the decaying position. In diffusion tracing method by ^8Li, however, a special attention is paid on the energy loss of α-particles subjected to the actual path length in the sample of interest from the decaying to the detection positions. For diffusion in a nano-scale, the diffusion length is too short to give a significant change in the energy loss of α-particles. We further apply a limitation in emission (detection) angles to the coincident measurement as shown in Fig. 12. With small emission angles, the actual path length of α-particles in the sample of interest can be made significantly longer than implanted depth, giving rise to a considerable energy loss difference against a nano-scale diffusion length toward the surface.

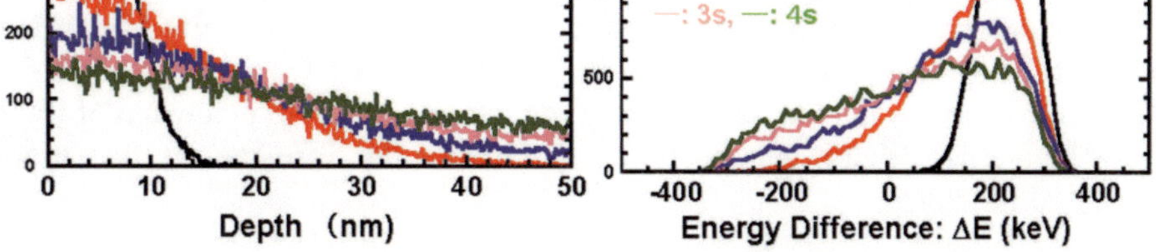

Fig. 13. (a) Simulated concentration profiles of ^8Li implanted into $LiCoO_2$ with a thickness of 50nm with an energy of 1keV and their evolution by diffusion simulated at 1, 2, 3 and 4 seconds after implantation with a diffusion coefficient of 10^{-12} cm^2/s. (b) Spectra of energy difference of α-particles coincidently emitted at angles of 5° relative to the surface of the sample. They are respectively simulated in the same time sequence as done in Fig.13 (a).

Along with the idea, we have performed a simulation to examine the feasibility and search for an observable most sensitive to diffusion profiles with a diffusion coefficient of 10^{-12} cm^2/s, a specific goal of present consideration. The simulation was performed in a similar way discussed in Jeong et al, 2003, by using the energy loss and straggling of α-particles provided by the SRIM-code. We assumed the diffusion coefficient as 10^{-12} cm^2/s, the sample thickness as 50nm, the emission angle as 5°, and the implantation energy as 1keV.

As a result of the simulation shown in Fig.13, we found that the energy difference between two coincidently measured α-particles could provide nano-scale sensitivity. The concentration profiles of ^8Li simulated sequentially with a condition given above are shown in Fig. 13 (a); the as-implanted profile is broadening with time by diffusion. Simulated in the same time sequence as done for profiles, the spectra of energy difference of two α-particles to be measured coincidently are compared in Fig. 13 (b). There is a good correspondence between the diffusion profiles and energy difference spectra, more specifically a clear one-to-one correspondence between decaying position of ^8Li and energy difference of two coincidently measured α-particles. For example, looking at the time evolution of the counts of coincident events with zero energy difference in the energy difference spectra should be a good measure of the time when ^8Li is across the middle of the sample, because the zero energy difference means that the path lengths experienced by two coincident α-particles are identical. We can conclude that the sensitivity against the diffusion of 10^{-12} cm^2/s could be easily achieved by the coincident measurement of two α-particles with the geometry assumed in the simulation.

5. Conclusions

A non-destructive and on-line radiotracer method for diffusion studies in lithium ionic conductors has been reviewed. As the tracer, the pulsed beam of the short-lived α-emitting radioisotope of ^8Li was implanted into a sample of interest. By analyzing the time-dependent yields of the α-particles from the diffusing ^8Li measured in coincident with the repetition cycle of the beam, the tracer diffusion coefficients were extracted with a good accuracy. The method has been successfully applied to measure the lithium diffusion coefficients in a typical defect-mediated lithium ionic conductor of LiGa, well demonstrating that the method is very efficient to measure the diffusion in the micro-meter regime per second. Anomalous composition-dependence of Li diffusion coefficients in β-LiGa was observed; the stoichiometric LiGa showed the highest diffusivity of Li. The anomaly was discussed qualitatively in terms of the formation of defect complex and the interaction between the constituent defects. The ordering of the Li vacancies in the Li-deficient LiGa was observed for the first time in terms of the Li diffusion by the present method, and its thermodynamic aspect was discussed. Further development, as an extension of the present method for higher sensitivity, was proposed to measure the diffusion on the nano-scale in lithium ionic conductors.

6. Acknowledgments

The author would like to thank all the members of the TRIAC collaboration between KEK (I. Katayama, H. Kawakami, Y. Hirayama, N. Imai, H. Ishiyama, H. Miyatake, Y.X. Watanabe, S. Arai, K. Niki, M. Okada, M. Oyaizu), JAEA (A. Osa, Y. Otokawa, M. Sataka, H. Sugai, S.

Ichikawa, S. Okayasu, K. Nishio, S. Mitsuoka, T. Nakanoya), and Aomori University (M. Yahagi, T. Hashimoto). The present studies have been greatly indebted to the staffs of the Tandem Accelerator Center at JAEA for ensuring stable delivery of primary beams for producing radiotracer beam of ^{8}Li at the TRIAC. The work is partly supported by a Grant-in-Aid for Scientific Research (B), No. 16360317 from the Japan Society for the Promotion of Science.

7. References

Bonner, P. W. et al. (1948). A study of breakup of ^{8}Li, *Physical Review*. Vol. 73, Issue 8, (April 1949), pp. 885-890, 0031-899X

Brun, T. O. et al. (1983). Ordering of Vacancies in LiAl, *Solid State Communications,* Vol. 45, Issue 8, (Feburary1983), pp. 721-724, 0038-1098

Chandra, S. (1981). *super-ionic solids: principle and applications,* North-Holland Publishing Company, ISBN 0-444-86039-8, New York

Cornell, J. (Ed.). (December 2003). Solid-Sate Physics at EURISOL, In: *The Physics Case for EURISOL,* GANIL, available from http://pro.ganil-spiral2.eu/eurisol/feasibility-study-reports/feasibility-study-appendix-a

Hamanaka, H. et al. (1998). Anomalous heat capacity and defect structure in β-LiGa, *Solid State Ionics,* Vol. 113-115, (December 1998), pp. 69-72, 0167-2738

Ichikawa, S. (2003). Ion source development for the JAERI on-line isotope separator, *Nuclear instruments & methods in physics research B,* Vol. 204, *(May 2003),* pp. 372-376, 0168-583X

Ehrenberg, H. et al. (2002). Phase Transition from the Cubic Zintl Phase LiIn into a Tetragonal Structure at Low Temperature, *Journal of solid state chemistry,* Vol. 167, Issue 1, (August 2002), pp. 1-6, 0022-4596

Jeong, S. C. et al. (2003). Simulation Study on the Measurements of Diffusion Coefficients in Solid Materials by Short-lived Radiotracer Beams, *Japanese journal of applied physics,* Vol. 42, No. 7A, (July 2003), pp. 4576-4583, 0021-4922

Jeong, S. C. et al, (2005a). Measurement of diffusion coefficients in solids by the short-lived radioactive beam of ^{8}Li, *Nuclear Instruments and Methods in Physics Research B,* Vol. 230, Issues 1-4, (April 2005), pp. 596-600, 0168-583X

Jeong, S.C. et al. (2005b) Measurement of self-diffusion coefficients in Li ionic conductors by using the short-lived radiotracer of 8Li, *Journal of Phase Equilibria and Diffusion,* Vol. 26, No. 5, (September 2005), pp. 472-476, 1547-7037

Jeong, S.C. et al. (2008) On-Line Diffusion Tracing in Li Ionic Conductors by the Short-Lived radioactive Beams of 8Li, *Japanese journal of applied physics,* Vol. 47, No. 8, (July 2008), pp. 6413-6415, 0021-4922

Jeong, S.C. et al. (2009). Abnormal Li diffusion in b-LiGa by the formation of defect complex, *Solid State Ionics,* Vol. 180, Issues 6-8, (May 2009), pp. 626-630, 0167-2738

Kishio, K & Britain, J.O. (1979). Defect structure of β-LiAl, *Journal of Physics and Chemistry of Solids,* Vol. 40, Issue 12, (1979), pp. 933-940, 0022-3697

Kittel, C. (2005) *Introduction to Solid State Physics* (8-th ed.), Wiely, New York, p71

Kuriyama, K. et al. (1996). Defect Structure and Li-Vacancy Ordering in β-LiGa, *Physical review. B; Condensed matter and materials physics,* Vol. 54, Issue 9, (September 1996), pp. 6015-6018, 1098-0121

Kuriyama, K. et al. (1986). Electrical-transport properties in the semimetallic compound LiGa, *Physical review. B; Condensed matter and materials physics*, Vol. 33, Issue 10, (May 1986), pp. 7291-7293, 1098-0121

Kuriyama, K., Kamijoh, K & Nozaki, T. (1980). Anomalous Electrical Resistivity in LiAl Near Critical Composition, *Physical review. B; Condensed matter and materials physics*, Vol. 22, Issue 1, (July 1980), pp. 470-471, 1098-0121

Prandolini, M. J. (2006). Magnetic nanostructure: radioactive probes and recent developments, *Reports on Progress in Physics*, Vol.69, No. 5, (May 2006), pp. 1235-1324, 0034-4885

Nakamura K. et al. (2007). Li+ ionic diffusion and vacancy ordering in β-LiGa, *Faraday Discussions*, Vol. 134, (2007), pp. 343-352, DOI: 10.1039/B602445A

Saint, J. et al. (2005). Exploring the Li-Ga room temperature phase diagram and the electrochemical performance of the Li_xGa_y alloys vs. Li, *Solid State Ionics*, Vol. 176, issues 1-2, (January 2005), pp. 189-197, 0167-2738

Sato, H. et al, (1997). Electrochemical characterization of thin-film $LiCoO_2$ electrodes in propylene carbonate solutions, *Journal of Power Sources*, Vol. 68, Issue 2, (October 1997), pp. 540-544, 0378-7753

Sugai, H. et al. (1995) Defect Structure in Neutron-Irradiated b-6LiAl and b-7LiAl: Electrical Resistivity and Li Diffusion, *Physical review. B; Condensed matter and materials physics*, Vol. 52, Issue 6, (August 1995), pp. 4050-4059, 1098-0121

Takai S. et al. (2005). Diffusion coefficients measurements of La2/3-xLi3xTiO3 using neutron radiography, *Solid States Ionics*, Vol. 176, Issues 39-40, (September 2005), pp. 2227-2233, 0167-2738

Tarczon, J. C. et al. (1988).Vacancy-Antistructure Defect Interaction Diffusion in β-LiAl and β-LiIn, *Materials Science &. Engineering A*, Vol. 101, (May 1988), pp. 99-108, 0921-5093

Tuijn, C. (1997). On the history of Solid-state diffusion. *Defect and Diffusion Forum*, Vol. 141-142, (1997), pp. 1-48, 1662-9507

Yahagi, M. (1980). Single crystal growth of LiAl, *Journal of Crystal Growth*, Vol. 49, Issue 2, (June 1980), pp. 396-398, 0022-0248

Watanabe, Y. X. et al. (2007). Tokai Radioactive Ion Accelerator Complex (TRIAC), *European Physical Journal - Special Topics*, Vol. 150, No. 1, (March 2007), pp. 259-262, 1951-6355

Wen, C. J. & Huggins, R. A. (1981). Electrochemical Investigation of the Lithium-Gallium System, *Journal of The Electrochemical Society*, Vol. 128, issue 8, (August 1981), pp. 1636-1641, 0013-4651

Wen, C. J. et al. (1979). Thermodynamic and Mass Transport properties of "LiAl", *Journal of The Electrochemical Society*, Vol. 126, Issue 12, (December 1979), pp. 2258-2266, 0013-4651

Wenwer, F. et al. (1996). A universal ion-beam-sputtering device for diffusion studies, *Measurement science & technology*. Vol. 7, No. 4, (April 1996), pp. 632-640, 0957-0233

Wichert, T. & Deicher, M. (2001). Studies of semiconductors, *Nuclear Physics A*, Vol. 693, No. 3-4, (October 2001), pp.327-357, 0375-9474

Ziegler, J. F., Biersack, J. F. & Littmark, U. (1985). *The Stopping and Range of Ions in Solids* (2003 Version), Pergamon Press, New York, Chap. 8, available from http://www.srim.org/

5

Determination of Chemical State and External Magnetic Field Effect on the Energy Shifts and X-Ray Intensity Ratios of Yttrium and Its Compounds

Sevil Porikli[1] and Yakup Kurucu[2]
Erzincan University, Faculty of Art and Sciences, Department of Physics
Atatürk University, Faculty of Sciences, Department of Physics
Turkey

1. Introduction

The term 'X-ray fluorescence analysis' (XRF) refers to the measurement of characteristic fluorescent emission resulting from the deexcitation of inner shell vacancies produced in the sample by means of a suitable source of radiation. For a particular energy (wavelength) of fluorescent light emitted by a sample, the number of photons per unit time (generally referred to as peak intensity or count rate) is related to the amount of that analyte in the sample. The counting rates for all detectable elements within a sample are usually calculated by counting, for a set amount of time, the number of photons that are detected for the various analytes' characteristic X-ray energy lines. It is important to note that these fluorescent lines are actually observed as peaks with a semi-Gaussian distribution because of the imperfect resolution of modern detector technology. Therefore, by determining the energy of the X-ray peaks in a sample's spectrum, and by calculating the count rate of the various elemental peaks, it is possible to qualitatively establish the elemental composition of the samples and to quantitatively measure the concentration of these elements.

XRF is an analytical method to determine the chemical composition of all kinds of materials. The materials can be in solid, liquid, powder, filtered or other form. XRF can also sometimes be used to determine the thickness and composition of layers and coatings. The method is fast, accurate and non-destructive, and usually requires only a minimum of sample preparation. Applications are very broad and include the metal, cement, oil, polymer, plastic and food industries, along with mining, mineralogy and geology, and environmental analysis is of water and waste materials. XRF is also a very useful analysis technique for research and pharmacy.

For routine XRF analysis, two major approaches are distinguishable based on the type of detector used to measure the characteristic X-ray emission spectra. Wavelength dispersive X-ray fluorescence (WDXRF) analyses depend upon the use of diffracting crystal to determine the characteristic wavelength of the emitted X-rays. Energy dispersive X-ray fluorescence (EDXRF) employs detectors that directly measure the energy of the X-rays by collecting the ionization produced in suitable detecting medium.

X-ray emission spectra are known to be influenced by chemical combination of X-ray emitting atoms with different ligands. The effect of the chemical combination, however are not large and a theoretical interpretation of these effects has not been established completely. Therefore, chemical effects have rarely been utilized in the characterization of materials. The purpose of this work was to study chemical effects and discuss their applications to Yttrium (Y) in various compounds. So much so that, this paper presents and discusses the measured spectra both energy dispersive and wave-length dispersive X-ray spectrometer. In the first part of the study, the effect of the 0.6T and 1.2T external magnetic field and chemical state on the $K\alpha$, $K\beta_{1,3}$ and $K\beta_{2,4}$ X-ray energies and relative intensity ratios for Y, YBr_3, YCl_3, YF_3, $Y(NO_3)_3.6H_2O$, Y_2O_3, YPO_4, $Y(SO_4)_3.8H_2O$ and Y_2S_3 have been investigated, using the 22.69 keV X-rays from a ^{109}Cd and 59.54 keV γ-ray from a ^{241}Am as photon sources. The measurements were done using an energy dispersive Si(Li) detector with photon excitation by radioisotopes. For B=0, the present experimental results were compared with the experimental and theoretical data in the literature.
The results show that Y_2O_3, YF_3 and Y_2S_3 can change owing to the applied magnetic field. In addition, we found that the energy of characteristic X-ray series is totally independent of the excitation source and mode. However, changes have been observed in X-ray spectra when the element studied in the sample is chemically bonded to others. The development of high resolution spectrometers allows for the characterization and study of these effects.
In the second part of the study, energies and full width at half maximum (FWHM) values of the $K\alpha$, $K\beta_{1,3}$ and $K\beta_{2,4}$ X-ray of Y and its compounds were measured by a wavelength dispersive spectrometer. An accurate analytical representation of each line, obtained by a fit to a minimal set of Gaussians, is presented. The absolute energies and FWHM values derived from the data, agree well with previous measurements. Possible origins of chemical shifts are discussed. It was found that the chemical shifts of Y $K\alpha$ line in pure Y and its some compounds relatively small (less than 0.1 eV with pure Y as reference). The influence of crystal symmetry on the energy shifts of X-ray lines is an interesting aspect of our study. The results demonstrate a clear dependence of the energy shifts on the chemical state of the element in the sample. The relative intensities are more susceptible to the chemical environment than the energy shifts.
It is well known that the chemical environment of an element affects and modifies the various characteristics of its X-ray emission spectrum. Most of the works suffer from neglecting chemical influences, and usually theoretical atomic values (Scofield, 1974a, 1974b) are used as a reference even for quite different chemical compounds of certain element. However, some papers deal with chemical effects (Berenyi et al., 1978; Rao et al., 1986), mostly in connection with X-ray emission after an electron capture process (EC) and partially after photoionisation (PI). Paic &Pecar (1976) found that for first-row transition elements the $K\beta/K\alpha$ ratio depends on the mode of excitation. The difference between the ratios for electron-capture decay and photoionization becomes almost 10%. Similar results were obtained by Arndt et al., (1982) and they pointed out that the difference comes from a strong shake-off process accompanying photoionization.
The 3*d* transition metals have played an important role in the development of modern technology, and knowledge of their valence electronic structure is very important for understanding their physical properties. X-ray spectroscopy is an established tool for probing the electronic structure of 3*d* transition metal compounds (Meisel et al., 1989). A number of techniques, such as photoemission spectroscopy, X-ray absorption and X-ray emission spectroscopy create a hole in an inner shell in order to investigate the valance

electron configuration. Although some investigations have been made to study their electronic structures individually, no systematic study has been made so far for understanding the valence electronic structure of all the 3*d* transition metals. With a deeper theoretical understanding of the underlying processes and further improving X-ray sources, sophisticated experiments have been developed (e.g., resonant inelastic scattering, magnetic dichroism (Groot, 1994a,b)) that give detailed information on the valance electron configuration.

In a number of X-ray spectral studies of 3*d* transition metals it has been observed that the $K\beta$-to-Ka X-ray intensity ratios are dependent on the physical and chemical environments of the elements in the sample. In the earlier studies of 3*d* metal compounds (Küçüköder et al., 1993; Padhi et al., 1993, 1995), the influence of chemical effects has shown difference in the $K\beta$-to-Ka X-ray intensity ratios up to nearly 10%. Such chemical effects can be caused either by a varying 3*d* electron population or by the admixture of *p* states from the ligand atoms to the 3*d* states of the metal or both. Brunner et al. (1982) explained their experimental results by the change in screening of 3*p* electron by 3*d* valance electrons as well as the polarization effect. They also pointed out that the chemical effect is almost the same order of magnitude as the effect of excitation mode and both effects should be studied separately. However, most of these measurements have been performed with solid-state X-ray detectors and the change in the satellite peaks in the $K\beta$ X-ray region has not been studied because of poor energy resolution. Urch (1979) discussed the chemical effect on the *K* X-ray spectra based on molecular-orbital (MO) theory. Similar studies on the chemical effect on the X-ray spectra have already been done extensively. However, these studies are concerned mostly on with transition energies and profiles of X-rays, and qualitative discussions on the intensities have not yet been made. Tamaki et al. (1979) studied Cr and 55Mn-labeled compounds and reported that the $K\beta/Ka$ ratio increases with increasing formal oxidation number of the element in the compound. Kataria et al. (1986) found deviations of up to 10% for the same ratio in the case of Mn compounds. Mukoyama et al. (1986) experimentally confirmed the theoretical predictions following Brunners' (1982) model in the case of Te and Mo compounds for $K\beta_{1,3}$ and $K\beta_2$ components.

Wide employed applications and the intriguing asymmetry of the Cu Kα and Kβ line shapes (Deutsch&Hart, 1982) along with those of all 3d transition elements, led in turn to a century of extensive spectrometric studies of the Cu Ka and $K\beta$ spectra. In spite of these extensive studied, recent studies reveal that surprises still lurk under the skewed $Ka_{1,2}$ and overlapping $K\beta_{1,3}$ lines, and the related multi-electronic satellite (S) and hypersatellite (HS) spectra. The asymmetric lineshape of the copper emission lines were attributed in the past to a number of different processes: Kondo-like interaction of the conduction electrons with the core holes, final state interactions between the core holes and the incomplete 3*d* shell, 2*p*/3*d* shell electrostatic exchange interaction, and most importantly, shake-up and shake-off of electrons from the 3*l* shells. The last process, in particular, received in the past strong experimental support.

Raj et al. (1998) were carried studies on CrB, CrB2 and FeB forms in order to look into the electronic structure of the transition metals in monoborides and diborides. In order to understand the valence electronic structure of the transition metals in the compounds, they have tried to compare the measured $K\beta$-to-Ka ratios with the multiconfiguration Dirac±Fock calculations assuming different electronic configurations for the transition metal. Such a comparison would provide information on the valence electronic structure of the transition metals in the compounds, which could in turn provide information on the rearrangement of electrons between 3*d* and 4*s* states of the metal or electron transfer from the 3*d* state of the metal to the ligand atoms or vice-versa.

The chemical environment has a strong effect on the transitions originated in valence band and its influence could clearly be observed in the emission spectrum structure. The P-$K\beta$ spectrum has been studied by many authors (Takashi, 1972; Taniguchi 1984; Torres Delluigi et al., 2003), who used both single-crystal and two-crystal spectrometers with conventional X-ray sources. These authors showed some modifications in the $K\beta$ spectra and its relation with P chemical environment. Compounds with oxygen as ligand atom, a relationship between the ratio of the $K\beta'$ line intensity to the total intensity of the $K\beta$ line and the energy shift of the $K\alpha 1,2$ lines was found by them. Fichter (1975) discussed the Ka-line shifts related to the oxidation number of the P-atom. The chemical shift of X-ray emission lines is usually interpreted with the effective charges or oxidation number of the X-ray emitting atom (Leonhardt&Meisel, 1970; Meisel et al., 1989). For example, the Al Ka lines shift to higher energy in going from the metal to the oxide (Nagel et al., 1974). By comparing the measured chemical shifts with those of the reference compounds, Gohshi et al. (1973, 1975) determined the chemical state of S, Cr and Sn. They obtained not only qualitative, but also quantitative results.

Theoretical studies of emission spectra were performed mostly to study atoms with simple electronic configurations (see, e.g., the review by Mukoyama et. al., 2004). Theoretical calculations for solids and molecules have been done mainly to predict transition energies and line profiles, but evaluation of transition probabilities is rather scarce. This is due to two reasons: Firstly, molecular orbital methods and band theories are originally developed for ground states and sometimes difficult to apply to excited states with an inner-shell vacancy. Secondly, matrix elements for absorption and emission processes in molecules include multi-center integrations, which are tedious and require long computing times. Most individual authors indicate that their results favor the Dirac-Hartree-Fock calculations of Scofield (1974a), rather than the significantly lower predictions of the same author's earlier Dirac-Hartree-Slater calculations (Scofield, 1969). Both of these describe the de-excitation of a single *K* vacancy in a neutral atom. However careful examinations (Salem et al., 1974; Khan&Karimi, 1980) of all available data reveal a tendency for $K\beta/Ka$ to fall somewhat below the DHF predictions in the atomic number region $21<Z<32$ where the 3*d* subshell is filling.

Band et al. (1985) applied the scattered-wave (SW) Xa MO method to calculate the chemical effect on the $K\beta/Ka$ intensity ratios. They performed the MO calculations for different chemical compounds of Mn and Cr using the cluster method and obtained the spherically averaged self-consistent potential and the total charge of the valance electrons in the central atom region. Chemical effect on the $K\beta/Ka$ X-ray intensity ratios or some Mn and Cr compounds has been studied both theoretically and experimentally by Mukoyama et al. (1986). The *K* X-ray spectra were measured by the use of a double crystal spectrometer with high energy resolution. The theoretical calculations were made with the use of the discrete-variational Xa molecular-orbital method and the X-ray intensities were evaluated in the dipole approximation using molecular wave functions. Mukoyoma et al. (2000) have calculated the electronic structures of tetraoxo complexes of 4*d* and 5*d* elements with the discrete-variational Xa (DV-Xa) MO method. They found that the for Tc compounds, the calculated values were in good agreement with the measured values. In the case of Mo *K* X-rays, the agreement theory and experiment is not as good as with Tc compounds. Yamoto et al. (1986) studied the variation of the relative *K* X-ray intensity ratios for compounds involving Tc isotopes, ^{95m}Tc, ^{97m}Tc and ^{99m}Tc. They found that the chemical effect on the $K\beta/K\alpha$ ratios for 4*d* elements is small but the dependence of the $K\beta_2/K\alpha$ ratios on the chemical environments is appreciable.

Mukoyoma et al. (1986) have calculated the $K\beta_2/K\alpha$ intensity ratios for chemical compounds of 4*d* transition elements by the use of the simple theoretical method of Brunner et al. (1982), originally developed for 3*d* elements. Although they obtained good agreement between theories and experimental, it was found that their model is inadequate for the metallic cases.

These investigations on the effect of 3*d* and 4*d* electrons were performed only to understand the chemical effect on the X-ray intensity ratios. However, if the dependence on the excitation mode is also caused by the difference in the number of 3*d* electrons, as shown in our previous work, both effects, i.e. the dependence on the chemical environment and on the excitation mode, can be treated simultaneously to estimate the $K\beta/K\alpha$ ratios in terms of the number of 3*d* electrons. However it may also be possible that these ratios are also expressed as a function of other parameters, such as bond length and effective number of 4*p* electrons. Considering these facts, it is interesting to study the dependence of the $K\beta_2/K\alpha$ ratio in 3*d* elements on various parameters of chemical compounds.

Iiahara et al. (1993) measured the *L* X-ray intensity ratios for some Nb and Mo compounds. When the measured $L\gamma_1/L\beta_1$ ratios were plotted as a function of the effective number of 4*d* electrons, they found that the experimental data are experimental data are almost on a straight line. However, it should be noted that the $4d\rightarrow 2p$ transitions are allowed dipole transition and the 4d electron is the valance shell electron which participates directly in the X-ray emission. In this case the X-ray emission rate is proportional to the number of 4*d* electrons and increases with increasing effective number of 4*d* electrons.

The chemical behavior of actinide atoms (in particular, that of uranium) is determined by valance nl-electrons of three types: 7s, 6d and 5f. Although the bond energies of these electrons are almost equal, their wave-function differs greatly in distribution in the radial direction (Katz et al., 1986; Balasubramanian et al., 1994). It can be said that the 5f electrons have an only core arrangement in the atom. Therefore, when actinides chemical bonding is studied, several questions should be raised: (1) the possibility and form of 5f electrons participation in chemical bonding; (2) the necessity for taking into account the splitting of valance levels of the atom into two sublevels nl_+ and nl_- with total angular momentum $j=1\pm1/2$ because of the relativistic effect of spin-orbital splitting (SOS) (Pyykko,1988; Pepper et al., 1991); (3) the energetic stabilization of the specific chemical state of the heavy atom due to fine effects of electron density redistribution on valance orbital; (4) the possibility of independent participation of split subshells in chemical bond formation. One of the methods of modern precise spectroscopy capable of providing a correct description of chemical bonding process is the chemical shift (CS) method of X-ray emission lines, i.e. the change in their energy when the chemical state of the emitting atom is changed (Gohshi&Ohtsuka, 1973; Makarov, 1999; Batrakov et al., 2004).

Atomic theory has shown that the magnetic dipole moments observed in bulk matter arise from one or two origins: one is the motion of the electrons about their atomic nucleus (orbital angular momentum) and the other is the rotation of the electron about its own axis (spin angular momentum). The nucleus itself has a magnetic moment. Except in special types of experiments, this moment is so small that it can be neglected in the consideration of the usual macroscopic magnetic properties of bulk matter. When the atom is placed in an external magnetic field, the magnetic field produces a torque on the magnetic dipole. The torque is tending to align the dipole with the field, associated with this torque; there is a potential energy of orientation:

$$\Delta E=-\mu_l B \quad (1)$$

μ_l is the orbital magnetic dipole moment of an electron. According to the quantum theory, all spectral lines arise from transitions of electrons between different allowed energy levels within the atom and the frequency of the spectral line is proportional to the energy difference between the initial and final levels. The slight difference in energy is associated with these different orientations in the magnetic field. In the presence of a magnetic field, the elementary magnetic dipoles, whether permanent or induced, will act to set up a field of induction of their own that will modify the original field.

Today investigations of magnetic effects on X-ray spectra became actual both from theoretical and experimental points of view. The numbers of works on this subject deal with magnetic circular dichroism (MCD) in X-ray absorption spectroscopy (XAS), that gives information on empty electron states in a valence band and their spin configurations (Thole et al., 1992, Stöhr&Wu, 1994). Several experiments have been performed on the external magnetic field effect on the *K* shell X-ray emission lines. Demir et al., (2006a) determined how the radiative transitions and the structures of the atoms in a strong magnetic field are affected, Kα and Kβ X-ray production cross sections, the *K*-shell fluorescence yields and $I(K\beta/K\alpha)$ intensity ratios for ferromagnetic Nd, Gd, and Dy and paramagnetic Eu and Ho were investigated using the 59.5 keV incident photon energy in the external magnetic fields intensities ± 0.75 T. On the other hand, Demir et al., (2006b) measured L_3 subshell fluorescence yields and level widths for Gd, Dy, Hg and Pb at 59.5 keV incident photon energy in the external magnetic field of intensities ± 0.75 T. Porikli et al. (2008a; 2008b; 2008c) conduct measurements using pure Ni, Co, Cu and Zn and their compounds. Characteristic quantities such as position of line maxima, full widths at half maximum (FWHM), indices of asymmetry and intensity ratio values were determined in the values of external magnetic field 0.6 T and 1.2 T. Several experiments have been performed on the external magnetic field effect on the *K* shell X-ray emission lines. Commonly, experimental *L* X-ray intensities are measured using radioisotopes as excitation sources (Han et al., 2010; Porikli, 2011b). They have the advantages of stable intensity and energy and of small sizes, which allow compact and efficient geometry, and they operate without any external power.

Our motivation in performing this experiment has been two fold. First, with the aim of a better understanding of the chemical effect and external magnetic field effect, we conduct measurements using pure yttrium (Y) and its compounds. Characteristic quantities such as position of line maxima, full widths at half maximum (FWHM), indices of asymmetry and $K\beta_1/K\alpha$, $K\beta_2/K\alpha$, $K\beta_2/K\beta_1$ and $K\beta/K\alpha$ intensity ratio values are determined in the values of external magnetic field 0.6 T and 1.2 T. In the present work, the measurements were done using a filtered 22.69 keV from Cd-109 and 59.54 keV from Am-241 point source and Si(Li) detector. Particle size effects were circumvented. Peak areas were determined using Gaussian fitting procedures and the errors in various corrections such as self-absorption and detector efficiency were minimized. The measured values were compared due to the external magnetic field and chemical effect. The measured values for B=0 were compared with other experimental and theoretical results. To our knowledge, these intensity ratio values of Y in the external magnetic field have not been reported in the literature and appear to have been measured here for the first time. Secondly, spectra of *K* X-rays emitted from a Y target were measured in high resolution wave-length dispersive X-ray spectrometer (WDXRF). After the measurement, characteristic quantitative such as peak energy, indices of asymmetry, FWHM are determined. The measured spectra were described in terms of a background function (a straight line) and peaks having Gaussian profiles. The Microcal Orgin 7.5 was used for peak resolving and background subtraction of *K* X-rays.

2. Experimental

2.1 Experimental set up (EDXRF)

Yttrium compounds can serve as host lattices for doping with different lanthanide cations and they used as a catalyst for ethylene polymerization. As a metal, it is used on the electrodes of some high-performance spark plugs. Yttrium is also used in the manufacturing of gas mantles for propane lanterns as a replacement for thorium, which is radioactive. Developing uses include yttrium-stabilized zirconia in particular as a solid electrolyte and as an oxygen sensor in automobile exhaust systems. Yttrium is used in the production of a large variety of synthetic garnets. Small amounts of yttrium (0.1 to 0.2%) have been used to reduce the grain sizes of chromium, molybdenum, titanium, and zirconium. It is also used to increase the strength of aluminium and magnesium alloys. The addition of yttrium to alloys generally improves workability, adds resistance to high-temperature recrystallization and significantly enhances resistance to high-temperature oxidation (see graphite nodule discussion below).

The studied elements were Y, YBr_3, YCl_3, YF_3, $Y(NO_3)_3.6H_2O$, Y_2O_3, YPO_4, $Y(SO_4)_3.8H_2O$ and Y_2S_3. The purity of commercially obtained materials was better than 99%. For powdered samples, particle size effects have a strong influence on the quantitative analysis of infinitely thick specimens. Even for specimens of intermediate thickness, in which category the specimens analyzed in the present study fall, these effects can be significant. Therefore, to circumvent particle size effects all samples were grounded and sieved through a -400 mesh (<37 μm) sieve. The powder was palletized to a uniform thickness of 0.05-0.15 g cm^{-2} range by a hydraulic press using 10 ton in^{-2} pressure. The diameter of the pellet was 13 mm.

All of the lines were excited using a 100 mCi Am-241 annular radioactive source and Cd-109 point source of 10 mCi strength (providing $5.0x10^3$ $steradian^{-1}$ photon flux of Ag X-radiation). The fluorescent X-rays emitted from the targets were analyzed by a Si(Li) detector (effective area 12.5 mm^2, thickness 3 mm, Be window thickness 0.025 mm).

For each sample three separate measurements have been made just to see the consistency of the results obtained from different measurements agreed with a deviation of less than 1%. The experimental setup consist of a Si(Li) detector and Cd-109 radioactive source as shown in Fig. 1. The mechanical arrangement to house the source-sample-detector combination in a definite geometry was shown in Fig. 1. An Al, Pb conical collimator was used between the sample and the detector for the excitation to obtain a large beam of emergent radiation and to avoid the interaction of the X-rays emitted by the component elements of the radioactive capsule and detector.An Al, Pb conical collimator was used between the sample and the detector for the excitation to obtain a large beam of emergent radiation and to avoid the interaction of the X-rays emitted by the component elements of the radioactive capsule and detector. This collimator has an external diameter of 13 mm and it was placed in the internal diameter of the radioactive source (8 mm). A graded filter of Pb, Fe and Al to obtain a thin beam of photons scattered from the sample and to absorb undesirable radiation shielded the detector. The sample-detector and excitation source-sample distances were optimized to get maximum count rate in the fluorescent peaks. The sample was placed approximately at 45° to the source-plane as well as to the detector-plane so that the intensity of scattered radiation could be minimized (Giauque et al., 1973). The count rate kept below 1000 counts s^{-1} in order to avoid peak broadening, energy shift and non-linearity. The data were collected into 16384 channels of a digital spectrum analyzer DSA-1000. The energy per channel was adjusted as 4 eV to determine the peak centroits and to discriminate the overlapped peaks. The samples were mounted in a sample holder placed between the pole pieces of an electromagnet capable of producing the magnetic field of approximately 2.66T at 2 mm pole range. During the study, the magnetic field intensities of, 0.6 T and 1.2 T were applied to the samples. An ammeter monitored the continuity and stability of the currents feeding the electromagnet. A typical *K* X-ray spectrum of Y at the 0.0 T, 0.6 T and 1.2 T is shown in Fig. 2. A typical *Ka*, $K\beta_{1,3}$ and $K\beta_{2,4}$ spectrum of Y, YBr_3, YCl_3, $Y(SO_4)_3.8H_2O$ and Y_2S_3 are shown in Fig. 3.

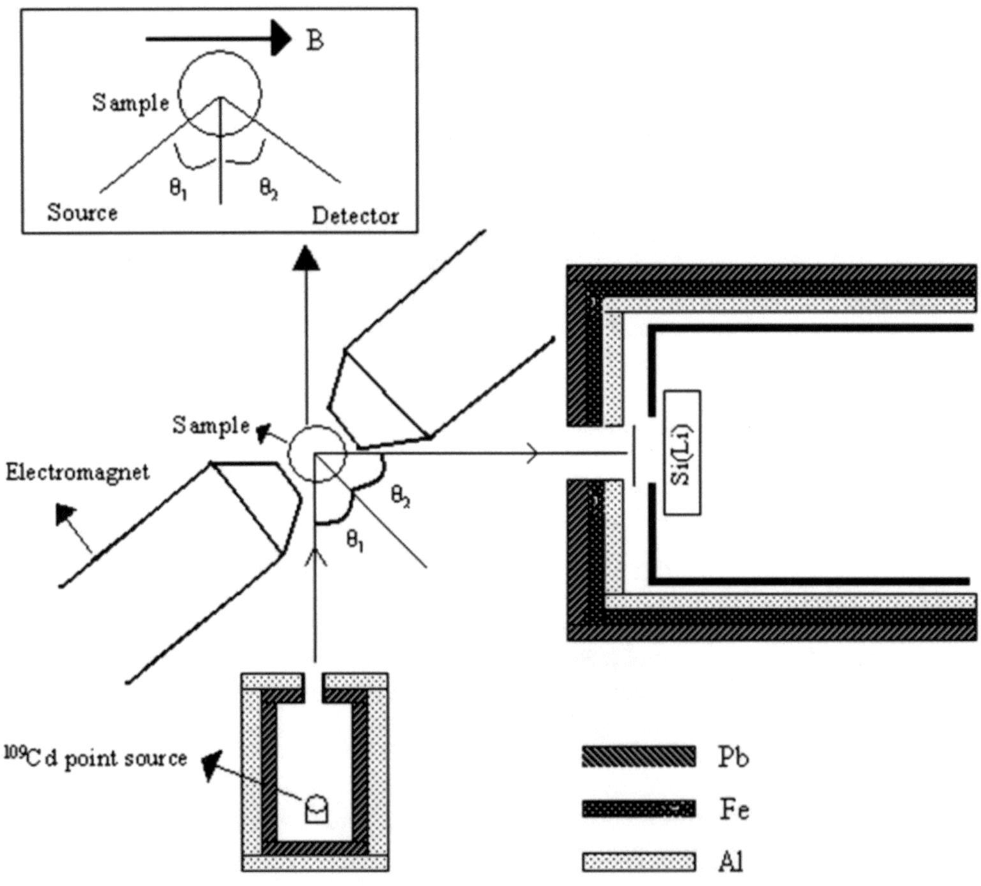

Fig. 1. Experimental set-up.

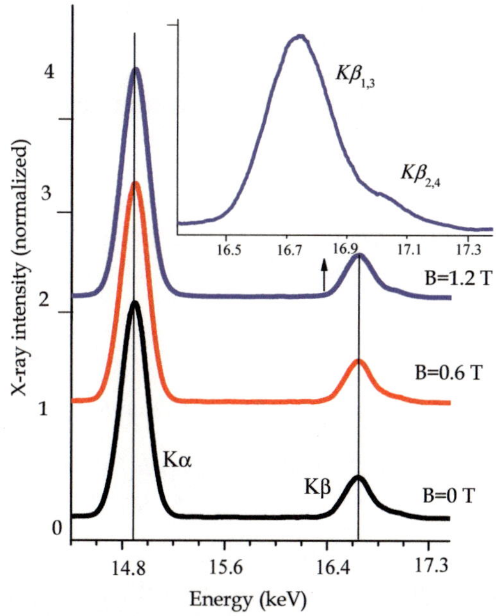

Fig. 2. A typical *K* X-ray spectrum of the Y target in B=0, B=0.6T and B=1.2 T magnetic field. The spectra were plotted after smoothing.

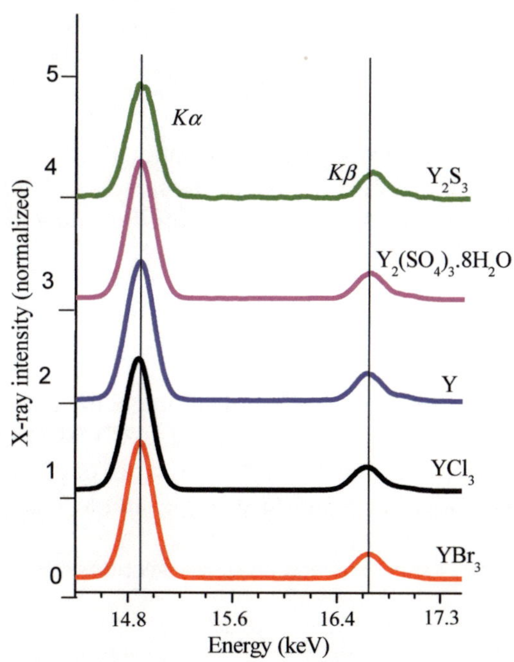

Fig. 3. Measured $K\alpha$, $K\beta_{1,3}$ and $K\beta_{2,4}$ spectra of Y, YBr_3, YCl_3, $Y(SO_4)_3.8H_2O$ and Y_2S_3. The spectra were plotted after smoothing.

Spectrum evaluation is a crucial step in X-ray analysis, as much as sample preparation and quantification. As with any analytical procedure, the final performance of X-ray analysis is determined by the weakest step in the process. The processing of ED spectra by means of computers has always been more evident because of their inherent digital nature. Due to relatively low resolving power of the employed Si(Li) detector, the process of evaluating XRF spectra is prone to many errors and requires dedicated software. For this purpose a software package called ORGIN was used for peak resolving background subtraction and determination of the net peak areas of *K* X-rays which is based on the non-linear least squares fitting of a mathematical model of the XRF spectrum.

2.2 Data analysis (EDXRF)

The $K\beta/K\alpha$ X-ray intensity ratio values have been calculated by using the relation

$$\frac{I(K\beta)}{I(K\alpha)} = \frac{N(K\beta)}{N(K\alpha)} \frac{\varepsilon(K\alpha)}{\varepsilon(K\beta)} \frac{\beta(K\alpha)}{\beta(K\beta)} \tag{2}$$

where $N(K\alpha)$ and $N(K\beta)$ are the net counts under the $K\alpha$ and $K\beta$ peaks, respectively. $\beta(K\alpha)$ and $\beta(K\beta)$ are the self-absorption correction factor of the target and $\varepsilon(K\alpha)$ and $\varepsilon(K\beta)$ are the detector efficiency for $K\alpha$ and $K\beta$ rays. The values of the factors, $I_0G\varepsilon$ which contain terms related to the incident photon flux, geometrical factor and the efficiency of the X-ray detector, were determined by collecting the $K\alpha$ and $K\beta$ X-ray spectra of Ti, As, Br, Sr, Y, Zr and Ru with the mass thickness 0.02-0.17 g/cm^2 in the same geometry and calculated by using the following equation

$$I_0 G\varepsilon_{Ki} = \frac{N_{Ki}}{\sigma_{Ki}\beta_{Ki}t_i} \tag{3}$$

where N_{Ki} and β_{Ki} ($i=\alpha, \beta$) have the same meaning as in Eq. (2). σ_{Ki} is X-ray fluorescence cross-section, G is a geometry factor and t is the mass of the sample in g/cm^2.
The self absorption correction factor β is calculated for both $K\alpha$ and $K\beta$ separately by using the following expression

$$\beta_{Ki} = \frac{1-\exp\{-[\mu(E_0)\sec\theta_1 + \mu_{Ki}(E)\sec\theta_2]t\}}{[\mu(E_0)\sec\theta_1 + \mu_{Ki}(E)\sec\theta_2]t} \tag{4}$$

where $\mu(E_0)$ and $\mu_{Ki}(E)$ are the total mass absorption coefficients taken from WinXCOM programme which is the Windows version of XCOM. XCOM is the electronic version of Berger and Hubbell's Tables (Berger et al., 1987). The angles of incident photons and emitted X-rays with respect to the normal at the surface of the sample θ_1 and θ_2 were equal to 45° in the present setup.
The term σ_{Ki} represents the *K* X-ray fluorescence cross-sections and is given by

$$\sigma_{Ki} = \sigma_K^P w_K f_{Ki} \tag{5}$$

σ_K^P is the *K* shell photo ionization cross-section (Scofield, 1973), w_K is the fluorescence yield (Krause et al., 1979) and f_{Ki} is fractional X-ray emission rate (Scofield, 1974a).

2.3 Experimental set up (WDXRF)

A commercial WDXRF spectrometer (Rigaku ZSX 100e) was used for analysis of the different samples. This instrument is usually equipped with a 3 kW Rh-anode tube working at a voltage range of 20–50 kV and a current from 20 to 50 mA. It is possible to use primary beam filters (made of Zr, Al, Ti or Cu) between the primary radiation and the sample holder to reduce the background continuum and to improve the signal-to-noise ratio. Energy resolution and efficiency for each analytical line also depend on the collimator aperture and the analyzer crystal in use. Several different collimators can be used to reduce the step/scan resolution, as well as up to ten analyzer crystals, to better enhance spectral data for a specific element. Detection can be performed using a flow proportional counter (light elements) or a scintillation counter (heavy elements). In this work, analyses were made in vacuum atmosphere. Moreover, to avoid possible problems with inhomogeneity when measuring the samples, a sample spinner facility was used in all cases.

To investigate the spectrometer sensitivity in measuring of intensity and energy shift, one sample at same conditions was measured for three times. Because of the use of instruments such as sieve weight and hydraulic press, errors are caused in the results of analysis. These errors were called manual and instrumental errors. Three samples were prepared and measured for same conditions to determine these errors.

2.4 Spectral profile analysis (WDXRF)

The common method for evaluation of spectra in WDXRF is by the use of net peak line intensity. This is due to the high efficiency in the analytical results from the scintillation and/or the flow counter detectors. These detectors can receive up to 2×106 cps. In contrast, the common spectra evaluation in EDXRF is based on integration of the gross or net peak area due to a lower efficiency in the solid state detectors, usually limited to a maximum count of 5×104 cps. Taking into account these facts and to improve the sensitivity of the signal, the spectral data obtained by the WDXRF equipment were treated using the deconvolution software (Microcal Orgin 7.5), traditionally used in EDXRF spectrometry, to obtain the peak areas. The total number of counts increases considering the total peak area instead of only the analytical line. This leads also to an improvement of sensibility and detection limits. Once samples were analyzed, the identification of elements from the WDXRF spectra was done by using the qualitative scanning mode linked to the equipment, which includes automatic peak and element identification. The principle of WDXRF spectrometry is the use of different analyzer crystals to diffract and separate the different characteristic wavelengths of the elements present in the sample. For that reason, in WDXRF measurements, a multi-spectrum was obtained resulting from the use of different analyzing crystals, excitation conditions, etc.

Rigaku has improved their semi-quantitative software package further with the introduction of SQX. It is capable of automatically correcting for all matrix effects, including line overlaps. SQX can also correct for secondary excitation effect by photoelectrons (light and ultra-light elements), varying atmospheres, impurities and different sample sizes.

The obtained multispectra were split into the different individual spectra and were converted to energies by inversion of the channels to be treated using the means of the SQX software to perform spectral deconvolution and fitting and to evaluate element net peak areas from the spectra. Peak fitting was done by iteration to better adjust the peak and the background to minimize the chi-square of the fitting on each spectra. Fig. 4 shows the spectrum of Y. Measured numbers of counts are shown as solid black circles, while the red line represents the overall fit. The background is shown as a blue line.

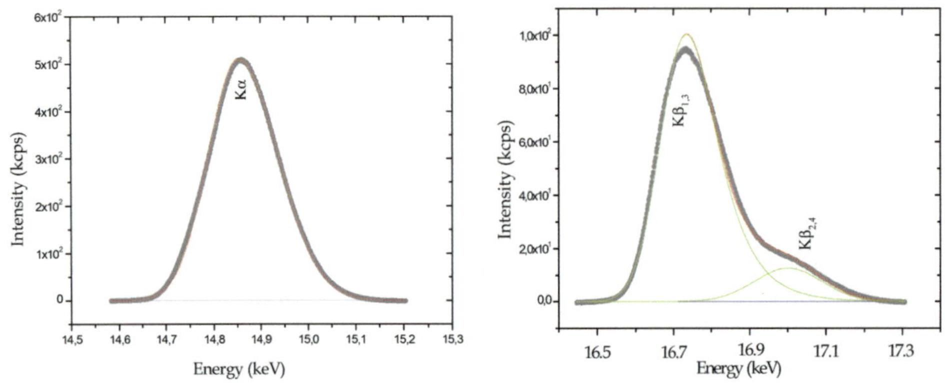

Fig. 4. Solid circles: Measured spectrum of Y *K* X-rays. Lines: Overall fitting function (red) and its components (green).

3. Results and discussion

A frequently used and convenient (but not very accurate) quantification of the line shape is by its full width at half maximum (FWHM) and index of asymmetry. These characteristics of lines (line parameters) are sensitive to change: line position (chemical shift), line shape (full width at half maximum (FWHM) and index of asymmetry) and additionally mutual ratios of line intensities. The peak position was determined at the center point of the 9/10 intensity of the smoothed line shape as illustrated in Fig. 5. It was known from our experience that the standard deviation of the peak position was determined using the peak top. Parameters such as FWHM and asymmetry index, defined in Fig. 5, were evaluated using the smoothed data. The Savitzky-Golay smoothing method was iteratively processed one time. Spectral smoothing was important for reducing the standard deviation of these parameters.

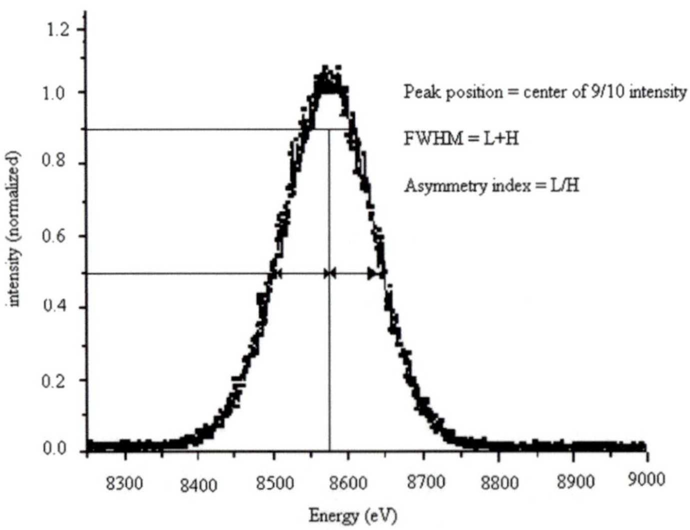

Fig. 5. Definition of asymmetry index, FWHM and peak position determined from 9/10 intensity.

The FWHM values for all compounds investigated for Cd-109 and Am-241 radioactive sources are given in Tables 1 and 2. Among all the Y compounds, the $Y(NO_3)_3.6H_2O$ *Ka* emission line shows large widths for both lines. Both the *Ka*, $K\beta_{1,3}$ and $K\beta_{2,4}$ lines become narrow when the compounds crystal system is monoclinic. In all cubic compounds such as Y_2O_3 and Y_2S_3 the lines FWHM values come closer to pure Y FWHM value. As can be seen from Table 1 and 2, for Zn compounds, the variation in the values of FWHM is relatively small. But when we compare the Y compounds with pure forms, we realize changes in both *Ka*, $K\beta_{1,3}$ and $K\beta_{2,4}$ FWHM values.
The nonmonotonic behavior of the *Ka* widths is probably due to the behavior of the *L* levels widths rather than the *K* level ones. The smaller overlap of the *M* and *K* wave functions, as compared to the *K* and *L* ones, may reduce the relative influence of possible similar sized nonmonotonic contributions originating in the final state level widths. Since experimental results FWHM and the index of asymmetry for B≠0 cannot be found in the literature, the comparison is not made with the other experimental values. As can be seen from Table 1 and 2, all FWHM values systematically decrease with increasing magnetic field intensity. Table 1 and 2 show that the experimental values of FWHM are unity within experimental uncertainties, suggesting the absence of chemical and external magnetic field effects.

Element	External Magnetic Field	Asymmetry Index (eV)			FWHM (eV)	
		Ka	$K\beta_{1,3}$	$K\beta_{2,4}$	*Ka*	$K\beta_{1,3}$
Y	**B=0**	**1.125**	**1.136**	**1.008**	**3.232**	**4.333**
	B=0.6T	1.079	1.112	1.006	3.223	4.268
	B=1.2T	0.991	0.109	0.996	3.057	4.016
$Y(NO_3)_3.6H_2O$	**B=0**	**1.122**	**1.127**	**1.011**	**3.211**	**4.228**
	B=0.6T	0.997	1.103	1.009	3.170	4.220
	B=1.2T	0.968	1.002	0.994	3.109	4.103
YCl_3	**B=0**	**1.110**	**1.119**	**0.994**	**3.197**	**4.267**
	B=0.6T	1.017	1.077	0.991	3.133	4.209
	B=1.2T	0.984	1.004	0.985	3.110	4.111
YPO_4	**B=0**	**1.114**	**1.111**	**0.987**	**3.199**	**4.337**
	B=0.6T	0.983	1.071	0.967	3.139	4.259
	B=1.2T	0.980	1.002	0.956	3.062	4.058
YBr_3	**B=0**	**1.098**	**1.078**	**0.971**	**3.201**	**4.284**
	B=0.6T	1.007	1.006	0.970	3.187	4.065
	B=1.2T	1.000	0.989	0.944	3.110	3.943
Y_2O_3	**B=0**	**1.071**	**1.05**	**0.938**	**3.266**	**4.121**
	B=0.6T	1.010	1.001	0.930	3.133	4.043
	B=1.2T	0.994	0.983	0.915	3.109	3.997
YF_3	**B=0**	**1.099**	**1.077**	**0.933**	**3.245**	**4.264**
	B=0.6T	1.022	1.011	0.929	3.120	4.001
	B=1.2T	0.988	0.984	0.886	3.100	3.966
$Y(SO_4)_3.8H_2O$	**B=0**	**1.055**	**1.035**	**0.921**	**3.189**	**4.166**
	B=0.6T	1.003	1.004	0.917	3.166	3.976
	B=1.2T	0.969	0.966	0.910	3.037	3.874
Y_2S_3	**B=0**	**1.024**	**1.031**	**0.910**	**3.337**	**4.494**
	B=0.6T	1.012	0.994	0.883	3.266	4.441
	B=1.2T	0.985	0.981	0.867	3.190	4.284

Table 1. Full width at half maximum (FWHM) and asymmetry index values of *Ka*, $K\beta_{1,3}$ and $K\beta_{2,4}$ emission lines in Y compounds for Cd-109 radioactive source.

To obtain more definite conclusions on FWHM dependency of the external magnetic field, more experimental data are clearly needed. The experimental uncertainties are always <0.05 eV for the FWHM.

According to Allinson (1933), the index of asymmetry of an X-ray emission line is defined as the ratio of the part of the FWHM lying to the long-wavelength side of the maximum ordinate to that on the short-wavelength side. In Table 1 and 2, the index of asymmetry for $K\alpha$, $K\beta_{1,3}$ and $K\beta_{2,4}$ emission lines are presented. The experimental uncertainties in the values cited in the table were determined taking into account multiple measurements and multiple fits of each spectrum. The errors for the index of asymmetry are ≤0.1 eV for $K\alpha$, $K\beta_{1,3}$ and $K\beta_{2,4}$.

Element	External Magnetic Field	Asymmetry Index (eV)			FWHM (eV)	
		$K\alpha$	$K\beta_{1,3}$	$K\beta_{2,4}$	$K\alpha$	$K\beta_{1,3}$
Y	**B=0**	**1.120**	**1.113**	**1.022**	**3.229**	**4.297**
	B=0.6T	1.089	1.100	1.016	3.212	4.203
	B=1.2T	1.009	0. 989	0.984	3.050	4.100
$Y(NO_3)_3.6H_2O$	**B=0**	**1.113**	**1.110**	**1.020**	**3.114**	**4.116**
	B=0.6T	1.028	1.104	1.013	3.017	4.111
	B=1.2T	0.996	0.993	1.001	3.091	4.075
YCl_3	**B=0**	**1.109**	**1.105**	**1.017**	**3.077**	**4.122**
	B=0.6T	1.003	1.007	1.009	3.041	4.110
	B=1.2T	0.978	0.987	0.987	3.000	4.004
YPO_4	**B=0**	**1.111**	**1.075**	**0.993**	**3.201**	**4.177**
	B=0.6T	0.989	1.032	0.984	3.126	4.170
	B=1.2T	0.969	1.008	0.953	3.013	4.005
YBr_3	**B=0**	**1.101**	**1.031**	**0.988**	**3.421**	**4.136**
	B=0.6T	1.017	1.004	0.980	3.234	4.065
	B=1.2T	0.994	0.971	0.965	3.048	3.993
Y_2O_3	**B=0**	**1.056**	**1.006**	**0.974**	**3.555**	**4.008**
	B=0.6T	1.031	0.999	0.972	3.229	3.989
	B=1.2T	0.987	0.977	0.954	3.096	3.974
YF_3	**B=0**	**1.077**	**1.001**	**0.970**	**3.301**	**4.123**
	B=0.6T	1.005	0.984	0.954	3.137	4.012
	B=1.2T	0.993	0.975	0.950	3.009	3.940
$Y(SO_4)_3.8H_2O$	**B=0**	**1.064**	**0.991**	**0.964**	**3.202**	**4.109**
	B=0.6T	1.001	0.990	0.966	3.113	3.989
	B=1.2T	0.978	0.946	0.956	3.074	3.866
Y_2S_3	**B=0**	**1.011**	**0.995**	**0.949**	**3.441**	**4.301**
	B=0.6T	1.010	0.994	0.937	3.368	4.039
	B=1.2T	0.974	0.961	0.921	3.222	3.974

Table 2. Full width at half maximum (FWHM) and asymmetry index values of $K\alpha$, $K\beta_{1,3}$ and $K\beta_{2,4}$ emission lines in Y compounds for Am-241 radioactive source.

When the crystal system of Y compounds is cubic, the *Ka* emission lines are almost symmetric for $Y_2O_3$and Y_2S_3. It is clear from results that, except for Y and $Y(NO_3)_3.6H_2O$,the asymmetry are generally larger for the *Ka* peak. So we can say that, the line shapes of *Ka* are not symmetric. It is also found from the Table 2, when the crystal system of Y compounds is trigonal (YBr_3), the *Ka*, $K\beta_{1,3}$ and $K\beta_{2,4}$ emission lines are almost symmetric too. For Am-241, line shapes are more symmetric than the Cd-109. As seen from Table 1 and 2, in the presence of an external magnetic field, the asymmetry index of the Y compounds change. The *Ka*, $K\beta_{1,3}$ and $K\beta_{2,4}$ emission lines asymmetry indices values decrease with external magnetic field. However, a more asymmetric structure is encountered for the elements of which their crystal symmetry is cubic. Also, it is observed that monoclinic group is more symmetric than the others.

Element	**External Magnetic Field**	**Chemical shift (ΔE) (eV)**			**Energy shift (δE) (eV)**		
		Ka	$K\beta_{1,3}$	$K\beta_{2,4}$	*Ka*	$K\beta_{1,3}$	$K\beta_{2,4}$
Y	**B=0**	**0**	**0**	**0**	**0**	**0**	**0**
	B=0.6T				0.074	0.109	0.101
	B=1.2T				0.161	0.260	0.237
$Y(NO_3)_3.6H_2O$	**B=0**	**-0.333**	**-0.441**	**-0.216**	**0**	**0**	**0**
	B=0.6T				0.086	0.098	0.115
	B=1.2T				0.141	0.137	0.227
YCl_3	**B=0**	**-0.238**	**-0.309**	**-0.115**	**0**	**0**	**0**
	B=0.6T				0.481	0.235	0.288
	B=1.2T				0.612	0.368	0.339
YPO_4	**B=0**	**-0.121**	**-0.235**	**-0.103**	**0**	**0**	**0**
	B=0.6T				0.335	0.279	0.336
	B=1.2T				0.455	0.355	0.399
YBr_3	**B=0**	**-0.033**	**-0.065**	**-0.065**	**0**	**0**	**0**
	B=0.6T				0.444	0.131	0.338
	B=1.2T				0.657	0.4	0.551
Y_2O_3	**B=0**	**0.135**	**0.017**	**0.056**	**0**	**0**	**0**
	B=0.6T				0.111	0.124	0.543
	B=1.2T				0.185	0.303	0.441
YF_3	**B=0**	**0.164**	**0.191**	**0.198**	**0**	**0**	**0**
	B=0.6T				0.167	0.166	0.112
	B=1.2T				0.533	0.529	0.144
$Y(SO_4)_3.8H_2O$	**B=0**	**0.609**	**0.387**	**0.33**	**0**	**0**	**0**
	B=0.6T				0.089	0.12	0.051
	B=1.2T				0.354	0.293	0.237
Y_2S_3	**B=0**	**0.899**	**0.872**	**0.808**	**0**	**0**	**0**
	B=0.6T				0.173	0.204	0.206
	B=1.2T				0.199	0.307	0.333

Table 3. Chemical shift (ΔE) and energy shift (δE) values of *Ka*, $K\beta_{1,3}$ and $K\beta_{2,4}$ emission lines in Y compounds for Cd-109 radioactive source.

The more unpaired 3d or 4d electrons the atom possesses, the more asymmetric will be the line observed. This kind observation led Tsutsumi (1959) to consider that the interaction between the hole created in the $2p_{3/2}$ or $2p_{1/2}$ shell (due to the transition of an electron from this shell to the 1s level) and the electrons in the incomplete 3d shell in the transition metal atoms is responsible for asymmetric nature of *Ka* lines. They proposed a theoretical model based on this idea to account for the asymmetry in the X-ray emission lines in the firs-row transition metal compounds. However, this is not the only consideration which can explain the origin of the asymmetry of the line; there are other considerations which are based on the relaxation effect of the inner state proposed by Parratt (1959) or on the interactions between 2p hole and electrons in the Fermi sea as proposed by Doniach and Sunjic (1970).

Element	**External Magnetic Field**	**Chemical shift (ΔE) (eV)**			**Energy shift (δE) (eV)**		
		$K\alpha$	$K\beta_{1,3}$	$K\beta_{2,4}$	$K\alpha$	$K\beta_{1,3}$	$K\beta_{2,4}$
Y	**B=0**	0	0	0	0	0	0
	B=0.6T				0.125	0.121	0.132
	B=1.2T				0.179	0.219	0.269
$Y(NO_3)_3.6H_2O$	**B=0**	**-0.401**	**-0.347**	**-0.316**	0	0	0
	B=0.6T				0.091	0.111	0.158
	B=1.2T				0.104	0.169	0.201
YCl_3	**B=0**	**-0.023**	**-0.109**	**-0.153**	0	0	0
	B=0.6T				0.226	0.254	0.259
	B=1.2T				0.559	0.472	0.318
YPO_4	**B=0**	**0.231**	**0.304**	**0.221**	0	0	0
	B=0.6T				0.553	0.602	0.436
	B=1.2T				0.598	0.556	0.511
YBr_3	**B=0**	**-0.133**	**-0.227**	**-0.194**	0	0	0
	B=0.6T				0.342	0.313	0.387
	B=1.2T				0.600	0.499	0.458
Y_2O_3	**B=0**	**0.440**	**0.316**	**0.241**	0	0	0
	B=0.6T				0.447	0.423	0.505
	B=1.2T				0.682	0.613	0.553
YF_3	**B=0**	**0.206**	**0.391**	**0.190**	0	0	0
	B=0.6T				0.367	0.276	0.211
	B=1.2T				0.533	0.590	0.438
$Y(SO_4)_3.8H_2O$	**B=0**	**0.194**	**0.276**	**0.133**	0	0	0
	B=0.6T				0.194	0.146	0.115
	B=1.2T				0.402	0.281	0.296
Y_2S_3	**B=0**	**0.575**	**0.503**	**0.421**	0	0	0
	B=0.6T				0.197	0.340	0.247
	B=1.2T				0.331	0.349	0.366

Table 4. Chemical shift (ΔE) and energy shift (δE) values of $K\alpha$, $K\beta_{1,3}$ and $K\beta_{2,4}$ emission lines in Y compounds for Am-241 radioactive source.

The chemical shift was the difference between the center point of the 9/10 peak intensity of a compound and that of pure Y measured before and after the measurement of the compound. When the environment of the emitting atom is changed, there are changes in the position of emission lines with respect to those in the pure metal. These changes are called chemical shifts. They are presented in Tables 3 and 4. Both the 4*d* electron configuration and crystal structure affect the chemical shift and energy shift with applied external magnetic field. There is a clear relationship with the external magnetic field values and the energy shift values of Y compounds, as is found in Table 3 and 4's last column. For higher values of external magnetic field, the values of energy shift increases systematically. But we do not find any relationship between external magnetic field and crystal structure of the compound. To obtain more definite conclusion, more experimental data for 4*d* compounds which crystal structure different are needed.
The errors of the chemical and energy shifts, originate mainly from the limited precision of our measurements, determined by repetitive measurements of all pure targets, which were prepared and placed in the same experimental geometry. The precision in the position of *Ka*, $K\beta_{1,3}$ and $K\beta_{2,4}$ lines was determined as 0.05 eV, whereas the maximum deviation of a single measurement from the average value was 0.1 eV.

Element	**Differences between FWHM values [ΔFWHM=FWHM$_{com.}$-FWHM$_{pure}$] (eV)**			**Chemical shift (ΔE) (eV)**		
	Ka	$K\beta_{1,3}$	$K\beta_{2,4}$	*Ka*	$K\beta_{1,3}$	$K\beta_{2,4}$
Y	0	0	0	0	0	0
$Y(NO_3)_3.6H_2O$	0.155	0.031	0.012	0.132	0.126	0.125
YCl_3	0.091	0.058	0.033	0.268	0.238	0.284
YPO_4	-0.223	-0.087	-0.107	-0.454	-0,444	-0.415
YBr_3	0.114	0.071	0.067	0.144	0.179	0.156
Y_2O_3	-0.077	-0.068	-0.021	-0.096	-0.126	-0.085
YF_3	-0.095	-0.034	-0.048	-0.115	-0.127	-0.167
$Y(SO_4)_3.8H_2O$	0.054	0.023	0.181	0.183	0.145	0.106
Y_2S_3	-0.013	-0.064	-0.079	-0.301	-0.397	-0.385

Table 5. Chemical shift *(ΔE)* and differences between FWHM values of *Ka*, $K\beta_{1,3}$ and $K\beta_{2,4}$ emission lines in Y compounds obtained for WDXRF.

It is also seen from Table 5 that the *Ka* line width of YPO_4 compound is wider than that of the other compounds. Compare with the *Ka* peak of the pure Y, that of cubic crystal structure Yttrium compounds shifted to lower energy, and the peak shift ordering was $YF_3<Y_2O_3<Y_2S_3<YPO_4$. The line shapes of *Ka* are generally symmetric. Y_2S_3 and Y_2O_3, where the $K\beta_{1,3}$ and $K\beta_{2,4}$ peak shifts are large, show prominent asymmetry.
The accurate knowledge of the $K\beta_1/Ka$, $K\beta_2/Ka$, $K\beta_2/K\beta_1$ and $K\beta/Ka$ intensity ratios is required for a number of practical applications of X-rays, e.g. molecular and radiation physics investigations, in non-destructive testing, elemental analysis, medical research etc. Therefore, these ratios depend sensitively on the atomic structure. Thus they have been widely used also for critical evaluation of atomic structure model calculations. We now discuss the values of these ratios as obtained in our measurements.
The relevant information in a spectrum is contained in its peaks whose position and area are linked respectively to the photon energy and the activity of the connected radionuclide. The peak areas can also be used to determine emission probabilities. In this work, peak areas

were determined after the *Ka*, $K\beta_{1,3}$ and $K\beta_{2,4}$ areas were separated by fitting the measured spectra with multi-Gaussian function plus polynomial backgrounds using Microcal Orgin 7.5 software program. Details of the experimental set up and data analysis have been reported earlier (Porikli et al., 2011b).
Table 6 lists the theoretical values which were calculated by Scofield (Scofield 1974a; Scofield, 1974b). Addition to this, the measured values of the $K\beta_1/Ka$, $K\beta_2/Ka$, $K\beta_2/K\beta_1$ and $K\beta/Ka$ intensity ratios in Y, and previous experimental and the other theoretical values of these ratios for pure elements and their compounds are listed in Table 6.

Element	External Magnetic Field	Intensity Ratio	This Work	Scofield (1974a)	Manson &Kennedy (1974)	Ertuğral et al. (2007)
Y	**B=0**	$K\beta_{1,3}/Ka$	0.2307±0.010	0.22910		
		$K\beta_{2,4}/Ka$	0.0317±0.008	0.02902		
		$K\beta_{2,4}/K\beta_{1,3}$	0.1981±0.011	0.19220		
		$K\beta/Ka$	**0.1822±0.008**	**0.16960**	**0.1685**	**0.1856±0.009**
	B=0.6T	$K\beta_{1,3}/Ka$	0.2289±0.007			
		$K\beta_{2,4}/Ka$	0.0311±0.008			
		$K\beta_{2,4}/K\beta_{1,3}$	0.1956±0.011			
		$K\beta/Ka$	**0.1753±0.011**			
	B=1.2T	$K\beta_{1,3}/Ka$	0.2275±0.007			
		$K\beta_{2,4}/Ka$	0.0304±0.008			
		$K\beta_{2,4}/K\beta_{1,3}$	0.1941±0.011			
		$K\beta/Ka$	**0.1712±0.011**			
$Y(NO_3)_3.6H_2O$	**B=0**	$K\beta_{1,3}/Ka$	0.2339±0.008			
		$K\beta_{2,4}/Ka$	0.0325±0.010			
		$K\beta_{2,4}/K\beta_{1,3}$	0.1987±0.011			
		$K\beta/Ka$	**0.1829±0.006**			
	B=0.6T	$K\beta_{1,3}/Ka$	0.2320±0.008			
		$K\beta_{2,4}/Ka$	0.0316±0.008			
		$K\beta_{2,4}/K\beta_{1,3}$	0.1977±0.010			
		$K\beta/Ka$	**0.1796±0.011**			
	B=1.2T	$K\beta_{1,3}/Ka$	0.2315±0.008			
		$K\beta_{2,4}/Ka$	0.0310±0.011			
		$K\beta_{2,4}/K\beta_{1,3}$	0.1954±0.011			
		$K\beta/Ka$	**0.1742±0.010**			
YCl_3	**B=0**	$K\beta_{1,3}/Ka$	0.2341±0.010			
		$K\beta_{2,4}/Ka$	0.0329±0.008			
		$K\beta_{2,4}/K\beta_{1,3}$	0.1992±0.009			
		$K\beta/Ka$	**0.1836±0.010**			
	B=0.6T	$K\beta_{1,3}/Ka$	0.2337±0.006			
		$K\beta_{2,4}/Ka$	0.0305±0.009			
		$K\beta_{2,4}/K\beta_{1,3}$	0.1952±0.009			

Element	External Magnetic Field	Intensity Ratio	This Work	Scofield (1974a)	Manson &Kennedy (1974)	Ertuğral et al. (2007)
		$K\beta/Ka$	**0.1821±0.011**			
	B=1.2T	$K\beta_{1,3}/Ka$	0.2322±0.010			
		$K\beta_{2,4}/Ka$	0.0324±0.008			
		$K\beta_{2,4}/K\beta_{1,3}$	0.1972±0.009			
		$K\beta/Ka$	**0.1818±0.010**			
YPO_4	**B=0**	$K\beta_{1,3}/Ka$	0.2355±0.010			
		$K\beta_{2,4}/Ka$	0.0333±0.010			
		$K\beta_{2,4}/K\beta_{1,3}$	0.1999±0.010			
		$K\beta/Ka$	**0.1840±0.009**			
	B=0.6T	$K\beta_{1,3}/Ka$	0.2336±0.007			
		$K\beta_{2,4}/Ka$	0.0320±0.009			
		$K\beta_{2,4}/K\beta_{1,3}$	0.1966±0.011			
		$K\beta/Ka$	**0.1773±0.008**			
	B=1.2T	$K\beta_{1,3}/Ka$	0.2322±0.011			
		$K\beta_{2,4}/Ka$	0.0304±0.009			
		$K\beta_{2,4}/K\beta_{1,3}$	0.1693±0.009			
		$K\beta/Ka$	**0.1738±0.011**			
YBr_3	**B=0**	$K\beta_{1,3}/Ka$	0.2359±0.008			
		$K\beta_{2,4}/Ka$	0.0341±0.007			
		$K\beta_{2,4}/K\beta_{1,3}$	0.2004±0.009			
		$K\beta/Ka$	**0.1848±0.010**			
	B=0.6T	$K\beta_{1,3}/Ka$	0.2342±0.008			
		$K\beta_{2,4}/Ka$	0.0321±0.009			
		$K\beta_{2,4}/K\beta_{1,3}$	0.1987±0.010			
		$K\beta/Ka$	0.1818±0.010			
	B=1.2T	$K\beta_{1,3}/Ka$	0.2321±0.009			
		$K\beta_{2,4}/Ka$	0.0303±0.011			
		$K\beta_{2,4}/K\beta_{1,3}$	0.1976±0.010			
		$K\beta/Ka$	0.1794±0.009			
Y_2O_3	**B=0**	$K\beta_{1,3}/Ka$	0.2364±0.011			
		$K\beta_{2,4}/Ka$	0.0349±0.010			
		$K\beta_{2,4}/K\beta_{1,3}$	0.2008±0.010			
		$K\beta/Ka$	0.1856±0.011			
	B=0.6T	$K\beta_{1,3}/Ka$	0.2351±0.008			
		$K\beta_{2,4}/Ka$	0.0340±0.008			
		$K\beta_{2,4}/K\beta_{1,3}$	0.1998±0.010			
		$K\beta/Ka$	0.1831±0.010			
	B=1.2T	$K\beta_{1,3}/Ka$	0.2302±0.010			
		$K\beta_{2,4}/Ka$	0.0307±0.010			
		$K\beta_{2,4}/K\beta_{1,3}$	0.1985±0.010			

Element	External Magnetic Field	Intensity Ratio	This Work	Scofield (1974a)	Manson &Kennedy (1974)	Ertuğral et al. (2007)
		$K\beta/K\alpha$	0.1824±0.009			
YF_3	**B=0**	$K\beta_{1,3}/K\alpha$	0.2371±0.008			
		$K\beta_{2,4}/K\alpha$	0.0352±0.009			
		$K\beta_{2,4}/K\beta_{1,3}$	0.2011±0.010			
		$K\beta/K\alpha$	0.1859±0.010			
	B=0.6T	$K\beta_{1,3}/K\alpha$	0.2338±0.006			
		$K\beta_{2,4}/K\alpha$	0.0330±0.008			
		$K\beta_{2,4}/K\beta_{1,3}$	0.2003±0.011			
		$K\beta/K\alpha$	0.1841±0.010			
	B=1.2T	$K\beta_{1,3}/K\alpha$	0.2307±0.009			
		$K\beta_{2,4}/K\alpha$	0.0325±0.010			
		$K\beta_{2,4}/K\beta_{1,3}$	0.1989±0.011			
		$K\beta/K\alpha$	0.1829±0.011			
$Y(SO_4)_3.8H_2O$	**B=0**	$K\beta_{1,3}/K\alpha$	0.2380±0.007			
		$K\beta_{2,4}/K\alpha$	0.0355±0.012			
		$K\beta_{2,4}/K\beta_{1,3}$	0.2015±0.010			
		$K\beta/K\alpha$	0.1867±0.011			
	B=0.6T	$K\beta_{1,3}/K\alpha$	0.2323±0.008			
		$K\beta_{2,4}/K\alpha$	0.0332±0.012			
		$K\beta_{2,4}/K\beta_{1,3}$	0.2006±0.011			
		$K\beta/K\alpha$	0.1844±0.012			
	B=1.2T	$K\beta_{1,3}/K\alpha$	0.2311±0.007			
		$K\beta_{2,4}/K\alpha$	0.0318±0.012			
		$K\beta_{2,4}/K\beta_{1,3}$	0.2001±0.010			
		$K\beta/K\alpha$	0.1816±0.011			
Y_2S_3	**B=0**	$K\beta_{1,3}/K\alpha$	0.2385±0.012			
		$K\beta_{2,4}/K\alpha$	0.0359±0.009			
		$K\beta_{2,4}/K\beta_{1,3}$	0.2019±0.009			
		$K\beta/K\alpha$	0.1876±0.008			
	B=0.6T	$K\beta_{1,3}/K\alpha$	0.2354±0.007			
		$K\beta_{2,4}/K\alpha$	0.0335±0.010			
		$K\beta_{2,4}/K\beta_{1,3}$	0.2009±0.009			
		$K\beta/K\alpha$	0.1867±0.011			
	B=1.2T	$K\beta_{1,3}/K\alpha$	0.2332±0.006			
		$K\beta_{2,4}/K\alpha$	0.0301±0.007			
		$K\beta_{2,4}/K\beta_{1,3}$	0.2002±0.008			
		$K\beta/K\alpha$	0.1849±0.012			

Table 6. $K\beta_{1,3}/K\alpha$, $K\beta_{2,4}/K\alpha$, $K\beta_{2,4}/K\beta_{1,3}$ and $K\beta/K\alpha$ X-ray intensity ratios of pure Y their compounds.

When you look at the Table 6, a serious difference between the $K\beta_{1,3}/K\alpha$ experimental and theoretical values can be seen. This situation is mainly because of the limited resolution of the detector. The $K\alpha_1$ and $K\alpha_2$ X-ray components appear as one line. In the most general case, chemical speciation is preferably performed via the analysis of the $K\beta_{1,3}$ or $K\beta_{2,4}$ lines. These lines, emitted after transition of valance electrons are more sensitive to the chemical environment.

As can be seen from Table 6, the $K\beta/K\alpha$ ratios of Y in all Y compounds are in close agreement with the ratios of corresponding pure metals. The greatest increase of the $K\beta/K\alpha$ ratio has been observed for Y_2S_3. We found a general increase of the $K\beta/K\alpha$ intensity ratios for different compounds. This situation is more complex because the $K\beta/K\alpha$ intensity ratio is affected by the chemical bonding type, (ionic, metallic, covalent), the individual characteristics of the structure of molecules, complexes and crystals (polarity, valency and electronegativity of atoms, co-ordination number, ionicities of covalent bond etc.).

We found that the chemical effect on the $K\beta/K\alpha$ ratios for 4d elements is small but the dependence of the $K\beta_2/K\alpha$ ratios on the chemical environments is appreciable. This can be understood by the fact that in 4d elements the valance state consists of the 4d, 5s and 5p electrons and the influence of the chemical state on the $K\beta_{1,3}$ (3p→1s) X-ray emission is negligible. Yamoto et al. (1986) found similar results for compounds involving Tc isotopes and Mukoyoma et al. (2000) found similar results theoretically for Mo and Tc compounds.

The overall error in the present measurements is estimated to be 3-8%. This error is attributed to the uncertainties in different parameters used to determine the $K\beta_1/K\alpha$, $K\beta_2/K\alpha$, $K\beta_2/K\beta_1$ and $K\beta/K\alpha$ values; such as, $I_0G\varepsilon$ product (1.0-2.5%), in the absorption correction factor (0.3-1.5%), the error in the area evaluation under the $K\alpha$, $K\beta_1$, $K\beta_2$ and $K\beta$ X-ray peak (0.5-3.0%) and the other systematic errors (1.0-2.0%).

4. Conclusion

There has been increasing interest in chemical speciation of the elements in recent years which can be attributed to the great alterations in the chemical and biological properties of the elements depending on their oxidation state, the type of chemical bonds etc. Usually, the influence of the chemical environment results in energy shifts of the characteristic X-ray lines, formation of satellite lines and changes in the emission linewidths and relative X-ray intensities. High resolution X-ray spectroscopy, employing crystal spectrometers of a few eV resolutions, can be applied to probe these phenomena efficiently, exploiting them for chemical state analysis. Measurements of the shapes and wavelengths of certain X-ray lines have been made by previous investigators with both EDXRF and WDXRF. It has been shown that the WDXRF spectrometer is capable of measuring X-ray wavelengths with a precision equal to or greater than that attained with the EDXRF system. Both EDXRF and WDXRF technique has been used to study the effect of chemical state of an element on characteristic X-rays.

We have presented and discussed the effect of chemical composition and external magnetic field on the $K\beta_{1,3}/K\alpha$, $K\beta_{2,4}/K\alpha$, $K\beta_{2,4}/K\beta_{1,3}$ and $K\beta/K\alpha$ intensity ratios for some Yttrium compounds. The experimental measurements have been performed with a Si(Li) detector. The observed spectral features, namely the asymmetry indices, FWHM values, chemical shifts, energy separations between $K\alpha$ and $K\beta$ lines and $K\beta/K\alpha$ intensity ratio values show an interesting correlation with crystal symmetries. Furthermore, these values change symmetrically with the external magnetic field. There is a relation between the crystal structures and *K* X-ray emission rate because of the change in bond distance, inter atomic distance, the interaction between ligand atoms and the central atom, and the Auger electron and dipole transition. These situations cause a redistribution of the electron configuration in the molecule.

A correlation between the $K\beta/K\alpha$ intensity ratio of 4d elements and chemical state was found in this work. Excluding the values for Y, we can generally state that $K\beta/K\alpha$ intensity ratio increases for different compounds. The $K\beta_1/K\alpha$, $K\beta_2/K\alpha$, $K\beta_2/K\beta_1$ and $K\beta/K\alpha$ intensity ratio values were obtained in the present work and listed in Table 6 and compared with other experimental and theoretical values. As a result, we can say that the uncertainties of the measured values are too large to allow any statement about the specific dependence of the $K\beta/K\alpha$ intensity ratio on the crystal symmetry, but small enough to show significant increase in the $K\beta/K\alpha$ intensity ratio with increasing external magnetic field values.

In general, our experimental values are qualitatively in agreement with the other experimental values. There are some differences between the results of this study and that of previous experimental work because these studies were carried out in different laboratories and different systems. We were not obtained researches interested in $K\beta_{1,3}/K\alpha$, $K\beta_{2,4}/K\alpha$, $K\beta_{2,4}/K\beta_{1,3}$ and $K\beta/K\alpha$ intensity ratio values for Y compounds. So we do not compare compounds these intensity ratio values in literature values. Rigorous systematic experiments and theoretical calculations are urgently needed for comparison with present experimental result. To obtain more definite conclusions on the magnetic field and crystal structure dependency of the atomic parameters, more experimental data are clearly needed, particularly for different symmetries and for chemical compounds.

5. Acknowledgment

This work was supported by the Scientific and Technological Research Council of Turkey (TUBITAK), under the project no 106T045.

6. References

Allinson, S.K. (1933). The Natural Widths of the *Ka* X-Ray Doublet from ^{26}Fe to ^{47}Ag. *Phys. Rev.* Vol.44, pp. 63-72.

Arndt, E.; Brunner, G. & Hartmann, E. (1982). *Kβ/Ka* Intensity Ratios for 3d Elements by Using Photoionisation and Electron Capture. *J. Phys. B: At. Mol. Opt. Phys.*, Vol.15, pp. 887-889.

Balasubramanian, K. (1994). Relativistic Effects and Electronic Structure of Lanthanide and Actinide Molecules, in: Gschneidner, K.A.; Eyring, L.; Choppin, G.R. & Lander G.H. (Ed.), *Handbook of Physics and Chemistry of Rare Earth*, 18, Elsevier, Amsterdam, Chap. 119, pp. 29–50.

Band, I.M.; Kovtun, A.P.; Listengarten, M.A. & Trzhaskovskaya, M.B. (1985). The Effect of the Chemical Environment of Manganese and Chromium Atoms on the *Kβ/Ka* X-Ray Intensity Ratio. *J. Electr. Spectr. and Relat. Phenom.*, Vol.36, pp. 59-68.

Batrakov, Y.F.; Krivitsky, A.G. & Puchkova E.V. (2004). Relativistic Component of Chemical Shift of Uranium X-Ray Emission Lines. *Spectrochim. Acta Part B,* Vol.59, pp. 345 351.

Berenyi, D.; Hock, G.; Ricz, S.; Schlenk, B. & Valek, A. (1978). *Kα/Kβ X*-Ray Intensity Ratios and *K*-Shell Ionisation Cross Sections for Bombardment by Electrons of 300-600 keV. *J. Phys. B,* Vol.11, pp. 709-713.

Berger, M.J. & Hubbell, J.H. (1987). XCOM: Photon Cross-Sections on a Personnel Computer with (Version 1.2).

Brunner, G.; Nagel, M.; Hartmann, E. & Arndt, E. (1982). Chemical Sensitivity of the *Kβ/Kα* X-Ray Intensity Ratio for 3d Elements. *J. Phys. B,* Vol.15, pp. 4517-4522.

Demir, D. & Şahin, Y. (2006a). Measurement of the *K* shell X-Ray Production Cross-Sections and Fluorescence Yields for Nd, Eu, Gd, Dy and Ho using Radioisotopes X-Ray Fluorescence in the External Magnetic Field. *Eur. Phys. J. D,* Vol.44, pp. 34-38.

Demir, D. & Şahin, Y. (2006b). The Effect of an External Magnetic Field on the L_3 Subshell Fluorescence Yields and Level Widths for Gd, Dy, Hg and Pb at 59,54 keV. *Nucl. Instr. and Meth.,* Vol.254, pp. 43-48.

Deutsch, M. & Hart, M. (1982). Wavelength, Energy Shape, and Structure of the Cu $K\alpha_1$ X Ray-Emission Line. *Phys. Rev. B,* Vol.26, pp. 5558-5567.

Ertuğral, B.; Apaydın, G.; Çevik, U.; Ertuğrul, M. & Kopya, A.İ. (2007). *Kβ/Kα* X-Ray Intensity Ratios for Elements in the Range 16<Z<92 Excited by 5.9, 59.5 and 123.6 keV Photons. *Radiation Phys. and Chem.,* Vol.76, pp. 15-22.

Fichter, M. (1975). Das K-Röntgenemissionsspektrum von Phosphor in Abhängigkeit von der Chemischen Bindung. *Spectrochim. Acta Part B,* Vol.30, pp.417-431.

Giauque, R.D.; Goulding, F.S.; Jaklevic, J.M. & Pehl, R.H. (1973). Trace Element Determination with Semiconductor Detector X-Ray Spectrometers. *Anal. Chem.,* Vol.45, pp. 671-681.

Gohshi, Y. & Ohtsuka, A. (1973). The Application of Chemical Effects in High Resolution X-Ray Spectrometry. *Adv. X-Ray Anal.,* Vol. 28, pp. 179-188.

Gohshi, Y.; Hirao, O. & Suzuki, I. (1975). Chemical State Analysis of Sulfur, Chromium and Tin by High Resolution X-Ray Spectrometry. *Adv. X-Ray Anal.,* Vol.18, pp. 406-414.

Groot, F.M.F. (1994a). X-Ray Absorption and Dichroism of Transition Metals and Their Compounds. *J. Electron Spectrosc. Relat. Phenom.,* Vol.676, pp. 529-622.

Groot, F.M.F. (1994b). New Directions in Research with Third-Generation Soft X-Ray Synchrotron Radiation Sources. *Applied Sciences,* Vol.254.

Han, I.; Porikli, S.; Şahin, M. & Demir. D. (2010). Measurementof *Lα, Lβ* and Total *L* X-ray Fluorescence Cross-Sections for Some Elements with 40<Z<53. *Radiation Physics and Chemistry,* Vol.79, pp. 393–396.

Iiahara, J.; Omorr, T., Yoshihara, K.; & Ishii, K. (1993). Chemical Effects on Chromium *L* X-Rays. *Nucl. Instr. Methods. B,* Vol.73, pp. 32-34.

Kataria, S.K.; Govil, R.; Saxena, A. & Bajpei, H.N. (1986). Chemical Effects in X-ray Fluorescence Analysis. *X-Ray Spectr.,* Vol.15, pp. 49-53.

Katz, J.J.; Seaborg, G.T. & Morss, L.R. (1986). 2nd ed, The Chemistry of the Actinide Elements, 2, *Chapman and Hall*, New York, pp. 1131–1165.

Khan, M.R. & Karimi, M. (1980). *Kβ/Ka* Ratıos ın Energy-Dispersive X-Ray Emission Analysis. *X-Ray Spectrom.*, Vol.9, pp. 32-35.

Krause, M.O. (1979). Atomic Radiative and Radiationless Yields for *K* and *L* Shells. *Chem. Ref. Data.*, Vol.8, pp. 307-327.

Küçükönder, A.; Şahin, Y.; Büyükkasap, E. & Kopya, A. (1993). Chemical effect on *Kβ/Ka* X Ray Intensity Ratios in Coordination Compounds of Some 3d Elements. *J. Phys. B: At. Mol. Opt. Phys.*, Vol.26, pp. 101-105

Leonhardt, G. & Meisel, A. (1970). Determination of Effective Atomic Changes From the Chemical Shifts of X-Ray Emission Lines. *J. Chem. Phys.* Vol.52, pp. 6189-6198.

Makarov, L.L. (1999). X-ray Emission Effects as a Tool to Study Light Actinides. *Czech. J. Phys.*, Vol.49, pp. 610–616.

Manson, S. T. (1974). X-Ray Emission Rates in the Hartree–Slater Approximation. *At. Data Nucl. Data Tables.* Vol.14, pp. 111-120.

Meisel, A.; Leonhardt, G. & Szargan, R. (1989). X-Ray Spectra and Chemical Binding, *Chemical Physics*, Vol.37, edited by F.P. Schafer, V.I. Goldanskii & J.P. Toennies (Springer-Verlag, Berlin).

Mukoyoma, T.; Taniguchi, K. & Adachi, H. (1986). Chemical Effect on *Kβ/Ka* X-Ray Intensity Ratios. *Phys. Rev. B*, Vol.34, pp. 3710-3716.

Mukoyoma, T.; Taniguchi, K. & Adachi, H. (2000). Variation of *Kβ/Ka* X-Ray Intensity Ratios in 3d elements. *X-Ray Spectr.*, Vol.29, pp. 426-429.

Mukoyama, T. (2004). Theory of X-ray Absorption and Emission Spectra. *Spectrochim. Acta Part B*, Vol.59, pp. 1107-1115.

Nagel, D.J. & Baun, W.L. (1974). In: L.V. Azaroff (Ed.), *X-Ray Spectroscopy*, McGraw-Hill, US, Ch. 9.

Padhi , H.C.; Bhuinya, C.R. & Dhal, B.B. (1993). Influence of Solid-State Effects on the *Kβ/Ka* Intensity Ratios of Ti and V in TiB_2, VB_2 and VN. *J. Phys. B: At. Mol. Opt. Phys.*, Vol.26, pp. 4465-4469.

Padhi , H.C. & Dhal, B.B. (1995). *Kβ/Ka* X-Ray-Intensity Ratios of Fe, Co, Ni, Cu, Mo, Ru, Rh and Pd in Equiatomic Aluminides. *Solid State Commun.*, Vol. 96, pp. 171-173.

Paic, G. & Pecar, V. (1976). Study of Anomalies in *Kβ/Ka* Ratios Observed Following *K* Electron Capture. *Phys. Rev. A*, Vol.14, pp. 2190-2192.

Pepper, M. & Bursten, B.E. (1991). The Electronic Structure of Actinide Containing Molecules: a Challenge to Applied Quantum Chemistry. *Chem. Rev.*, Vol.91, pp. 719–741.

Porikli, S. & Kurucu, Y. (2008a). The Effect of an External Magnetic Field on the *Ka* and *Kβ* X-Ray Emission Lines of the 3*d* Transition Metals. *Instr. Science and Tech.*, Vol.36:4, pp. 341-354.

Porikli, S. & Kurucu, Y. (2008b). Effects of the External Magnetic Field and Chemical Combination on *Kβ/Ka* X-Ray Intensity Ratios of Some Nickel and Cobalt Compounds. *Appl. Rad. and Isot.*, Vol.66, pp. 1381-1386.

Porikli, S.; Demir, D. & Kurucu, Y. (2008c). Variation of *Kβ/Ka* X-Ray Intensity Ratio and Lineshape with the Effects of External Magnetic Field and Chemical Combination. *Eur. Phys. J. D,* Vol.47, pp. 315–323.

Porikli, S.; Han, İ.; Yalçın, P. & Kurucu, Y. (2011a). Determination of Chemical Effect on the $K\beta_1/K\alpha$, $K\beta_2/K\alpha$, $K\beta_2/K\beta_1$ and $K\beta/K\alpha$ X-Ray Intensity Ratios of 4d Transition Metals. *Spectroscopy Letters,* Vol.44, pp. 38–46.

Porikli, S. (2011b). Influence of the Chemical Environment Changes on the Line Shape and Intensity Ratio Values for La, Ce and Pr L Lines *Spectra. Chem. Phys. Lett.,* Vol.508, pp. 165-170.DOI information: 10.1016/j.cplett.2011.04.021

Pyykko, P. (1988). Relativistic Effects in Structural Chemistry. *Chem. Rev.,* Vol.88, pp. 563-94.

Raj, S.; Padhi, H.C. & Polasik, M. (1998). Influence of Chemical Effect on the *Kβ*-to-*Ka* X-Ray Intensity ratios of Ti, V, Cr, and Fe in TiC, VC, CrB, CrB_2 and FeB. *Nucl. Instrum. Meth. in Phys. Res. B,* Vol.145, pp. 485–491.

Rao, N.V.; Reddy, S.B; Satyanarayana, G. & Sastry, D.L. (1986). *Kβ/Ka* X-Ray Intensity Ratios. *Physica C,* Vol.142, pp. 375-380.

Salem, S.I.; Panossian, S.L. & Krause, R.A. (1974). Quantum Mechanics of Atomic Spectra and Atomic Structure. *At. Data Nucl. Data Tables,* Vol.14, pp.91-109.

Scofield, J.H. (1969). Radiative Decay Rates of Vacancies in *K* and *L* Shells. *Phys. Rev.,* Vol.179, pp. 9-16.

Scofield, J.H. (1973). Theoretical Photoionization Cross Sections from 1 to 1500 keV, Unpublished *Lawrence Livermore Laboratory Report* UCRL-51326, Livermore, California

Scofield, J.H. (1974a). Relativistic Hartree–Slater Values for *K* and *L* Shell X-Ray Emission Rates. *Atomic Data and Nuclear Data Tables,* Vol.14, pp. 121-137.

Scofield, J.H. (1974b). Exchange Corrections of *K* X-ray Emission Rates. *Phys. Rev. A,* Vol.9, pp. 1041–1049.

Stöhr, Y. & Wu, Y. (1994). X-Ray Magnetic Circular Dichroism: Basic Concepts and Theory for 3d Transition Metal Atoms. *New Directions in Research with 3rd Generation Soft X-Ray Synchrotron Radiation Sources,* Editors F. Schlachter and F. Wuilleumier (Kluwer, Netherlands, 1993), p. 221.

Tamaki, Y.; Omori, T. & Shiokawa, T. (1979). Chemical Effect on the *Kβ/Ka* Intensity Ratios of the Daughter Atoms Formed by the EC decay of ^{51}Cr and ^{54}Mn. *Radiochem. Radioanal. Lett.,* Vol.37, pp. 39-44.

Taniguchi, K. (1984). Chemical-State Analysis by Means of Soft X-Ray Spectroscopy. 2. *Kβ* Spectra for Phosphorus, Sulfur, and Chlorine in Various Compounds. *Bull. Chem. Soc. Jpn.,* Vol.57, pp. 915-920.

Thole, B.T.; Carra, P.; Sette, F. & Van der Lean, G. (1992). X-Ray Circular-Dichroism as a Probe of Orbital Magnetization. *Phys. Rev. Lett.,* Vol.68, pp. 1943-1946.

Torres Deluigi, M.; Perino, E.; Olsina R. & Riveros de la Vega, A. (2003). Sulfur and Phosphorus *Kβ* Spectra Analyses in Sulfite, Sulfate and Phosphate Compounds by X-Ray Fluorescence Spectrometry. *Spectrochim. Acta Part B,* Vol.58, pp- 1699-1707.

Urch, D.S. (1979). Theory, Techniques, and Application; Brundle, C. R., Baker, A. D., Eds.; Academic Press: New York, *Electron Spectrosc.*, Vol.3, pp. 1-39.

Yamoto, I.; Kaji, H. & Yoshihara, K. (1986). Studies on Chemical Effects on X-Ray Intensity Ratios of $K\beta/K\alpha$ in Nuclear Decay of Technetium Nuclides ^{99m}Tc, ^{97m}Tc and ^{95m}Tc. *J. Chem. Phys.*, Vol.84, pp- 522-527.

6

Angular Dependence of Fluorescence X-Rays and Alignment of Vacancy State Induced by Radioisotopes

İbrahim Han
Ağrı İbrahim Çeçen University
Turkey

1. Introduction

This chapter concerns angular distribution measurements for fluorescence X-ray and the alignments of atoms with inner-shells vacancy resulting from ionization by radioisotope sources. The discussion on this topic is done by evaluating measurements of X-ray fluorescence parameters (such as cross-section, alignment parameter, polarization degree) from sample in various emission angles.

When an atom is ionized in one of its inner shells, the electrons rearrange themselves to fill the vacancy, with the transition energy released as a photon or transferred to another electron. The following X-ray or Auger electron may have an isotropic or non-isotropic angular distribution. The study of alignment of the inner-shell vacancy in ions can provide information about ionization process and the wave functions of inner-shell electrons, and calculations showed that the alignment was a sensitive testing parameter for theoretical models. For the last five decades there have been both theoretically and experimentally renewed efforts towards better understanding of the physics concerned with alignment of atoms with inner-shells vacancy and/or angular dependence of fluorescent X-rays emitted atoms induced photons or charged particle (electrons, protons, heavy ions). Generally, the alignments of atoms with inner-shells vacancy resulting from ionization by photons are investigated by measuring the anisotropic emission of X-ray lines using a detector (such as Si(Li) or Ge(Li)) and radioisotope photon source in various emission angles.

2. Historical background and current status of topic

The aim of paper interested in this topic is to determine the relationship between the angular distributions of X-rays with respect to total angular momentum values (J) of vacancy states. It is well-known that when radioisotope source, X-ray tube or charged particles produce vacancies in atoms at energy levels with $J>1/2$, the resulting ions will be aligned. The signature of this alignment is the anisotropic angular distribution of the emitted characteristic X-ray radiation, or the degree of polarization of the X-ray radiation. Total angular momentum (J) of vacancy states after photoionization is greater than 1/2, the population of its magnetic sub-states is non-statistical by the ionized atoms and this is reason of this anisotropic behavior. A lot of theoretical studies have been reported so far along this topic (Mehlhorn 1968; Mc Farlene, 1972; Berezhko and Kabachnik 1977; Sizov and Kabachnik, 1980, 1983) and the predictions of these researchers have been experimentally supported by some researchers (Schöler and Bell, 1978; Pálinkás, 1979, 1982; Wigger et al., 1984; Jesus et al., 1989; Mitra et al., 1996). The experimental study of alignment generally involves measurements of the angular distribution or polarization of the induced X-rays (Hardy et al., 1970; Döbelin et al., 1974; Jamison and Richard, 1977; Jistchin et al., 1979, 1983; Pálinkás, et al., 1981; Stachura et al., 1984; Bhalla, 1990; Mehlhorn, 1994; Papp 1999). In 1969, Cooper and Zare, (1969) first suggested a theoretical model relevant to aligned photon induced atoms. According to calculation by Cooper and Zare, (1969), after photoionization the inner-shell vacancy states have statistical population of magnetic substates. The vacancies produced after photoionizationin sub-shells are not be aligned at all and so the angular distribution of the fluorescent X-rays subsequent to photoionization will be isotropic. In 1972, 3 years after Cooper and Zare, the predictions of Flügge et al., (1972) showed that when vacancies are created in states with J>1 2, the population of its magnetic sub-states are non-statistical and therefore the resulting ions will be aligned. Mc Farlane (1972) calculated the polarization of X-rays from the decay of a vacancy in the $2p_{2/3}$ sub-shell using hydrogenic wave- functions in the Bethe approximation and the first Born approximation. After Caldwell and Zare (1977) first made an experimental investigation of the photon-induced alignment of Cd and they measured the degree of polarization of the emitted radiation from Cd. Since then, many experiments and calculations have been done to study the alignment of atoms and angular dependence characteristic X-rays by measuring either the angular distribution or the degree of polarization of the emitted X-rays. All these studies confirmed either alignment or not-alignment of the atoms after photoionization. The angular correlation between ionizing and fluorescent X-rays has been calculated relativistically, including all the radiation multipoles using single particle wavefunctions calculated in the Hartree-Slater model, by (Scofield, 1976). More recently, Scofield, (1989) used a relativistic model to study the angular distribution of the photoelectrons produced from photo- ionization by linear polarize photons and its inverse process (radiative recombination) in the energy region of 1–100keV. Scofield, (1989) found that the cross-section has a maximum at 90° compared to the direction of the incoming photons in the x–z plane (polarization plane) while the cross-section is independent of the angle between the incoming photon and the ejected electron in the y–z plane (normal to the polarization plane). Kamiya et al., (1979) measured L X-rays of Ho and Sm produced by protons and ^{3}He impacts with Si(Li) detector over the incident energy ranges E_p = 0.75–4.75MeV and $E_{3_{He}}$ = 1,5–9,4 MeV in the direction of 90° to the projectile. Kamiya, et al., (1979) reported that the ratios of X-ray production cross-sections for the La and Ll lines depend clearly on projectile energy, but are independent of the projectile charge. Theoretical values of the alignment parameter for different states of various atoms calculated using the Herman-Skillman wave functions, have been reported by Berezhko and Kabachnik, (1977). The very strong anisotropy was reported for the emission of L lines for various elements by several scientists (Kahlon, et al., 1990a,b, 1991a,b; Ertuğrul, et al., 1995, 1996; Ertuğrul, 1996, 2002; Kumar, et al., 1999 Sharma and Allawadhi, 1999; Seven and Koçak, 2001,2002; Seven, 2004; Demir, et al., 2003). However, in all these investigations, the observed anisotropy is much higher than the predicted theoretical values of Scofield, (1976) and Berezhko and Kabachnik, (1977). On the other hand, anisotropic emission for L X-rays of Pb, Th and U was reported by some scientists (Mehta, et al., 1999; Kumar, et al 1999, 2001). Recently, Yamaoka et al., (2002, 2003) performed experiments using synchrotron radiation to determine the angular distribution of

L X-ray photons of Pb and Au. Although they found an isotropic distribution of the Pb L_3 lines within the experimental errors, non-isotropic angular distribution of the Au L_3 lines have been obtained. Papp and Campbell, (1992) reported the magnitude of the anisotropy and the alignment parameter for the *L* lines of Er. The alignment parameter of the ions of Xe was obtained by Küst, et al., (2003).

Kahlon et al., (1990a) reported experimental investigation of the alignment of the L_3 subshell vacancy state produced after photoionization in lead by 59.57 keV photons. The values of differential cross sections for the emission of the Ll, $L\alpha$, $L\beta$ and $L\gamma$ X-ray lines were determined at different emission angles varying from 40° to 120°. It was seen from the results that the Ll, and $L\alpha$ peaks show anisotropic emission, while the $L\beta$ and $L\gamma$ peaks are emitted isotropically. The angular dependence of emission intensity of L shell X rays induced by 59.57 keV photons in Pb and U was investigated by Kahlon et al., (1990b) measuring the normalized intensities of the resolved L X-ray peaks at different angles varying from 40° to 140°. It was observed that while the Ll and $L\alpha$ peaks (originating from J=3/2 state) show some anisotropic angular distribution, the emission of the $L\beta$ and $L\gamma$ peaks are emitted isotropically. Kahlon et al., (1991a) measured the angular distribution and polarization of the L shell fluorescent X-rays excited by 59.54 keV photons in Th and U. It was found that the $L\gamma$ group of L X-rays is isotropic in spatial distribution and unpolarized but, the Ll and $L\alpha$ groups are anisotropically distributed and polarized. Although no anisotropy of the $L\beta$ group is detected, it was slightly polarized. Kahlon et al., (1991b) investigated the differential cross sections for emission of Ll, $L\alpha_2$, $L\alpha$, $L\beta$ and $L\gamma$ groups of L X-ray lines induced in Au by 59.54 keV photons at different angles varying from 40° to 120°. The L X-rays represented byLl, $L\alpha_2$ and $L\alpha$ peaks were found to be anisotropic in the spatial distribution while those in $L\beta$ and $L\gamma$ peaks were isotropic. Papp and Campbell, (1992) measured angular distributions of the L_l, $L\alpha_{1,2}$ and $L\beta_{2,15}$ transitions of erbium in the angular range of 70°–150° following photoionization by 8.904 keV photons. A Johansson-type monochromatic was used to select the Cu $K\beta_1$ line for ionization. Anisotropy parameters for Ll, $L\alpha_{1,2}$ and $L\beta_{2,15}$ were found as 0.052±0.016, 0.16±0.022 and 0.012±0.015, respectively. Ertugrul et al., (1995, 1996a, 1996b) measured differential cross-sections for the emission of Ll, $L\alpha$, $L\beta$ and $L\gamma$ X-rays of Au, Hg, Tl, Pb, Bi, Tb and U at different emission angles varying from 45° to 135°. They found that Ll and $L\alpha$ peaks are emitted isotropically, while $L\beta$ and $L\gamma$ peaks show anisotropic emission. Sharma and Allawadhi, (1999) measured values of Ll, $L\alpha$ and $L\beta$ differential X-ray production cross sections in Th and U at 16.896 and 17.781 keV at emission angles 60°, 70°, 80° and 90°. From the results of the measurements it was evident that, in the present case, all the three Ll, $L\alpha$ and $L\beta$ differential X-ray production cross sections depend on the emission angle and thus, the emission is anisotropic. Demir et al., (2000) indicated differential cross-sections for the emission of M shell fluorescence X-rays from Pt, Au and Hg by 5.96 keV photons at seven angles ranging from 50° to 110° at. The differential cross-sections were found to decrease with increase in the emission angle, showing an anisotropic spatial distribution of M shell fluorescence X-rays. Seven and Koçak (2001, 2002) measured the Ll, $L\alpha$, $L\beta$ and $L\gamma$ X-ray production cross-sections in U, Th, Bi, Pb, Tl, Hg, Au, Pt, Re,W, Ta, Hf, Lu, and Yb using 59.5 keV incident photon energies in the angular range 40°-130°. Although differential cross sections for $L\beta$ and $L\gamma$ X-rays were found to be angle independent within experimental error, those for the Ll and La X-rays were found to be angle dependent. Ertugrul et al., (2002) measured the alignment parameter the $I_{L\alpha}/\,I_{Ll}$ intensity ratio. The Ll and $L\alpha$ X-rays of the elements were measured with a Si (Li) detector at a direction of 90° to the projectile. The L3 edges of Nd,

Gd, Tb, Dy, Ho, Er, Yb, Hf, Ta, W, Au, Hg, Tl, Pb, Bi, Th and U the elements were excited with the K X-ray energy of 17.781(Mo$K_{\alpha,\beta}$), 16.896(Nb$K_{\alpha,\beta}$), 14.980(RbK_{β}), 13.300(BrK_{β}), 12.503(SeK_{β}), 12.158(Br$K_{\alpha,\beta}$), 10.983(GeK_{β}), 10.073(Ge$K\alpha_{1,\beta}$), 9,572(ZnK_{β}), 8.976(Cu$K_{\beta 2}$), 8.907(CuK_{β}), 8.265(NiK_{β}), 7.649(Co$K_{\beta 1}$), 6.490(Mn$K_{\beta 1}$) keV from the selected elements, respectively. They noticed that the L_3 X-rays show large anisotropy, the measured alignment parameter varying from -0.115 to +0.355. Demir et al., (2003) reported Ll, $L\alpha$, $L\beta$ and $L\gamma$ X-ray differential cross-sections, fluorescence cross-sections and σ_{L1}, σ_{L2} and σ_{L3} subshell fluorescence cross-sections for Er, Ta, W, Au, Hg and Tl at an excitation energy of 59.6 keV. The differential cross-sections for these elements have been measured at different angles varying from 54° to 153°. The Ll and $L\alpha$ groups in the L X-ray lines were found to be spatially anisotropic, while those in the $L\beta$ and $L\gamma$ peaks are isotropic. The Ll, $L\alpha$, $L\beta_{2,4}$, $L\beta_{1,3}$ and $L\gamma$ X-ray production cross-sections and L-subshell fluorescence yields ω_1 and ω_2 in Th and U have been determined by Seven (2004) at an incident photon energy of 59.54 keV by measuring differential cross-sections with angles changing from 40° to 130°. The Ll, $L\alpha$ and $L\beta_{2,4}$ X-rays have an anisotropic spatial distribution while $L\beta_{1,3}$ and $L\gamma$ X-rays have isotropic spatial distributions. Özdemir et al., (2005) measured the angular dependence of L_3 subshell to M-shell vacancy transfer probabilities for the elements Lu, Hf, Ta, W, Os and Pt at the excitation energies of 5.96 keV and K X-rays of Zn, Ga, Ge, and As, respectively, at seven angles varying from 120° to 150°. It was observed that angular dependence from L_3 subshell to M-shell vacancy transfer probabilities increase with increasing cosθ. The angular dependence of M X-ray production differential cross-sections for selected heavy elements between Lu and Pt have been measured by Durak (2006) at 5.59 keV of incident photon energy and at seven emission angles in the range of 120°-150°. Angular dependence of M X-ray production differential cross sections has been derived, using the M-shell fluorescence yields, experimental total M X-ray production cross sections and theoretical M-shell photoionization cross sections. M X-ray production differential cross-sections were found to decrease with increase in the emission angle, showing an anisotropic spatial distribution of M X-rays. Angular dependence from L_3 subshell to M-shell vacancy transfer probabilities for selected heavy elements from Au to U were measured by Özdemir and Durak (2008) at different angles varying from 120° to 150°. It was observed that angular dependence from L_3-subshell to M-shell vacancy transfer probabilities increase with increasing cosθ. Apaydın et al., (2008) measured Mi ($i = a + \beta$) X-ray production differential cross sections for Re, Bi and U elements at the 5.96 keV incident photon energy in an angular range 135°-155°. They found that the angular dependence M X-rays production cross sections decrease with increase in the emission angle, showing anisotropic spatial distribution

Kumar et al., (1999) investigated the angular dependence of emission of L x-rays following photoionization at 22.6 and 59.5 keV in 82Pb by measuring the intensity ratios $I_{Ll}/I_{L\gamma}$, $I_{L\alpha}/I_{L\gamma}$ and $I_{L\beta}/I_{L\gamma}$ at different angles varying from 50° to 140°. The measured intensity ratios for various L x-rays were found to be angle independent within experimental error. Mehta et al., (1999) measured the L_l, L_α, L_η, $L\beta_6$, $L\beta_{2,4}$, $L\beta_{1,3}$, $L\beta_{9,10}$ and L_γ x-ray production differential cross sections in 92U using the 22.6- and 59.5-keV incident photon energies in an angular range 43°–140°. Differential cross sections for various L x rays were found to be angle independent within experimental error. Puri et al., (1999) measured The Ll, $L\alpha$, $L\beta_{2,4}$, $L\beta_{1,3}$ and $L\gamma_{1,5}$ X-ray production differential cross sections in 90Th have at 22.6 keV incident photon energy in an angular range 50° -130° The measured differential cross sections for various L X-rays were found to be angle-independent within experimental error. Kumar et al., (2001a) measured the the Ll, $L\alpha$ and $L\beta_{2,5,6,715}$ X-ray fluorescence (XRF) differential cross-sections in Pb at the 13.6 keV incident photon energy ($E_{L3} < E_{inc} < E_{L2}$, E_{Li} being the Li sub-shell binding energy) and in the angular range 90-160°. At this incident photon energy, the L_3 sub-shell vacancies (J = 3/2) are produced only due to the direct ionization and the reduction in the observed anisotropy in the emission of the Ll, $L\alpha$ and $L\beta_{2,5,6,715}$ X-rays due to the transfer of unaligned L_1 and L_2 subshell vacancies (J = 1/2) to the L_3 sub-shell through Coster–Kronig transitions was eliminated. The differential cross-sections for various x-rays were found to be angle-independent within experimental error. The L X-ray production (XRP) differential cross sections in Th and U have been measured by Kumar et al., (2001b) at the 17.8 keV incident photon energy ($E_{L3} < E_{inc} < E_{L2}$, E_{Li} is the Li subshell ionization threshold) in an angular range 90°-160°and at the 25.8 and 46.9 keV incident photon energies ($E_{L1} < E_{inc} < E_K$) at an angle of 130°. The present measurements rule out the possibility of a strong angular dependence of differential cross sections for various L_3 subshell X-rays following selective photoionization of the L_3 subshell. Tartari et al., (2003) investigated the anisotropy of L X-ray fluorescence induced by 59.54 keV unpolarized photons by means of an experimental procedure which allows the relative L X-ray production cross section to be evaluated without taking account of the angular set-up and the instrumental efficiency. Thick targets of Yb, Hf, Ta, W and Pb are considered, and the angular trend of the relative experimental ratios, $I_{L\alpha}/I_{L\beta}$, is calculated by simple evaluations of the peak area alone. Within the experimental uncertainties, which were found to be of the order of 1.6% in the worst cases, the results do not show any significant angular dependence of the $L\alpha$ emission lines. Santra et al., (2007) measured the angular distribution of the L X-ray fluorescent lines from Au and U induced by 22.6-keV X-rays in the angular range of 70°–150°. No strong anisotropy was observed as mentioned by some groups. In the case of Au, a maximum anisotropy of 5% was observed while for U it was within experimental errors 2%. From the angular distribution of the L_l line of Au, the alignment parameter was obtained and its value was found to be 0.10±0.14. Kumar et al., (2008) investigated alignment of the M_3(J= 3/2), M_4(J= 3/2) and M_5(J= 5/2) subshell vacancy states produced following photoionization in the M_i (i=1-5) subshells of Au, Bi, Th and U through angular distribution of the subsequently emitted M X-rays. The unpolarized Mn K X rays (E_{KX}=5.97 keV) from the 55Fe radioisotope were used to ionize the Mi subshells in an angular range 90°-160°and the emitted M X-rays were measured under vacuum using a low energy Ge detector. The M X-ray spectra taken at different emission angles were normalized using the isotropically emitted K shell (J= 1/2) X-rays measured simultaneously from a 23V thin target placed adjoining the M X-ray target. The present precision measurements infer that anisotropy in the $M_{\alpha\beta\gamma}$ X-ray emission shows trends and order of magnitude predicted by theoretical calculations, i.e., anisotropy parameter (β_2)~0.01.

In the recent experimental study (Han et al., 2008), the angular distribution of characteristic K and L X-rays, emitted from Sm, Eu, Gd Tb, Dy, Ho, and Er as a result of K and L shell vacancies produced by 59.54 keV photon impact was investigated. Thus, K and L X-rays emitted from these elements were simultaneously measured in the same experimental geometry. In this study, Sm, Eu, Gd, Tb, Dy, Ho, and Er lanthanides were chosen since both K shell and L shell electrons of these elements can be excited simultaneously by an Am-241 point source. Also, K and L peaks of the chosen elements are well resolved. Earlier experimental investigations have been only performed on the K X-ray cross sections or on the angular distribution of L X-rays. This is the first report of the angular distributions of Li X-ray and Ki X-ray (i = α, β) cross sections for Sm, Eu, Gd, Tb, Dy, Ho, and Er at different angles. It is well known that K X-ray cross sections have no angular dependency (J = 1/2). The experimental investigation on K X-ray cross sections at different angles was made to check the validity of the angular dependency of experimental L X-ray cross sections. The experimental K X-ray cross sections were compared with theoretically calculated values and fairly good correspondence was observed. This means that the present measurements regarding angular dependency of L X-rays are reliable.

In following the work of us (Han et al., 2009) experimental results of the angular distribution of characteristic X-rays were introduced. We preferred to use of $I_{La} / I_{Ll}(\theta)$ intensity ratios to obtain the values of alignment parameters (A_2). In that case, the background subtraction problem is considerably reduced and statistical errors are significantly less. It was observed from measured intensities that L*a* and L*l* X-ray intensities for the L_3 sub-state depended on the emission angle, meaning that L*a* and L*l* X-rays had an anisotropic spatial distribution. Thus, the L*a* to L*l* intensity ratios for a set of elements was determined and alignment parameters for each element were obtained using these ratios. In this study, three L sub-shells electrons were excited. Therefore, alignment parameter values are influenced by Coster–Kronig transitions from vacancies induced in the L_1 or L_2 sub-shells. L_1 and L_2 sub-shells have the same J= 1/2 value therefore the transferred vacancies are not-aligned and the observed anisotropy of the X-rays is attenuated. For this reason, corrected value of the alignment parameter was calculated using attenuation factor F. If photon energies exciting only L_3 sub-shell electrons are chosen, the alignment parameter will be independent from Coster–Kronig transitions

In more recently study (Han and Demir, 2011a), we investigated the angular distribution of characteristic L X-rays emitted from heavy elements (Pt, Au, Pb, Bi, Th and U) as a result of L shell vacancy production by 59.54 keV photon impact and angular distribution of Compton scattering photons from the same elements. Thus, emitted fluorescent L X-rays and Compton scattering photons from elements were simultaneously measured in the same experimental geometry. Earlier experimental investigations have been only performed on the angular distribution of L X-rays or Compton scattering photons. This is the first report of the angular distribution of L*i* (*i*= *l*, *a*, β and γ) X-rays fluorescent and Compton scattering differential cross sections for Pt, Au, Pb, Bi, Th and U at different angles in the same experimental geometry. It is well known that Compton scattering differential cross sections have angular distribution. The experimental investigation on Compton scattering differential cross sections at different angles was made to check the validity of angular distribution of experimental L X-rays fluorescent differential cross sections. The experimental Compton scattering differential cross sections were compared with theoretically calculated values and fairly good correspondence was observed. This means that the present measurements regarding angular distribution of L X-rays are reliable. In the meantime, L_3-subshell alignment of Th and U ionized by 59.5 keV photons has been investigated by evaluating the angular dependence of L*i* (*i*=*l*, *a*, η, β and γ) X-ray lines. The angular dependence measurements were performed by measuring the fluorescence cross section, σ_{Li} (*i*= *l*, *a*, η, β and γ) and $\sigma_{Ll}/\sigma_{L\gamma}$, $\sigma_{L\eta}/\sigma_{L\gamma}$, $\sigma_{La} / \sigma_{L\gamma}$ and $\sigma_{L\beta}/\sigma_{L\gamma}$ ratios at different angles. It was observed from the measurements that L*i* (i=*l* and *a*) X rays for the L_3-subshell depended on the emission angle and had an anisotropic spatial distribution. On the other hand, there was no dependence of emission angle and any significant anisotropy for other L X rays. The both L*l* and L*a* X-rays originate from the filling of vacancies in states L_3-subshell with *J* = 3/2. The results of measurements indicate that the L_3-subshell vacancy states with *J* =3/2 are aligned, whereas L_1, and L_2 vacancy states with *J* =1/2 are non-aligned. Integral cross-sections for the L*i* (*i*= *l*, *a*, η, β and γ) X-rays and L subshell fluorescence yields ω_i (*i*= 1, 2 and 3) were also determined and results were compared with theoretically calculated values and results of others and fairly good correspondence was observed. The Lγ X-rays, originating purely from the L_1 and L_2 subshells, having isotropic emission were used to normalize the intensities of the anisotropic L*l* and the L*α* X-rays originating from the L_3 subshell. It was observed from measurements that L*l* and L*α* X-ray for the L_3 sub-state depended on the emission angle, meaning that L*l* and L *α* X-rays had an anisotropic spatial distribution. On the other hand, the L*β* and L*γ* X-rays don't show any significant anisotropy. The fluorescence cross sections for L*l* and L*α* X-rays are decreased with increased emission angles (Han and Demir, 2011b).

3. Conclusion

In the light of all these, above; data from different researchers show contradictory and the existing results on the angular dependence of fluorescence X-ray and the alignment of atoms with inner-shells vacancy following ionization are still controversial and quite confusing. Therefore, more experimental and theoretical investigations should be required to settle the present discrepancies

4. Acknowledgment

I thank to M.R. Kacal for his help and advice during the preparation of this chapter.

5. References

Apaydın, G., Tırasoglu, E., Sogut, O., 2008. Measurement of angular dependence of M X-ray production cross-sections in Re, Bi and U at 5.96 keV Eur. Phys. J. D 46, 487–492

Berezhko, E.G., Kabachnik, N.M., 1977. Theoretical study of inner-shell alignment of atoms in electron impact ionisation: angular distribution and polarization of X-rays and Auger electrons. J. Phys. B At. Mol. Opt. Phys. 10, 2467–2477.

Bhalla, C.P., 1990. Angular distribution of Auger electrons and photons in resonant transfer and excitation in collisions of ions with light targets. Phys. Rev. Lett. 64, 1103–1106.

Caldwell, C.D., Zare, R.N., 1977.Alignment of Cd atoms by photoionization. Phys. Rev. A 16, 255–262.

Cooper, J., Zare, N., 1969. Potoelectron angular distributions. In:Geltman, S., Mahanthappa, K.T., Brittin, W.E.(Eds.), Lecturesin Theoretical Physics: Atomic Collision Processes, vol. X1C. Gordon and Breach, NewYork, pp. 317–337.

Demir, L., Şahin, M., Kurucu, Y., Karabulut, A., Şahin, Y., 2000. Measurement of angular dependence of photon-induced differential cross-sections of M X-rays from Pt, Au and Hg at 5.96 keV. Radiat. Phys. Chem. 59 355-359

Demir, L., Şahin, M., Kurucu, Y., Karabulut, A., Şahin, Y., 2003. Angular dependence of *Ll*, *Lα*, *Lβ* and *Lγ* X-ray differential and fluorescence cross-sections for Er, Ta, W, Au, Hg and Tl Radiat. Phys. Chem. 67, 605–612

Döbelin, E., Sandner, W., Mehlhorn, W., 1974. Experimental study of inner shell alignment of atoms in electron impact ionization. Phys. Lett. A 49, 7–8.

Durak, R., 2006. Measurement of angular dependence of M X-ray production differential cross-sections in heavy elements at 5.96 keV. Can. J Anal. Sci. Spectrosc. 51, No. 2.

Ertuğrul, M., Büyükkasap, E., Erdoğan, H., 1996a. Experimental investigation of the angular dependence of photon-induced differential cross-sections of L X-rays from U, Th and Bi at 59.5 keV. Il Nuovo Cimento D, 18, 671–676.

Ertuğrul, M., Büyükkasap, E., Küçükönder, A., Kopya, A.İ., Erdoğan, H., 1995. Anisotropy of *L* -shell X-rays in Au and Hg excited by 59.5 keV photons. Il Nuovo Cimento D, 17, 993-998.

Ertuğrul, M., Öz, E., Şahin, Y., 2002. Measurement of alignment parameters for photon induced L_3 vacancies in the elements 59≤Z≤92. Physica Scripta 66, 289–292.

Ertuğrul, M.,1996b. Measurement of cross-sections and Coster–Kronig transition effect on L sub-shell X-rays of some heavy elements in the atomic range 79≤Z≤92 at 59.5 keV. Nucl. Instrum. Methods Phys. Res. B 119, 345–351.

Flügge, S.,Mehlhorn, W.,Schmidt,V.,1972. Angular distribution of auger electrons following photoionization. Phys.Rev.Lett. 29, 7–9 Erratum: Phys.Rev.Lett. 29, 1288.

Han, I., Demir, L., 2011a. Angular distribution of fluorescent L X-rays and Compton scattering photons Spectrosc. Lett. 44, 95–102.

Han, I., Demir, L., 2011b. Angular dependence of L_3-subshell X-ray emission following photoionisation. J X-Ray Sci. Technol. 19, 13–21.

Han, I., Sahin, M., Demir, L., 2008. Angular variations of K and L X-ray fluorescence cross sections for some lanthanides Can. J. Phys. 86, 361–367.

Han, I., Sahin, M., Demir, L., 2009. The polarization of X-rays and magnetic photoionization cross-sections for L_3 sub-shell. Appl. Radiat. Isot. 67, 1027–1032.

Hardy, J., Henins, A., Bearden, J.A., 1970. Polarization of the L_{a1} X-rays of mercury. Phys. Rev. A 2, 1708–1710.

Jamison, K.A., Richard, P., 1977. Polarization of target K X-rays. Phys. Rev. Lett. 38, 484–487.

Jesus, A.P., Ribeiro, J.P., Niza, I.B., Lopes, J.S., 1989. L_3-subshell alignment of Au induced by proton, deuteron and alpha-particle impact. J. Phys. B At. Mol. Opt. Phys. 22, 65.

Jitschin, W., Hippler, R., Shanker, R., Kleinpoppen, H., Schuch, R., Lutz, H.O., 1983. L X-ray anisotropy and L_3-sub-shell alignment of heavy atoms induced by ion impact. J. Phys. B At. Mol. Opt. Phys. 16, 1417–1431.

Jitschin, W., Kleinpoppen, H., Hippler, R., Lutz, H.O., 1979. L-shell alignment of heavy atoms induced by proton impact ionization. J. Phys. B At. Mol. Opt. Phys. 12, 4077–4084.

Kahlon, K.S., Aulakh, H.S., Singh, N., Mittal, R., Allawadhi, K.L., Sood, B.S., 1990a. Experimental investigation of alignment of the L_3 sub-shell vacancy state produced after photoionization in lead by 59.57 keV photons. J. Phys. B At. Mol. Opt. Phys. 23, 2733–2743.

Kahlon, K.S., Aulakh, H.S., Singh, N., Mittal, R., Allawadhi, K.L., Sood, B.S., 1991a. Measurement of angular distribution and polarization of photon-induced fluorescent x rays in thorium and uranium. Phys. Rev. A 43, 1455–1460.

Kahlon, K.S., Shatendra, K., Allawadhi, K.L., Sood, B.S., 1990b. Experimental investigation of angular dependence of photon induced L shell X-ray emission intensity. Pramana 35, 105–114.

Kahlon, K.S., Singh, N., Mittal, R., Allawadhi, K.L., Sood, B.S., 1991b. L_3-sub-shell vacancy state alignment in photon–atom collisions. Phys. Rev. A 44, 4379–4385.

Kamiya,M., Kinefuchi, Y., Endo, H., Kuwako, A., Ishii, K., Morita, S., 1979. Projectile- energy dependence of intensity ratio of La to Ll X-rays produced by proton and ^{3}He impacts on Ho and Sm. Phys. Rev. A 20, 1820–1827.

Kumar, A., Garg, M.L., Puri, S., Mehta, D., Singh, N., 2001a. Angular dependence of L_3 x-ray emission following L_3 sub-shell photoionization in Pb. X-Ray Spectrom. 30, 287–291

Kumar, A., Puri, S., Mehta, D., Garg, M.L., Singh, N., 1999. Angular dependence of L x-ray emission in Pb following photoionization at 22.6 and 59.5 keV. J. Phys. B At. Mol. Opt. Phys. 32, 3701–3709.

Kumar, A., Puri, S., Shahi, J.S., Garg, M.L., Mehta, D., Singh, N., 2001b. L X-ray productioncross-sections in Th and U at 17.8, 25.8 and 46.9 keV photon energies. J. Phys. B At. Mol. Opt. Phys. 34, 613–623.

Kumar, S., Sharma, V., Mehta, D., Singh, N., 2008. Alignment of *Mi* (*i*=3–5) subshell vacancy states in 79Au, 83Bi, 90Th, and 92U following photoionization by unpolarized Mn *K* x rays. Phys. Rev. A 77, 032510

Küst, H., Kleiman, U., Mehlhorn, W., 2003. Alignment after Xe L_3 photoionization by synchrotron radiation. J. Phys. B At. Mol. Opt. Phys. 36, 2073.

Mc Farlane, S.C., 1972. The polarization of characteristic X radiation excited by electron impact. J. Phys. B At. Mol. Opt. Phys. 5, 1906–1915.

Mehlhorn, W., 1968. On the polarization of characteristic X radiation. Phys. Lett. A 26, 166–167.

Mehlhorn,W., 1994. Alignment after inner-shell ionization by electron impact near and at threshold. Nucl. Instrum. Methods Phys. Res. B. 8, 7227–7233.

Mehta, D., Puri, S., Singh, N., Garg, M.L., Trehan, P.N., 1999. Angular dependence of L X-ray production cross-sections in U at 22.6 and 59.5 keV photon energies. Phys. Rev. A 59, 2723–2731.

Mitra, D., Sarkar, M., Bhattacharya, D., Chatterjee, M.B., Sen, P., Kuri, G., Mahapatra, D.P., Lapicki, G., 1996. L_3-subshell alignment in gold and bismuth induced by low-velocity carbon ions. Phys. Rev. A 53, 2309–2313.

Ozdemir, Y., Durak R., 2008. Angular dependence from L_3-subshell to M-shell vacancy transfer probabilities for heavy elements using EDXRF technique. Annals of Nuclear Energy 35 1335–1339.

Ozdemir, Y.,Durak, R., Esmer, K., Ertugrul, M., 2005. Measurement of angular dependence from L_3-subshell to M-shell vacancy transfer probabilities for the elements in the atomic region $71 \leq Z \leq 78$. J Quant. Spectrosc. Radiat. Transf. 90 161–168.

Pálinkás, J., Sarkadi, L., Schlenk, B., Török, I., Kálmán, Gy., 1982. L_3-subshell alignment of gold by C^+ and N^+ impact ionisation. J. Phys. B At. Mol. Opt. Phys. 15, L451.

Pálinkás, J., Schlenk, B., Valek, A., 1979. Experimental investigation of the angular distribution of characteristic X-radiation following electron impact ionisation. J. Phys. B At. Mol. Opt. Phys. 12, 3273.

Pálinkás, J., Schlenk, B., Valek, A., 1981. The Coulomb deflection effect on the L_3-sub-shell alignment in low-velocity proton impact ionization. J. Phys. B At. Mol. Opt. Phys. 14, 1157–1159.

Papp, T., 1999. On the angular distribution of X-rays of multiply ionized atoms. Nucl. Instrum. Methods Phys. Res. B 154, 300–306.

Papp, T., Campbell, J.L., 1992. Non-statistical population of magnetic substates of the erbium L_3 subshell in photoionization. J. Phys. B At. Mol. Opt. Phys. 25, 3765.

Puri, S., Mehta, D., Shahid, J.S., Garg, M.L., Singh, N., Trehan, P.N., 1999. Photon-induced L X-ray production di€erential cross sections in thorium at 22.6 keV. Nucl. Instrum. Methods Phys. Res. B 152 19-26

Santra, S., Mitra, D., Sarkar, M., Bhattacharya, D., 2007. Angular distribution of Au and U *L* x rays induced by 22.6-keV photons. Phys. Rev. A 75, 022901.

Schöler, A., Bell, F., 1978. Angular distribution and polarization fraction of characteristic X-radiation after proton impact. Z. Phys. A 286, 163–168.

Scofield, J.H., 1976. Angular dependence of fluorescent X-rays. Phys. Rev. A 14, 1418–1420.

Scofield, J.H., 1989. Angular and polarization correlations in photoionization and radiative recombination. Phys. Rev. A 40, 3054–3060.

Seven, S., 2004. Measurement of angular distribution of fluorescent X-rays and L subshell fluorescence yields in thorium and uranium. Radiat. Phys. Chem. 69, 451–460

Seven, S., Koçak, K., 2001. Angular dependence of L x-ray production cross sections in seven elements from Yb to Pt at a photon energy of 59.5 keV. J. Phys. B At. Mol. Opt. Phys. 34, 202.

Seven, S., Koçak, K., 2002. Angular dependence of L x-ray production cross-section in seven elements from Au to U at 59.5 keV photon energy. X-Ray Spectrom. 31, 75–83.

Sharma, J.K., Allawadhi, K.L., 1999. Angular distribution of Lβ X-rays from decay of L_3 sub-shell vacancies in uranium and thorium following photoionization. J. Phys. B At. Mol. Opt. Phys. 32, 2343–2349.

Sizov, V.V., and Kabachnik, N.M., 1980. Inner-shell alignment of atoms in ion-atom collisions. I. Impact ionisation. J. Phys. B At. Mol. Opt. Phys. 13, 1601.

Sizov, V.V., and Kabachnik, N.M., 1983. Inner-shell alignment of atoms in ion-atom collisions. III. Light target atoms. J. Phys. B At. Mol. Opt. Phys. 16, 1565.

Stachura, Z., Bosch, F., Hambsch, F.J., Liu, B., Maor, D., Mokler, P.H., Schonfeldt, W.A., Wahl, H., Cleff, B., Brussermann, M., Wigger,J., 1984. Anisotropy of Ll X-ray transition observed in 1.4 MeV N^{-1} heavy ion-atom collisions. J. Phys. B At. Mol. Opt. Phys. 17, 835–847.

Tartari, A., Baraldi, C., Casnati, E., Re, A.D., Fernandez, J.E., Simone, T., 2003. On the angular dependence of L x-ray production cross sections following photoionization at an energy of 59.54 keV. J. Phys. B: At. Mol. Opt. Phys. 36 843–851.

Wigger, J., Altevogt, H., Brüssermann, M., Richter, G., Cleff, B., 1984. M_3, M_4 and M_5 alignment of thorium by proton impact ionisation. J. Phys. B At. Mol. Opt. Phys. 17, 4721.

7

Determination of Actinides Using Digital Pulse Processing Analysis

B. de Celis[1], V. del Canto[1], R. de la Fuente[1], J.M. Lumbreras[1], J. Mundo[2], and B. de Celis Alonso[3]
[1]*Escuela de Ingenieria Industrial, Universidad de Leon*
[2]*Universidad Autonoma de Puebla*
[3]*Universidad de Erlangen,*
[1]*Spain*
[2]*Mexico*
[3]*Germany*

1. Introduction

The determination of low concentrations of actinides (Pu, U, Am, Np, etc) in environmental samples is vital for evaluating radioactive contamination caused by nuclear reactors, atomic bomb tests, and any type of nuclear incident. In an emergency situation, rapid analytical methods are essential to provide timely information to the authorities working to protect the environment and the population from the consequences of possible contamination.

Actinides disintegrate by emission of alpha particles which are difficult to detect because they are absorbed by the samples themselves. They are very dangerous to man because of the high relative biological effectiveness of alpha particles and their tendency to accumulate in several parts of the body (bones, kidney, liver, etc.) for many years because of their long biological and physical half-lives. Their determination usually demands lengthy analytical procedures which usually employ radiochemical pre-treatment of the sample followed by measurements using alpha spectrometry.

Alpha-emitting radioisotopes produce alpha particles at characteristic energies between 4 and 7 MeV, which can differ by as little as 10 keV, close to the smallest resolution of the silicon detectors used in alpha spectrometers. Alpha particles are heavy charged particles. Therefore, any physical medium between the alpha-emitting radionuclide and the detector will strongly absorb most of the alpha particle energy. These attenuations are produced by the sample itself or by any material between the sample and the detector. To solve this problem two different analytical procedures are commonly applied. A separation and concentration step is necessary in both procedures to avoid possible energy interferences between radionuclides and to concentrate in a small volume a significant amount of radioactivity, making it easier to measure. In order to account for any loss of the sample during separation, a known quantity of a specific isotope or tracer is usually added. The tracer is an isotope of the element under study with similar chemical behaviour. The radionuclide and tracer behaviour during the chemical treatment is the same, assuming that the tracer is homogeneously mixed and brought into chemical equilibrium with the sample.

The first procedure applies radiochemical sample treatment in an aqueous medium to separate and concentrate the different actinides before their measurement using high resolution passivated implanted planar silicon (PIPS) detectors (Holm, 1984; Aggarwal et al., 1985). This procedure allows for the alpha energies of the nuclides present in the sample to be identified with almost no interference. Thus very low minimum detectable activities (MDA) can be reached, but this process does require long analysis times. In this work we summarize this procedure and present results obtained in our laboratory using PIPS technology.

The second procedure using liquid scintillation (LS) is accurate and reproducible and considerably faster than other methods (Basson and Steyn, 1954; McKliven and McDowell, 1984; McDowell, 1986; Abuzwrda et al., 1987; Miglio and Willis, 1988; Lucas, 1980). The radionuclides are in a solution of an organic scintillator and there is no risk of sample self-absorption, giving a counting efficiency of nearly 100%. Recent improvements have been made using organic cocktails which act simultaneously as extractors and scintillators, enabling elements to be separated in the relatively short time of several hours. However, the identification of the different radioisotopes is not simple because the resolution is low.

The latest developments have focused on obtaining higher energy resolution by optimizing the chemical separation techniques and the extracting cocktails to provide the maximum amount of light emission and reduce quenching (McDowell, 1986). The electronics associated with these techniques have also been improved to reduce electronic noise and discriminate between beta and alpha pulses (McDowell, 1986; Reboli et al., 2005). The technique called photon-electron rejecting alpha liquid scintillation extracts the actinides from an aqueous solution into an organic phase containing the extractant, an energy transfer reagent, and a light emitting fluor. The counting efficiency reached is nearly 100% and the energy resolution is 5% for 4 to 6 MeV alpha particles. Using analogue pulse shape discrimination (PSD), it is possible to reduce the background from photo-electrons produced by external gamma-rays and to eliminate interference from beta emitters. PSD electronically selects pulses produced by alpha particles based on the longer decay time of their light emission (30 nsec), which is due to their much higher linear energy transfer to the scintillation solution.

The LS technique offers the advantage of a short analysis time but the resolution and the possibility of reducing interferences is limited. The initial way of increasing energy resolution was to use scintillation cocktails that produced more photons and improved light collection by using mirrors and a liquid between the sample and the photocathode to reduce light reflection (Hanschke, 1972; McDowell, 1994). The use of solid state detectors has also been used to improve resolution. Avalanche photodiodes (APD) have a larger gain than common photodiodes and are more sensitive to the frequency of the scintillator due to their greater spectral response. The replacement of photomultiplier tubes (PMTs) by APDs was successful for gamma-ray scintillation detection with energy resolution improvements and more compact detectors. Large area avalanche photodiodes have been applied to alpha liquid scintillation spectroscopy (Reboli et al., 2005). The influence of several parameters on energy resolution was studied: temperature, bias voltage, nature of the scintillating cocktail, and geometry of the counting vial. The improvement in energy resolution was attributed to the higher quantum efficiency and the more uniform active area of the photocathode compared to PMTs. The silicon photodiodes have a larger spectral response than PMTs, making it easier to match the fluorescence spectra of the scintillators with the APD spectral

response. However, APDs have the typical limitations of photodiodes: a limited gain and a higher noise contribution.
In this review we describe the experimental work performed in our laboratory in applying digital cards to the analysis of actinides in environmental samples and we compare our methodology with the classical procedures described above. The first section shows some results obtained using PIPS detectors. The use of PSD with fast digital cards was applied to LS spectrometry. Finally, we studied the results obtained by applying coincident liquid scintillation/high resolution gamma spectrometry to reduce interference and increase the limits of detection.

2. High resolution alpha spectroscopy with passivated implanted planar silicon detectors

This procedure consists of the radiochemical separation of the radioisotopes of interest and their electro-deposition to form a few atomic layers in order to avoid the loss of alpha particle energy. Such attenuation can be produced by the sample itself or by any material between the sample and the detector. The result is a characteristic tail in the alpha peak (Fig.1). To reduce the size of the tail the samples are counted in a vacuum camera and the samples have to be as thin as possible to avoid self absorption. Electro-deposition, evaporation and precipitation are the most commonly used procedures to produce the thinnest samples possible. The sample is placed in front of the detector inside a vacuum camera and data are acquired for a preset period of time. Achieving the desired lower limit of detection requires very long times because of the low activities involved. Count times of one or two days are common, and so the total analysis time including sample preparation can be several days.
The proper preparation of the sample is an important step in achieving high quality results in alpha spectroscopy. Different methodologies have been published for specific applications (Glover, 1984; Greeman et al., 1990; Gomez et al., 1998; Sarin et al., 1990). There are three principal steps in the preparation of an alpha spectroscopy sample: preliminary treatment of the sample, chemical separation and preparation of the source to be measured.
The preliminary treatment is performed to homogenize the sample and to prepare it for the chemical processing. Different procedures are used for solid samples (e.g., food, soils, plants), liquid samples (water, blood, urine, etc.), and absorbents of gaseous or liquid samples (air filters, wipe type samples, etc.). In this early step the tracer nuclides are added to the samples.
Chemical separation is used to isolate and concentrate the elements of interest. Techniques used for separation include co-precipitation, solvent extraction, ion exchange and extraction chromatography.
The last step is required to prepare a sample suitable for being measured. The source preparation is an important step in obtaining the maximum resolution with an alpha spectrometer. The techniques used must be able to produce a very thin and uniform deposit to minimize the energy attenuation of alpha particles. The sources must be stable, and free of liquid, solvent and acid residues that could damage the vacuum camera or the detector. There are three main procedures for source preparation: electro-deposition (Singh et al.,1979), in which case the source is supposed to have a few atomic layers of the alpha emitter and the attenuation is extremely small; evaporation from an aqueous or organic solvent (Talvmtie, 1972; Aggarwal et al., 1985), which is sometimes followed by flaming

under vacuum (Glover, 1984) to remove the organic material; and co-precipitation (Singh et al., 1979 and 1983).

Fig.1 shows an alpha spectrum of a solution containing Ra-226 taken with a Canberra A450-18AM PIPS detector, with $450mm^2$ of active area, 18 keV of alpha resolution, and a background of 6 counts/day. The sample was obtained after the evaporation of a few microliter droplets of the radioactive solution on a metal disk. Despite the small amount of residue left, about $0.1\ mg/cm^2$, the attenuation was still significant as can be seen by the long tails of the alpha peaks. The peak energies of the alpha radionuclides could be easily determined by looking at the peak edge position but the large tails make their quantitative determination difficult.

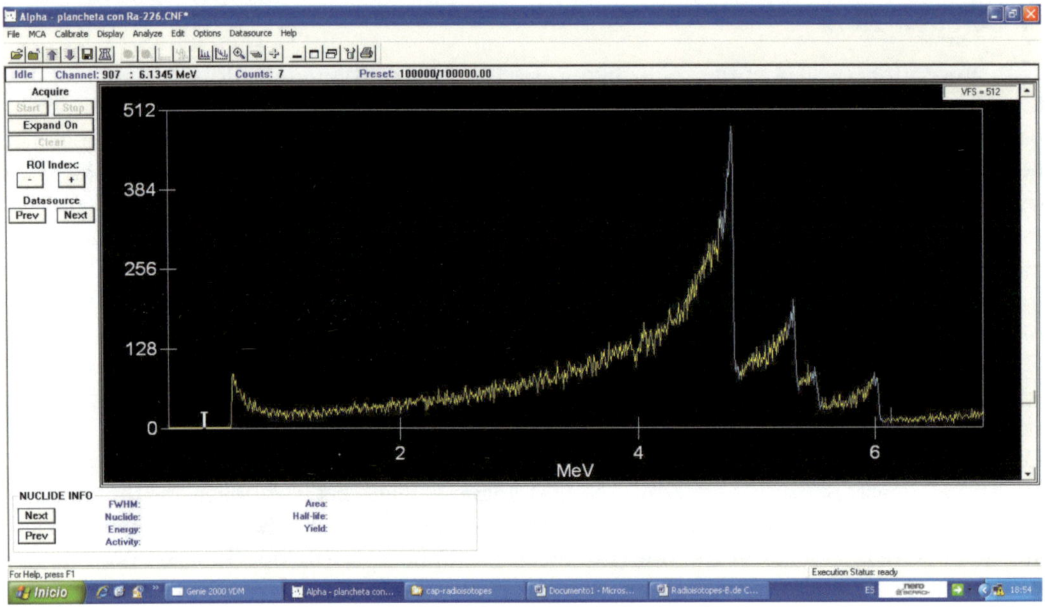

Fig. 1. Alpha spectrum of an evaporated solution of Ra-226 taken with a passivated implanted planar silicon detector. The software employed was Genie-2000, Canberra. The first peak situated to the left of the spectrum corresponds to Ra-226 (4.7 MeV), while subsequent peaks correspond to Po-210 (5.3MeV), Rn-222 (5.49MeV), and Pb-218 (6.0MeV).

Fig.2 shows the spectrum of a mineral sample containing U-238 and U-234. The sample was prepared by evaporation of the organic cocktail on a metal plate. The peaks of U-238 (4.2 MeV) and U-234 (4.7 MeV) are clearly visible but not the U-235 peak (4.5 MeV) that lies between both of them. The energy resolution was worse than in the case of aqueous evaporation because there was more organic residue ($1\ mg/cm^2$). The size of both peaks was the same because the sample contained natural uranium, and the U-238 and U-234 were approximately in radioactive equilibrium. The energy of the alpha peaks could be determined but a quantitative analysis was not possible because of the large peak tails. Chemicals to spread the sample solution can be added prior to its evaporation to obtain more uniform deposition, and the organic deposits burned off before counting, but this causes poor adherence of the sample to the backing and poor resolution. In some cases it is possible to obtain a nearly solid-free deposit source sample. The procedure consists of evaporating the organic or aqueous sample to dryness, then treating with perchloric and nitric acids to oxidize the residual organic matter. The purified ion in aqueous solution is then extracted into thenoyltrifluoroacetone (TTA), deposited onto a stainless steel disc, and evaporated.

Sources are usually prepared by depositing micrograms of the element onto a flat, polished metal disk by electroplating from an aqueous medium. Properly prepared sources weigh less than 50 μg/cm². In this case a resolution of 50-100 keV for 4-7 MeV alpha-particles can be reached. Fig.3 shows the alpha spectrum of a soil sample containing U-238, U-234 , U-235, and U-232 (the yield tracer). A small peak corresponding to U-235 can also be observed between the U-238 and U-234 peaks. The energy resolution of the spectrum enables all the uranium isotopes to be identified. Quantitative analysis can also be performed because the peaks are clearly defined and their area can be easily evaluated. The background is very low and it is possible to reach very small MDAs.

Cooprecipitation offers an alternative to electrodeposition which is too time consuming. The actinides can co-precipitate with small amounts of a rare-earth element carrier. The carrier elements may be precipitated as a fluoride by addition of hydrofluoric acid. The precipitate is filtered onto a 0.1 μm membrane filter which is dried, mounted onto a support backing, and used for alpha spectroscopy. The method is fast, inexpensive, and produces resolution nearly as good as electrodeposition.

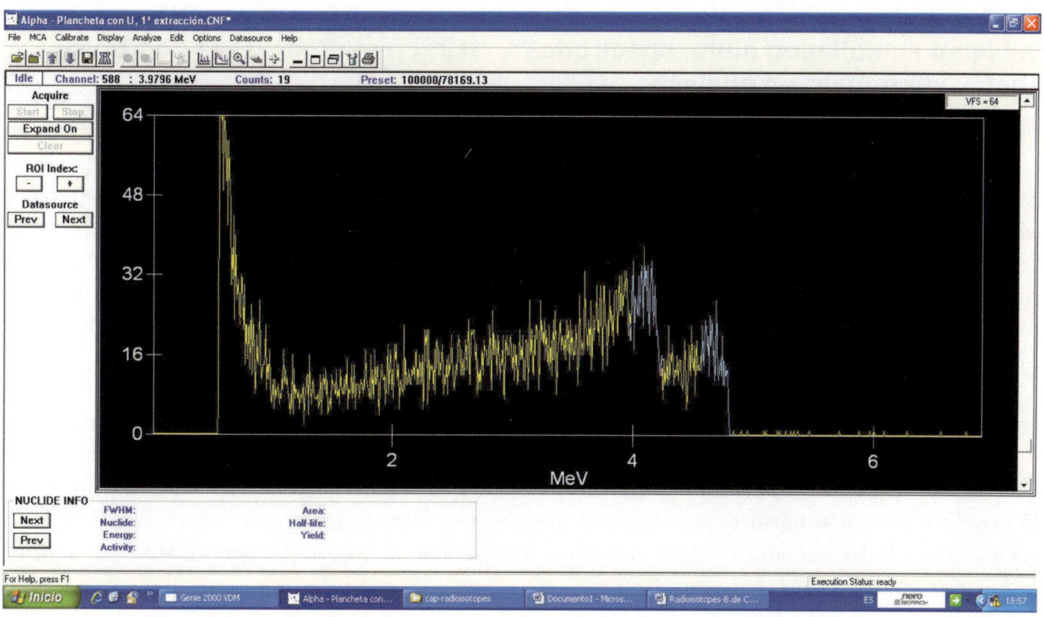

Fig. 2. Alpha spectrum of an evaporated organic extractant containing uranium. The U-238 (4.2 MeV) and U-234 (4.7 MeV) peaks (in blue) are clearly visible but not the U-235 peak (4.5 MeV) that lies between the two of them.

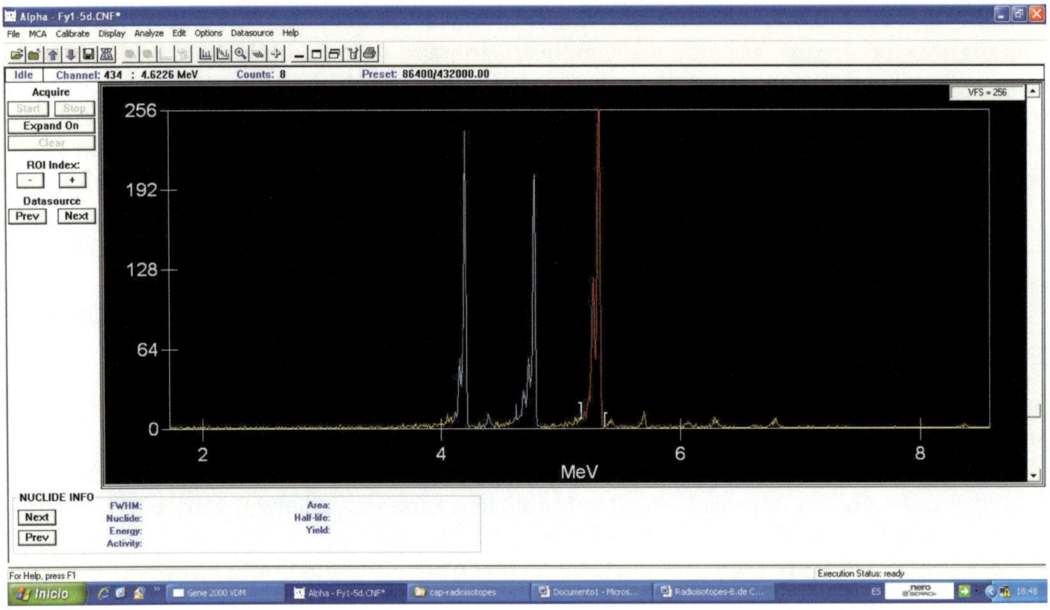

Fig. 3. Alpha spectrum of a soil sample containing U-238 (4.2 MeV), U-234 (4.7 MeV) (blue peaks), and U-232 (5.3 MeV) added as a yield tracer (red peak). A small peak corresponding to U-235 (4.4 MeV) can also be observed between the large peaks of U-238 and U-234.

3. Liquid scintillation alpha spectrometry using fast digital card and pulse shape discrimination analysis

The experimental equipment used in our laboratory to analyse actinides by LS spectrometry consisted of a standard photomultiplier tube with its pre-amplifier output directly connected to a fast digital card installed in a desktop computer. Two coincident photomultipliers were not needed, because the thermal electron noise in the phototubes was below the alpha pulse level. However, electronic noise, when superposed on the signal, could affect final resolution of the energy peaks. Usually, samples are immersed in a liquid to improve the transmission of light to the photomultiplier tube and they are surrounded by a light reflector to increase the efficiency of light collection. In our case the sample was placed directly above the photomultiplier tube. However, it has been observed that the geometry and reflector shape are important if good energy and pulse shape resolution are desired (McKlveen and McDowell, 1975).

The photomultiplier tube was a Saint Gobain RCA XP2412B with an AS07 preamplifier. It is possible to use several types of solid and liquid scintillators by placing them directly above the photomultiplier tube, but it is necessary to keep the phototube in a dark environment to avoid causing damage to the photocathode when a high voltage is applied. In our case the scintillator was the organic extraction cocktail containing the radionuclides of interest.

The typical analytical procedures to treat the solid samples were drying, grinding, and combustion of the sample to remove organic matter. Next, the sample was dissolved in a strong acid solution (nitric, perchloric and hydrofluoric acids). When samples contained silicate material, hydrofluoric acid was required, and repeated evaporation of the sample dissolved in acid was needed to remove the silicates as silicon tetrafluoride. For the general case of samples without significant amounts of thorium or iron a two-stage extraction is a fast, simple separation procedure (Cadieux et al., 1994; McDowell, 19869). A first extraction from 0.8M nitric acid separated U and Pu into the organic phase. After the organic phase was removed, the aqueous phase was adjusted to a pH of 2.5 to 3.0 by the addition of a formate buffer solution. The second extraction is then performed to remove americium and curium. The extractive scintillator mixture contains 120g/l of HDEHP, 180g/l of scintillation grade naphthalene and 4.0g/l of 2-(4'-biphenylyl-6-phenyl-benzoxazole) [PBBO] dissolved in spectroscopic grade p-xylene. After the phases were separated the oxygen was purged to reduce quenching and improve energy resolution. Tests of the two-step extraction on water samples with known actinide activities have shown recoveries higher than 95% with a precision of 2% to 3% (Cadieux et al. 1994). In our case we found recoveries of 95% for the first extraction and about 80% for the second one with a precision of 5%.

PSD analysis is an essential part of the procedure and can be performed with analog or digital electronics. The recent availability of fast digital cards offers the advantage of greater reliability and simplicity of use. This is achieved because it removes the need for the electronic equipment necessary when employing conventional analog procedures, i.e. multichannel analysers (MCAs), amplifiers, coincidence electronics, single-channel analysers, etc. The direct analog-to-digital conversion of the detector signal and the use of software techniques allows to simplify tasks usually performed by the analog amplifier, i.e. electronic noise reduction, the treatment of pile-up pulses, trigger level restoration, and the reduction of ballistic deficit. MCAs are not needed because the digitized signal has all the necessary information to determine the pulse amplitude, which is proportional to the particle energy, and to obtain the energy spectrum. The direct analog-to-digital conversion of the detector signal offers many other possibilities (White and Miller, 1999; Warburton el al., 2000), such as alpha/beta/gamma discrimination by analysis of the electronic pulse shape (PSD), precise detection time determination of pulses by using time stamping information, the implementation of coincidence techniques, and dead-time reduction.

In order to test the accuracy of the procedure a standard liquid containing known activities of uranium (U-238, U-234, U-235), Am-241, and Sr-90 was employed for the analysis. The results from the first extraction, containing U-238, U-234, and U-235, and from the second extraction, containing Am-241, were place in two glass vials to be measured. The vials were situated directly in front of the photomultiplier tube to allow the maximum entrance of the light emitted by the sample. Other arrangements, adding reflectors to the vial or immersing it in liquid to improve light transmission, were also studied, but it was observed that different light paths produced a broadening of the energy peak and these procedures were abandoned.

The electronic signal at the pre-amplifier output was connected to a digital card (AlazarTechs ATS330 PCI) with the following technical specifications: two channels independently sampled at 12-bit resolution and a sampling rate of 50 million samples per second [Ms/sec], with multiple possibilities for triggering, multiple records and time stamping. The two channels are useful if coincidence measurements are performed with two independent electronic circuits and detectors. The device had been previously tested with different types of scintillation cocktails containing radioactive sources of known radioactivity and different emitting radionuclides (Ra-226, Sr-90, Am-141), to establish their pulse shape characteristics, the optimum bias voltage, and the card parameters (sampling rate, samples per pulse, triggering thresholds, voltage range, input impedance, and noise level). Those hardware conditions determined the optimum parameters for particle detection when the organic liquid scintillator was coupled with the photomultiplier tube.

The digital card was supplied with basic software to set up the acquisition hardware. It contained a software development kit to allow full control of the card. In order to develop a complete system able to perform alpha/beta/gamma spectrometry and coincidence experiments, many other functions were developed and incorporated in the original software. The system developed acts like a digital oscilloscope that visualizes each individual pulse, and also like a conventional MCA. Several selection functions were incorporated, such as voltage amplitude, record length, sampling rate, trigger level, input impedance. The characteristics incorporated into the MCA were identification of a region of interest, peak area determination, calibration of energies, dead-time determination, and loading and recording of spectra.

Several algorithms were also developed to perform PSD analysis. They separated the signals that came from different scintillator types or were caused by different type of particles, allowing simultaneous acquisition of alpha/beta/gamma spectra from a radioactive sample. In addition, AND and OR relations could be used during acquisition to obtain a compound spectrum from the sample.

Fig. 4 shows the spectrum of a soil sample containing uranium. The spectrum was obtained with the organic cocktail from the first chemical extraction. It was difficult to differentiate the peaks for U-238 (4.2 MeV) and U-234 (4.7 MeV), because the resolution is 10% (0.5 MeV), larger than the resolution (5%) reported by other researchers using specific instrumentation (McKlveen and McDowell, 1984; Cadieux, 1990) or employing avalanche photodiodes (Reboli et al., 2005). Different procedures were used to reduce electronic noise (i.e. elimination of noise harmonics by using fast Fourier transform), but the best energy resolution reached was 8%.

Fig. 5 shows the spectrum of a sample containing Ra-226. The peaks for Ra-226 (4.7 MeV) and its descendants, Rn-222 (5.49 MeV), Pb-218 (6.0 MeV), and Po-214 (7.69 MeV) interfered and formed a unique peak. The small peak to the right corresponded to Po-214 (7.69 MeV). The sample was immersed in an aqueous medium containing Packard alpha/beta Ultima-Gold scintillator and a few drops of Ultima-Gold F scintillator to improve the energy resolution which was 8%.

The counting efficiency was 80%. Efficiencies near 100% could probably be reached with better energy resolution and light collection. A disperse background of counts was observed to the left of the peaks. That was attributed to pulses of lower amplitude produced by more attenuated light paths. The background was measured with a sample of the pure cocktail extractor. Its value was 10^{-4} counts per minute [cpm] below the energy peak area. It could provide an MDA close to 10 Becquerels per kilogram [Bq/kg] in the case of a ten-gram sample and a measurement time of one day, assuming the absence of energy interference with other radionuclides. In any case, the results obtained in terms of energy resolution and MDA are very similar to those obtained by using a Packard TRICARB LSC 2900TR.

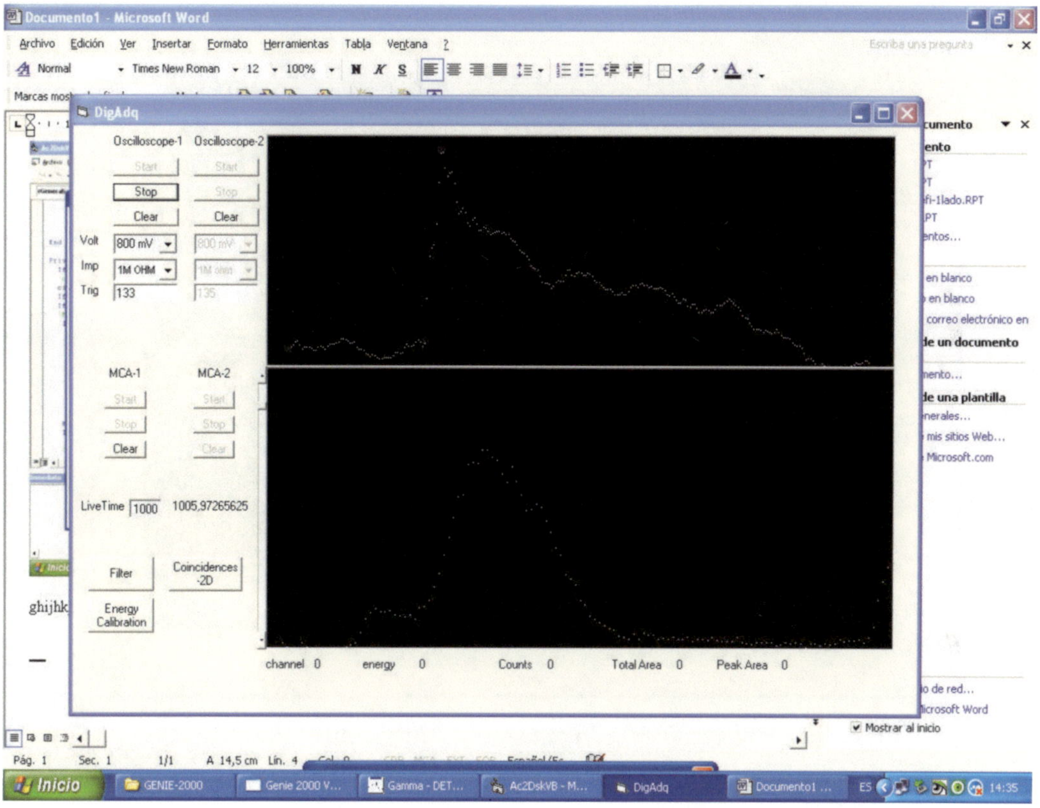

Fig. 4. Alpha spectrum of a uranium sample obtained by liquid-liquid extraction with organic solvents. The energy resolution was about 10% (0.5 MeV) and the peaks of U-238 (4.2 MeV) and U-234 (4.7 MeV) interfered. The upper part of the figure shows the shape of the electronic pulse at the preamplifier output. Each point of the pulse graph corresponds to 20 nsec.

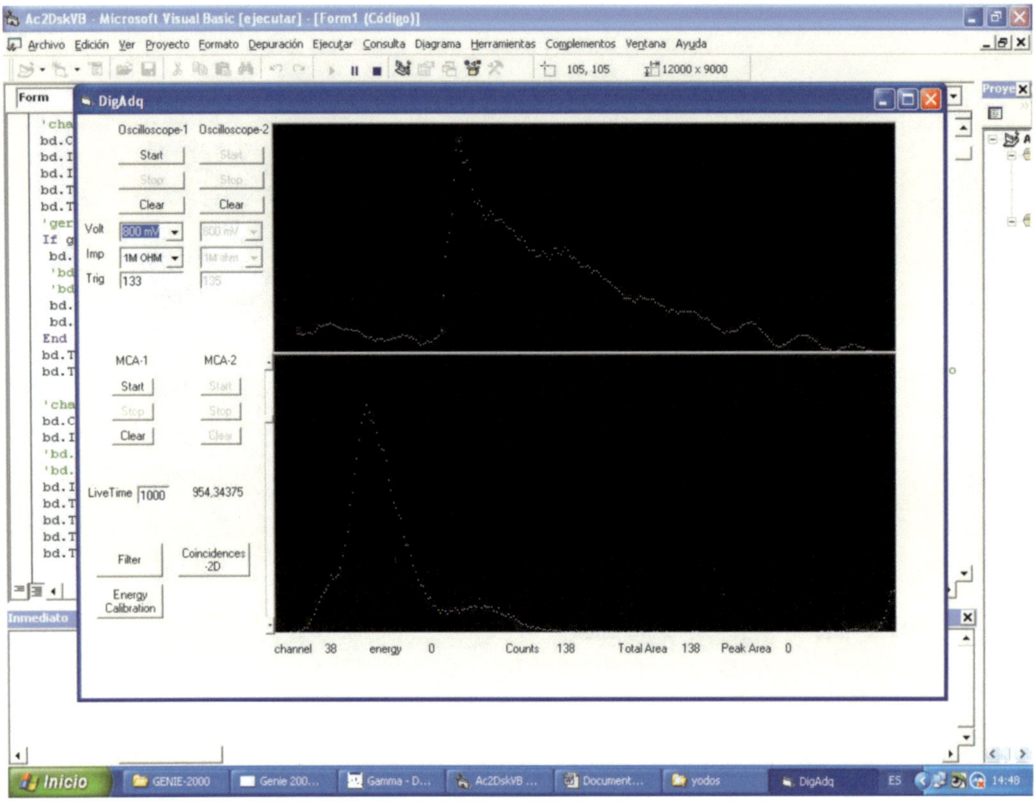

Fig. 5. Alpha spectrum of a radium sample in an aqueous medium using alpha/beta Ultima-Gold scintillator. The large peak corresponds to Ra-226 (4.7 MeV) and its descendents Po-210 (5.3 MeV), Rn-222 (5.49 MeV), and Pb-218 (6.0 MeV). They interfered and formed a unique peak. The small peak to the right is Po-214 (7.69 MeV). The upper part of the figure shows the shape of the electronic pulse at the preamplifier output. Each point of the pulse graph corresponds to 20 nsec.

4. Coincidence alpha X-ray in actinides using fast digital cards and pulse shape discrimination analysis

In order to increase the sensitivity of the procedure and reduce energy interference, gamma pulses emitted by actinides were measured in coincidence with alpha pulses. Many actinides decay by alpha particle emission in coincidence with low energy gamma/X-rays. When the alpha energies of two radionuclides interfere because their energy separation is below the energy resolution limit of LS spectroscopy, the coincidence procedure allows to obtain additional information by measuring the energy of the gamma/X-ray radiation coincident with the alpha emission.

The coincidence procedure is easy to implement by using a two channel digital card. The electronic procedure consists of recording on the second channel the X or gamma radiation coinciding with the alpha particles recorded on the first one. Using this method, it is possible to obtain X and gamma spectra with a significant reduction in background and the typical lead shield is not needed (de Celis, de la Fuente et al., 2007; de la Fuente, de Celis et al., 2008).

The detector employed for gamma radiation was a low-energy germanium, Canberra GL1515 with a 0.5mm Be window and a FWHM resolution of 304 eV at 5.9 keV and 551 eV at 122 keV. The detector area was 1500 square millimetres and its thickness 15 mm, which guarantees a high gamma-collection efficiency and a low energy resolution.

Tests were conducted with an Am-241 liquid source of known activity dissolved in 1ml of liquid scintillator. Am-241 emits 5.4 MeV alpha particles in coincidence with 59 keV gamma-rays. A plastic vial containing the radioactive source was placed directly over the photomultiplier tube and the germanium detector was situated above the sample. Small vials about 1cm in height were fabricated to reduce the distance between the liquid sample and the germanium detector. To implement the procedure in the digital card, the alpha trigger of the first channel is set "on" and the gamma trigger of the second channel is set to "off", in order to record gamma/X ray radiation only when the alpha channel is active. The pulses from the two detectors were stored jointly with the detection times and coincidences are identified by comparing the arrival time of the pulses proceeding from the liquid detector (alpha particles) and germanium detector (gamma/X rays). A delay time between pulses smaller than 1μs was considered to indicate a coincidence event.

Experimental tests were performed with a soil sample of known activity, previously reduced to ash and dissolved in nitric acid. The sample contained Am-241 and Pu-238 with activities below 1 Bq/kg. In conventional analysis most of these radionuclides are only detectable if radiochemical separation is combined with alpha spectrometry using PIPS detectors. With this procedure it was possible to identify and determine quantitatively the Am-241 activity. The MDA of the procedure was 0.01Bq/kg, which is very low compared with the 10 Bq/kg recorded when using directly gamma spectrometry and lower than the 0.1 Bq/kg achieved when using PIPS detectors. This is due to the high background reduction of the coincidence procedure. Other alpha/X-ray radionuclides, such as Pu-238, were not identified because of the efficiency reduction in the 10-20 keV region by the Ge absorption edge and X-ray self-absorption in the liquid sample. Tests to increase the MDA at these energies could be carried out with Si(PIN) detectors or using a phosphor sandwich [phoswich] detector with a higher resolution scintillator. Fig. 6 shows a coincidence event (alpha/gamma) and the coincidence gamma spectrum of a soil sample. The gamma pulse was delayed with respect to the alpha pulse because an analog amplifier with a shaping constant of 1μs was employed to reduce electronic noise and obtain a better resolution in the germanium detector. Electronic noise reduction in the germanium chain could also be achieved using digital procedures.

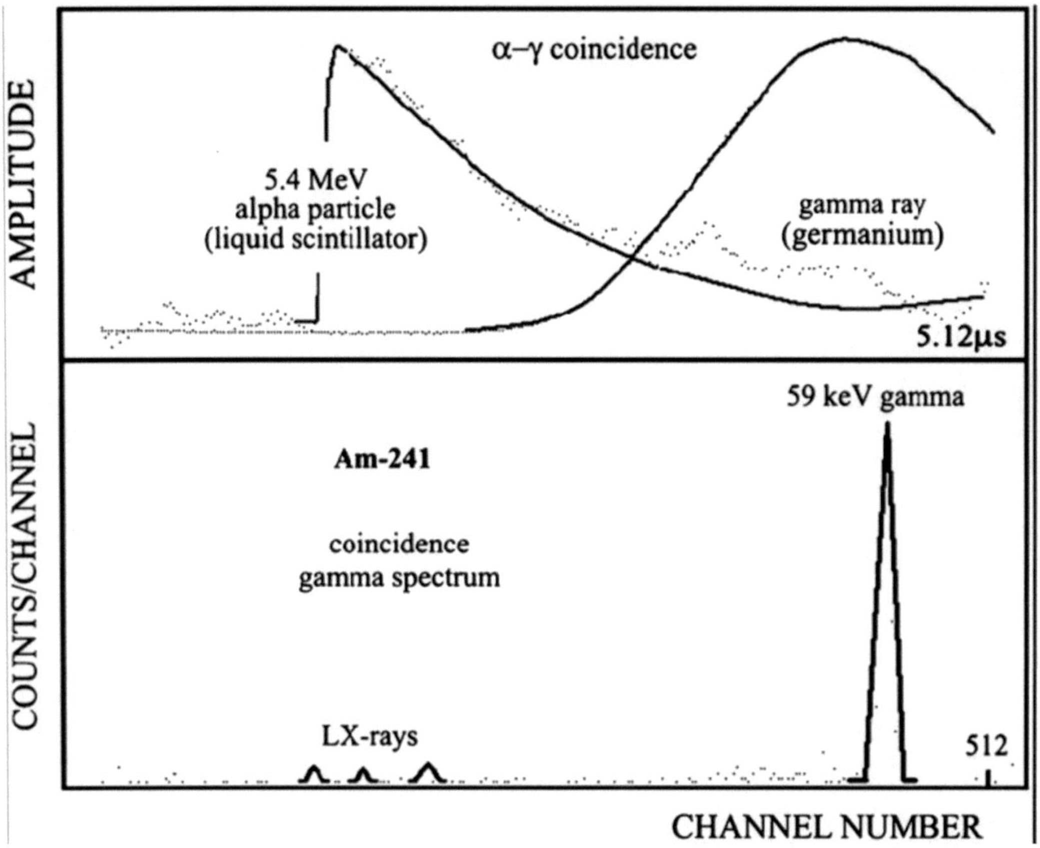

Fig. 6. Am-241 alpha/gamma coincident pulses (above) and coincident gamma spectrum (below) using liquid scintillation and a low energy germanium detector.

The procedure has the advantage of using the same geometry and chemical preparation for samples and standards, reducing the need for corrections for quenching. Interferences by beta/gamma coincidences are easily avoided, as beta particle pulses detected in the liquid scintillator are usually of smaller amplitude than alpha particle pulses. However, PSD, using the different scintillation decay time of alpha and beta particles, is intended for use in future. The use of a digital PSD system offers many other advantages, particularly the fact that the background of the detection system, which ultimately determines the sensitivity of the procedure, can be further reduced by studying the shape and detection time of the coincident particles, excluding those pulses which do not meet certain conditions.

5. Conclusions

Analytical procedures used to determine actinides in environmental samples were reviewed. The standard procedure using PIPS detectors is the most often used. It allows very small MDA to be reached due to the high resolution and very low backgrounds of these detectors. However, the analysis time is long (several days) and impossible to apply when many samples need to be processed or it is necessary to know the results in record time. LS spectrometry overcomes this inconvenience but at the expense of a poor resolution which makes it difficult to eliminate some interferences and thus determine quantitatively certain radionuclides. The LS analysis time is very short and improvements could be made to increase the light emission from the cocktail, reducing quenching and improving energy resolution. Future photodiode detectors could also help to improve energy resolution. A digital system is the natural way to implement many of these new developments due to its advantages in treating the digitized pulses. The MDA is ultimately dependent on the background of the technique. In our case it was possible to reduce the background by eliminating electronic signals which did not correspond with the correct pulse shape and also random electronic noise using standard digital procedures. Discrimination between different types of particles could be implemented by measuring the pulse light decay time without the need to resort to analogue electronic equipment.

Coincidence experiments could be easily set up using digital cards with two or more channels. The coincidence time could be determined with resolution times between 20 nsec and 1 nsec for cards of 50 Ms/sec to 1 Gs/sec sampling rates. Two-parameter studies using coincidence and 2-D diagrams can help to determine certain types of actinides. The technique is usually applied with analogue electronic equipment to determine radio-xenons resulting from nuclear subterranean tests but the setup presents more difficulties than when using digital equipment. In the case of actinide identification the technique could be applied to eliminate interferences between radioisotopes of similar alpha energies but different gamma emissions.

In some cases, for instance, the analysis of certain plutonium isotopes (Pu-239 and Pu-240), the alpha energies are similar and also equal the X-ray energy emissions. In this case conversion electron spectrometry with PIPs detectors using the same electro-deposited samples prepared for alpha spectrometry could help in obtaining a quantitative determination of both radioisotopes.

X-ray spectrometry could also be used to determine the total activity of the different radioisotopes of an element by using a high-resolution germanium detector in coincidence with LS spectrometry. The coincidence experiment is necessary to eliminate the background of the germanium detector and improve the MDA limit. Si (PIN) detectors, cooled by the Peltier effect, could offer a simple alternative if detectors of large area and small electronic noise are built in the future. Modern Si (PIN) detectors can reach energy resolutions of 125 eV at 5 keV with very low backgrounds but at the cost of a low counting efficiency because of their small detector area.

6. Acknowledgments

This research was supported by the Consejo de Seguridad Nuclear (CSN) of Spain, and the REM (Red de Estaciones de Muestreo) and PVRAIN (Plan de Vigilancia Radiológica Ambiental en la Central Nuclear de Sta. Mª Garoña) programmes.

7. References

Abuzwrda M., Abouzreba S., Almedhem B., Zolotarev Yu.A. and Komarov N.A. (1987). Evaluation of a photoelectron-rejecting alpha liquid-scintillation, *Radioanal. Nucl. Chem.,* 111 (1987) 11.

Aggarwal S.K., Chourasiya G., Duggal R.K., Singh C.P., Rawat A.S., and H.C. Jain. (1985). A comparative study of different methods of preparation of sources for alpha spectrometry of plutonium. *Nuclear Instruments and Methods in Physics Research,* Section A, Volume 238, Issues 2-3,1 August 1985, Pages 463-468.

Cadieux, J.R. (1990). Evaluation of a photoelectron-rejecting alpha liquid-scintillation (PERALS) spectrometer for the measurement of alpha-emitting radionuclides. *Nuclear Instruments and Methods in Physics Research* Section A: Accelerators, Spectrometers, Detectors and Associated Equipment, Volume 299, Issues 1-3, 20 December 1990, Pages 119-122.

Cadieux, J.R., Clark S., Fjeld R.A., Reboul S. and Sowder A. (1994). Measurement of actinides in environmental samples by photon-electron rejecting alpha liquid scintillation. *Nuclear Instruments and Methods in Physics Research* Section A: Accelerators, Spectrometers, Detectors and Associated Equipment, Volume 353, Issues 1-3, 30 December 1994, Pages 534-538.

de Celis, B., de la Fuente, R., Williart, A., de Celis Alonso, B., (2006). Coincidence measurements in a/b/g spectrometry with phoswich detectors using digital pulse shape discrimination analysis. In: *ISRP10* Symposium, Coimbra, Portugal, 17-22 September 2006.

de Celis, B., de la Fuente, R., Williart, A. and de Celis Alonso, B. (2007). Coincidence measurements in α/β/γ spectrometry with phoswich detectors using digital pulse shape discrimination analysis, *Nuclear Instruments and Methods in Physics Research Section A: Accelerators, Spectrometers, Detectors and Associated Equipment,* Volume 580, Issue 1, 21 September 2007, Pages 206-209. ISSN 0168-9002.

de la Fuente R., de Celis B., del Canto V., Lumbreras J.M., de Celis Alonso B., Martín-Martín A. and Gutierrez-Villanueva, J.L. (2008). Low level radioactivity measurements with phoswich detectors using coincident techniques and digital pulse processing analysis *Journal of Environmental Radioactivity,* Volume 99, Issue 10, October 2008, Pages 1553-1557

Glover, K. M. (1984). Alpha-particle spectrometry and its applications. *Int. J. Appl. Radiat. Isot.* 35, 239, 1984.

Gomez Escobar, V., Vera Tome, F., Lozano and J.C., Martin Sanchez, A., (1998). Extractive procedure for uranium determination in water samples by liquid scintillation counting, *Appl. Radiat. Isot.* 49, 875-883.

Greeman, D.J. and Rose, A.W. (1990). Form and behaviour of radium, uranium and thorium in central Pennsylvania soils derived from dolomite. *Geophys. Res. Lett.* 17, 833-836., 1990.

Hanschke, T. (1972). High-resolution alpha spectroscopy with liquid scintillators by optimisation of the geometry, Ph.D. Thesis, Hannover Technical University, 1972.

Holm E. , (1984). Review of Alpha-Particle Spectrometric Measurements of Actinides, *Int. J Appl . Radiat . Isot .* 35 285.

Lucas L.L . (1980). The Standardization of Alpha-Particle Sources, ASTM STP 698 (1980) 342.

McDowell W.J. and McDowell L.B., *Liquid Scintillation Alpha Spectrometry,* CRC Press, Boca Raton, FL, USA, 1994.

McDowell W.J., Nucl. Sci. Ser. on Radiochem. Tech. NAS-NS 3116 (1986).

McDowell, W.J. (1980). Alpha liquid scintillation counting: Past, present, and future *The International Journal of Applied Radiation and Isotopes,* Volume 31, Issue 1, January 1980, page 23.

McKliven, J.W. and McDowell, W.J. (1984). Liquid scintillation alpha spectrometry techniques, *Nuclear Instruments and Methods in Physics Research,* Volume 223, Issues 2-3, 15 June 1984, Pages 372-376.

Miglio J.J. and Willis L.C ., (1988). Simultaneous liquid scintillation determination of ^{239}Pu and^{241}Am in tissue, *J. Radioanal. Nucl . Chem.* 12 3 (1988) 517.

Reboli, A., Aupiais, J. and Mialocq, J.C. (2005). Application of large area avalanche photodiodes for alpha liquid scintillation counting, *Nuclear Instruments and Methods in Physics Research* Section A: Accelerators, Spectrometers, Detectors and Associated Equipment, Volume 550, Issue 3, 21 September 2005, Pages 593-602.

Sarin, M.M., Krishnaswami, S., Somayajulu, B.L.K. and Moore, W.S. (1990). Chemistry of uranium, thorium and radium isotopes in the Ganga-Brahmaputra river system. *Geochim. Cosmochim. Acta* 5, 1990.

Singh N.P. , Ibraim S.H., Cohen N. and Wrenn M.E.,(1979). Determination of plutonium in sediments by solvent extraction and α-spectrometry, *Anal. Chem.* 50, 357 (1979) Pages 265-274.

Singh N.P. and Wrenn M.E. (1983). Determination of alpha-emitting uranium isotopes in soft tissues by solvent extraction and alpha-spectrometry, *Talanta,* Volume 30, Issue 4, April 1983, Pages 271-274.

Singh N.P., McDonald and Wrenn, E. (1988). Determinations of actinides in biological and environmental samples,*Science of The Total Environment,*Volume 70, March 1988, Pages 187-203.

Talvitie N.A . (1972). Electrodeposition of actinides for alpha spectrometric determination. *Anal .Chem.* 44 (1972) 280-283.

Warburton, W.K., Momayezi, M., Hubbard-Nelson, B., Skulski, W., (2000). Digital pulse processing: new possibilities in nuclear spectroscopy. *Applied Radiation and Isotopes* 53, 913-920. 2000.

White, T. and Miller, W. (1999). A triple crystal phoswich detector with digital pulse shape discrimination for alpha/beta/gamma spectroscopy, *Nuclear Instruments and Methods in Physics Research* Section A: Accelerators, Spectrometers, Detectors and Associated Equipment 422, 144-147.

8

The Newly Calculations of Production Cross Sections for Some Positron Emitting and Single Photon Emitting Radioisotopes in Proton Cyclotrons

E. Tel[1], M. Sahan[1], A. Aydin[2], H. Sahan[1], F. A.Ugur[1] and A. Kaplan[3]
[1] *Faculty of Arts and Science, Osmaniye Korkut Ata University,*
[2] *Faculty of Arts and Science, Kirikkale University*
[3]*Faculty of Arts and Science, Süleyman Demirel University*
Turkey

1. Introduction

Nowadays, radioisotopes are produced using both nuclear reactors and cyclotrons. Especially, the induced by intermediate and high energy protons nuclear reactions are very important because of a wide range technical applications. These reactions are required for advanced nuclear systems, such as spallation reaction for production of neutrons in spallation neutron source (capable of incinerating nuclear waste and producing energy), high energy proton induced fission for the radioisotope production alternatives etc. [1,2]. By using the intermediate proton induced reactions, we can directly produce radionuclides used in medicine and industry.

In the last decade, a big success has been provided on production and usage of the radionuclides. The radioisotopes obtained from using charged particles (proton, deuteron, alpha etc.) play an important role in medical applications [3-6]. A medical radioisotope can be classified as a diagnostic or a therapeutic radionuclide, depending on its decay properties. Radionuclides are used in diagnostic studies via emission tomography, i.e. Positron Emission Tomography (PET), Single Photon Emission Computed Tomography (SPECT), and Endoradiotheraphy (internal therapy with radio nuclides). In general, the diagnostic radioisotopes can also be classified into two groups; namely β^+–emitters (^{11}C, ^{13}N, ^{15}O, ^{18}F, ^{62}Cu, ^{68}Ga, etc.) and γ - emitters (^{67}Ga, ^{75}Se, ^{123}I, etc.). The use of positron emitting radioisotopes such as ^{11}C, ^{13}N, ^{15}O, and ^{18}F together with PET offers a highly selective and quantitative means for investigating regional tissue biochemistry, physiology and pharmacology [7]. The positron emitting nuclei which are neutron deficient isotopes are important for PET studies. Positrons annihilate with electrons emitting two photons ($E\gamma$=511 keV) in opposite direction. Most of the positron emitters are still being studied in terms of their applicability for diagnostic purposes. PET has been developing with the increasing number of clinical facilities raising interest in the use of PET in routine practice [8,9].

In the radioisotope production procedure, the nuclear reaction data are mainly needed for optimization of production rates. This process involves a selection of the projectile energy range that will maximize the yield of the product cross section for nuclear reaction and minimize that of the radioactive impurities [5,6]. The total cross section of production yields are also important in accelerator technology from the point of view of radiation protection safety. The nuclear reaction calculations based on standard nuclear reaction models can be helpful for determining the accuracy of various parameters of nuclear models and experimental measurements. Today, experimental cross-sections are available in EXFOR file [7]. The theoretical calculations of production rates for different medical nucleus reactions were calculated in statistical equilibrium (compound) and pre-equilibrium model in literature [10-14].

In this study, the newly calculations of proton cyclotron production cross sections in some PET , SPECT and others used in medical applications radioisotopes used in medical applications were investigated in a range of 5–100 MeV incident proton energy range. Excitation functions for pre-equilibrium calculations were newly calculated by using hybrid model, geometry dependent hybrid (GDH) model. The reaction equilibrium component was calculated with a traditional compound nucleus model developed by Weisskopf-Ewing (W-E) model [16]. We have investigated the optimum energy range for proton cyclotron production cross sections. We have presented the decay data taken from the NUDAT database [23]. Calculated results were also compared with the available excitation functions measurements in EXFOR file [7]. The optimum energy range and the decay data for the investigated radionuclides are given in Table 1.

2. The nuclear reaction cross-section calculations

Calculations based on nuclear reaction models play an important role in the development of reaction cross sections [15]. For many years, it has been customary to divide nuclear reactions into two extreme categories. Firstly, there are very fast, direct reactions which on a time scale comparable to the time ($\cong 10^{-22}$ s) necessary for the projectile to traverse a nuclear diameter, involve simple nuclear excitations, and are non-statistical in nature. Secondly, there are equilibrium nucleus reactions which occur on a very much longer time scale ($\cong 10^{-16}$ to 10^{-18}s) where emissions can be treated by the nuclear statistical model. This second process can be described adequately with equilibrium nucleus theories developed by the Weisskopf-Ewing (W-E) [16] and Hauser-Feshbach [17]. Equilibrium nucleus wave function is very complicated, involving a large number of particle-hole excitations to which statistical considerations are applicable. The spectra of the emitted particles of equilibrium nucleus are approximately Maxwellian, and angular distributions of emitted particles are symmetric about 90 degrees. During the nineteen-fifties and sixties, evidence accumulated suggesting that in some nuclear reactions, it is not possible to understand all emission processes in terms of equilibrium nucleus and direct processes. This reaction is known as pre-equilibrium reaction. The pre-equilibrium reactions occur on time scale about 10^{-18} to 10^{-20} s. Deviations from a Maxwellian shape for the emission spectra were observed for intermediate to high emission energies, with the theory under predicting data. The first developments were made to understand these observations by Griffin [18], who proposed the pre-equilibrium 'exciton model'. Pre-equilibrium processes are important mechanisms in nuclear reactions induced by light projectiles with incident energies above about 8-10 MeV. After Griffin introduced the exciton model, a series of semi-classical models [19-21] of varying complexities have been developed for calculating and evaluating particle emissions in the continuum. More recently, researchers have formulated several quantum–mechanical reaction theories [22] that are based on multi-step concepts and in which statistical evaporation at lower energies is connected to direct reactions at higher energies. The hybrid model for pre-compound decay is formulated by Blann [19] as

$$\frac{d\sigma_v(\varepsilon)}{d\varepsilon} = \sigma_R P_v(\varepsilon)$$

$$P_v(\varepsilon)d\varepsilon = \sum_{\substack{n=n_0 \\ \Delta n=+2}}^{\bar{n}} [\ {}_n\chi_v N_n(\varepsilon,U)/N_n(E)]\, g\, d\varepsilon[\lambda_c(\varepsilon)/(\lambda_c(\varepsilon)+\lambda_+(\varepsilon))]D_n \quad (1)$$

where $P_v(\varepsilon)d\varepsilon$ represents number of particles of the ν type (neutron or proton) emitted into the unbound continuum with channel energy between ε and $\varepsilon + d\varepsilon$. The quantity in the first set of square brackets of Eq.(1) represents the number of particles to be found (per MeV) at a given energy (with respect to the continuum) for all scattering processes leading to an "n" exciton configuration. It has been demonstrated that the nucleon-nucleon scattering energy partition function Nn(E) is identical to the exciton state density ρn(E). The second set of square brackets in Eq. (1) represents the fraction of the ν type particles at energy, which should undergo emission into the continuum, rather than making an intra-nuclear transition. The Dn represents the average fraction of the initial population surviving to the exciton number being treated. Early, $\lambda_c(\varepsilon)$ is emission rate of a particle into the continuum with channel energy ε and $\lambda_+(\varepsilon)$ is the intranuclear transition rate of a particle. Nn (ε ,U) is the number of ways. comparisons between experimental results, pre-compound exciton model calculations, and intra-nuclear cascade calculations indicated that the exciton model gave too few pre-compound particles and that these were too soft in spectral distribution for the expected initial exciton configurations. The intra-nuclear cascade calculation results indicated that the exciton model deficiency resulted from a failure to properly reproduce enhanced emission from the nuclear surface. In order to provide a first order correction for this deficiency the hybrid model was reformulated by Blann and Vonach [24]. In this way the diffuse surface properties sampled by the higher impact parameters were crudely incorporated into the pre-compound decay formalism, in the geometry dependent hybrid (GDH) model. The differential emission spectrum is given in the GDH model as

$$\frac{d\sigma_v(\varepsilon)}{d\varepsilon} = \pi \lambda\!\!\!^{-2} \sum_{l=0}^{\infty}(2l+1)T_l\, P_v(l,\varepsilon) \quad (2)$$

where $\lambda\!\!\!^{-}$ is reduced de Broglie wavelength of the projectile and T_l represents transmission coefficient for l th partial wave ℓ is orbital angular momentum in n unit $\hbar$.

3. Results and discussion

This work describes new calculations on the excitation functions of $^{18}O(p,n)^{18}F$, $^{57}Fe(p,n)^{57}Co$, $^{57}Fe(p,\alpha)^{54}Mn$, $^{68}Zn(p,2n)^{67}Ga$, $^{68}Zn(p,n)^{68}Ga$, $^{93}Nb(p,4n)^{90}Mo$, $^{112}Cd(p,2n)^{111}In$, $^{127}I(p,3n)^{125}Xe$,$^{133}I(p,6n)^{128}Ba$ and $^{203}Tl(p,3n)^{201}Pb$ reactions carried out in the 5-100 MeV proton energy range. The pre-equilibrium calculations involve the hybrid model and the geometry dependent hybrid (GDH) model. Equilibrium reactions have been calculated according to Weisskopf-Ewing (W-E) model. The ALICE/ASH code was used in the calculations of all models described above. The ALICE/ASH code is an advanced and modified version of the ALICE codes. The modifications concern the implementation in the code of models describing the pre-compound composite particle emission, fast γ-emission, different approaches for the nuclear level density calculation, and the model for the fission fragment yield calculation. The ALICE/ASH code can be applied for the calculation of excitation functions, energy and angular distribution of secondary particles in nuclear reactions induced by nucleons and nuclei with the energy up to 300 MeV. The initial exciton number as n_o =3 and the exciton numbers (for protons and neutrons) in the calculations for proton induced reactions as,

$$ {}_3X_p = 2\frac{(\sigma_{pn}/\sigma_{pp})N + 2Z}{2(\sigma_{pn}/\sigma_{pp})N + 2Z}, \quad {}_3X_n = 2 - {}_3X_p \tag{3} $$

where σ_{xy} is the nucleon-nucleon interaction cross-section in the nucleus. Z and N are the proton and neutron numbers, respectively, of the target nuclei. The ratio of nucleon-nucleon cross-sections calculated taking into account to Pauli principle and the nucleon motion is parameterized

$$ \sigma_{pn}/\sigma_{pp} = \sigma_{np}/\sigma_{nn} = 1.375\times10^{-5}T^2 - 8.734\times10^{-3}T + 2.776 \tag{4} $$

where T is the kinetic energy of the projectile outside the nucleus. The super-fluid model [26] has been applied for nuclear level density calculations in the ALICE/ASH code. The results of the calculations are plotted in [Fig. 1], [Fig. 2], [Fig. 3], [Fig. 4], [Fig. 5], [Fig. 6], [Fig. 7], [Fig. 8], [Fig. 9], and [Fig. 10]. Calculated results based on hybrid model, geometry dependent hybrid model and equilibrium model have been compared with the experimental data. The experimental taken from EXFOR [7] are shown with different symbols (such as +, o, Δ) symbols in all figures. The results are given below.

3.1 $^{18}O(p,n)^{18}F$ reaction process

The calculation for the excitation function of $^{18}O(p,n)^{18}F$ reaction has been compared with the experimental values in Fig.1. In general, the hybrid model calculations are the best agreement with the measurements of $^{18}O(p,n)^{18}F$ reaction up to 30 MeV the incident proton energy. The equilibrium W-E model calculations are only in agreement with in energy region lower than 20 MeV. The optimum energy range for production of ^{18}F is E_p= 10 → 5 MeV.

3.2 $^{57}Fe(p,n)^{57}Co$ reaction process

The calculation cross section of $^{57}Fe(p,n)^{57}Co$ reaction has been compared with the experimental values in Fig.2. The hybrid model calculations are the best agreement with the measurements of Levkovskij [7] at above the 20 MeV energie regions. The W-E model calculations are only in agreement with in energy region lower than 20 MeV. The GDH model calculations are a little higher than the measurements of Levkovskij [7] at above the 20 MeV for incident proton energies. The optimum energy range for production of ^{57}Co is E_p= 15 → 5 MeV.

3.3 $^{57}Fe(p,\alpha)^{54}Mn$ reaction process

The calculated $^{57}Fe(p,\alpha)^{54}Mn$ reaction has been compared with the experimental values in Fig.3. The hybrid model and GDH model calculations are in good agreement with the measurements of Levkovskij [7] at above the 20 MeV energie regions. The W-E model calculations are good in agreement with in energy region lower than 20 MeV. The optimum energy range for production of ^{54}Mn is E_p= 20 → 10 MeV.

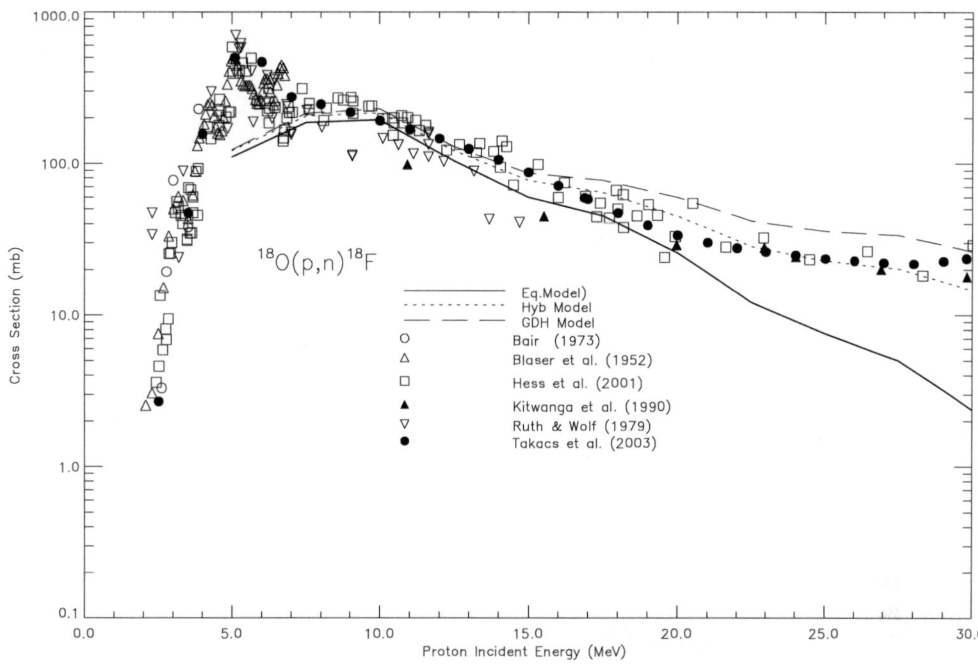

Fig. 1. The comparison of the calculated cross section of $^{18}O(p,n)^{18}F$ reaction with the values reported in Ref. [7].

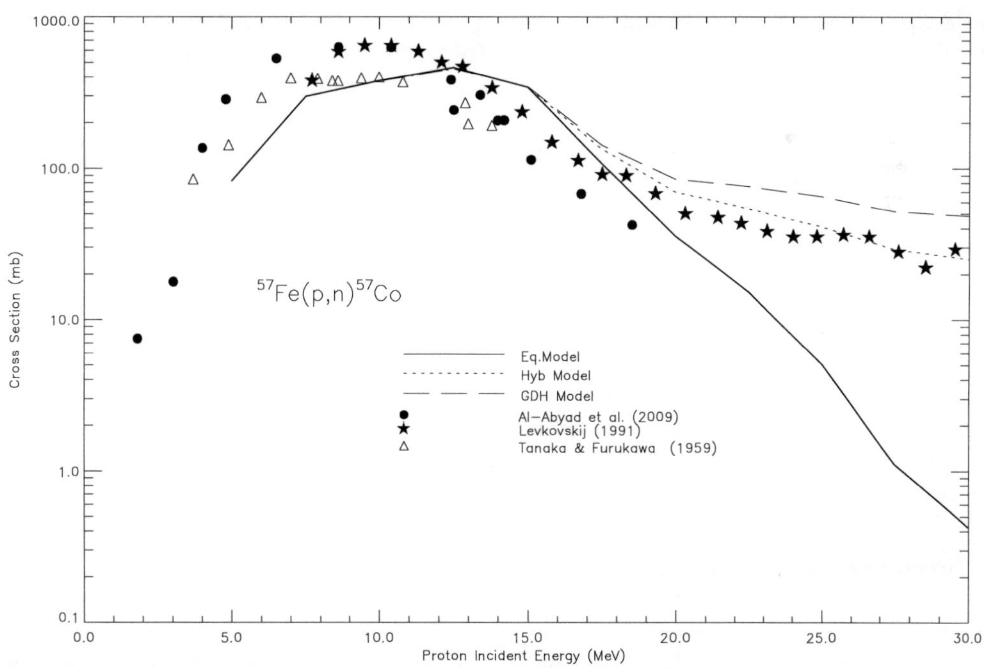

Fig. 2. The comparison of the calculated cross section of $^{57}Fe(p,n)^{57}Co$ reaction with the values reported in Ref. [7].

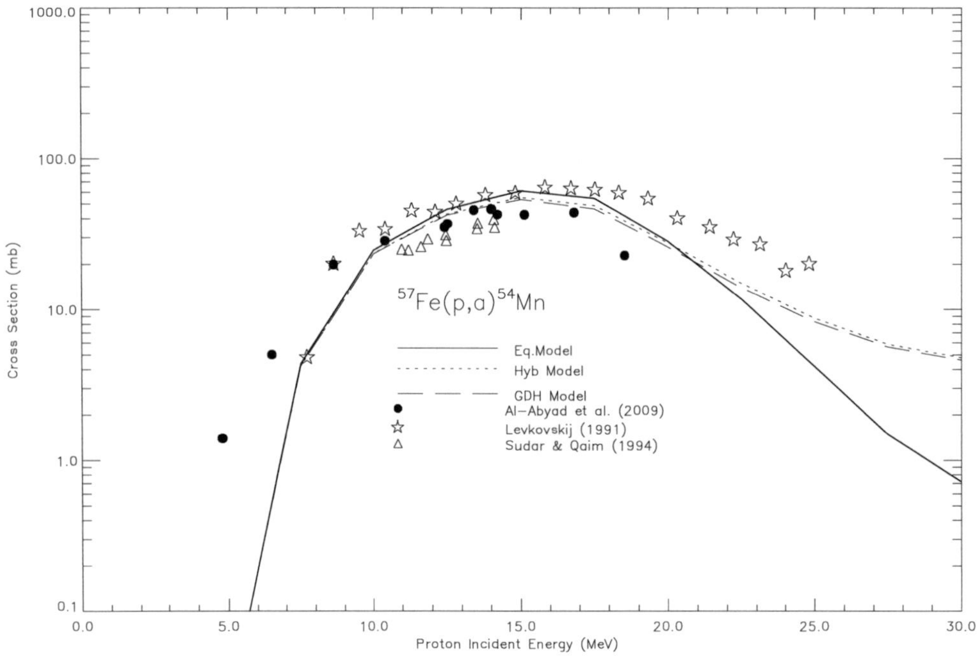

Fig. 3. The comparison of the calculated cross section of $^{57}Fe(p,\alpha)^{54}Mn$ reaction with the values reported in Ref. [7].

3.4 $^{68}Zn(p,2n)^{67}Ga$ reaction process

The calculation on the excitation function of $^{68}Zn(p,n)^{67}Ga$ reaction has been compared with the experimental values in Fig.4. The W-E model calculations are in agreement with the measurements up to 25 MeV. Also, the GDH and hybrid model calculations are in very good harmony with the experimental data. The optimum energy range for production of ^{67}Ga is E_p= 30 →15 MeV.

3.5 $^{68}Zn(p,n)^{68}Ga$ reaction process

The calculation on the excitation function of $^{68}Zn(p,n)^{68}Ga$ reaction has been compared with the experimental values in Fig.5. The experimental data of the measurements are in good agreement with each other. The W-E model calculations are in agreement with the measurements up to 15 MeV. The GDH model calculations are the best agreement with the experimental data of 5 - 30 MeV energy range. Also, the hybrid model calculations are in very good harmony with the experimental data. The optimum energy range for production of ^{68}Ga is E_p= 15 → 5 MeV.

3.6 $^{93}Nb(p,4n)^{90}$ Mo reaction process

The calculation on the excitation function of $^{93}Nb(p,4n)^{90}Mo$ reaction has been compared with the experimental values of Ditroi *et al.* [7] in Fig.6. While the GDH model calculations are the best agreement with the experimental data of 35 - 60 MeV energy range, the other model calculations are higher than the experimental data. The optimum energy range for production of ^{90}Mo is E_p= 55 →45 MeV.

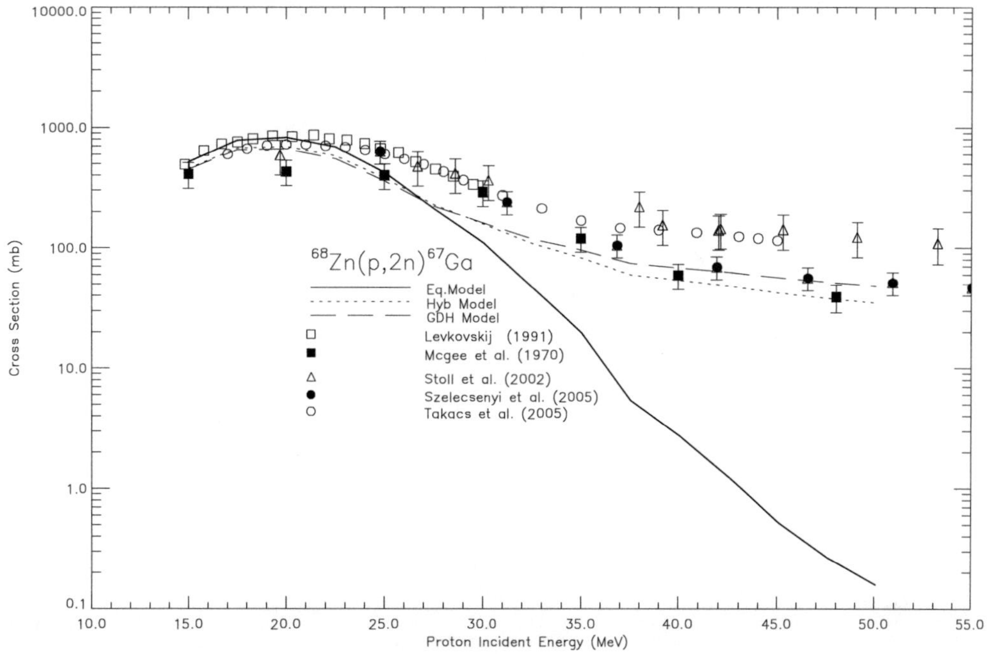

Fig. 4. The comparison of the calculated cross section of $^{68}Zn(p,2n)^{67}Ga$ reaction with the values reported in Ref. [7].

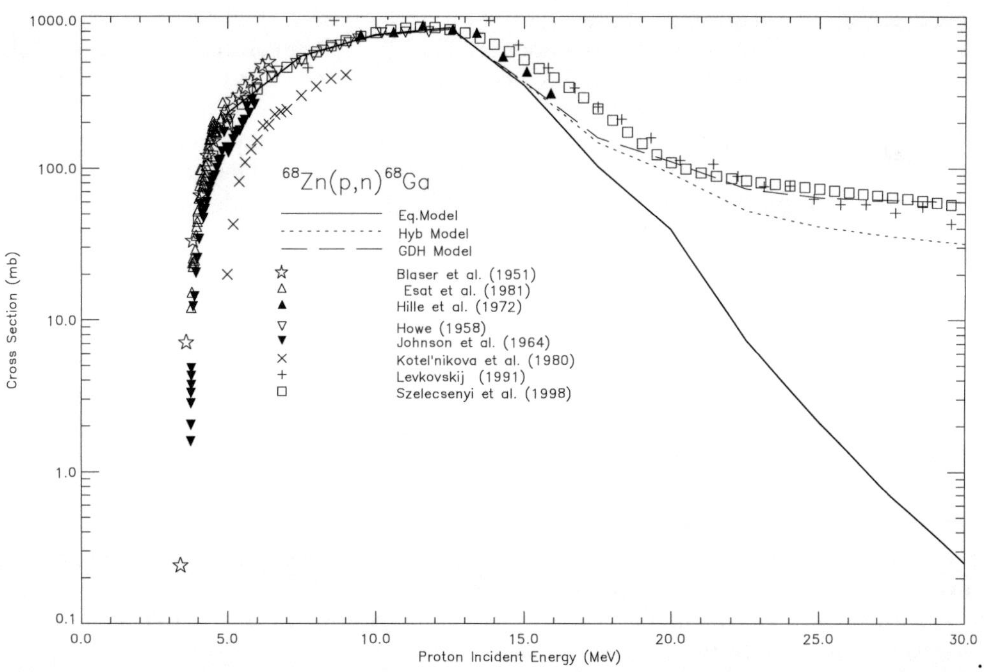

Fig. 5. The comparison of the calculated cross section of $^{68}Zn(p,n)^{68}Ga$ reaction with the values reported in Ref. [7].

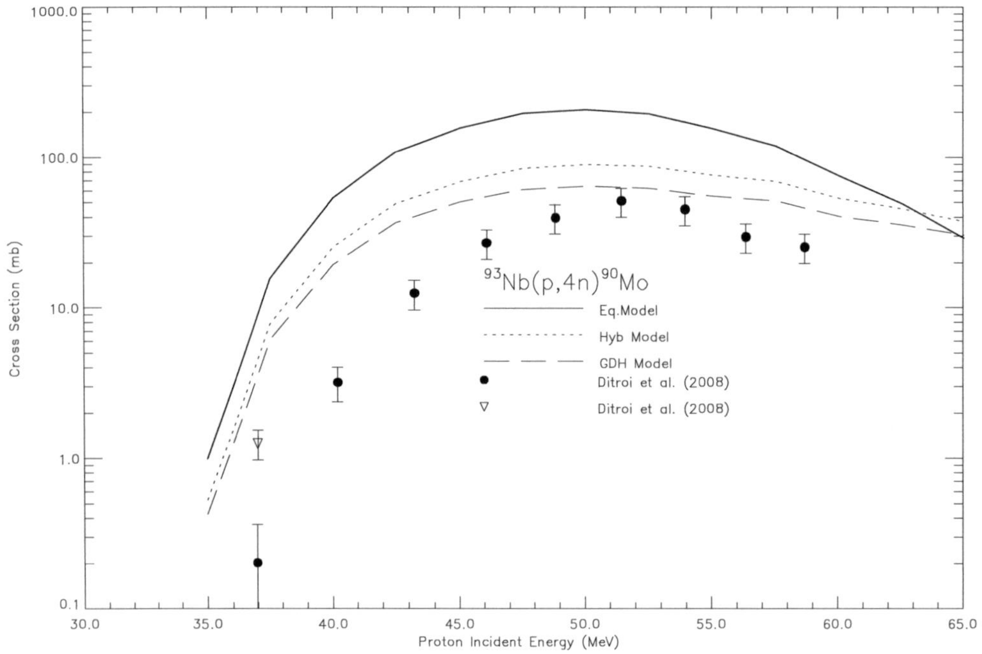

Fig. 6. The comparison of the calculated cross section of ^{93}Nb(p,4n)^{90}Mo reaction with the values reported in Ref. [7].

3.7 ^{112}Cd(p,2n)^{111}In reaction process

Especially, the ^{111}In radionuclei are very important for SPECT. The calculation on the excitation function of ^{112}Cd(p,2n)^{111}In reaction has been compared with the experimental values in Fig.7. Generally, the experimental data of the measurements are in good agreement with each other. The GDH and hybrid model calculations are in very good harmony with the experimental data. The equilibrium W-E model calculations are in agreement with the measurements up to 30 MeV. The optimum energy range for production of ^{111}In is E_p= 25 →15 MeV.

3.8 ^{127}I(p,3n)^{125}Xe reaction process

The calculation on the excitation function of ^{127}I(p,3n)^{125}Xe reaction has been compared with the experimental values in Fig.8. We can say that the experimental data of the measurements are in good agreement with each other. The equilibrium W-E model calculations are in agreement with experimental values up to 35 MeV The GDH and hybrid model calculations are in very good harmony with the experimental data. The optimum energy range for production of ^{125}Xe is E_p= 35 →25 MeV.

3.9 ^{133}Cs(p,6n)^{128}Ba reaction process

The calculation on the excitation function of ^{133}Cs(p,6n)^{128}Ba reaction has been compared with the experimental values in Fig.9. The equilibrium W-E model calculations are not agreement with experimental values. The GDH model calculations are the best agreement with the experimental data of Deptula *et al.* [7] for 45-100 MeV energy range. While the hybrid model calculations are the good agreement with the experimental data of 45-65 MeV energy range, for above the 65 MeV proton incident energy this model calculations are higher than the experimental data.The optimum energy range for production of ^{128}Ba is E_p= 70 →50 MeV.

3.10 ^{203}Tl(p,3n)^{201}Pb reaction process

The ^{201}Tl radioisotopes are very important for using in the SPECT. The reaction process for ^{201}Tl is ^{203}Tl(p,3n)^{201}Pb, $^{201}Pb \xrightarrow{9.33h} {}^{201}Tl$. The calculation on the excitation function of ^{203}Tl(p,3n)^{201}Pb reaction has been compared with the experimental values of Blue *et al* [7], Lebowitz *et al.* [7] and Al-saleh *et al.* [7] Takacs *et al.* [7] in Fig.8. The Weisskopf-Ewing model calculations are in agreement with the measurements up to 35 MeV. The GDH model and hybrid model calculations are in good agreement with the experimental data of 20 - 35 MeV energy range. The optimum energy range for production of ^{201}Tl is E_p= 30→ 20 MeV.

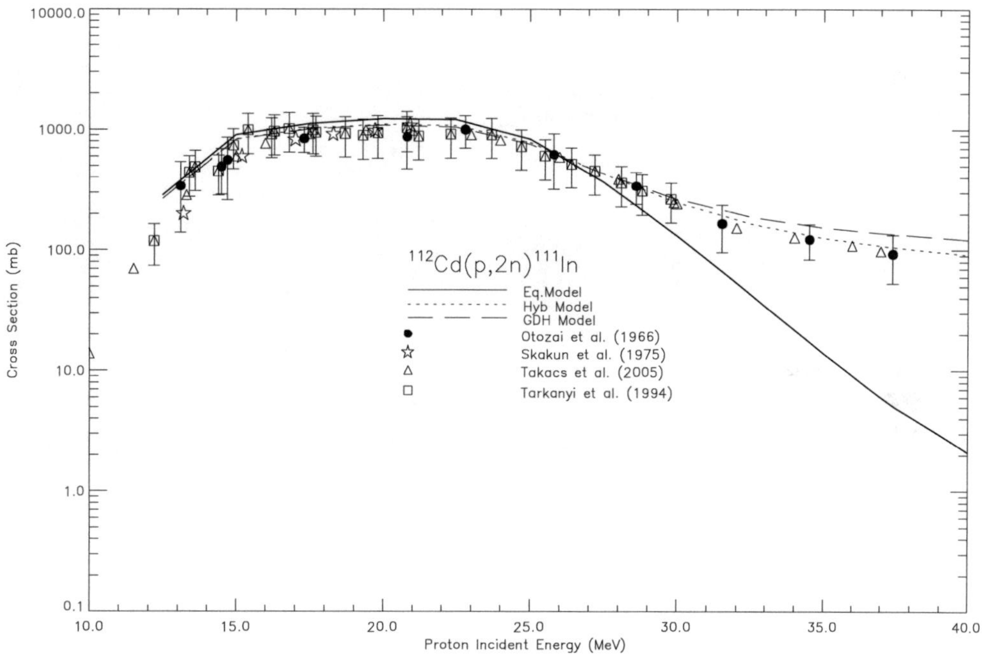

Fig. 7. The comparison of the calculated cross section of ^{112}Cd(p,2n)^{111}In reaction with the values reported in Ref. [7].

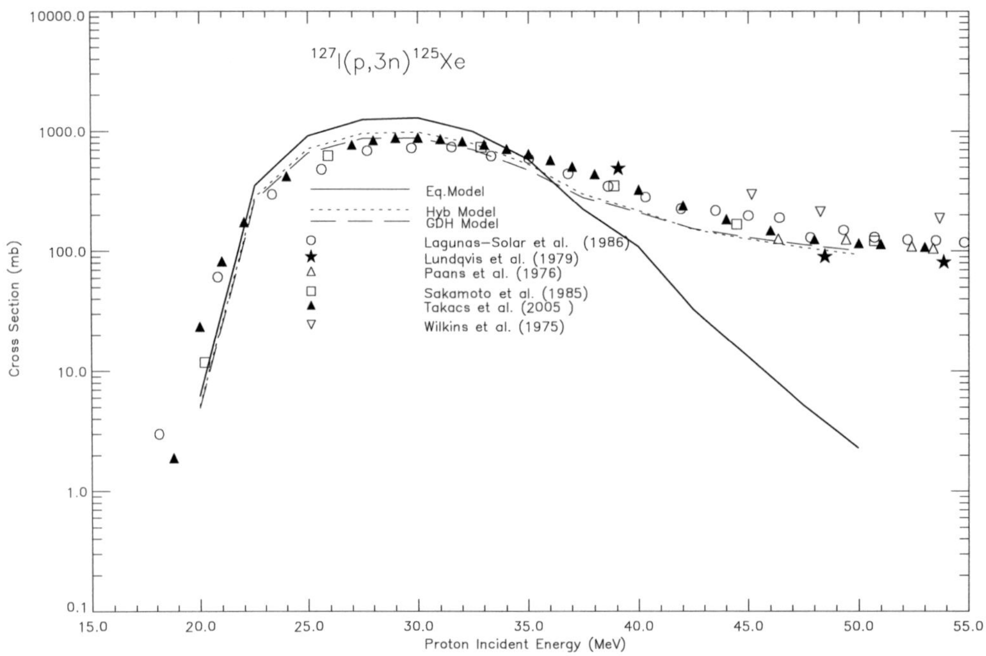

Fig. 8. The comparison of the calculated cross section of ^{127}I(p,3n)^{125}Xe reaction with the values reported in Ref. [7].

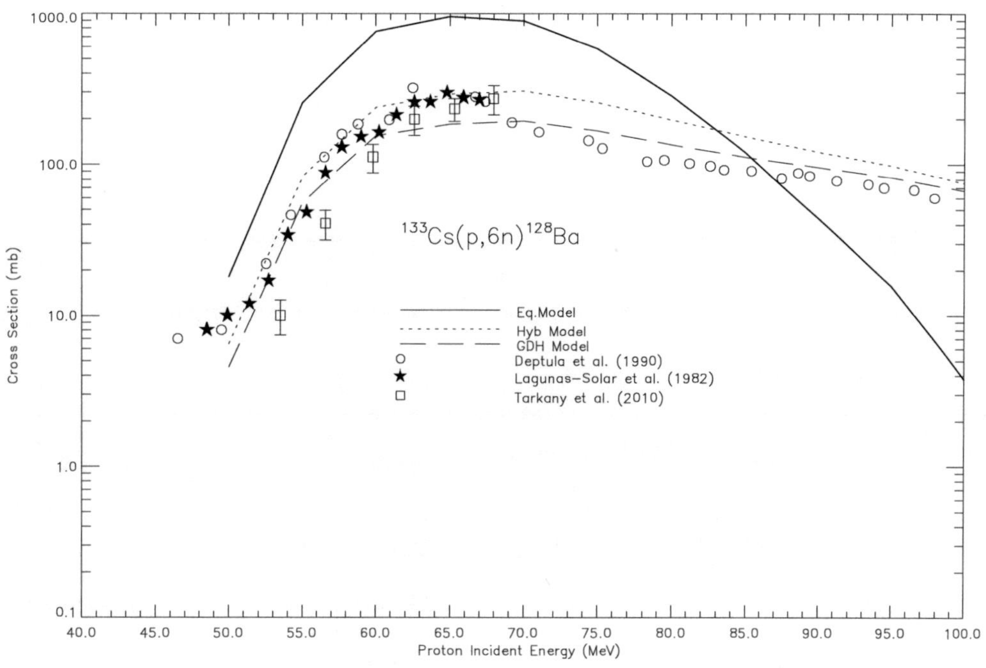

Fig. 9. The comparison of the calculated cross section of $^{133}Cs(p,6n)^{128}Ba$ reaction with the values reported in Ref. [7].

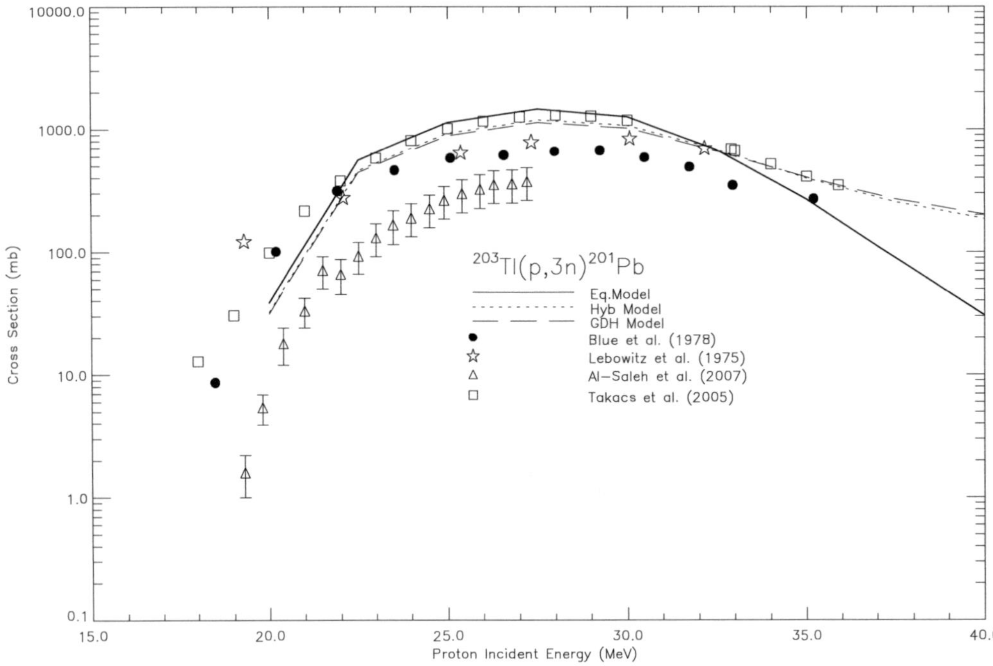

Fig. 10. The comparison of the calculated cross section of $^{203}Tl(p,3n)^{201}Pb$ reaction with the values reported in Ref. [7].

Producted radioisotopes	Half life	Mode of decay (%)	E□ (keV)	I□□(%)	Optimum Energy Range (MeV)
^{18}F	1.83 h	EC + β^+ (100)	0.52	0.01795	E_p= 10 → 5
^{57}Co	271.74 d	EC + β^+ (100)	122.06065	85.60	E_p= 15 → 5
^{54}Mn	312.05d	EC + β^+ (100)	834.848	99.97	E_p= 20 → 10
^{67}Ga	3.2617 d	EC + β^+ (100)	393.527	4.56	E_p= 30 →15
^{68}Ga	67.63m	EC+ β^+ (100)	1077.35	3	E_p= 15 → 5
^{90}Mo	5.67h	EC + β^+ (100)	257.34	78	E_p= 55 →45
^{111}In	2.8047 d	EC + β^+ (100)	245.350	94.10	E_p= 25 →15
^{125}Xe	56.9s	EC + β^+ (100)	111.3	60.2	E_p= 35 →25
^{128}Ba	2.43d	EC + β^+ (100)	273.44	14.5	E_p= 70 →50
^{201}Pb	9.33 h	EC+ β^+ (100)	546.280	0.279	E_p= 30→ 20

Table 1. The decay data and optimum energy range for investigated radionuclides.

4. Conclusions

The new calculations on the excitation functions of $^{18}O(p,n)^{18}F$, $^{57}Fe(p,n)^{57}Co$, $^{57}Fe(p,\alpha)^{54}Mn$, $^{68}Zn(p,2n)^{67}Ga$, $^{68}Zn(p,n)^{68}Ga$, $^{93}Nb(p,4n)^{90}Mo$, $^{112}Cd(p,2n)^{111}In$, $^{127}I(p,3n)^{125}Xe$,$^{133}I(p,6n)^{128}Ba$ and $^{203}Tl(p,3n)^{201}Pb$ reactions have been carried out using nuclear reaction models. Although there are some discrepancies between the calculations and the experimental data, in generally, the new evaluated hybrid and GDH model calculations (with ALICE/ASH) are good agreement with the experimental data above the incident proton energy with 5-100 MeV in Figs. 1-10. While the Weisskopf-Ewing model calculations are only in agreement with the measurements for lower incident proton energy regions, hybrid model calculations are in good harmony with the experimental data for higher incident proton energy regions. Some nuclei used in this study were examined and compared in previous paper written by Tel et al.[13,14]. Detailed informations can be found in these papers. And also new developed semi-empirical formulas for proton incident reaction cross-sections can be found in Ref. [27,28].

When Comparing the experimental data and theoretical calculations, the production of ^{18}F,^{57}Co, ^{54}Mn, $^{67,68}Ga$, ^{90}Mo, ^{111}In, ^{125}Xe , ^{128}Ba and ^{201}Pb radioisotopes can be employed at a medium-sized proton cyclotron since the optimum energy ranges are smaller than 50 MeV, except for ^{128}Ba. We gave the optimum energy range and the decay data for the investigated radionuclides in Table 1.

5. References

[1] M B Chadwick, P G Young, S Chiba, S C Frankle, G M Hale, H G Hughes, A J Koning, R C Little, R E MacFarlane, R E Prael, L S Waters, *Nucl. Sci. Engin.* 131 293 (1999)

[2] C. Rubbia, J A Rubio, S Buorno, F Carminati, N Fitier, J Galvez, C Gels, Y Kadi, R Klapisch, P Mandrillon, J P Revol, and Ch. Roche, *European Organization for Nuclear Research,* CERN/AT/95-44 (ET) (1995).

[3] S M Qaim *Radiat. Phys. Chem.* 71 917 (2004).

[4] S M Qaim *Radiochim. Acta* 89 297 (2001).

[5] B Scholten, E Hess, S Takacs, Z Kovacs, F Tarkanyi, H H Coenen and S M Qaim *J. Nucl. Sci. and Tech.* 2 1278 (2002).

[6] S M Qaim Radiochim. *Acta* 89 223 (2001).

[7] EXFOR/CSISRS (Experimental Nuclear Reaction Data File), Brookhaven National Laboratory, National Nuclear Data Center, (http://www.nndc.bnl.gov/exfor/) (2009) .

[8] A P Wolf, J S Fowler *Positron Emitter Labeled Radiotracers, Chemical Considerations in Positron Emission Tomography* (Alan R. Liss, Inc. Pub.) (1985) .

[9] A P Wolf and W B Jones *Radiochim. Acta* 34 1 (1983).

[10] K Gul *Appl. Radiat. Isotopes* 54 147 (2001).

[11] K Gul *Appl. Radiat. Isotopes* 54 311 (2001).

[12] A Aydın, B Sarer, E Tel *Appl. Radiat. Isotopes* 65 365 (2007).

[13] E Tel, E G Aydin, A. Kaplan and A. Aydin, *Indian J. Phys.* 83 (2) 1-20 (2009).

[14] E G Aydin, E Tel, A Kaplan and A Aydin, *Kerntechnic,* 73, 4, (2008).

[15] M B Chadwick *Radiochim. Acta* 89 325 (2001).

[16] V F Weisskopf and D H Ewing *Phys. Rev.* 57 472 (1940) .

[17] W Hauser and H Feshbach *Phys. Rev.* 87 366 (1952) .

[18] J J Griffin, *Phys. Rev. Lett.* 17 478 (1966) .

[19] M Blann *Annu. Rev. Nucl. Sci.* 25 123 (1975) .
[20] E Betak *Comp. Phys. Com.* 9 92 (1975).
[21] H. Gruppelaar, P Nagel, P E Hodgson and LaRivasta Del *Nuovo Cim.* 9 1 (1986).
[22] H Feshbach, A Kerman and S Koonin *Annu.Phys.* (NY), 125 429 (1980) .
[23] NUDAT - Decay Radiation Database, http://www.nndc.bnl.gov/nudat2
[24] M Blann and H K Vonach *Phys. Rev.* C28 1475 (1983).
[25] C. H. M. Broeders, A. Yu. Konobeyev, Yu. A. Korovin, et al., http://bibliothek.fzk.de/zb/berichte/FZKA7183.pdf.
[26] A. V. Ignatyuk, K. K. Istekov, G. N. Smirenkin, *Yad. Fiz.* 29, 875 (1979).
[27] Tel, E., Aydın, E. G., Aydın, A., Kaplan, A., *Appl. Radiat. Isotopes* 67 (2), 272 (2009)
[28] Tel, E., Aydın, A., Aydın, E. G., Kaplan, A., Ö. Yavaş, İ. Reyhancan, , *Pramana-J. Phys.*, 74 (6), 931-944, (2010).

9

History of Applications of Radioactive Sources in Analytical Instruments for Planetary Exploration

Thanasis E. Economou
Laboratory for Astrophysics and Space Research, Enrico Fermi Institute, University of Chicago, Chicago, USA

1. Introduction

Space age started with the launch of Sputnik-1 more than 50 years ago. Since then we have visited all the planets, some of them many times with very capable spacecrafts that are equipped with sophisticated payload that are returning significant information about the composition, physical condition of their surfaces and atmospheres and much more information. However, at the beginning of the space age, there were no techniques and instruments available to be used in space and had to be invented, designed, built and tested for the harsh environmental conditions in space. For the first analytical instruments in space, transuranium artificial radioisotopes produced in the national laboratories, proved to be very useful for applications in space. In early sixties Anthony Turkevich and his group at the University of Chicago applied a novel technique -- the Rutherford backscattering -- that is based on the interaction of the alpha particles with matter, to devise an instrument to obtain in-situ the chemical composition of the lunar surface [1]. In addition to measuring the backscattered alpha particles, the instrument also measures the proton energy spectra derived from the (α,p) reaction of the alpha particles with some light elements in the analyzed sample. Since the bombardment of a sample with a beam of alpha particles and x-rays from the same source also produces specific characteristic x-rays that results in additional compositional information, an additional x-ray detector was added to the ASI instrument to detect the produced x-rays. Based on the successful performance of this instrument on the lunar missions, more advanced and miniaturized versions complimented with an x-ray mode were developed and used in many NASA, ESA and Russian missions to several planetary bodies.

2. The alpha scattering instrument for the lunar missions

The technique of alpha backscattering for obtaining the chemical composition of planetary bodies was described for the first time and in detail by A. Turkevich, 1961[1], 1968[2]; Patterson et al., 1965[3]; Economou et al., 1970[4], 1973[5]. The compositional information is obtained from the energy spectra of scattered alpha particles and protons generated in (α,p) reactions of alpha particles with the matter in the analyzed sample. As it is shown in Fig. 1, a beam of monochromatic alpha particles with energy of 6.1 MeV from ^{242}Cm alpha source is bombarding the sample to be analyzed. The resulting scattered alpha particles and protons from (α,p) reactions are detected by solid state Si detectors situated as close to 180° as possible for maximum separation of individual elements. The energy of scattered alpha particles depends mostly on the scattered angle and the mass of the nucleus A, from which it is scattered. The sensor head of the Alpha Scattering Instrument (ASI) containing all the detectors, sources and the first stages of the electronic preamplifiers is deployed to the surface of the Moon and it is exposed to the harsh lunar environmental conditions, as it is shown in Fig. 2. The rest of the ASI electronics is placed inside the spacecraft compartment that is more controlled for temperature extremes.

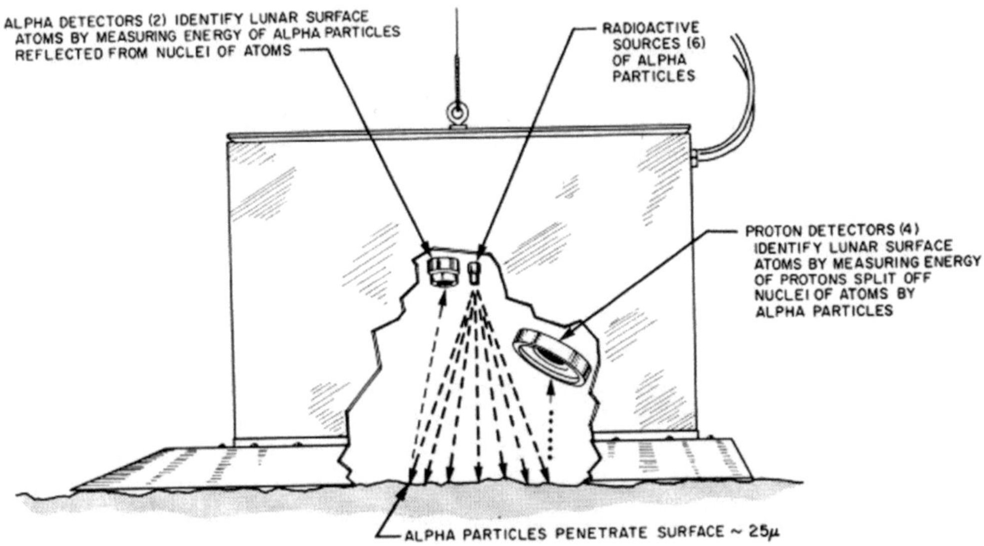

Fig. 1. The Alpha Scattering Instrument sensor head [2]

Fig. 2. The Alpha Scattering instrument deployed on the surface of the Moon

The energy E of scattered alpha particles in respect to its initial energy E_0 is a function of scattering angle θ and the mass A, of the target atom.

$$\frac{E}{E_0}=\left(\frac{4\cos\Theta+(A^2-16\sin^2\Theta)^{1/2}}{A+4}\right)^2$$

Which for θ=180° converts to

$$\frac{E}{E_0}=\left(\frac{A-4}{A+4}\right)^2$$

Since we know the initial source energy, E_0, the mass A, from which an alpha particle is scattered, can be calculated from the measured energy E from the above equation. For the Surveyor lunar missions in the 1960's the ASI used several hundred millicuries of ^{242}Cm alpha particle source of 6.1 MeV. Figure 3 shows some alpha spectra from pure elements and the alpha and proton spectra from lunar samples from the Surveyor 7 lunar mission. From these spectra, the elemental composition of the lunar surface material has been derived. The ASI on three Surveyor series lunar missions in 1967-68 provided the first detailed and complete elemental composition of the lunar surface. The comparison of the lunar analyses results provided by the ASI on the Surveyor 5 and Surveyor 7 with those analyses of lunar material brought back from the Moon by the Apollo 11 astronauts is shown in Fig. 4 [6].

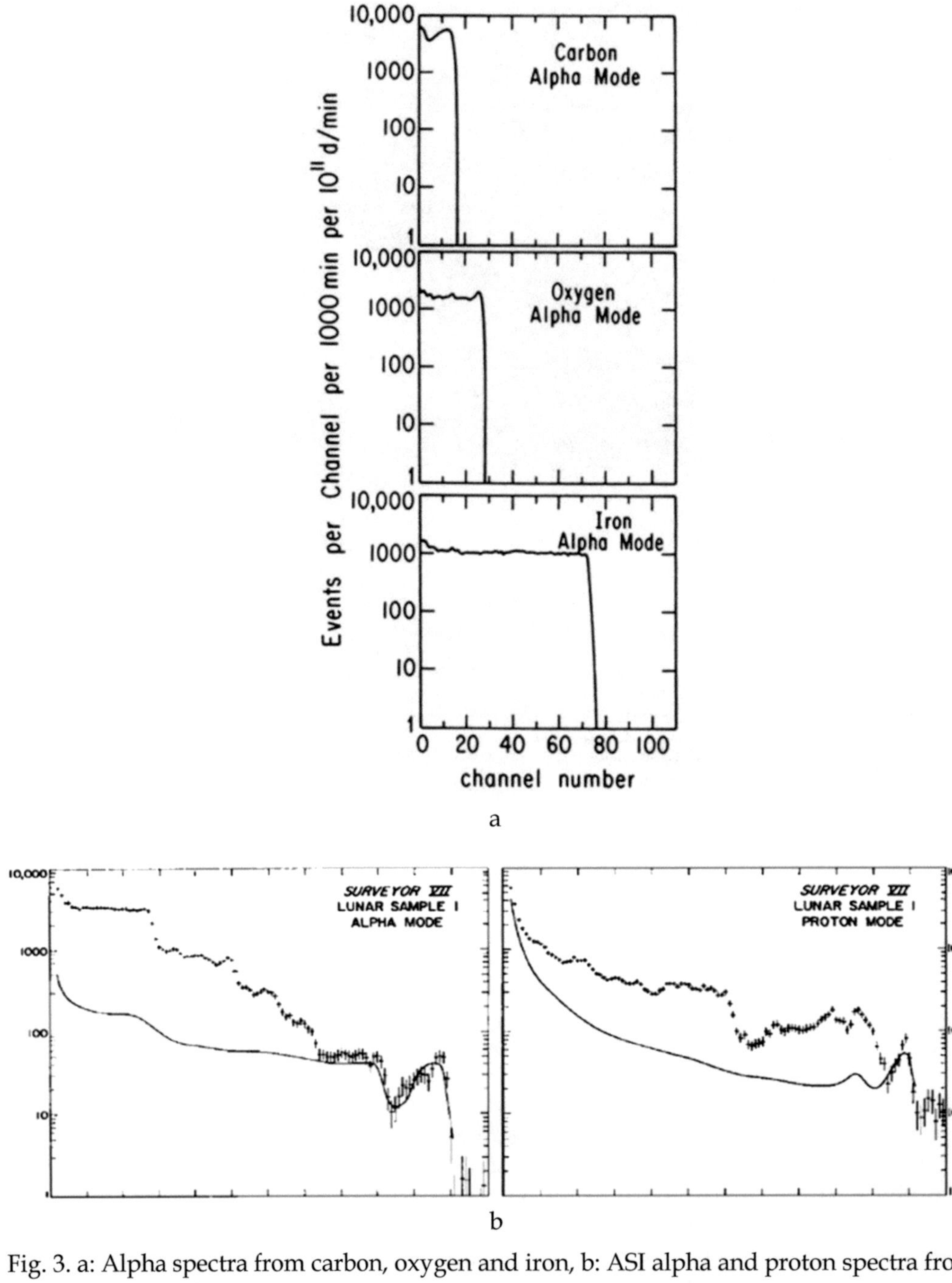

Fig. 3. a: Alpha spectra from carbon, oxygen and iron, b: ASI alpha and proton spectra from Lunar sample 1 of Surveyor 7.

This experiment was credited as the beginning of the Alpha Backscattering Spectroscopy (ABS) which has become a common analytical technique in many terrestrial laboratories. The ABS is using, however, a beam of alpha particles from a particle accelerator instead from radioisotopes.

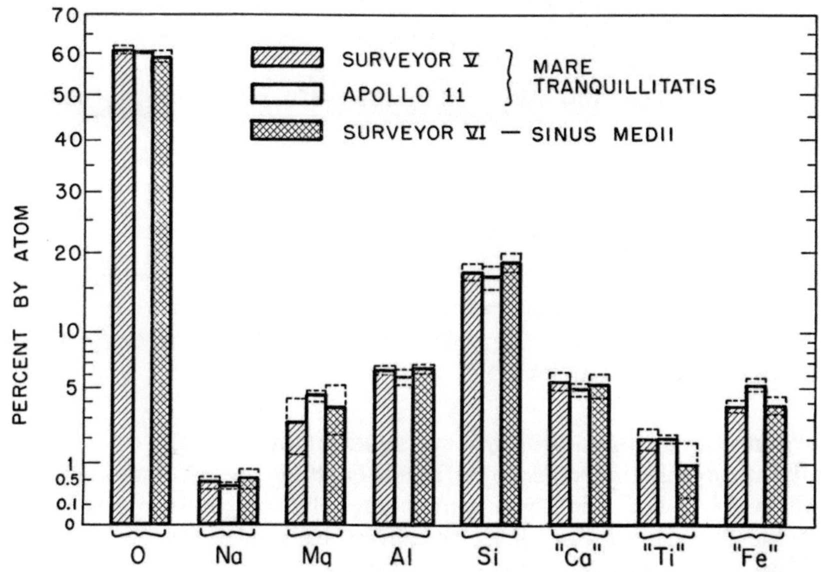

Fig. 4. Comparison of the ASI Surveyor 5 and Surveyor 7 results of the chemical composition of lunar surface material with those brought back by the Apollo 11 astronauts.

2.1 The ^{242}Cm alpha sources

The Alpha Scattering Instrument on the Surveyor lunar missions that provided the first chemical analysis of lunar material required alpha particle sources of high intensity and quality. In choosing the right isotope several criteria had to be taken into consideration. The decay half-live time has to be short in order to have essentially weightless sources, but long enough so that the sources will not decay appreciably before the completion of the measurements. Also, the source had to emit monoenergetic alpha particles with energy as high as possible and any decay daughter products should grow very slowly.

It was found that ^{242}Cm isotope ($T_{1/2}$ = 163 days, E_{α}=6.11 MeV) fits all the criteria best. This isotope emits essentially monoenergetic alpha particles and the decay half-life time of its daughter product ^{238}Pu is 86 yrs. So, it was ideal for the short missions to the Moon.

The procedure of preparing ^{242}Cm is long and complicated and it is described in details by Paterson et al., [7]. It starts with irradiation of ^{241}Am with neutrons in nuclear reactors which converts it to ^{242}Am by neutron capture followed by EC decay into ^{242}Cm and ^{242}Pu. After that, substantial purification yielded pure ^{242}Cm that was deposited on stainless steel plates of an active area of 2.8 mm in diameter and loaded into the ASI instrument.

Although ^{241}Am is wildly abundant because it is the decay product of reactor produced ^{241}Pu, the availability of ^{242}Cm is very scarce, especially in large quantities, due to its short half-life time. For the Surveyor lunar missions, a special program supported by the Atomic Energy Commission was undertaken at Argonne National Laboratory to produce adequate quantities of ^{242}Cm alpha sources with high qualities required for the experiment. Table 1 lists the main source characteristics of ^{242}Cm alpha sources used in the ASI on the Surveyor lunar missions in the late 1960's.

Radioisotope	^{242}Cm
Number of Sources	6
Total Intensity	**120-470 millicuries**
E_{α}	**6.11 MeV**
$T_{1/2}$	**162.8 days**
FWHM	**1.5-1.9 %**
FW0.1M	**2.5-3.0%**
FW0.01M	**5.0-8.0%**

Table 1. Characteristics of the ^{242}Cm alpha radioactive source used in the ASI instrument on the Surveyor 5-7 missions to the Moon.

2.2 The "Mini-Alpha" alpha proton X-ray spectrometer (APXS)

Most of the alpha emitting radioisotopes, beside emitting alpha particles, also emit x-rays in the energy range 14-20 keV and some gamma rays that are useful in exciting with different efficiency the characteristic XRF lines from every element. The higher Z elements are more efficiently excited by the X-ray lines emitted by the Curium isotopes alpha sources, while the lower Z elements are better excited by the alpha particles of 5-6 MeV (see Fig. 5a). These generated x-rays contain additional compositional information and were used to dramatically improve the analytical information of the ASI instrument, extending the performance to the presence of minor elements down to several tens of ppm range. On the other hand, the alpha mode is capable of better separating the light elements, while the x-ray mode has a better separation resolution for the heavier Z elements (see Fig. 5b). Therefore, an additional mode has been added to the original ASI instrument to detect the resulting characteristic x-rays that are produced when a sample is bombarded with a beam of alpha particles and x-rays from the same radioactive source. It is the combination of all three modes, the Rutherford alpha backscattering, the x-ray fluorescence (XRF) and the particle PIXE techniques that resulted in one integrated low power, low volume, but very powerful Alpha Proton X-ray Spectrometer (APXS) analytical instrument [8] that has been used in so many space missions [9-20].

A refined and miniaturized instrument with alpha, proton and x-ray modes (the "Alpha-Proton X-ray Spectrometer by Economou et al., 1976[8]) was proposed for analysis of Martian surface material during preparation for the Viking missions, but it was not selected for that mission. But the development of the "Mini-Alpha" APXS instrument was essential for future missions to Mars and other planetary bodies.

Figure 6 shows the diagram of the "Mini-Alpha" instrument. Its sensor head contains two telescopes with a combination of a thin dE/dx and thicker Si solid state detectors for the alpha and proton modes and a separate x-ray detector for the x-ray mode. The big challenge of miniaturization and enabling the use of such an instrument for space missions was the requirement of a good energy resolution for the x-ray detector. In the terrestrial laboratories every x-ray detector is kept at very low temperatures by cooling it with liquid nitrogen, which is for space application prohibitive. An enormous effort was undertaken to replace the Si-cooled x-ray detector with one that can operate at ambient Mars environment. Varieties of detectors made from exotic materials like Ge, HgI_2, CdTe, CdZn and other materials were tried with a different degree of success. Finally, we used a Si PIN detector that could provide sufficient energy resolution at martian ambient temperatures, and space qualify it at the last moment for the Pathfinder missions to Mars in 1996.

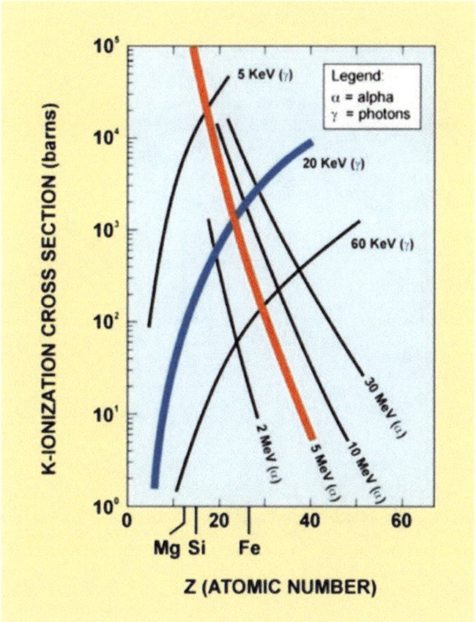

Fig. 5a. The characteristic X-rays are the result of two different mechanisms: X-ray fluorescence (in this particular case by the Pu L- lines in the 14-20 keV range) which is most effective for high Z elements and by particle induced X-ray emission (PIXE) that is most effective for the low Z elements.

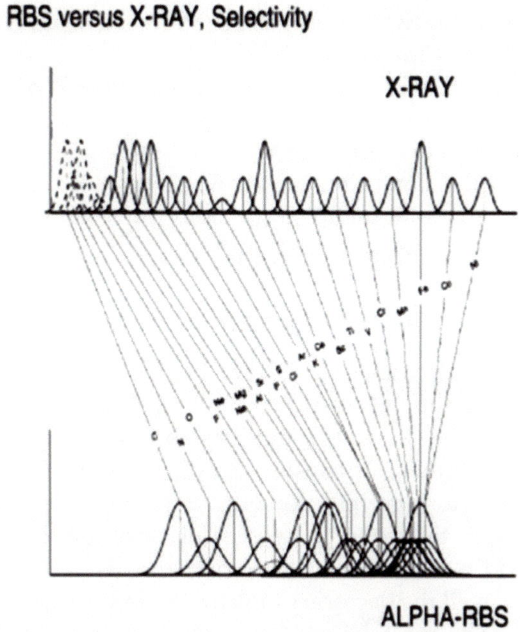

Fig. 5b. The X-ray mode of the APXS separates better the heavier Z elements while the alpha mode has much better resolution power for the light elements. The combination of these two modes makes the APXS a very powerful analytical instrument.

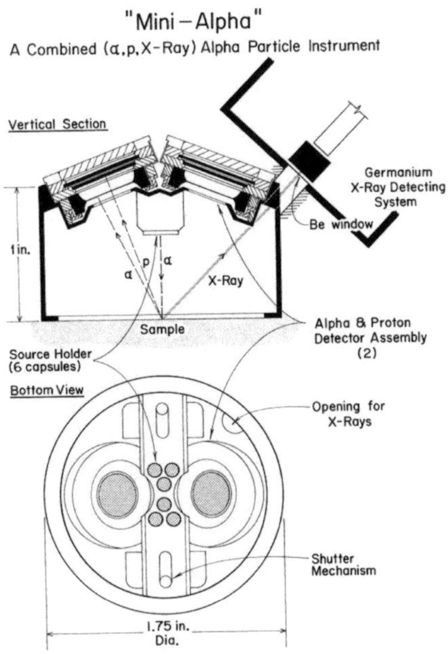

Fig. 6. An Alpha Proton X-ray Spectrometer "Mini-alpha" combining the alpha, proton and x-ray modes in one low power, low volume but high performance analytical instrument on basis of which several version have been used in many planetary missions. The big challenge was to replace the LN cooled x-ray detector with one able to operate at the ambient temperatures.

Based on the "Mini-Alpha" instrument several APXS instruments have been proposed and selected for the following space missions:

1. Soviet Phobos 1 and Phobos 2 missions to martian satellite Phobos in 1988.
2. Russian Mars-96 mission to Mars in 1994 (1996)
3. Pathfinder NASA mission to Mars in 1996
4. ESA's Rosetta mission to Comet Cheryumov-Gerasimenko in 2002
5. APXS on NASA Mars Exploration Rover missions (Spirit and Opportunity) in 2003
6. APXS on NASA Mars Science Laboratory in 2011.

3. Mars pathfinder alpha proton X-ray spectrometer (APXS)

The Alpha Proton X-ray Spectrometer (APXS) for the Mars Pathfinder mission in 1996 [12] was based on the design of the "Mini-Alpha" APXS [8] and used about 45 millicuries of ^{244}Cm alpha source instead the previously used ^{242}Cm isotope. Fig.7 is a photograph of the Pathfinder APXS flight instrument. It consists of two parts: 1/ the sensor head containing nine ^{244}Cm alpha sources in a ring-type geometry and three detectors for the measurement of the three components: A telescope of two Si-detectors for the measurement of alpha-particles and protons and a Si-PIN X-ray detector with its preamplifier, and 2/the main electronic box that contains all the necessary electronic components of a spectrometer and its interface with the spacecraft. The total weight of the Pathfinder APXS is 570 grams and it needed only 370 mWatts of power for its operation. The sensor head of the APXS is mounted in rear of the Sojourner microrover and it is deployed to the surface by its deployment mechanism as it is shown in Figure 8.

The Mars Pathfinder APXS performed very well on Mars and provided us for the first time the chemical composition of martian rocks [13-15]. It was also found that the soil chemical composition was very similar to that found by Viking XRF spectrometer in 1976[15].

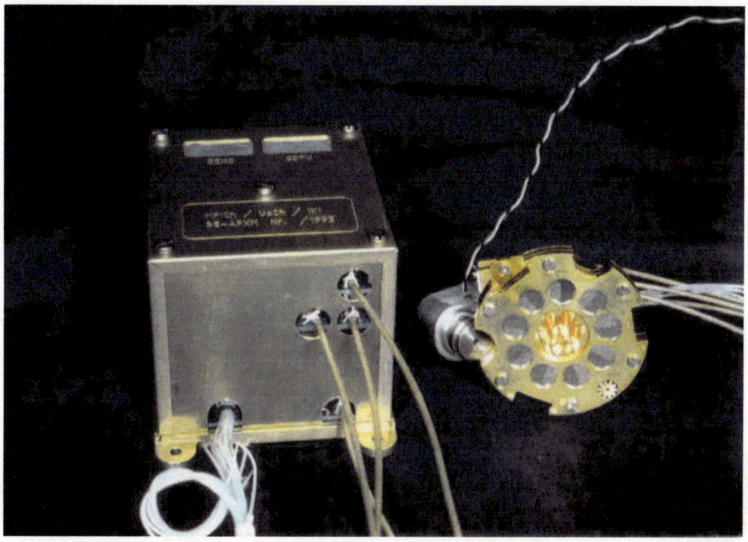

Fig. 7. A photograph of the Pathfinder APXS flight instrument

Fig. 8. APXS is mounted on the rear of Sojourner microrover and it is deployed to the surface by its own deployment mechanism.

3.1 Cm-244 alpha sources

For longer mission to distant planetary bodies, ^{242}Cm has too short a half-life decay time (163 days) and it would be too weak by the time the spacecraft reaches the target planets. ^{244}Cm with a half-life time of 18.1 years and E_{α}=5.8 MeV seems to be a much better choice for such missions because it has similar characteristics of 242Cm but has a much longer half-life time. The disadvantage of a longer half-life isotope is that for the same intensity one must pack more radioactive material per unit area, which deteriorates the energy resolution of the source by self absorption in its thickness of the material. Earlier, the ^{244}Cm was readily available in the US from the Livermore National Laboratory, but presently it is available only from the State Scientific Centre Research Institute of Atomic Reactors Dimitrovgrad, Russia.[21] There, the sources are prepared by high temperature condensation of metal curium vapor onto silicon substrates. For source production, the initial fraction of ^{244}Cm content was about 93%. Then, before the stage of source preparation, the curium was purified from daughter ^{240}Pu nuclide and additionally purified from americium, microquantities of ^{252}Cf and other impurities. The final deposition was on the silicon disks with ^{244}Cm fixed on their surfaces as a silicide. The sources for the Mars Pathfinder mission have overall dimensions as follows: disk diameter 8 mm; thickness 0.3 mm; and 6 mm diameter active spot. The source activities are 5 ± 1 mCi and the alpha-line half-widths are equal to (1.7–2.5) and (2.9–4.5)% of full width at half maximum energy of 5.8 MeV. The sources and their spectral characteristics stability were studied at wide intervals of physical and chemical parameters of the environment simulating real conditions of storage and maintenance. Thermo–vacuum (from -60 °C up to 1000 °C), mechanical, and vibrational tests were performed to demonstrate that the sources maintained their characteristics. Table 2 lists the main characteristics of the alpha sources used for the Pathfinder APXS instrument.

Radioisotope	^{244}Cm
Number of Sources	9
Total Intensity	**50 millicuries (1.85*109 Becq.)**
Ea	**5.807 MeV**
$T_{1/2}$	**18.1 years**
FWHM	**2.3 %**
FW0.1M	**3.5%**
FW0.01M	**10.0%**

Table 2. Characteristics of the ^{244}Cm alpha radioactive source used for the Mars Pathfinder APXS instrument.

Also, an important factor in determining the resulting energy spread of a source is the chemical composition of the source material: the ideal case is to use the source material in elemental form. In the case of curium, however, the metal becomes chemically unstable and the sources deteriorate rapidly. More recent work [21] concentrated on the formation of curium silicides on the surface of semiconductor grade silicon. This technology has yielded the best results so far and sources for the current or future APXS instruments will be produced by this technique.

The ^{242}Cm and ^{244}Cm alpha sources as used for the lunar and earlier Mars missions are called "open" alpha sources without any cover in order not to degrade the energy resolution of the alpha sources, which is one of the requirements for a good alpha spectrometer. However, using "open" sources present some challenges in handling radioactive sources and preventing source contamination of the spectrometer and its environment. During the fission decay, some recoil product can get out of the source housing and contaminate the instrument and the targets as well. In these cases, a very thin film of Al_2O_3 and VYNS (polypropylene) combination of a total thickness of 1200 Å was used in front of the source collimators. This thickness did not not affect the energy resolution of the alpha particles much but it is thick enough to stop the recoil products and prevent any contamination.
For the most recent APXS, the 244Cm are placed in a sealed housing covered with a thin 2 micron Ti foil. This somewhat degrades the alpha source energy resolution, however without degrading significantly the analytical performance of the APXS. First, for convenience, aluminum foils have been used for that purpose, but their use was discontinued when it was realized that the aluminum is easily oxidizing in the air with time causing continuous decreasing of the energy of the alpha particles.
The big advantage of sealed alpha sources is in much easier handling of the radioactive sources, easier way to prevent source contamination, but mainly it makes the transportation of radioactive material much easier and without a requirement for a special transport containers approved by the IAEA.

4. The Mars exploration rover APXS

For the Mars Rover Exploration (MER) mission in 2003 a more advanced x-ray sensor was developed which uses a silicon drift x-ray detector with a 5-μm beryllium entrance window and has an energy resolution of about 155 eV at 5.49 keV that rivals the energy resolution of the best terrestrial XRF laboratory spectrometers [16]. The rest of the APXS is almost similar to the one used on the Pathfinder. The sensor head of the APXS that contains the alpha sources and all the detectors together with the first stage of amplification electronics was mounted on the Instrument Deployment Device (IDD) of the MER rovers Spirit and Opportunity (robotic arm) and could be deployed to any selected target on the martian surface in sequence with other analytical instruments on the robotic arm, as can be seen in Fig. 9. The rest of the APXS electronics is located inside the temperature controlled compartment of the rover.
The APXS on the MER mission performed very well throughout the entire mission. Although the MER mission was designed to operate only for 3 months on the surface of Mars, we are now still operating successfully after 7 years on the surface of Mars without any degradation in its performance. Actually, due to the decay of radioactive sources used on Mössbauer instrument that was raising the APXS background, the signal to noise ratio in the APXS is better now, 7 years later in the mission, rather than at the beginning of the MER mission in 2004.
Fig. 10 shows the spectra obtained by the APXS on Mars Explorer mission of soil on two different landing sites, on Meridiani planum and Gusev crater. As it can be seen from the spectra, except for some small differences, the soil composition on both sites is very similar. The APXS has analyzed hundreds of martian samples and found many different lithologies, starting with basalt-like rocks, altered basalts, to rocks containing large amount sulfur, phosphor, magnesium, etc., on both landing sites. One such lithology with almost half of the sample in the form of sulfates is shown in Figure 11.

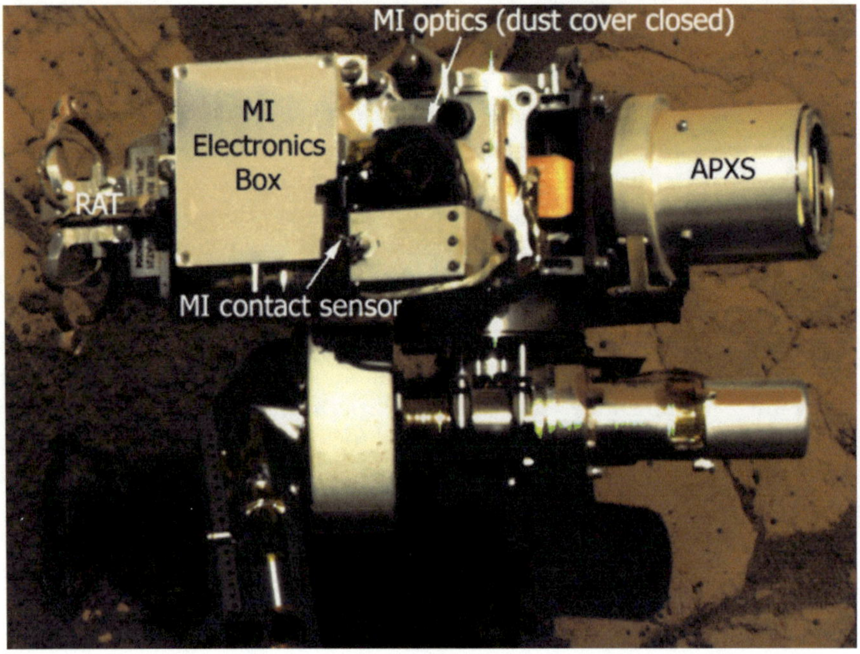

Fig. 9. The Instrument Deployment Device (IDD) of the MER rovers Spirit and Opportunity (robotic arm) showing the mounting of the APXS. The deployment of the APXS to the surface is done by commanding whenever it is desirable.

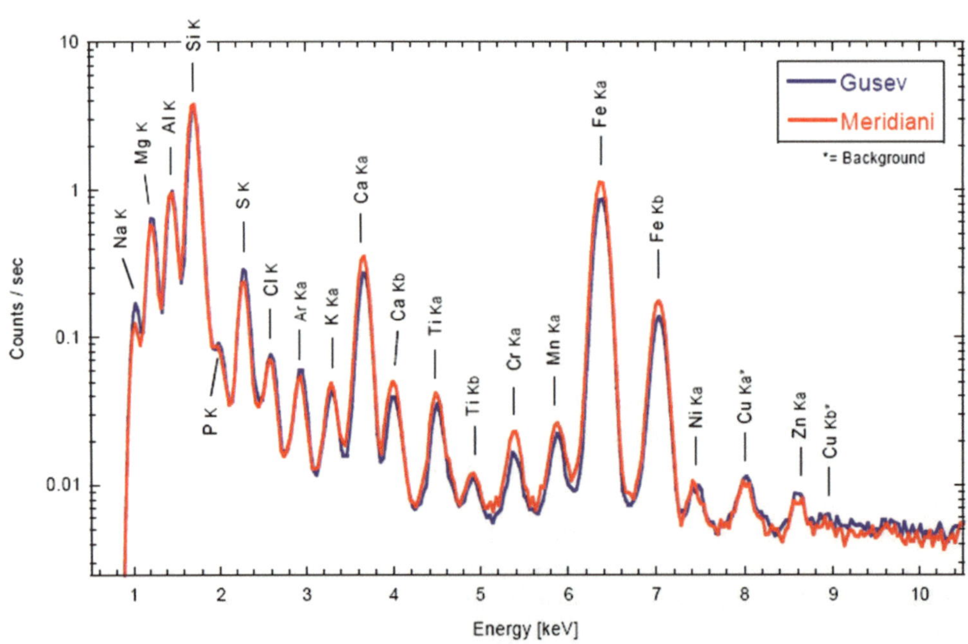

Fig. 10. APXS X-ray spectra of Martian soils measured at Gusev Crater (blue) and Meridiani Planum (red).

Sample Name:	A401_sulf_soil
Spacecraft:	MER2
Sol #:	401
Measurement Start (LST):	05:58:54
Measurement Duration:	6:50:10
DAT.-file #:	2A161961338EDRA600N1438N0M1
LBL.-file #:	2A161961338EDRA600N1438N0M1
Time created:	2005-02-19 06:59:26

Fe^{2+}/Fe_{tot} from MB:	0.7
Relative Geometric Yield :	0.85
Ratio of Pu-Lα scatter lines (coh/incoh):	0.62
Intensity of Pu-Lα coh. scatter line (cps):	0.40
Geometric Yield from scatter lines:	0.67

Ratios:	Mg/Si	Al/Si	FeO/MgO[4]
by wt	0.43	0.29	3.00

APXS Preliminary Analysis

Elements	wt%	Oxides	wt%	mol%
Na[3]	1.60	Na_2O	2.20	2.60
Mg	4.00	MgO	6.60	11.90
Al	2.70	Al_2O_3	5.10	3.60
Si	9.20	SiO_2	19.70	23.90
P	2.53	P_2O_5	5.80	2.97
S	12.30	SO_3	30.70	27.90
Cl	0.45	Cl	0.45	0.90
K	0.07	K_2O	0.09	0.07
Ca	4.60	CaO	6.40	8.30
Ti	0.27	TiO_2	0.44	0.40
Cr	0.04	Cr_2O_3	0.05	0.02
Mn	0.17	MnO	0.21	0.22
Fe	15.50	(FeO)[2]	14.00	14.10
Ni (ppm)	70	(Fe_2O_3)[2]	6.60	3.00
Zn (ppm)	90	Fe_2O_3T	22.20	
Br (ppm)	460	Σ Oxides[1]	99.9	100.0

Notes:

(1) sample matrix (for matrix correction and geometric yield) by closure, assuming all Fe as Fe_2O_3, and matrix free of H_2O, CO_2, etc.

(2) calculated with Fe^{2+}/Fe_{tot} from MB (default assumption: Fe^{2+}/Fe_{tot} =0.7).

(3) Na too low (by ~30 %) due to faulty algorithm. Will be corrected asap.

(4) assumes all Fe as FeO

Fig. 11. The chemical analyses results from a soil sample A_0401, obtained by the APXS on Spirit on Martian sol 401 showing very high concentration of sulfates.

5. The nano alpha-X instrument for MUSES-C mission

In the year of 2000, NASA, with collaboration with Japanese JAXA, were getting ready for the MUSES-C mission to bring back to Earth a sample from an asteroid. On that mission, it was also proposed to study in-situ the surface characteristics of that asteroid with a nanorover populated with several analytical instruments, including an Alpha-X Spectrometer (AXS) [22] to obtain the chemical composition of the asteroid surface material. For that mission, a further miniaturization of the AXS was required to fit it inside the Jet Propulsion Laboratory's nanorover with the overall dimensions of 14x15x6 cm. A massive hybridization and elimination of the proton mode resulted in a miniaturized Alpha-X Spectrometer shown in Fig. 12 and with specifications listed in Table 3. The AXS, although drastically miniaturized, is a complete spectrometer capable of providing highly accurate analytical results, similar to other APXS instruments.

Fig. 12. A photograph of AXS laboratory instrument for MUSES-C mission.

APXS Characteristics for MUSES-C Mission

Weight:	95 g
Volume: a/ Sensor head	15 cm^3
b/ Electronics	50 cm^3
Total Volume	65 cm^3
Power :	<200 mW (~25 mA @7.5 V)
Voltages:	+7.5 V DC
	-7.5 V DC
Radioactive Sources:	30 mCi of Cm-244 radioisotope
E_α:	5.8 MeV
$T_{½}$:	18.1 years
Accumulation Time:	3-5 hours for each sample
Data Requirements:	100 kb per sample

Table 3. Characteristics of the AXS instrument for MUSES-C mission.

The MUSES-C mission suffered several long delays and NASA eventually cancelled its participation on this mission. The Japanese, however, went ahead with the mission that was renamed Hayabusa after the launch in 2003 and succeeded in bringing back a small amount of asteroid material from asteroid Itokawa in June, 2010. Despite the cancellation, the AXS was designed, built and fully flight qualified. It is now available for some potential future mission.

6. Summary

We have described here some of the techniques and analytical instruments that have been developed in the past half century and used for space applications. Many other space analytical instruments have used radioactive sources in one way or another that have not been described here. Some of the developed techniques had a profound influence in developing analytical instruments that are used today in many terrestrial laboratories around the world.

Table 4 shows some of the radioisotopes used in different analytical instruments on space missions. Besides the ^{242}Cm and ^{244}Cm that were used in the APXS, the Mössbauer experiment on the MER mission used about 400 millicuries of ^{57}Co at the start of the mission, to obtain the mineralogy of the iron bearing rocks [10]. Similarly, ^{55}Fe has been used in an XRF instrument on Viking mission in 1976 [11] and on the Beagle 2 mission in 2003 [12].

	Isotope	**Half Live Time**	**Energy, Type**	**Space Mission**
1.	Cm-242	161 days	6.1 MeV, α	ASI Surveyor Lunar Missions, 1967-1968
2.	Cm-244	18.1 years	5.8 MeV, α	Phobos1@2('88), Mars96('96), Pathfinder('97), MER(2004), MSL(2011), Rosetta (2002)
3.	Fe-55	2.7 y	5.9 keV, x	Viking 1&2 (1976), Beagle 2 (2003)
4.	Co-57	271 days	14, 122 keV, γ	Viking 1&2 (1976), MER (2004)
5.	Am-241	432 y	5.49 MeV,α,γ	XRF
6.	Pu-238	87.74 y	5.5 MeV, α	Radioisotope Thermoelectric Generators (RTG), Radioactive Heating Units (RHU), many space missions
7.	Pu-239	2.4x104 y	5.15 MeV,α	Calibrations

Table 4. The most common radioisotopes used in the analytical instrument for space applications

7. References

[1] Turkevich, A., "Chemical analysis of surfaces by use of large-angle scattering of heavy charged particles", Science, 134, 672, 1961.

[2] A. L. Turkevich, W. A. Anderson, T. E. Economou, E. J. Franzgrote, H. E. Griffin, S. L. Crotch, J. H. Patterson and K. P. Sowinski. "The Alpha-Scattering Chemical Analysis Experiment on the Surveyor Lunar Missions". Jet Propulsion Laboratory Technical Report 52-1265, pp. 505-82, June 15, 1968.

[3] Patterson, J.H, A. L. Turkevich and E. J. Franzgrote, Chemical analysis of surfaces using alpha particles, J. Geophys.Res., 70, 1311, 1965.

[4] Thanasis E. Economou, Anthony L. Turkevich, Keneth P. Sowinski, James H. Patterson and Ernest J. Franzgrote, The Alpha-Scattering Technique of Chemical Analysis, J. Geophys. Res., 75, No 32, 6514, 1970.

[5] Thanasis E. Economou,Anthony L. Turkevich and James H. Patterson, An Alpha Particle Experiment for Chemical Analysis of the Martian Surface and Atmosphere, J. Geophys. Res., 78, No 5, 781, 1973.

[6] Patterson, J.H, Ernest J. Franzgrote,A. L. Turkevich, W.A. Anderson, T.E. Economou, H.E. Griffin, S.L. Groach and K.P. Sowinski, Alpha Scattering Experiment on Surveyor 7: Comparison with Surveyor 5 and 6, J. Geophys. Res., 74, 6120, 1969.

[7] James H. Patterson, Harry E. Griffin, E. Philip Horwitz, and Carol A. Bloomquist, Preparation of High Level Alpha-Particle Sources for the Surveyor Alpha Scattering Experiment, Nuclear Technology, Vol. 18, 277-285, 1073.

[8] Thanasis E. Economou and Anthony L. Turkevich. "An Alpha Particle Instrument with Alpha, Proton and X-ray Modes for Planetary Chemical Analyses". Nucl. Instr. & Methods 134, 1976, p. 391-399.

[9] Hovestadt, D., et al., PHOBOS, Proceedings of the International Workshop, Moscow, 1986, p 302

[10] Economou, T.E, J.S. Iwanczyk and R. Rieder, A HgI_2 X-ray Instrument for the Soviet Mars '94 Mission, Nucl. Instr. & Methods A322,633-638, 1992.

[11] Linkin, V., et al., "A Sophisticated Lander for Scientific Exploration of Mars : Scientific Objective and Implementation", Planetary Space Science Journal, Vol. 46, Issues 6-7, 1998, pp 717-737.

[12] Rieder, R., H. Wänke, T. Economou and A. Turkevich; "Determination of the Chemical Composition of Martian Soil and Rocks: The Alpha-Proton-X-Ray Spectrometer", J. Geophys. Res. 1997 pp. 4027-4044.

[13] Rieder, R., T. Economou, J. Brückner, G. Dreibus, H. Wänke and A. Turkevich, First Elemental Analysis of Martian Surface by the Mobile Alpha Proton X-ray Spectrometer Attached to Mars Pathfinder Rover Sojourner. Meteoretics and Planetary Science, Vol. 32, No. 4, A107, 1997.

[14] Rieder, R., T. Economou, H. Wänke, A. Turkevich, J. Crisp, J. Brückner, G. Dreibus,H.Y. McSween, Jr., The Chemical Composition of Martian Soil and Rocks Returned by the Mobile Alpha Proton X-ray Spectrometer: Preliminary Results from the X-ray Mode, Science Vol. 278, (1997)1771-1774.

[15] C. Nicole Foley, Thanasis E. Economou and Robert N. Clayton, Final Chemical Results from the Mars Pathfinder Alpha Proton X-ray Spectrometer, J. Geoph. Res. Vol. 108, No 12, doi:10.1029/2003JE002019, 2003.

[16] Rieder, R., et al., *J. Geophys. Res., 108(E12), doi:10.1029/2003JE 002150, 2003*

[17] http://www.esa.int/esaMI/Rosetta

[18] Klingelhoefer, G., J. Geophys. Res., 108, p. 8067, 2003

[19] Clark,B.C., et al., J. Geophys. Res., 87, p.10059, 1982

[20] Beagle 2 mission to Mars in 2003, http://www.beagle2.com/index.htm

[21] V. Radchenko, B. Andreichikov, H. Wanke, V. Gavrilov, B. Korchuganov, R. Rieder, M Ryabinin, and T. Economou, Curium-244 alpha-sources for space applications, Applied Radiation and Isotopes, 2000 Oct., 53(4-5):821-824.

[22] Economou, T.E, The Chemical Composition of an Asteroid Surface by the Alpha X-Ray Spectrometer on the MUSES-C Mission Lander, In Lunar and Planetary Science XXXI, 2000, 1861.pdf

Part 2

Radioisotopes in Environment

10

Environmental Dosimetry – Measurements and Calculations

Mats Isaksson
*Department of Radiation Physics, Institute of Clinical Sciences, The Sahlgrenska Academy, University of Gothenburg
Sweden*

1. Introduction

According to UNSCEAR (2008) the largest contribution to external exposure comes from naturally occurring, gamma-emitting, radioactive elements in the ground. Apart from ^{40}K, these elements are members of the decay chains starting with ^{238}U and ^{232}Th, respectively. Both of these radionuclides have half-lives comparable to the age of the earth and have been present in the earth's crust since its formation. The half-life of ^{40}K is of the same order of magnitude. However, nuclear weapons fallout and debris from accidents at nuclear facilities may at some sites contribute even more to the exposure.

Calculation of the external dose rate from field gamma spectrometric measurements requires knowledge of how the primary fluence rate, *i.e.* the number of unscattered photons per unit area and time, from different distributions of radioactive elements, at or in the ground, affects the detector response. Likewise, determination of external dose rate based on gamma spectrometry of soil samples must include a calculation of the primary fluence rate from the actual soil inventory.

The main topic of this chapter is to describe, and compare, some of the methods that can be used to estimate external dose rate from environmental sources of ionizing radiation from radioactive elements. The external dose rate from radioactive sources at or below ground may be estimated in several ways. One method is to simply measure the dose rate with a properly calibrated intensimeter or other type of dose meter (*e.g.* ionization chamber or thermoluminescent dosimeter, TLD). Another method is to calculate the dose rate based on gamma spectrometric measurements, *e.g.* field gamma spectrometry or soil sampling. However, due to assumptions of the relation between primary and scattered radiation, as well as calibration requirements, the dose rate estimations from these methods seldom agree in practical situations. The topics mentioned above will be exemplified with reference to actual measurements by field gamma spectrometry and intensimeter, as well as model calculations of fluence rates from different environmental geometries.

2. Primary fluence rate from different source geometries

The quantity fluence has been defined by the International Commission on Radiation Units and Measurements, ICRU, (ICRU, 1998) as $\Phi = \mathrm{d}N/\mathrm{d}a$, *i.e.* the number of particles per unit area. According to the definition, dN is the number of particles (here photons) incident on

the area da, which should be the cross-sectional area of a sphere. This definition of the area is chosen because the definition of fluence should be independent of the direction of the radiation field. The fluence rate is defined in terms of the increment of the fluence, dΦ, during the time interval dt as dΦ/dt.
Photons emitted from a radioactive source in a medium may undergo different kinds of interactions. Some of the photons pass through the medium without interacting and these will constitute the primary fluence at the detector position. Other photons, originally not directed towards the detector, may undergo Compton scattering and hit the detector. These latter photons, or scattered radiation, will also contribute to the detector signal. The relation between primary and scattered radiation is given by the build-up factor, which depends on several parameters (*e.g.* photon energy and material between source and detector) and must be determined by experiments or Monte Carlo-simulations.

2.1 Volume source

Figure 1 shows the geometry used for calculating the fluence rate at a reference point outside the active volume. In this case we consider a homogeneously distributed source in the ground, as an example. The number of photons emitted per unit time from a volume element dV at the depth z in the ground is given by $S_V(z,r,\eta)\cdot$dV, where dV equals d$r\cdot$d$R\cdot$dρ.

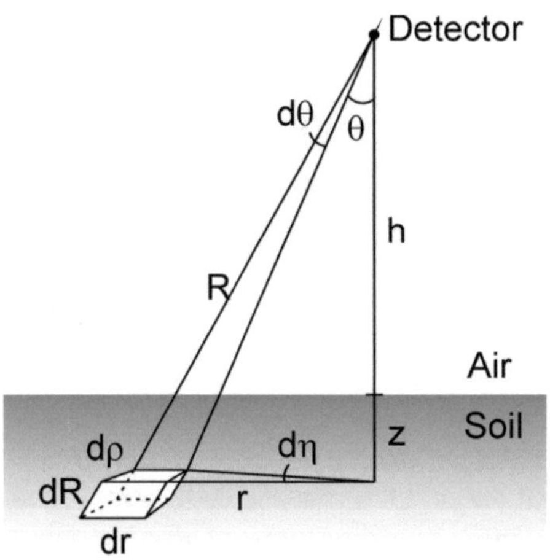

Fig. 1. The geometry used for calculating the primary fluence rate at the position of the detector, from a volume element in the ground. The volume element is considered as a point source.

The differential (primary) photon fluence rate at the detector position, taking into account self-attenuation in soil (μ_s) and attenuation in air (μ_a) is given by

$$d\dot{\phi}_p = \frac{S_V\, dV \cdot e^{-\mu_s(R-h/\cos\theta)-\mu_a \cdot h/\cos\theta}}{4\pi R^2} \qquad (1)$$

The contribution from the volume element dV is thus simply treated as the primary fluence rate from a point source at the distance of R from the detector. The exponential function accounts for the attenuation in the materials between the source and the detector (soil and air). The source strength, S_V, is related to the activity, A, through the relation $S_V \cdot dV = A \cdot \Sigma f_i$, where f is the probability for emission of a photon (gamma-ray) of a certain energy in each decay of the radionuclide.

The total primary photon fluence rate can now be determined by integrating the differential fluence rate over the whole volume, *i.e.* summing the contributions from each volume element dV. Changing to spherical coordinates will facilitate the integration and dV is then expressed as $dV = dr \cdot dR \cdot d\rho = R^2 \cdot \sin\theta \cdot d\theta \cdot d\eta \cdot dR$. The primary photon fluence is thus given by

$$\dot{\phi}_p = \int_0^{2\pi} \int_{h/\cos\theta}^{\infty} \int_0^{\pi/2} \frac{S_V \cdot R^2 \cdot \sin\theta \cdot e^{-\mu_s(R-h/\cos\theta)-\mu_a \cdot h/\cos\theta}}{4\pi R^2} d\eta\, dR\, d\theta \tag{2}$$

The above expression can be calculated analytically for a few cases of distributions and these will be exemplified by two kinds of homogeneous distributions: an infinite volume source and an infinite slab source.

2.1.1 Infinite volume source

An infinite volume source can often be assumed when considering naturally occurring radionuclides in the ground. These are the members of the uranium and thorium decay chains, together with ^{40}K. Since the mean free path of photons from naturally occurring radionuclides is about 10 cm (based on 3 MeV photons and a soil density of 1.5 $g \cdot cm^{-3}$) a soil volume of this thickness may be considered as an infinite source.

Integrating Eq. 2, we find that

$$\int_0^{2\pi} d\eta = [\eta]_0^{2\pi} = 2\pi \tag{3}$$

and

$$\int_{h/\cos\theta}^{\infty} e^{-\mu_s \cdot (R-h/\cos\theta)} dR = e^{\mu_s \cdot h/\cos\theta} \cdot \left[-\frac{1}{\mu_s} \cdot e^{-\mu_s \cdot R} \right]_{h/\cos\theta}^{\infty} = \frac{1}{\mu_s} \tag{4}$$

The expression for the primary fluence rate then reduces to

$$\dot{\phi}_p = \frac{S_V}{2\mu_s} \int_0^{\pi/2} \sin\theta \cdot e^{-\mu_a \cdot h/\cos\theta} d\theta \tag{5}$$

This integral can be solved by variable substitution and by using the properties of the exponential integrals, $E_n(x)$, defined as

$$E_n(x) = x^{n-1} \int_x^{\infty} \frac{e^{-t}}{t^n} dt \tag{6}$$

The substitution $t = \mu_a \cdot h/\cos\theta$ enables us to rewrite Eq. 5 as

$$\dot{\phi}_p = \int_0^{\pi/2} \frac{S_V \cdot \sin\theta \cdot e^{-\mu_a \cdot h/\cos\theta}}{2 \cdot \mu_s} d\theta = \frac{S_V}{2 \cdot \mu_s} \int_{\mu_a \cdot h/\cos 0}^{\mu_a \cdot h/\cos(\pi/2)} \sin\theta \cdot e^{-t} \cdot \frac{\cos^2\theta}{\mu_a \cdot h \cdot \sin\theta} dt =$$
$$= \frac{S_V}{2 \cdot \mu_s} \cdot \mu_a \cdot h \int_{\mu_a \cdot h}^{\infty} \sin\theta \cdot e^{-t} \cdot \frac{\cos^2\theta}{(\mu_a \cdot h)^2 \cdot \sin\theta} dt = \frac{S_V}{2 \cdot \mu_s} \mu_a \cdot h \int_{\mu_a \cdot h}^{\infty} \frac{e^{-t}}{t^2} dt \quad (7)$$

By using Eq. 6 with $x = \mu_a \cdot h$ and $n = 2$, the primary fluence rate can be written

$$\dot{\phi}_p = \frac{S_V}{2 \cdot \mu_s} E_2(\mu_a \cdot h) \quad (8)$$

The exponential integral $E_2(x)$ equals 1 for $x = 0$ and decreases monotonically to 0 as x approaches infinity; the value of $E_2(x)$ can be found in standard mathematical tables and graphs. A special case occurs when the reference point (detector) is at the surface of an infinite volume source, *i.e.* $h = 0$. Hence there is no attenuation in air and $\mu_a \cdot h = 0$, giving $E_2(0) = 1$. Eq. 8 then reduces to

$$\dot{\phi}_p = \frac{S_V}{2 \cdot \mu_s} \quad (9)$$

2.1.2 Infinite slab source

Ploughing after fallout of radionuclides on arable land may lead to a homogeneous distribution of the radionuclides in the upper part of the soil, whereas the deeper soil may be considered inactive with regard to the radionuclides in the fallout. The fluence rate from such a slab source (of infinite lateral extension) of thickness z can be found by taking the difference between two volume sources: one extending from the ground surface to infinite depth and one extending from depth z to infinity (Fig. 2).

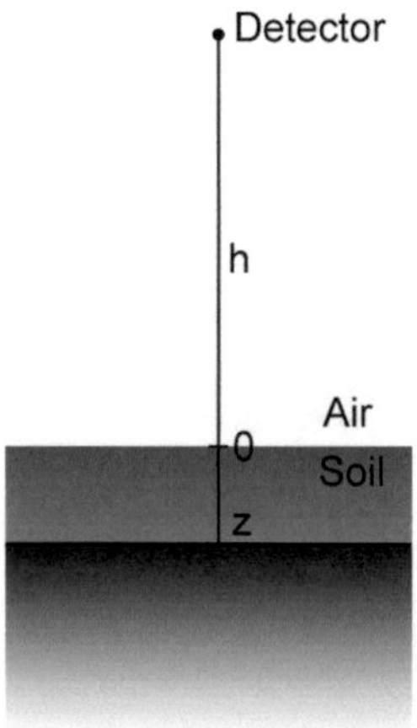

Fig. 2. The geometry of a slab source, extending from the soil surface to depth z.

The fluence rate from each of the two volume sources is given by Eq. 8 and the difference is then given by

$$\dot{\phi}_p = \frac{S_V}{2 \cdot \mu_s}\left[E_2(\mu_a \cdot h) - E_2(\mu_s \cdot z + \mu_a \cdot h)\right] \tag{10}$$

At the surface the attenuation in the air vanishes and Eq. 10 reduces to

$$\dot{\phi}_p = \frac{S_V}{2 \cdot \mu_s}\left[1 - E_2(\mu_s \cdot z)\right] \tag{11}$$

2.2 Plane source

Fresh fallout may be reasonably well described by a plane, or surface, source. The number of photons emitted per unit time from an area element dA at the surface of the ground is given by $S_A(r,\eta)\cdot$dA, where dA is given by d$r\cdot$dρ = d$r\cdot r\cdot$dη. The geometry of the calculations is shown in Figure 3.

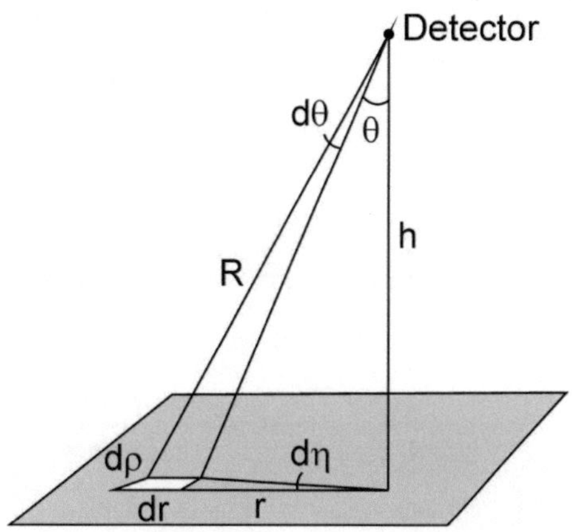

Fig. 3. The geometry used for calculating the primary fluence rate at the position of the detector, from an area element on the ground. The area element is considered as a point source.

The differential (primary) photon fluence rate at the detector position, taking attenuation in air (μ_a) into account is given by

$$d\dot{\phi}_p = \frac{S_A\, dA \cdot e^{-\mu_a \cdot h/\cos\theta}}{4\pi R^2} = \frac{S_A \cdot dr \cdot r \cdot d\eta \cdot e^{-\mu_a \cdot h/\cos\theta}}{4\pi R^2} \tag{12}$$

By integrating over all azimuthal angles, η, from 0 to 2π, according to Eq. 3 we have

$$d\dot{\phi}_p = \frac{S_A \cdot dr \cdot r \cdot e^{-\mu_a \cdot h/\cos\theta}}{2R^2} \tag{13}$$

2.2.1 Infinite plane source

The total primary photon fluence rate will again be found by integrating, this time over the whole plane, which we here will assume is of infinite extent. However, in order to be able to integrate, we need to find a relation between r and R. By looking at Figure 4 we see that $r = R\cdot\sin\theta$ and d$r\cdot\cos\theta = R\cdot$d$\theta$.

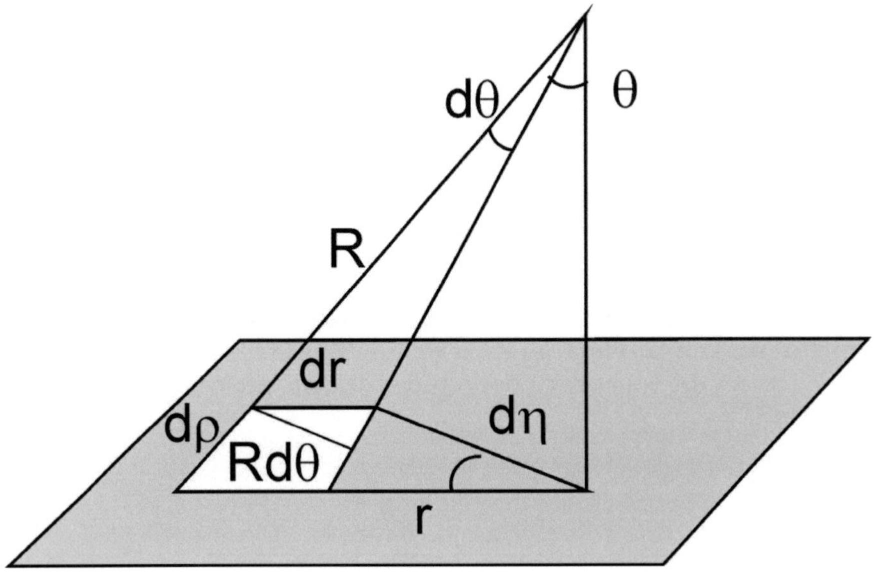

Fig. 4. Relations between distances and angles for the area element.

Substituting these relations into Eq. 13 gives an expression that may be integrated over the vertical angle θ.

$$d\dot{\phi}_p = \frac{S_A \cdot dr \cdot r \cdot e^{-\mu_a \cdot h/\cos\theta}}{2R^2} = \frac{S_A \cdot \sin\theta \cdot e^{-\mu_a \cdot h/\cos\theta}}{2 \cdot \cos\theta} d\theta \tag{14}$$

By making the same variable substitution as for the volume source, $t = \mu_a \cdot h/\cos\theta$, the primary fluence rate may again be expressed as an exponential integral (see Eq. 6).

$$\begin{aligned} \dot{\phi}_p &= \int_0^{\pi/2} \frac{S_A \cdot \sin\theta \cdot e^{-\mu_a \cdot h/\cos\theta}}{2 \cdot \cos\theta} d\theta = \frac{S_A}{2} \int_{\mu_a \cdot h/\cos 0}^{\mu_a \cdot h/\cos(\pi/2)} \frac{\sin\theta}{\cos\theta} \cdot e^{-t} \cdot \frac{\cos^2\theta}{\mu_a \cdot h \cdot \sin\theta} dt = \\ &= \frac{S_A}{2} \int_{\mu_a \cdot h}^{\infty} e^{-t} \cdot \frac{\cos\theta}{\mu_a \cdot h} dt = \frac{S_A}{2} \int_{\mu_a \cdot h}^{\infty} \frac{e^{-t}}{t} dt \end{aligned} \tag{15}$$

The exponential integral in this case is $E_1(\mu_a \cdot h)$ and the final expression for the primary fluence rate from an infinite plane source is

$$\dot{\phi}_p = \frac{S_A}{2} \cdot E_1(\mu_a h) \tag{16}$$

2.2.2 Disc source – Without attenuating media between source and detector

The primary fluence rate from a plane source of limited extent can be calculated by assuming a circular source, see Figure 5. The radionuclide is assumed to be homogeneously distributed over the circular area. The number of photons emitted per unit time from an area element dA is given by $S_A(r,\eta) \cdot dA$, where dA is given by $dr \cdot d\rho = dr \cdot r \cdot d\eta$ as shown in Figure 3.

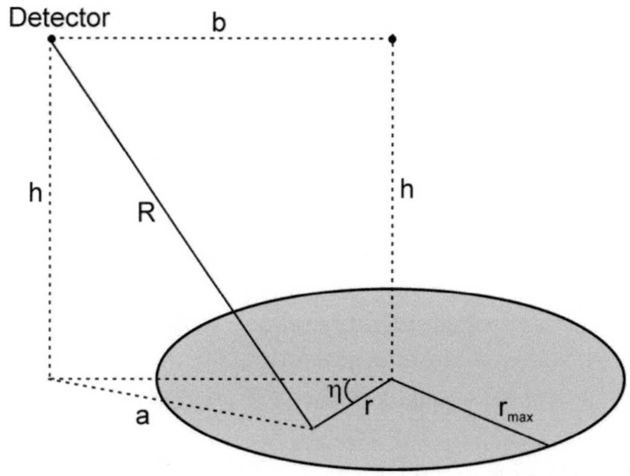

Fig. 5. Geometry for calculating the primary fluence rate from a circular area source of radius r_{max}.

Neglecting attenuation in the air and assuming that the area element acts as a point source, the primary fluence rate at the detector position is given by.

$$d\dot{\phi}_p = \frac{S_A \cdot dA}{4\pi \cdot R^2} \tag{17}$$

In order to carry out the integration to find the primary fluence rate, we need to express the distance R in more "fundamental" parameters. From Figure 5 we find that

$$a^2 = b^2 + r^2 - 2 \cdot r \cdot b \cdot \cos\eta \tag{18}$$

$$R^2 = h^2 + a^2 = h^2 + b^2 + r^2 - 2 \cdot r \cdot b \cdot \cos\eta \tag{19}$$

The expression for the primary fluence rate is then

$$\dot{\phi}_p = \frac{S_A}{4\pi} \int_0^{r_{max}} \int_0^{2\pi} \frac{r}{h^2 + b^2 + r^2 - 2 \cdot r \cdot b \cdot \cos\eta} dr\, d\eta \tag{20}$$

The integration over η is rather complicated, but could be solved using tables of standard integrals. Recognizing that the integral is of the form

$$\int_0^{2\pi} \frac{d\eta}{\alpha + \beta \cdot \cos\eta} = \frac{2\pi}{\sqrt{\alpha^2 - \beta^2}} \quad (\alpha > \beta \geq 0) \tag{21}$$

Eq. 20 now reduces to

$$\dot{\phi}_p = \frac{S_A}{2} \int_0^{r_{max}} \frac{r}{\sqrt{\left(h^2 + b^2 + r^2\right)^2 - 4 \cdot r^2 \cdot b^2}} dr \tag{22}$$

The expression within the square root in the denominator could be rewritten as

$$\left(h^2+b^2+r^2\right)^2-4\cdot r^2\cdot b^2=r^4+2r^2h^2-2r^2b^2+2h^2b^2+b^4+h^4 \tag{23}$$

and substituting $t = r^2$ yields

$$\left(h^2+b^2+r^2\right)^2-4\cdot r^2\cdot b^2=t^2+2t\cdot h^2-2t\cdot b^2+2h^2b^2+b^4+h^4 \tag{24}$$

By collecting terms we could write the right hand part of Eq. 24 as

$$t^2+2t\cdot h^2-2t\cdot b^2+2h^2b^2+b^4+h^4=t^2+2t\cdot\left(h^2-b^2\right)+\left(h^2+b^2\right)^2 \tag{25}$$

Now, since $r{\cdot}\mathrm{d}r = \mathrm{d}t/2$, Eq. 22 is given by

$$\dot{\phi}_p=\frac{S_A}{4}\int_0^{r_{\max}^2}\frac{dr}{\sqrt{t^2+2t\cdot\left(h^2-b^2\right)+\left(h^2+b^2\right)^2}} \tag{26}$$

The denominator is the square root of a polynomial and using standard integral tables we can find a solution since

$$\int\frac{dx}{\sqrt{\alpha x^2+\beta t+\gamma}}=\frac{1}{\sqrt{\alpha}}\ln\left(2\sqrt{\alpha\left(\alpha x^2+\beta t+\gamma\right)}+2\alpha x+\beta\right) \tag{27}$$

Identifying $\alpha = 1$, $\beta = 2(h^2\text{-}b^2)$, $\gamma = (h^2+b^2)^2$ and $x = t$ the integral in Eq. 26 is evaluated as

$$\left|\frac{1}{\sqrt{1}}\ln\left(2\sqrt{t^2+2t\cdot\left(h^2-b^2\right)+\left(h^2+b^2\right)^2}+2t+2\left(h^2-b^2\right)\right)\right|_0^{r_{\max}^2}=$$
$$=\ln\left(2\sqrt{r_{\max}^4+2r_{\max}^2\cdot\left(h^2-b^2\right)+\left(h^2+b^2\right)^2}+2r_{\max}^2+2\left(h^2-b^2\right)\right)-\ln\left(2\left(h^2+b^2\right)+2\left(h^2-b^2\right)\right)=$$
$$=\ln\frac{\sqrt{r_{\max}^4+2r_{\max}^2\cdot\left(h^2-b^2\right)+\left(h^2+b^2\right)^2}+r_{\max}^2+\left(h^2-b^2\right)}{2h^2} \tag{28}$$

And finally, the unattenuated fluence rate is given by

$$\dot{\phi}_p=\frac{S_A}{4}\ln\left|\frac{r_{\max}^2+h^2-b^2+\sqrt{\left(r_{\max}^2+h^2-b^2\right)^2+4h^2b^2}}{2h^2}\right| \tag{29}$$

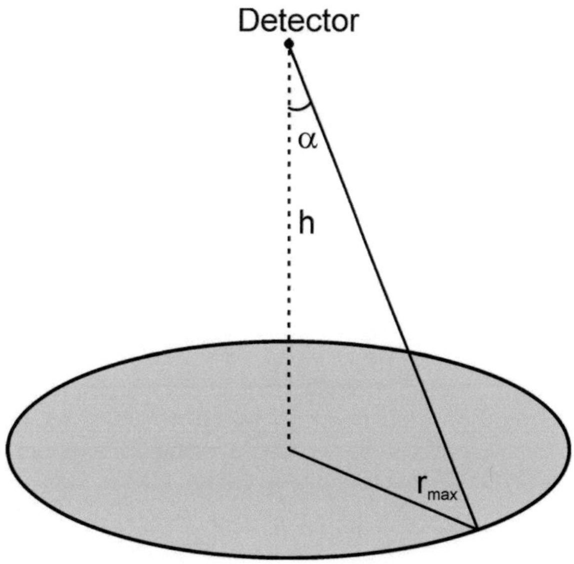

Fig. 6. Geometry for calculating the primary fluence rate at the axis of a circular area source of radius $r_{\max}$.

At a point centrally over the disc (Fig. 6), b equals zero and the expression for the fluence rate simplifies to

$$\dot{\phi}_p = \frac{S_A}{4}\ln\left(\frac{r_{\max}^2 + h^2}{h^2}\right) \tag{30}$$

In some measurements a plane source is approximated by a point source to simplify calculations, especially when trying to make a fast first estimation of the activity of the source. It should be noted that the commonly used "inverse square law", *i.e.* that the fluence rate decreases inversely proportional to the square of the distance between source and detector, is not generally valid for an extended source. The agreement depends on the radius of the source. It is clear, however, that beyond a certain distance an extended plane source may be replaced by a point source in the calculations.

This distance may be found by putting the fluence rate from a point source equal to the fluence rate from a disc source with the same activity (remember that S_A is the activity per unit area):

$$\frac{S_A \cdot \pi \cdot r_{\max}^2}{4\pi \cdot h^2} \approx \frac{S_A}{4}\ln\left(\frac{r_{\max}^2 + h^2}{h^2}\right) \quad \Rightarrow \quad \frac{r_{\max}^2}{h^2} \approx \ln\left(\frac{r_{\max}^2 + h^2}{h^2}\right) \tag{31}$$

A calculation shows that

$$\frac{r_{\max}^2}{h^2} \le 0.2 \quad \Rightarrow \quad \frac{\dfrac{r_{\max}^2}{h^2}}{\ln\left(\dfrac{r_{\max}^2 + h^2}{h^2}\right)} < 1.1 \tag{32}$$

meaning that the error we make is less than 10 % if $r_{max}/h < 0.45$ (since $0.45^2 = 0.2$). The disc source may hence be approximated by a point source if h is larger than 2.2 times the radius of the disc, r_{max}. In short, it is rather safe to make the approximation when the distance from the source is greater than the diameter of the source.

2.2.3 Disc source – With attenuating media between source and detector

If we want to take attenuation in the air into account we have to consider the exponential attenuation factor, which depends on the distance, R, (Fig. 5) and the material between the source and the detector position through the linear attenuation coefficient μ. Eq. 17 is then given by

$$d\dot{\phi}_p = \frac{S_A \cdot dA}{4\pi \cdot R^2} \cdot e^{-\mu \cdot R} \tag{33}$$

and the integral will be complicated to evaluate. However, if we again consider a point centrally over the disc the integral can be evaluated, using the exponential integral, as

$$\dot{\phi}_p = \frac{S_A}{2} \cdot \left[E_1(\mu_a h) - E_1\left(\frac{\mu_a h}{\cos\alpha}\right)\right] \tag{34}$$

where α is given by Figure. 6. If the radius of the source is very large we get, in the limit were r_{max} approaches infinity, $\cos\alpha = 0$ and hence $1/\cos\alpha \to \infty$. In the limit the exponential integral E_1 will be equal to zero and Eq. 34 reduces to

$$\dot{\phi}_p = \frac{S_A}{2} \cdot E_1(\mu_a h) \tag{35}$$

which is what we found for a plane source of infinite extent (Eq. 16).
A question that may arise is how large a plane source has to be before it can be treated as a source of infinite extent. Calculations with Eq. 34 and 35 for ^{137}Cs, assuming $S_A = 2$ m^{-2} s^{-1} (giving a unit multiplicative constant), are shown in Table 1. The calculations are made with $h = 100$ cm and $\mu_a = 9.3 \cdot 10^{-5}$ cm^{-1}, then $\mu_a \cdot h$ equals $9.3 \cdot 10^{-3}$ and $E_1(\mu_a \cdot h) = 4.1$.

α degrees	r_{max} m	$E_1(\mu_a \cdot h/\cos\alpha)$	$\dot{\phi}_p$ m^{-2} s^{-1}
10	0.18	4.10	0.015
20	0.36	4.05	0.062
30	0.58	3.97	0.14
40	0.84	3.85	0.26
50	1.2	3.67	0.44
60	1.7	3.43	0.68
70	2.7	3.05	1.1
80	5.7	2.40	1.7
89	57	0.52	3.6
89.9	570	$7.8 \cdot 10^{-4}$	4.1
90	∞	0	4.1

Table 1. Calculated primary fluence rate for disc sources of different radii, corresponding to vertical angles of 10° - 90°. The radii of the source for each vertical angle, assuming a detector height of 1 m, are also shown in the table. In this example the source strength $S_A = 2$ m^{-2} s^{-1}.

The primary fluence rate from a source of infinite extent is thus, in this example, equal to 4.1 m^{-2} s^{-1} and we see that the same value is reached for a limited source with a radius of about 500 metres. The table also shows that 90 % of the photons that reach the detector come from an area within a radius of about 60 metres.

2.3 Spherical source – Dose point inside the sphere

As an example of this geometry we can consider the release of radioactive material in the atmosphere. The radiation dose to a person submerged in a radioactive plume could be estimated using calculations based on this geometry. The same applies for a person surrounded by water containing radioactive elements. Figure 7 depicts the geometry, where the contribution to the primary fluence rate at a detector position in the centre of the sphere is due to a volume element dV. The number of photons emitted per unit time from the volume element dV is given by $S_V \cdot dV$.

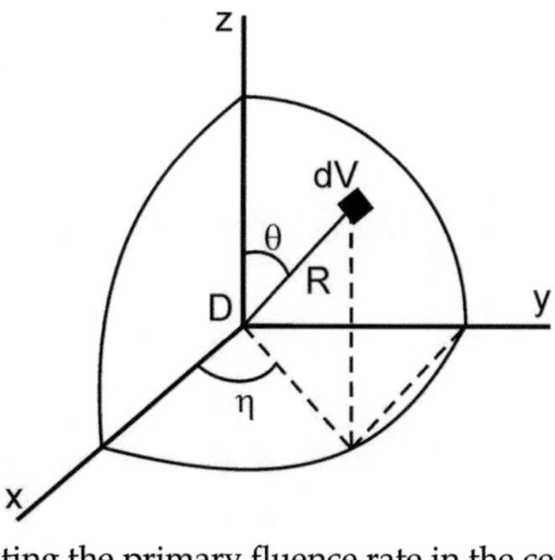

Fig. 7. Geometry for calculating the primary fluence rate in the centre of a spherical volume source. The source extends to the radius r_{max}.

Just as in the previous examples dV is treated as a point source and if we also take the attenuation into account we get

$$d\dot{\phi}_p = \frac{S_V \, dV \cdot e^{-\mu_s R}}{4\pi R^2} \tag{36}$$

where μ_s is the linear attenuation coefficient of the source medium. Recognizing that $dV = d\eta \cdot dR \cdot d\theta$ can be written in spherical coordinates as $dV = R^2 \cdot \sin\theta \cdot d\eta \cdot dR \cdot d\theta$ yields the expression for the total primary fluence rate at the detector position

$$\dot{\phi}_p = \int_0^{2\pi} \int_0^{r_{max}} \int_0^{\pi} \frac{S_V \cdot R^2 \cdot \sin\theta \cdot e^{-\mu_s R}}{4\pi R^2} d\eta \, dR \, d\theta \tag{37}$$

The integral is then evaluated as

$$\dot{\phi}_p = \frac{S_V}{4\pi} \cdot 2\pi \cdot \left| -\frac{1}{\mu_s} e^{-\mu_s R} \right|_0^{r_{max}} \cdot \left| -\cos\theta \right|_0^{\pi} = \frac{S_V}{\mu_s}\left(1 - e^{-\mu_s r_{max}}\right) \tag{38}$$

Assume that the source is a cloud or plume of air containing radioactive elements in the form of gasses or particles and the dose point is at the ground surface. Because of symmetry, the primary fluence rate is then given by half the value calculated from Eq. 38. This result is also consistent with what we found for a point at the surface of an infinite volume source, Eq. 9.

It could be interesting to see how large a radioactive plume can be before the primary fluence rate reaches equilibrium. The linear attenuation coefficient for 662 keV (gamma radiation from ^{137}Cs) in air is $9.3 \cdot 10^{-5}$ cm^{-1} and for a plume radius of 495 m the factor within the parenthesis is 0.99. Figure 8 shows the primary fluence rate as a function of plume radius for a spherical source in air and for some different photon energies, as well as the corresponding situation in water. Thus, for a plume of radius greater than about a few hundred metres, approximate calculations may be performed without knowledge of the linear attenuation coefficient for air. In water the radius is instead of the order of a few tens of centimetres.

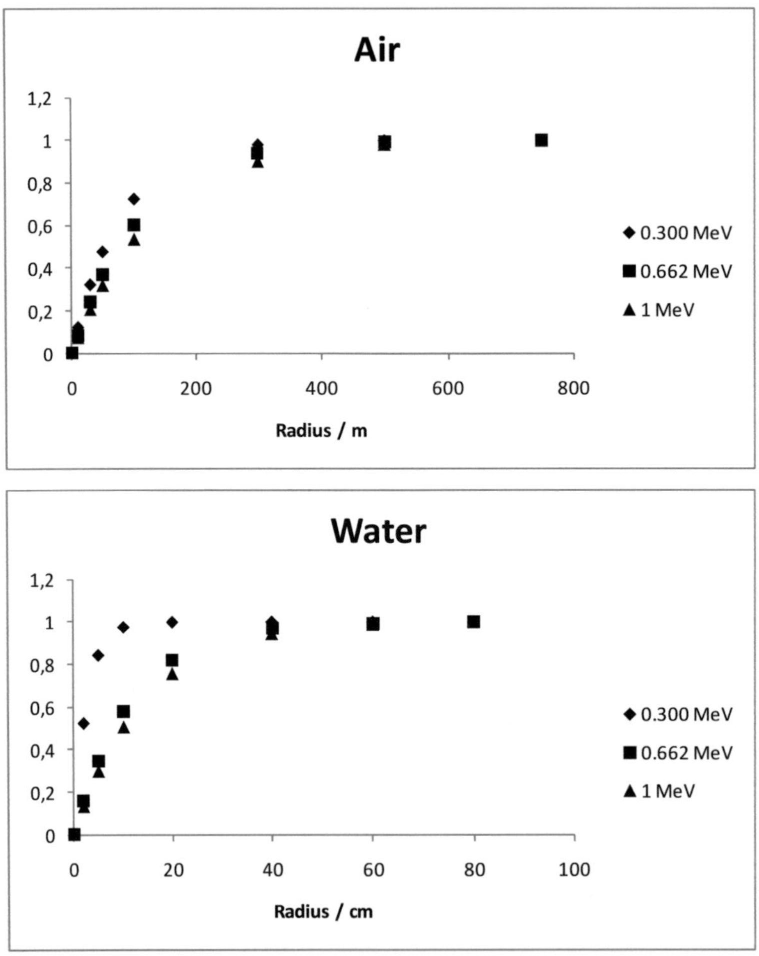

Fig. 8. Primary fluence rate at the centre of a spherical source of different radius, r_{max}, normalized to S_V/μ_s, for air and water, respectively, at some different photon energies.

3. Dose calculations

An important application of the calculations of primary fluence rate is to determine the effective dose to humans or absorbed dose to other biota. These calculations are often simplified by using conversions coefficients available from the literature. The concepts of absorbed dose, kerma and effective dose will be discussed below.

3.1 Absorbed dose rate and air kerma rate

Absorbed dose is defined by the ICRU (ICRU, 1998) and is basically a measure of how much of the energy in the radiation field that is retained in a small volume. The SI-unit of absorbed dose is 1 Gy (gray), which in the fundamental SI-units equals 1 J kg^{-1}. Absorbed dose rate to air in free air can be defined in terms of the fluence rate, photon energy, E_γ, and mass energy-absorption coefficient, μ_{en}/ρ, for air according to Eq. 39. The relation is valid for monoenergetic photons of energy E and if the radiation field consists of photons of different energies the contribution from each occurring photon energy has to be included in the calculation of the absorbed dose. In addition, the requirement of charged particle equilibrium, CPE, has to be fulfilled. This concept is further discussed in several dosimetry books (e.g. Attix, 1991, McParland, 2010).

$$\dot{D}_{p,E} = \dot{\phi}_{p,E} \cdot \left(\frac{\mu_{en}}{\rho}\right)_E \cdot E_\gamma \tag{39}$$

The total dose rate from primary and scattered photons is then given by

$$\dot{D}_{t,E} = \dot{D}_{p,E} + \dot{D}_{s,E} = \dot{D}_{p,E} \cdot B_E \tag{40}$$

where B_E is a build-up factor, which depends on the photon energy, and the distance and material between the source and the detector. Values of B_E can be determined by Monte Carlo-methods, found in tables and graphs, or calculated by analytical approximations (e.g. Shultis & Faw, 2000). The analytical expressions are convenient when the radiation field consists of photons of several energies, such as in environmental measurements. The contribution from each energy can then be integrated to yield the total absorbed dose. These calculations can of course be applied directly on the calculations of primary fluence rate instead.

Calculation of dose rate conversion factors, relating the activity per unit mass in the ground to dose rate at 1 m above ground, have been made by *e.g.* Clouvas *et al.* (2000). These factors can be applied when the activity in the ground has been determined from an *in situ* (field gamma spectrometry) measurement or gamma spectrometric soil sample analysis. As an example of the use of these conversion factors, consider the naturally occurring uranium decay series starting with ^{238}U and ending up with the stable lead isotope ^{206}Pb. To determine the activity of the radioactive elements in the series the bismuth isotope ^{214}Bi is often used due to the easily identified peak at 609.4 keV in the gamma spectrum. For this isotope a factor of 0.05348 nGy h^{-1} per Bq kg^{-1} is given by Clouvas *et al.* (2000). If secular equilibrium can be assumed, the activities of all members of the series are equal and the dose rate from all radioactive elements in the series can be calculated by using the factor 0.38092 nGy h^{-1} per Bq kg^{-1}, using the activity of the measured ^{214}Bi.

One member of the thorium series (^{232}Th to ^{208}Pb) is ^{208}Tl, which also has a convenient gamma line at 583.1 keV. The activity of this radionuclide, however, can not be used in the same straight-forward way as ^{214}Bi in the uranium series. The previous radionuclide in the series, ^{212}Bi, decays both by α (35.9 %) and β-decay (64.1 %) to ^{208}Tl and ^{212}Po, respectively and hence the activity of ^{208}Tl is only 35.9 % of the activity at equilibrium. The thorium series, as well as the uranium series, contains isotopes of radon that easily escape from the ground or from a sample. The degree of equilibrium therefore ought to be checked by measuring the activity of some member of each series, above radon.

Related to absorbed dose is the concept of kerma, *K*. The unit of this quantity is also 1 Gy, but kerma is a measure of how much of the photon energy in the radiation field that is transferred, via interactions, to kinetic energy of charged particles in a small volume (ICRU, 1998). Part of this kinetic energy may be radiated as breaking radiation, thus leaving the volume, and will not be a part of the absorbed dose. To account for this energy loss, kerma is defined in terms of the mass energy-transfer coefficient, μ_{tr}/ρ, insted of μ_{en}/ρ. An expression for air kerma rate in free air can thus be written

$$\dot{K}_{p,E} = \dot{\phi}_{p,E} \cdot \left(\frac{\mu_{tr}}{\rho}\right)_E \cdot E_\gamma \tag{41}$$

For a point source, an air kerma rate constant can be defined, which enables calculation of the air kerma rate from knowledge of the source activity. Such constants can be found in *e.g.* Ninkovic *et al.* (2005). The air kerma rate free in air is a useful concept in environmental dosimetry, since published dose conversion coefficients to determine effective dose are often expressed in this quantity, which will be discussed below.

3.2 Exposure of humans – Effective dose

The quantity absorbed dose is basically a measure of the energy imparted in a small volume due to irradiation. In some applications, such as radiation therapy, absorbed dose is used to determine the biological effect on (tumour) tissue but in radiation protection its use is limited. The effects on tissue depend, apart from the absorbed dose, on the type of radiation, *i.e.* whether the energy is transferred to tissue by α-particles, β-particles, γ-radiation or neutrons. For example, to cause the same degree of biological damage to cells the absorbed dose from γ-radiation has to be twenty times as large as the absorbed dose from α-particles. This dependence is accounted for by multiplying the absorbed dose with a so called radiation weighting factor (ICRP, 2007). The weighted quantity is then given a new name, Equivalent dose, H_T, and a new unit, 1 Sv (sievert). Equivalent dose can be used in regulations to limit, for example, the exposure to the hands and feet in radiological work.

However, in environmental radiation fields a large number of different irradiation geometries may occur and to determine the risk for an exposed individual would require knowledge of the equivalent dose to each exposed organ in the body. Furthermore, risk estimates for all possible combinations of irradiated organs for each possible equivalent dose would be needed. It is obvious that such a table would be quite cumbersome (mildly speaking) and instead ICRP has defined the quantity Effective dose (Jacobi, 1975; ICRP, 1991), with the unit 1 Sv. The effective dose, *E*, is a so called risk related quantity that can be assigned a risk coefficient for different detriments (*e.g.* cancer incidence) and be used for regulatory purposes concerning populations. The quantity is not applicable to individuals.

Effective dose is calculated by multiplying the equivalent dose to each exposed organ by a tissue weighting factor, w_T, and summing the contributions from each organ or tissue, T, according to Eq. 42. The effective dose can thus be described as the dose that, if given uniformly to the whole body, corresponds to the same risk as the individual organ doses together.

$$E = \sum_{T} w_T \cdot H_T \tag{42}$$

The effective dose, E, has replaced the earlier used quantity effective dose equivalent, H_E (ICRP, 1977; ICRP, 1991). Although the effective dose has been in use for many years, dose conversion factors published before the replacement and hence given in terms of effective dose equivalent are still useful (*e.g.* Jacob *et al.*, 1988). Relations between E and H_E has been published by ICRP (1996) and the difference is less than 12 % for all photon energies above 100 keV and for all irradiation geometries (see Fig. 9). For rotational symmetry E/H_E is between 0.95 and 1.0 for photon energies larger than 100 keV. Thus, for environmental applications the two quantities may often be interchangeable.

3.3 Relation between air kerma and effective dose

The effective dose can be related to the physical quantities photon fluence, absorbed dose and kerma by conversion coefficients and ICRP (1996) reports conversion coefficients from both photon fluence, ϕ, and air kerma, K_a, to effective dose. Conversion coefficients derived from measurements in a contaminated environment have also been reported (*e.g.* Golikov et al., 2007). The determination of effective dose from air kerma requires detailed knowledge about the energy deposition in the human body, *i.e.* the absorbed dose to each organ connected to an organ weighting factor. This can be achieved through Monte Carlo-simulations using different kinds of models or body-like phantoms. The effective dose will thus depend on the irradiation geometry and hence the simulations will yield different conversion factors for different irradiation geometries. Since the actual number of geometries can be very large, conversion coefficients are given for some idealised geometries, which approximates real situations (Fig. 9).

The AP & PA geometry approximates the irradiation from a single source in front of or behind the person, respectively and LAT may be used when a single source is placed to the left (or right) of the body. When the source is planar and widely dispersed the ROT geometry will be a good approximation, whereas a body suspended in a cloud of a radioactive gas can be approximated by the ISO geometry. The most usable geometries for environmental applications are ROT and ISO and Figure 10 shows the value of the conversion coefficient E/K_a for these geometries for different photon energies.

According to Figure 10 the conversion coefficient between effective dose rate and air kerma rate in rotational invariant irradiation situations concerning naturally occurring radionuclides is around 0.8 Sv Gy^{-1}, since the mean value of the gamma energy of those is close to 1 MeV. However, Golikov et al. (2007) reports a value of 0.71 Sv Gy^{-1} for adults and 1.05 Sv Gy^{-1} for a 1-year old child, based on phantom measurements in an environment contaminated with ^{137}Cs. Since the conversion coefficient is based on the absorbed dose to several organs in the body it will depend on the size of the body (distance from the source and self-shielding) as well as the energy distribution of the photons from the source. The photon fluence from a surface source will consist mainly of primary photons whereas the contribution from scattered photons (of lower energies) increases when the source is distributed with depth.

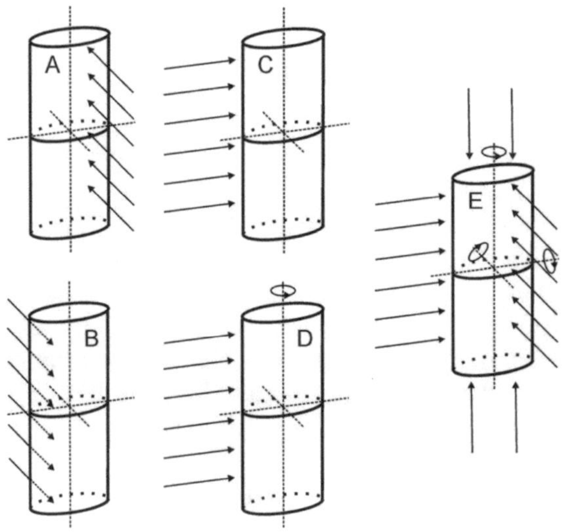

Fig. 9. Irradiation geometries used for simulation of real exposure situations: AP (Anterior-Posterior); B: PA (Posterior-Anterior); C: LAT (Lateral, in some applications specified as RLAT or LLAT depending on the direction of the radiation: from the left or from the right); D: ROT (Rotationally symmetric) and E: ISO (Isotropic radiation field).

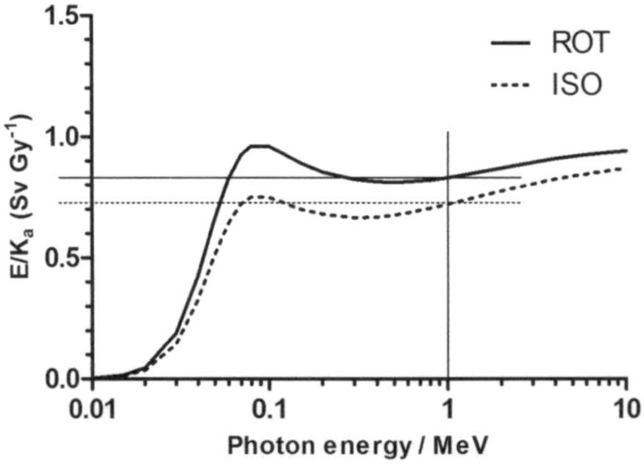

Fig. 10. Effective dose per unit air kerma for irradiation geometries ROT and ISO. Rotational symmetry (ROT) is applicable to a planar and widely dispersed source; isotropic symmetry may be used for a body suspended in a cloud of a radioactive gas. Data from ICRP (1996).

Also the angular distribution of the photons plays a role in this aspect. Figure 11 shows, schematically, the angular distribution of 662-keV photons from two different sources: an infinite volume source and an infinite plane source; the detector is placed one metre above ground. The expressions for the angular distribution of primary photons can be derived from Eq. 5 and Eq. 14, respectively. For angles of incidence less than about 75° the fluence increases with increasing angle of incidence due to the increasing source volume and area, respectively. At larger angles the attenuation in air limits the fluence at the detector and both curves in Figure 11 has a maximum. Detailed calculations for this photon energy show that the maximum for the infinite volume source occurs at 77.91° and the maximum for the infinite plane source at 89.45° (Finck, 1992). Thus most of the primary photons hitting the detector are nearly parallel to the ground surface.

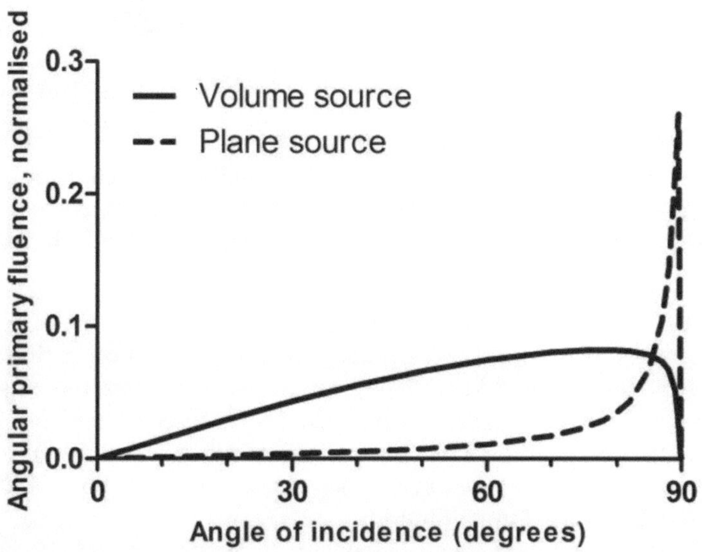

Fig. 11. Angular distribution of primary 662-keV photons from an infinite volume source and an infinite plane source as a function of angle of incidence on a detector placed one metre above ground.

4. Measurements

Since the risk related quantities, *e.g.* effective dose, could not be measured, some measurable quantities need to be defined. These measurable quantities should then be an estimation of the risk related quantities. As an example of the use of data from field gamma spectrometry to estimate the radiation dose, some results from repeated measurements are presented.

4.1 Measurable quantities

The relations between the different categories of quantities, risk related and measurable, are shown in Figure 12. The physical quantities, such as the absorbed dose, can be used to calculate risk related quantities by the use of weighting factors (ICRP 1991; ICRP, 2007), but also by conversion coefficients (*e.g.* ICRP, 1996). Operational quantities are derived from the physical quantities through definitions given by the ICRU (ICRU 1993; ICRU, 2001), for photon as well as for neutron irradiation. These definitions utilises the so called ICRU-sphere - a tissue equivalent sphere of 30 cm diameter - and the absorbed dose at different depths in the sphere. Furthermore, relations between risk related and measurable quantities can be found from data given by the ICRP (ICRP, 1996).

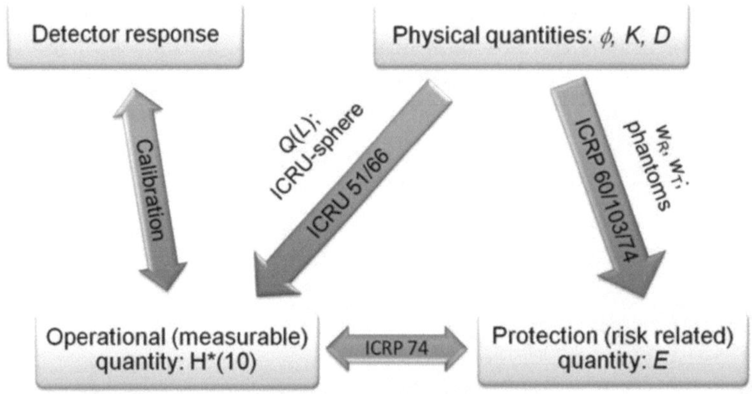

Fig. 12. Relations between the quantities of interest in radiation protection and measurements. Only one of the measurable and risk related quantities, respectively are shown in the figure; several other operational quantities are defined, which are used to monitor the radiation dose to individuals.

Instruments used for radiation protection purposes, *e.g.* intensimeters, are often calibrated to directly show a measurable quantity and guidelines for the calibration of those instruments have been issued by the IAEA (IAEA, 2000). Other types of detectors, *e.g.* high-purity germanium detectors (for determination of activity or fluence rate by field gamma spectrometry) or ionisation chambers (for determination of absorbed dose), can be used to measure a physical quantity. The relationship between detector response and the physical quantity of interest is then found by calibration.

Several measurable quantities have been defined for different purposes. Some are used to monitor personal exposure and others to monitor the radiation environment. The quantity of interest in environmental measurements is often the ambient dose equivalent, *H**(10), since it is an estimate of the effective dose. From Figure 13 it is obvious that an instrument calibrated to show ambient dose equivalent should never underestimate the effective dose. This may cause deviations if effective dose is estimated from field gamma spectrometry and compared to the reading of an intensimeter calibrated to show ambient dose equivalent.

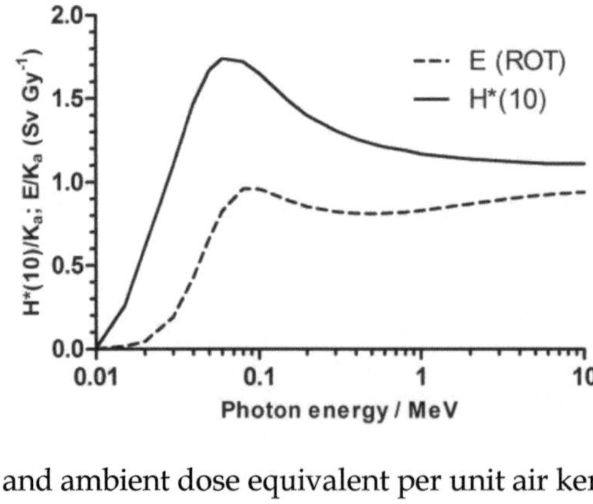

Fig. 13. Effective dose and ambient dose equivalent per unit air kerma, respectively. Data from ICRP (1996).

When the effective dose is estimated by detectors worn by individuals, special attention ought to be given to how the detector is calibrated. In these measurements the detector is often a TL-dosimeter (Thermo Luminescence Dosimetry, TLD), which is commonly used for monitoring workers at hospitals or in the nuclear industry. These detectors are calibrated to show the personal dose equivalent and if the reading is to be used for estimation of effective dose, the relation between the reading and air kerma must be known. Thereafter the effective dose can be estimated by the relations given above.

4.2 Measurements in western Sweden

We have made repeated field gamma measurements at 34 predetermined sites in western Sweden (Almgren & Isaksson, 2009) and the results are shown here to give an example of the use of these data to estimate the ambient dose equivalent. By a proper calibration of the field gamma detector, the amount of different radioactive elements in the ground can be determined. For the naturally occurring radionuclides with an assumed homogeneous depth distribution the inventory is given as Bq kg^{-1}. However, for ^{137}Cs a plane source is assumed and the activity given as Bq m^{-2}. The latter is a common procedure when the depth distribution is unknown and the reported "equivalent surface deposition" (Finck, 1992) will underestimate the true inventory due to the absorption of photons in the ground. The quantity is, however, still a good measure of the photon fluence rate above the ground.
Using published dose rate conversion factors and the relation between absorbed dose and ambient dose equivalent the field gamma measurements may be compared to intensimeter measurements made in connection to the field gamma measurements. Figure 14 shows the sum of the contribution to the ambient dose equivalent from the radionuclides in the uranium series, the thorium series and ^{40}K, as well as the contribution from ^{137}Cs. The figure also shows the results from intensimeter measurements, corrected for the contribution from cosmic radiation. Although a correction has been made to compensate for the fact that the intensimeter is calibrated for ^{137}Cs (0.662 MeV), whereas the mean energy of the naturally

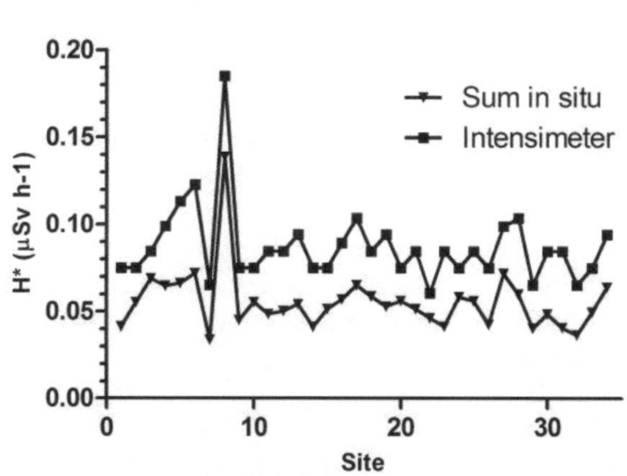

Fig. 14. Ambient dose equivalent rate at 34 reference sites shown as the sum of the contribution from the radionuclides in the uranium series, the thorium series and ^{40}K, as well as the contribution from ^{137}Cs. Also shown are the results from intensimeter measurements, corrected for the contribution from cosmic radiation and photon energy used in the calibration.

occurring radionuclides are slightly higher, a deviation between the results remains. One explanation may be that the measured area differs due to different angular sensitivity of the two measurements systems and that the correction of the intensimeter reading is insufficient. Still there is a good agreement between the two methods to estimate the ambient dose equivalent.

Figure 15 shows the contribution from each of the terms in the sum depicted in Figure 14. The main contributor to the ambient dose equivalent is ^{40}K and the contribution from ^{137}Cs is practically negligible in this area.

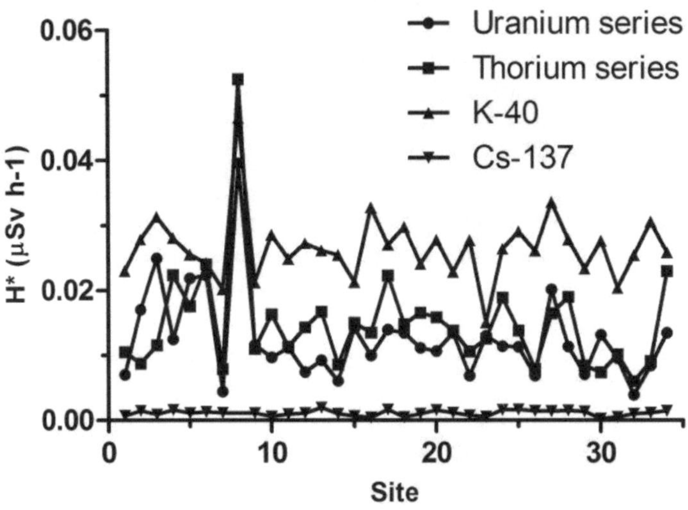

Fig. 15. Ambient dose equivalent rate at 34 reference sites from the radionuclides in the uranium series, the thorium series and ^{40}K, as well as from ^{137}Cs.

5. Conclusion

This chapter includes the basic relations for calculating the primary photon fluence rate from environmental sources of different shapes. Such calculations may be used for calibrating field equipment and also for estimating the exposure to people in the vicinity of the source. The chapter also dealt with practical environmental measurements and the importance to keep in mind the relation between effective dose and ambient dose equivalent. Measurements made by intensimeters tend to overestimate the effective dose due to the calibration requirements.

6. References

Almgren, S. & Isaksson, M. (2009). Long-term investigation of anthropogenic and naturally occurring radionuclides at reference sites in western Sweden, *Journal of Environmental Radioactivity*, Vol.100, pp.599-604.

Attix, F H. (1991). *Introduction to radiological physics and radiation dosimetry*, ISBN 0471011460, Wiley-VCH Verlag GmbH, Germany.

Clouvas, A.; Xanthos, S.; Antonopoulos-Domis, M. & Silva, J. (2000). Monte Carlo Calculation of Dose Rate Conversion Factors for External Exposure to Photon Emitters in Soil, *Health Physics*, Vol.78, No.3, pp.295-302.

Finck, R R. (1992). *High resolution field gamma spectrometry and its application to problems in environmental radiology*. Thesis. University of Lund, Department of Radiation Physics, Malmö, Sweden.

Golikov, V. ; Wallström, E. ; Wöhni, T. ; Tanaka, K. ; Endo, S. & Hoshi, M. (2007). Evaluation of conversion coefficients from measurable to risk quantities for external exposure over contaminated soil by use of physical human phantoms, *Radiation and environmental biophysics*, Vol.46, pp.375-382.

International Commission on Radiation Units and Measurements (1993). *Quantities and Units in Radiation Protection Dosimetry,* ICRU Publications 51, ICRU Publications, Bethesda, USA

International Commission on Radiation Units and Measurements (1998). *Fundamental Quantities and Units for Ionizing Radiation,* ICRU Publications 60 ,ICRU Publications, Bethesda, USA

International Commission on Radiation Units and Measurements (2001). *Determination of Operational Dose Equivalent Quantities for Neutrons,* ICRU Publications 66, ICRU Publications, Bethesda, USA

International Commission on Radiological Protection (1977). *Recommendations of the International Commission on Radiologicul Protection*. ICRP Publication 26. Annuals of the ICRP l(3). Pergamon Press, Oxford.

International Commission on Radiological Protection (1991). *1990 Recommendations of the International Commission on Radiological Protection*. ICRP Publication 60. Ann. ICRP 21 (1–3).

International Commission on Radiological Protection (1996). *Conversion coefficients for use in radiological protection against external radiation*. ICRP Publication 74. Ann. ICRP 26 (3/4).

International Commission on Radiological Protection (2007). *The 2007 Recommendations of the International Commission on Radiological Protection*, ICRP Publication 103, Ann. ICRP 37 (2-4)

International Atomic Energy Agency (2000). *Calibration of Radiation Protection Monitoring Instruments*, IAEA Safety Reports Series No.16.

Jacob, P.; Paretzke, H G.; Rosenbaum, H. & Zankl, M. (1988). Organ Doses from Radionuclides on the Ground. Part I. Simple Time Dependencies, *Health Physics*. Vol.54, No.6, pp. 617-633.

Jacobi, W. (1975). The Concept of the Effective Dose - A Proposal for the Combination of Organ Doses, *Rad. and Environm. Biophys*. Vol.12, pp.101-109.

McParland, B J. (2010). *Nuclear medicine radiation dosimetry*, ISBN 978-1-84882-125-5, Springer-Verlag London, GB.

Ninkovic, M M.; Raicevic, J J. & Adrovic, F. (2005). Air Kerma Rate Constants for Gamma Emitters used most often in Practice, *Radiation Protection Dosimetry*, Vol.115, No.1–4, pp. 247–250.

Shultis, J K & Faw, R E. (2000). *Radiation Shielding*, ISBN 0-89448-456-7, American Nuclear Society, La Grange Park, IL, USA.

United Nations Scientific Committee on the Effects of Atomic Radiation. (2008). *UNSCEAR 2008 Report to the General Assembly, with scientific annexes. Annex B.*

11

Radiological Survey in Soil of South America

María Luciana Montes and Judith Desimoni
Departamento de Física, Facultad de Ciencias Exactas, Universidad Nacional de La Plata
Instituto de Física La Plata – CONICET,
Argentina

1. Introduction

When the Earth was formed, the crust and consequently the soil and water were conformed by a wide variety of chemical elements with different concentrations; being some of these radioactives. There are different activity levels of natural radionuclides, as those of the ^{238}U and ^{232}Th decay chains, ^{40}K, ^{7}Be and ^{14}C, etc. along the planet [Cooper et al., 2003]. Among the 80 nuclides found in the environment, the more relevant concerning the radiobiological significance are ^{40}K, and the nuclides belonging to the ^{238}U and ^{232}Th decay chains. The human activities can strongly modify the natural concentrations due to the presence of residues or accumulation of elements caused by the release of effluents to the environment. In the 60´s the nuclear power production and nuclear weapon testing discharge to the environment anthropogenic nuclides. In particular, the Southern Hemisphere was mainly polluted by the debris originated in the South Pacific and middle Atlantic nuclear weapon tests [UNSCEAR, 2008]. Along with the class of anthropogenic gamma emitter nuclides releases, the ^{137}Cs is the most prominent isotope in the Earth crust originated by fission process. It is considered as one of the hazardous environmental contaminant due to the contribution to the external irradiation exposure and its incorporation to the human food chain [Singh et al., 2009].

Regardless, both natural and man-made nuclides have radiobiological implication because they significantly contribute to human external radiation dose and to the internal dose by inhalation and ingestion [Cooper et al., 2003; UNSCEAR, 2008]. The United Nations Scientific Committee on the Effects of Atomic Radiation (UNSCEAR) has estimated that exposure to natural sources is approximately 98% of the total radiation dose (excluding medical exposure) [UNSCEAR, 2000; UNSCEAR, 2008]. The dose arising from natural nuclides varies worldwide depending upon factors such as height above sea level, the amount and type of radionuclides in the air, food and water, as well as the concentration of the natural nuclides in the soil and rocks, which in turn depend on the local geology of each region, etc.

The information about the presence and migration anthropogenic radionuclides is crucial to fully understand the long-term behaviour in the environment, the uptake by flora and fauna including the human food chain, as well as potential contribution to groundwater. In consequence, before assessing the radiation dose to the population, a precise knowledge of the activity of a number of radionuclides is required [UNSCEAR, 2000;

UNSCEAR, 2008]. The mobility of the radionuclide in the ecosystem involves a number of complex mechanisms [Velasco et al., 2006; IAEA, 2010; Salbu, 2009; Cooper et al., 2003; Sawhney, 1972; Cornell, 1993; Staunton et al. , 2002; Bellenguer et al., 2008], and their transfer through the environmental compartments implies multiple interactions between the biotic and abiotic components of the ecosystem, as well as human interferences like the use of fertilizer [Tomazini da Conceic & Bonotto, 2006] or the overexploitation of the natural resources. For the identification of these interactions it is necessary to develop and test predictive models describing the radionuclide fluxes from the environment to the man.

In South America, the soil resource is extensively used in agriculture, stockbreeding and for building materials. Baselines of natural and anthropogenic activity nuclides in several countries are not established ye, as well regulations concerning the natural and anthropogenic activity and chemical restrictions in freshwater and food accordingly to the local situations. These facts and the scattered of the activity dataset put in relevance the present review on nuclide activity determinations in soils of South America, that could be considered as the first attempt in this direction.

A systematic compilation of radionuclide activity data of soil of Argentina, Brazil, Chile, Venezuela and Uruguay are presented. Radionuclide activity data concern to the natural ^{40}K, ^{238}U, and ^{232}Th and to the anthropogenic ^{137}Cs nuclides. These different pieces of information are put together, the quality of the environmental compartments is provided and the impact on the population is evaluated throughout the exposure dose. The migration of ^{137}Cs in soil is also analysed in the frame of different approaches [Kirchner, 1998; Schuller et al., 1997], and the transport parameters are discussed. Moreover, the caesium inventories are compared with the latitudinal UNSCEAR predictions [UNSCEAR 2000, UNSCEAR 2008].

2. Radionuclides in the environment

The man is continuously exposed to natural radiation since radioactive material is present in throughout nature. It occurs naturally in the soil, rocks, water, air, and vegetation. The components of the natural radioactive background are the cosmic radiation and the natural radioactivity of ground, atmosphere and water. Natural environmental radioactivity arises mainly from primordial radionuclides, such as ^{40}K and the nuclides from the ^{232}Th and ^{238}U series, which are at trace levels in all ground formations. Natural environmental radioactivity and the associated external exposure due to gamma radiation are primarily up to the geological and geographical conditions [UNSCEAR, 2000]. The specific concentrations of terrestrial environmental radiation are related to the composition of each lithologically separated area, and to the type of parental material from which the soils originate.

The high geochemical mobility of radionuclides in the environment allows them to move easily throughout the environmental matrixes. Rivers erode soil which contains radionuclides, and they reach lakes and oceans; atmospheric depositions can also occur on their surfaces; and groundwater containing some radionuclides can reach them.

Concerning the presence of artificial nuclides in the environment, after bombarding Hiroshima and Nagasaki in 1945, USA, USSR, France, England and Chine deserved to be a nuclear potency. In this frame, 543 underground and atmospheric nuclear weapon essays were carried from 1945 to 1980 in different regions of the globe. URSS, Chine and USA performed the tests in the North Hemisphere, while England and France in the South Hemisphere. The underground essays were the more numerous; however, the global environmental impact resulted small because the radioactive material remains in the essay area. On the contrary, the atmospheric ones delivered to the atmosphere huge amounts of radioactive detritus causing a big impact on the environment [UNSCEAR, 2008; Valkovic, 2000]. It is worth to mention that because of the atmospheric circulation, approximately the 82 % of the debris remain in the hemisphere of injection [UNSCEAR, 2008; Valkovic, 2000]. The relevant nuclides originate in the essays were ^{3}H, ^{14}C, ^{54}Mn, ^{55}Fe, ^{85}Kr, ^{89}Sr, ^{90}Sr, ^{95}Zr, ^{103}Ru, ^{106}Ru, ^{131}I, ^{137}Cs, ^{131}Ce and ^{144}Ce, among others [UNSCEAR, 1982]. Due to ^{90}Sr and ^{137}Cs are volatiles and have large half-life (28.6 years and 30.2 years, respectively) they are dispersed in the atmosphere, comprising the stratospheric global fallout, contributing to the residual background.

When analyzing the total annual effective dose received by human from natural sources, the dose received by the cosmic ray, terrestrial exposure, ingestion and inhalation of long-lived natural radionuclides needs consideration. Each environmental matrix, e.g. soil, air and water, has several associated pathways. These three environmental media cannot be thought as isolated and so, nuclide transfers are produced from one to the other. The different pathways exposure routes are schematized in the Fig. 1. The importance of these paths depends upon the particular radionuclide or radionuclides present in each compartment. The starting point to evaluate the people doses is to determine the nuclide concentrations in the environmental matrixes [USNCEAR, 2000; UNSCEAR, 2008].

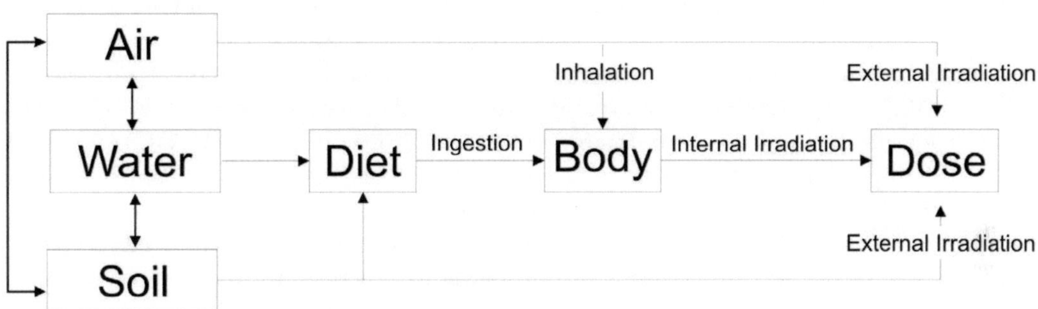

Fig. 1. Schematic terrestrial pathways of nuclide transfers and dose to humans.

3. Monitored regions and dataset

Two kinds of surveys have been performed, some of them deal with the determination of nuclide activity concentrations in depth, while others only reported single values of surface activity concentrations. Argentina, Brazil and Chile are the most studied countries, while there are reported a few data of Venezuela and Uruguay. The location of the monitored places, type of survey and monitored nuclides are summarized in the Table 1. Regarding the natural nuclides in South America, the reports of UNSCEAR only account values of the activity concentration of the natural nuclide ^{40}K for Argentina [UNSCEAR, 2000; UNSCEAR, 2008]. In San Luis Province, Argentina, two sites have been studied [Juri Ayub et al., 2008]. Recently, the first systematic studies to establish baseline activities for the naturally occurring radionuclides in unperturbed soils around La Plata city, Province of Buenos Aires, have been settled on samples taken from the surface down to a depth of 50 cm [Montes et al. 2010a, 2010b]. Moreover, in four superficial soils in the Ezeiza region, Argentina, the activities of ^{40}K and of natural chains of ^{238}U and ^{232}Th have been determined [Montes et al. 2011]. In the Brazilian State of Rio Grande do Norte the average concentrations of ^{226}Ra, ^{232}Th and ^{40}K in unperturbed soils have been determined [Malanka et al., 1996]. Samples of soils were also studied in different departments of Uruguay since 2004 to determine the activity concentrations of ^{40}K, ^{226}Ra and ^{232}Th up to 5 cm of depth [Odino Moure, 2010]. Regarding the anthropogenic nuclides, in Argentina ^{137}Cs reference activity profile was determined in the Pampa Ondulada region [Bujan et al., 2000, 2003] and in the central part of the country in natural and semi-natural grassland regions [Juri Ayub et al., 2007, 2008]. Beside the natural chains values, the profiles of ^{137}Cs in the region of Buenos Aires Province have been settled [Montes et al. 2010a, 2010b]. Some studies have been performed in Brazil, dealing with the determination of the activity of the ^{137}Cs globally presented on the soil because of nuclear weapon tests [Correchel et al., 2005; Handl et al., 2008]. Total inventories and depth distributions of ^{137}Cs were established in agricultural and sheep-farming regions of Chile [Schuller et al., 1997, 2002, 2004]. In Uruguay, surface soil ^{137}Cs activity has been determined in different regions since 2004 [Odino Moure, 2010]. In Venezuela, the ^{137}Cs concentration at two different depth (0 cm -20 cm and 20 cm - 40cm) were measured [Sajó-Bohus et al., 1999].

3.1 Natural radionuclides

According to the UNSCEAR [UNSCEAR, 2000], in South America only the activity concentration of ^{40}K in unperturbed soils has been measured in Argentina (UN in Table 1 and 2), being the activity concentration range 540 Bq/kg -750 Bq/kg. Later, data profiles of ^{226}Ra and ^{40}K of semi-natural grassland soils of the central part of the country, Province of San Luis (AS23 and AS24) have been reported down to 25 cm depth [Juri Ayub, 2008]. The activity concentrations of ^{40}K were determined to vary from 720 Bq/kg to 750 Bq/kg very close to the upper limit of the values reported by UNSCEAR [UNSCEAR, 2000; UNSCEAR, 2008] , while ^{226}Ra activities were in the range 64 Bq/kg to 73 Bq/kg, as observed in Fig. 2. The profiles recorded down to 22.5 cm indicated that both nuclides activity concentrations are constant in depth (see Fig.2). Activity concentrations down to 50 cm of natural nuclides (^{238}U and ^{232}Th chains and ^{40}K) have been determined in soil samples collected from inland (AS1 and AS2) and coastal (AS3 and AS4) areas of the La Plata River, located in the North eastern region of the Province of Buenos Aires, Argentina [Montes et al., 2010a; Montes et al., 2010b]. The main observed activity resulted originated from the decay of the ^{40}K with following in importance those of the natural ^{238}U (obtained from the ^{226}Ra activity) and ^{232}Th (obtained from the ^{228}Ac, ^{212}Pb, ^{212}Bi and ^{208}Tl activities) chains, as shown in Fig. 2. While the activity of ^{235}U was, in all the cases, lower than the detection limit (L_D= 0.02Bq/kg), the activity values of the ^{238}U and ^{232}Th chains lay in the intervals 52 Bq/kg – 104 Bq/kg and 32 Bq/kg - 50 Bq/kg, respectively. In the case of the ^{238}U, the activities resulted to some extent high when comparing with data from Uruguay [Odino Moure, 2010]. It was also observed that the coastal soils without magnetite and lower hematite relative fraction presented a higher U probably related to the geological origin of the soils [Montes et al., 2010a; Montes et al., 2010b]. The ^{40}K activity profiles were quite different when comparing the monitored soils, ranging the surface activities values from 531 Bq/kg to 873 Bq/kg as observed in Fig. 2. In the inland profiles, the activity increased with depth and the depletion of the activity was detected in the approximately first 20 cm of the inland soils. The Fe^{3+} relative fractions determined from Mössbauer spectroscopy [Vandenberghe, 1991] and the ^{40}K distribution had quite similar behaviour. This correlation could be ascribed to the soil pedogenic and edaphic properties [Montes et al, 2010b], as well as to the presence of plant roots that use both ions as nutrients.

The other studied region of the Buenos Aires Province is located in the neighbourhood of the Centro Atómico Ezeiza [Valdés et al, 2011]. In this case the monitoring dealt with surface samples and ^{238}U, ^{232}Th and ^{40}K activities were determined down to 10 cm (AS5- AS11) [Montes, 2011]. The activities, quoted in Table 2 and Fig. 3, ranged from 52 Bq/kg to 65 Bq/kg, 24 Bq/kg to 35 Bq/kg and from 470 Bq/kg to 644 Bq/kg for ^{238}U, ^{232}Th and ^{40}K, respectively.

In a frame of a survey program to study the environmental radioactivity in the Brazilian State of Rio Grande do Norte (BS1), the average concentrations of ^{226}Ra (29.2 Bq/kg), ^{232}Th (47.8 Bq/kg) and ^{40}K (704 Bq/kg) in fifty-two soil samples down 20 cm in areas with homogeneous lithology of the eastern and central regions of this states were determined. These values were higher than the world average and consistent with the predominance of granites and other Precambrian igneous rocks in the region [Malanca et al., 1996].

In Uruguay, the surface (down to 5 cm) ^{40}K activity values (US1-US8) ranged from 89.9 Bq/kg up to 1054 Bq/kg while activity values of ^{226}Ra and ^{232}Th were from 7.2 Bq/kg to 23.2 Bq/kg and 5.5 Bq/kg to 75.4 Bq/kg, respectively [Odino Moure, 2010].

The data of the all determined surface activity concentrations are compiled in Table 2 and Fig. 3 together with the worldwide average data reported by the UNSCEAR [UNSCEAR, 2008]. The mean and range worldwide values have been included by completeness [UNSCEAR, 2000; UNSCEAR, 2008]. It is clear that the reported data for ^{238}U for Brazil and Argentina are higher than the worldwide mean values. The observed ^{232}Th activities of Argentina are close to the worldwide mean values, while the Brazilian ones are quite higher than the worldwide average values. Due to the scattering and the scarcity of the data of Uruguay, it is not possible yet to extract a general conclusion. Finally, the ^{40}K data are higher than the mean values in most of the cases, and fit into the worldwide range with some exceptions.

Location	*Code*	*^{238}U*	*^{232}Th*	*^{40}K*	*^{137}Cs*	*Reference*
Argentina						
34°54.45' S; 58° 8.37' W	AS1	P	P	P	P	Montes et al., 2010b
35° 3.26' S; 57°51.21' W	AS2	P	P	P	P	
34°54.14' S; 57°55.10' W	AS3	P	P	P	P	
34°48.46' S; 58° 5.25' W	AS4	P	P	P	P	
34°48.08' S; 58° 5.04' W	AS5	S	S	S	S	Valdés, M. E. et al., 2011 Montes, 2011
35° 0.70' S; 57°44.29' W	AS6	S	S	S	S	
34°57.85'S; 57°45.66' W	AS7	S	S	S	S	
34°49.67' S; 58°35.14 W	AS8	S	S	S	S	

Location	*Code*	^{238}U	^{232}Th	^{40}K	^{137}Cs	*Reference*
34°49.30' S; 58°35.14' W	AS9	S	S	S	S	
34°50.69' S; 58°34.73' W	AS10	S	S	S	S	
34°50.46' S; 58°45.13' W	AS11	S	S	S	S	
33° 50.00′ S; 59°52.00′ W	AS12				P	Bujan et al., 2000; 2003;
33° 50.00′ S; 59°52.00′ W	AS13				P	
33° 50.00′ S; 59°52.00 W	AS14				P	
33°50.00′ S; 59°52.00′ W	AS15				P	
33°40.17' S; 65°23.45' W	AS16				P	Jury Ayub et al., 2007
33°40.17' S; 65°23.45' W	AS17				P	
33°40.17' S; 65°23.45' W	AS18				P	
33°40.17' S; 65°23.45' W	AS19				P	
33°39.93' S; 65°23.27' W	AS20				P	
33°39.93' S; 65°23.27' W	AS21				P	
33°39.93' S; 65°23.27' W	AS22				P	
33°40.17' S; 65°23.45' W	AS23	P	P	P	P	Jury Ayub et al., 2008
33°39.93' S; 65°23.27' W	AS24	P	P	P	P	
UNSCEAR	UN			S		UNSCEAR, 2008
Brazil						
Rio Grande do Norte	BS1	S	S	S		Malanca et al, 1996
22°42.00′ S; 47°38.00′ W	BS2				P	Corrochel et al, 2005
22°47.00′ S; 47°19.00′ W	BS3				P	
22°09.00′S; 47°01.00′ W	BS4				P	
22°40.00′ S; 48°10.00′ W	BS5				P	

Location	*Code*	^{238}U	^{232}Th	^{40}K	^{137}Cs	*Reference*
01°57.00′ S; 54°12.00′ W	BS6				P	
03°08.00′ S; 60°01.00′ W	BS7				P	
03°08.00′ S; 60°01.00′ W	BS8				P	
08°10.00′ S; 34°54.00′ W	BS9				P	
09°26.00′ S; 38°08.00′ W	BS10				P	
09°26.00′ S; 38°08.00′ W	BS11				P	
15°58.00′ S; 47°59.00′ W	BS12				P	
16°42.00′ S; 47°40.00′ W	BS13				P	
19°29.00′ S; 57°25.00′ W	BS14				P	
20°43.00′ S; 54°31.00′ W	BS15				P	Handl et al, 2008
20°22.00′ S; 43°24.00′ W	BS16				P	
20°21.00′ S; 43°29.00′ W	BS17				P	
22°20.00′ S; 43°37.00′ W	BS18				P	
22°30.00′ S; 44°30.00′ W	BS19				P	Handl et al, 2008
22°20.00′ S; 44°40.00′ W	BS20				P	
23°10.00′ S; 44°11.00′ W	BS21				P	
23°07.00′ S; 44°10.00′ W	BS22				P	
25°17.00′ S; 48°55.00′ W	BS23				P	
26°39.00′ S; 48°41.00′ W	BS24				P	
29°21.00′ S; 50°51.00′ W	BS25				P	
30°05.00′ S; 51°36.00′ W	BS26				P	

Location	*Code*	^{238}U	^{232}Th	^{40}K	^{137}Cs	*Reference*
Uruguay						
32°26.00' S; 54°19.00' W	US1	S			S	Odino Moure, 2010
31°18.00' S; 57°02.00' W	US2	S			S	
34°20.00' S; 56°43.00' W	US3	S			S	
34°10.00' S; 57°41.00' W (2004)	US4	S			S	
34°10.00' S; 57°41.00' W (2005)	US5	S			S	
34°10.00' S; 57°41.00' W (2006)	US6	S			S	
34°10.00' S; 57°41.00' W (2007)	US7	S			S	
34°10.00' S; 57°41.00' W (2009)	US8	S			S	
Venezuela						
Guaña	VS1				S	Sajó-Bohus et al, 1999
Chile						
39°44.00′ S; 73°22.80′ W	CS1				P	Schuller et al., 1997
38°41.50′ S; 72°53.00′ W	CS2				P	
39°41.30′ S; 72°57.10′ W	CS3				P	
40°23.00′ S; 72°57.50′ W	CS4				P	
1	CS5				P	Schuller et al., 2002
2	CS6				P	
3	CS7				P	
4	CS8				P	
5	CS9				P	
6	CS10				P	
7	CS11				P	
8	CS12				P	
9	CS13				P	
10	CS14				P	
11	CS15				P	
12	CS16				P	
13	CS17				P	

Location	*Code*	^{238}U	^{232}Th	^{40}K	^{137}Cs	*Reference*
14	CS18				P	
15	CS19				P	
16	CS20				P	
17	CS21				P	
18	CS22				P	
19	CS23				P	
20	CS24				P	
21	CS25				P	
22	CS26				P	
23	CS27				P	
24	CS28				P	
25	CS29				P	
26	CS30				P	
27	CS31				P	
28	CS32				P	
29	CS33				P	
50°53.00′ S; 72°40.00′ W	CS34				P	Schuller et al., 2004
51°08.00′ S; 53°10.00′ W	CS35				P	
51°10.00′ S; 73°05.00′ W	CS36				P	
51°12.00′ S; 73°00.00′ W	CS37				P	
52°20.00′ S; 68°25.00′ W	CS38				P	
52°16.00′ S; 68°50.00′ W	CS39				P	
52°35.00′ S; 69°50.00′ W	CS40				P	
52°38.00′ S; 70°15.00′ W	CS41				P	
52°40.00′ S; 70°50.00′ W	CS42				P	
52°25.00′ S; 71°25.00′ W	CS43				P	
52°35.00′ S; 71°33.00′ W	CS44				P	
51°55.00′ S; 72°00.00′ W	CS45				P	
52°35.00′ S; 71°42.00′ W	CS46				P	
53°36.00′ S; 70°50.00′ W	CS47				P	

Table 1. Location, type of survey and monitored nuclides in soils of South America. S: surface activity determinations and P: profile activity determination.

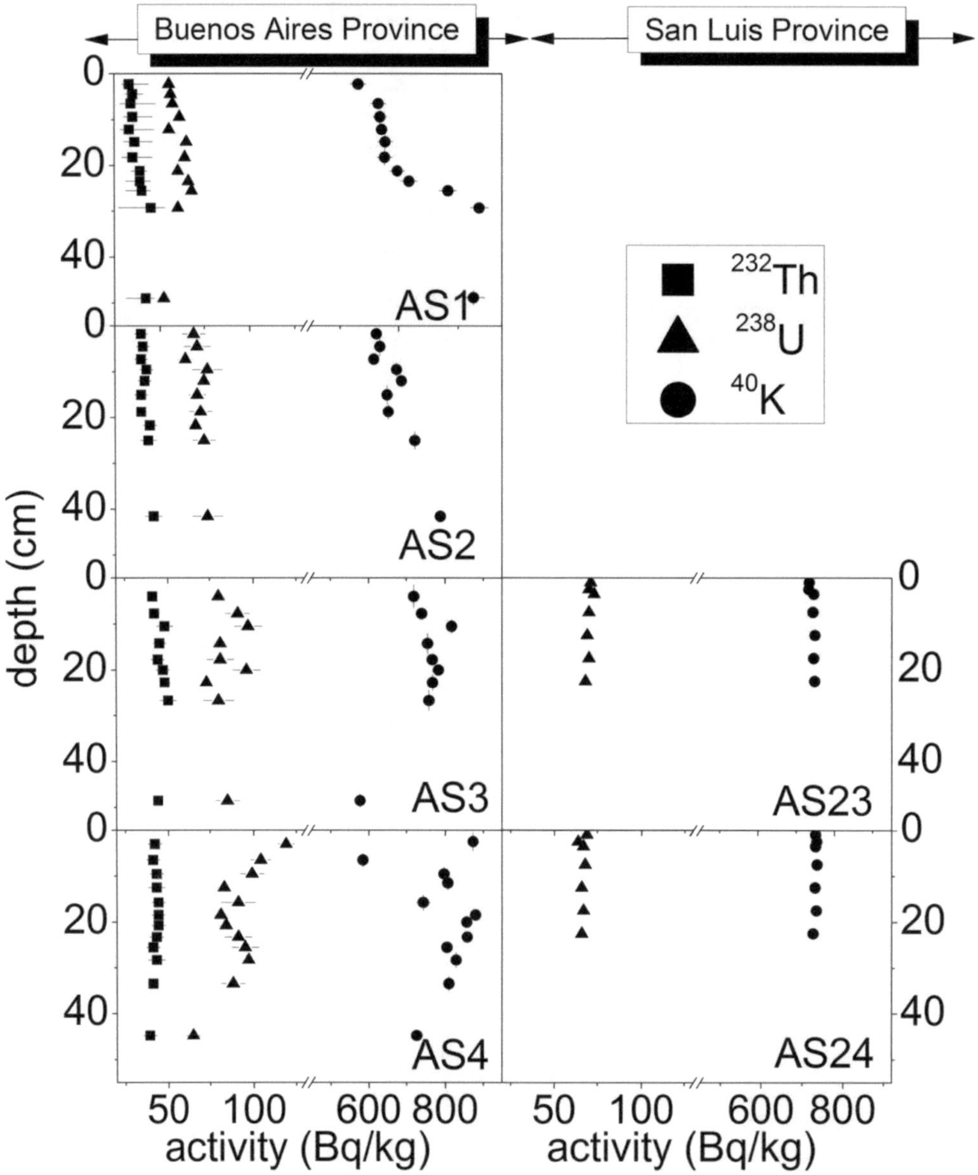

Fig. 2. Depth distribution activity of ^{232}The, ^{238}U and ^{40}K in Argentina, labelled with the sample code.

code	^{238}U *(Bq/kg)*	^{232}Th *(Bq/kg)*	^{40}K *(Bq/kg)*	*Code*	^{238}U *(Bq/kg)*	^{232}Th *(Bq/kg)*	^{40}K *(Bq/kg)*
AS1	55±6	33±4	531±13	BS1	10-136.7	12-191	56-1972
AS2	66±7	35±4	622±15	US1	19±2	75±7	1054±100
AS3	80±4	41±2	720±14	US2	7.2±0.5	11±1	90±5
AS4	119±5	42±4	717±15	US3	23±2	51±5	440±40
AS5	106±10	43±4	873±18	US4	19±2	8.6±0.5	492±45
AS6	61±8	35±2	576±15	US5	22±2	36±31	560±51
AS7	52±9	30±4	658±17	US6	21±2	35±30	495±45
AS8	65±18	35±17	644±29	US7	7.7±0.5	9.4±0.5	255±21
AS9	57±11	27±16	498±25	US8	14±1	19±2	340±31
AS10	53±13	24±12	470±22	**WA**	**35**	**30**	**400**
AS11	52±10	32±7	547±23	**WR**	**16-110**	**11-64**	**140-850**
AS23	71±4	-	733±11				
AS24	69±4	-	734±20				
UN	-	-	540-750				

Table 2. Natural surface activity concentrations in soils. WA: worldwide average and WR: worldwide range [UNSCEAR, 2000; UNSCEAR, 2008].

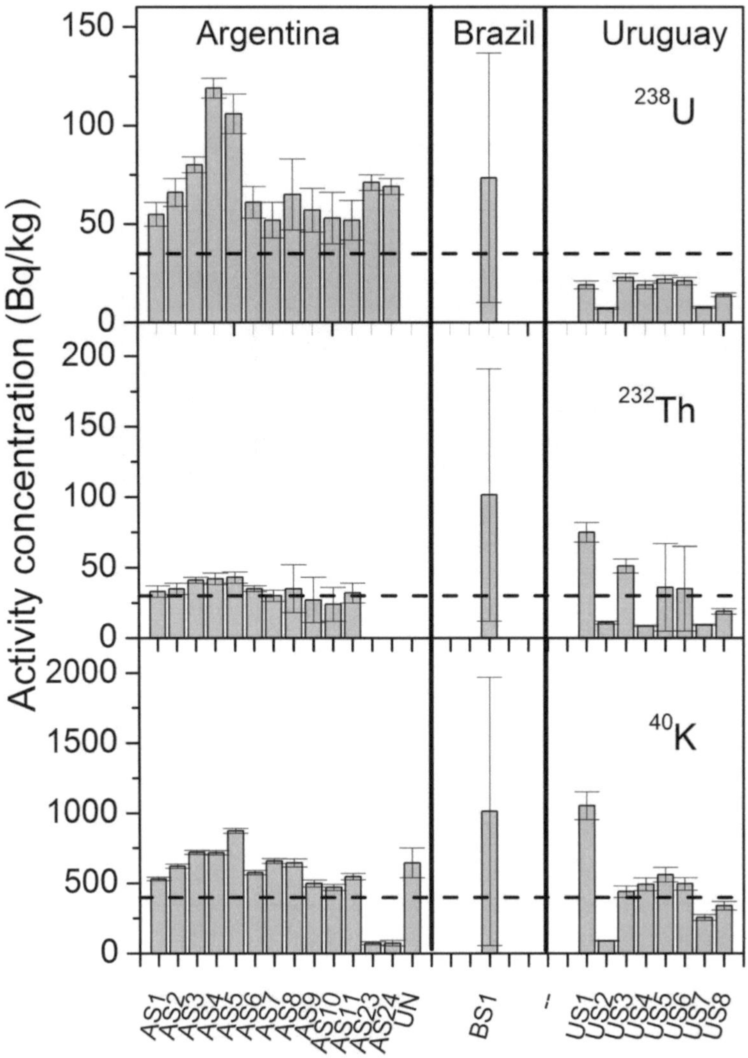

Fig. 3. Natural activity of ^{40}K, ^{232}Th and ^{238}U in soil surface. The dash line corresponds to the UNSCEAR average worldwide values [UNSCEAR, 2000]. The vertical bars correspond to the experimental errors in the case of Argentina and Uruguay, and to the standard deviation of a set of determinations in the case of Brazil.

3.2 Annual committed effective dose by external irradiation calculations

The contribution of natural nuclides to the absorbed dose rate at 1m above the ground depends on the concentration of radionuclides in the soil. There is a direct relationship between terrestrial gamma radiation dose and radionuclide natural concentrations in soils. The exposure dose rate can be evaluated accounting for the activity values of the nuclides (A_i) and the conversion factors (f_i). These coefficients are reported in Table 3 [UNSCEAR, 2000; UNSCEAR, 2008]. Based in the analysis of the UNSCEAR 1982 report [UNSCEAR, 1982], the International Committee of Radiation Protection (ICRP) used a coefficient (C_i) to convert the absorbed dose in air to annual committed effective dose (*aced*). Monte Carlo Calculations radiation- transport codes indicate that higher values should be used for infant and children. These values are quoted in the Table 3. To calculate the annual effective dose it has also considered that the spent time outdoors is 20% of total time [USNCEAR, 2008], i.e.:

$$\text{aced (Sv)}=10^{-9}\times 24\times 365\times C_c\times 0.2\times \sum_i f_i A_i C_i \tag{1}$$

nuclide	f_i *(nGyh⁻¹/Bqkg⁻¹)*	C_i *(Sv/Gy)*		
		infants	children	adults
^{40}K	0.0417	0.926	0.803	0.709
^{232}Th	0.604	0.907	0.798	0.695
^{238}U	0.462	0.899	0.766	0.672
Average		0.91	0.79	0.69

Table 3. Conversion factors (f_i) and absorbed dose to effective dose equivalent conversion coefficients (C_i) [UNSCEAR 2008, UNSCEAR 2000].

	aced (mSv)				*aced (mSv)*		
code	*infants*	*children*	*adults*	*code*	*infants*	*children*	*adults*
AS1	0.108±0.006	0.093±0.005	0.082±0.005	BS1	0.03-0.42	0.02-0.36	0.02-0.32
AS2	0.134±0.007	0.107±0.006	0.094±0.005	US1	0.16±0.01	0.137±0.009	0.120±0.008
AS3	0.146±0.004	0.126±0.003	0.111±0.003	US2	0.022±0.001	0.019±0.001	0.0667±0.0009
AS4	0.175±0.006	0.151±0.005	0.133±0.004	US3	0.095±0.006	0.083±0.005	0.073±0.005
AS5	0.177±0.009	0.153±0.007	0.135±0.007	US4	0.055±0.004	0.048±0.003	0.042±0.003
AS6	0.117±0.007	0.101±0.006	0.089±0.005	US5	0.09±0.03	0.08±0.03	0.07±0.03
AS7	0.111±0.008	0.096±0.007	0.084±0.006	US6	0.08±0.03	0.07±0.03	0.06±0.02
AS8	0.12±0.02	0.11±0.02	0.09±0.02	US7	0.032±0.002	0.028±0.002	0.024±0.001
AS9	0.10±0.02	0.09±0.02	0.08±0.02	US8	0.051±0.003	0.0447±0.003	0.039±0.003
AS10	0.09±0.02	0.08±0.01	0.07±0.01				
AS11	0.11±0.01	0.091±0.009	0.080±0.008				

Table 4. Calculated annual committed effective terrestrial exposure dose for infants, children and adults.

It is worth to mention that in the case of adults, the calculated annual committed effective doses due to terrestrial external exposure resulted slightly higher than the UNSCEAR reported values [UNSCEAR, 2000], as observed in Fig. 4.

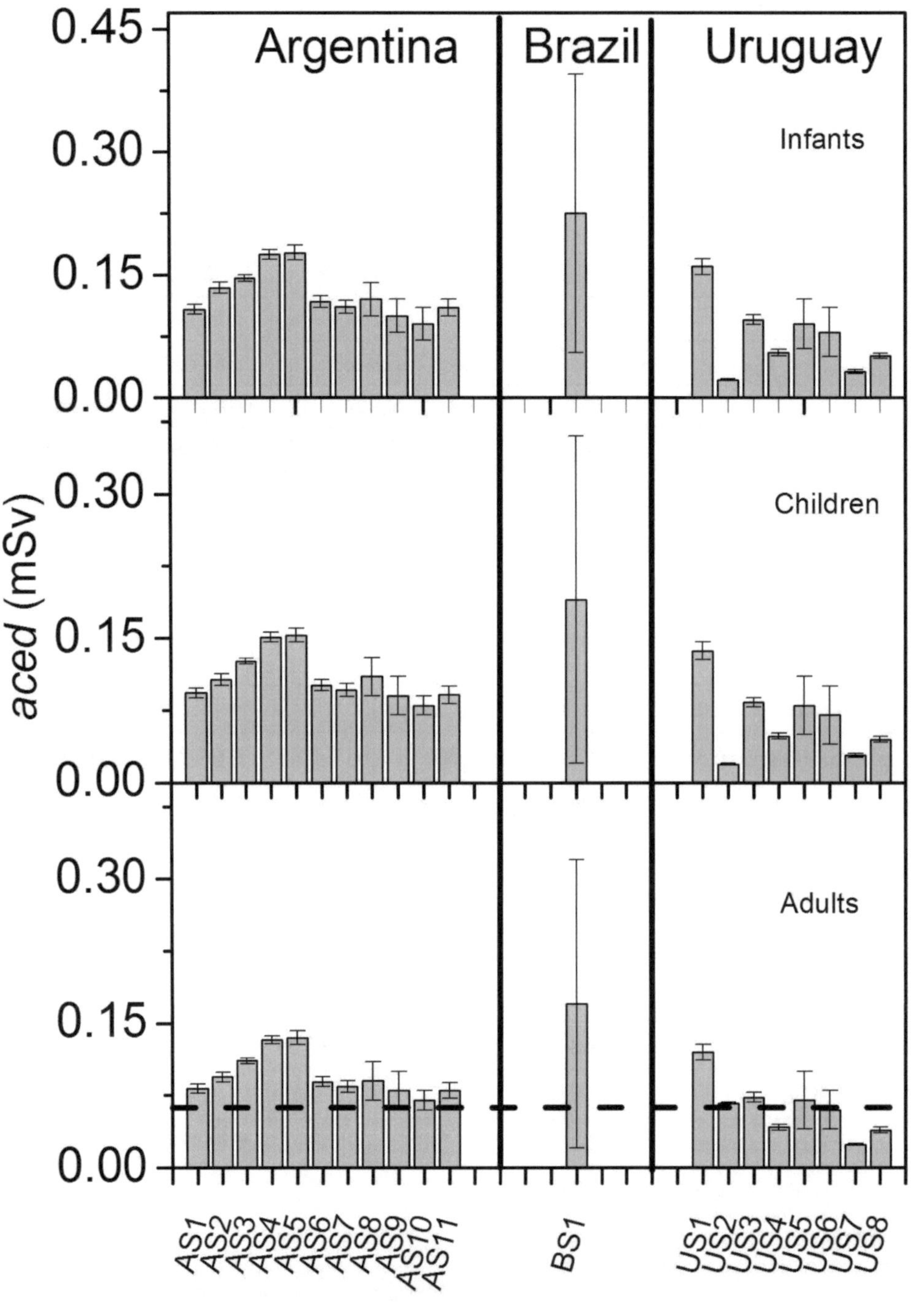

Fig. 4. Calculated annual committed effective dose for infants, children and adults. The dash line corresponds to the UNSCEAR reported values [UNSCEAR, 2000]. The vertical bars correspond to the experimental errors in the case of Argentina and Uruguay, and to the standard deviation of a set of determinations in the case of Brazil.

3.3 Anthropogenic nuclides

In the last decades, ^{137}Cs was the only monitored anthropogenic nuclide (gamma emitter) in the Southern Hemisphere. Aimed in the study of soil erosion, ^{137}Cs reference activity profiles (Fig. 5) were determined in the Pampa Ondulada of the Buenos Aires Province region, Argentina (AS12-AS15) [Bujan et al., 2000; Bujan et al., 2003]. The ^{137}Cs activities determined down to 90 cm declined sharply from the surface to the first 20 cm, the maxima activity was observed at the top layer. An average value of 1108 Bq/m^2 was obtained for the local inventory. In the La Plata city region, in spite that the ^{137}Cs integrated activities of the profiles obtained down to 50 cm were similar in all soils, differences in the ^{137}Cs depth distributions were detected (Fig. 5). The profiles of AS1 and AS3 sites followed a Gaussian-type feature, typical of a convective-diffusive process [Likar et al., 2001; Bossen & Kirchner, 2004]. The profile of AS2 was quite different since a Gaussian-shape was established down to 7 cm in depth. In the case of the AS4 soil, placed at 5 km from the La Plata river coast, the activity values were high at the surface and then suddenly decreased. Both facts, the high values at the surface and the deviation from the Gaussian shape [Likar et al., 2001;Bossen & Kirchner, 2004], could be explained considering the fine texture and the flat relief of the region which induce water-logging, i.e., this area shows a low permeability of the underlying horizons, and the phreatic water affects the deepest horizons [Imbellone, 2009]. It has been claimed that Cs is sorbed by Fe_3O_4 [Singh et al., 2009; Catallette et al., 1998; Marnier & Fromage, 2000]. However, by comparing the ^{137}Cs profiles and the Mössbauer relative fraction of Fe_3O_4 as well as with the other iron species [Montes et al.; 2010b], it was not observed an apparent correlation. A series of surface studies were also performed in the Buenos Aires Province in the neighbourhood of the Centro Atómico Ezeiza (AS5-AS11) showing that the activity concentration values down to 10 cm ranged between 0.9 Bq/kg and 2.6 Bq/kg [Vadés et al., 2011]. These values are consistent with the top layer activity data obtained from the profiles AS1-AS4 [Montes et al.; 2010b]. Vertical migration of ^{137}Cs was studied in soils of natural and semi-natural grassland areas of San Luis Province (AS16- AS21, AS23 and AS24) [Juri Ayub et al., 2007; Juri Ayub et al., 2008]. The inventories ranged from 330 Bq/m^2 to 730 Bq/m^2, while depth profiles had different shapes (see Fig. 5).

As observed in Fig. 6, differences in the patterns of ^{137}Cs depth distribution in the soil profiles of the different regions were found in the four studied sites of the South-Central region of Brazil (BS2-BS5), ascribed to chemical, physical, mineralogical and biological differences of the soils [Correchel et al., 2005]. The variability of the soil characteristics was not able to explain the spatial variability of the profiles. The average inventories of the four studied sites were 268 Bq/m^2, and the maximum activity value was detected at the top layer. The spatial distribution and behaviour of the ^{137}Cs in tropical, subtropical and equatorial unperturbed Brazilian soils have been investigated up to 40 cm (BS6-BS26) [Handl et al., 2008]. The shape of all 23 sampled sites depth profiles varied between the two ones showed in Fig. 6. The majority of the Cs content was observed in the 10-15 cm top layer while minor quantities were detected down to 35 cm. Low deposition densities were observed at the Amazon region where ascendant convection of water vapour is intense, while the south area exhibited considerable large concentrations. No correlation was observed between altitude and ^{137}Cs concentration. On the contrary, the results were correlated with the climatic de Martonne index, suggesting that the process can not be explained with single meteorological parameters [Handl et al., 2008].

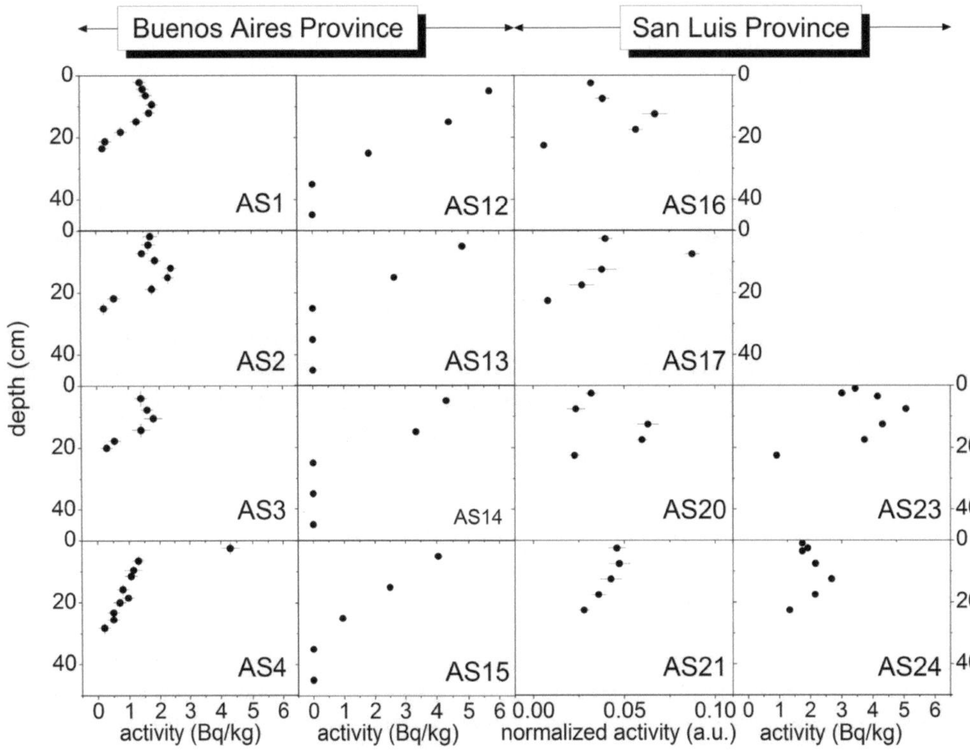

Fig. 5. ^{137}Cs depth profiles recorded in Argentina, labelled with the site code.

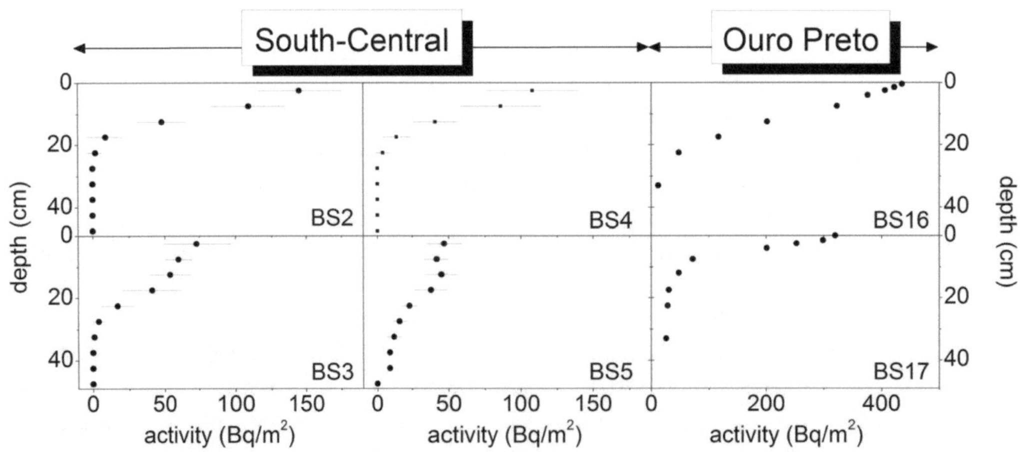

Fig. 6. ^{137}Cs depth profiles recorded in Brazil, labelled with the site code.

In Chile, total inventories and depth distributions of ^{137}Cs (Fig. 7) were determined at four sites of two agriculturally used soil types (CS1-CS4). The inventories were always higher than previously estimated for the Southern Hemisphere and depend on annual rainfall [Schuller et al, (1997)]. The depth distribution of ^{137}Cs in well-developed agricultural soil at 28 sites in different southern regions (CS5-CS33) was also studied [Schuller et al, 2002]. The profiles in most of the sites followed no systematic pattern in the upper few centimetres (Fig. 7), but below this depth an exponential behaviour was observed. The calculated relaxation depth [Schuller et al, 2002] ranged from 4.4 cm in Palehumults to 8.4 cm and 9.7 cm in Hapludands and Psamments soil types, respectively. The relaxation depth increased with decreasing clay content and increasing volume of coarse pores. Activity densities ranged from 450 Bq/m^2 to 5410 Bq/m^2, correlating with the mean annual rainfall rate of the sampling sites. The South Patagonia sheep-farming region (CS34-CS47) was studied (Fig. 7). The areal activity density varied from 222 Bq/m^2 to 858 Bq/m^2, positively correlated with the mean annual precipitation rate [Schuller et al., 2004].
In Venezuela, the ^{137}Cs concentration at two different depths, 0 cm - 20 cm and 20 cm - 40 cm, (VS1) was measured being the activity concentration around 0.5 Bq/kg and 10 Bq/kg at 20 cm in depth for the littoral and central regions, respectively [Sajó-Bohus et al., 1999].
Finally, the obtained values of ^{137}Cs surface activity concentration in different Uruguayan departments varied from 1.2 Bq/kg to 2.3 Bq/kg in the 2004 to 2009 period (US1-US8) [Odino Moure, 2010]. A spatial variation was also observed.

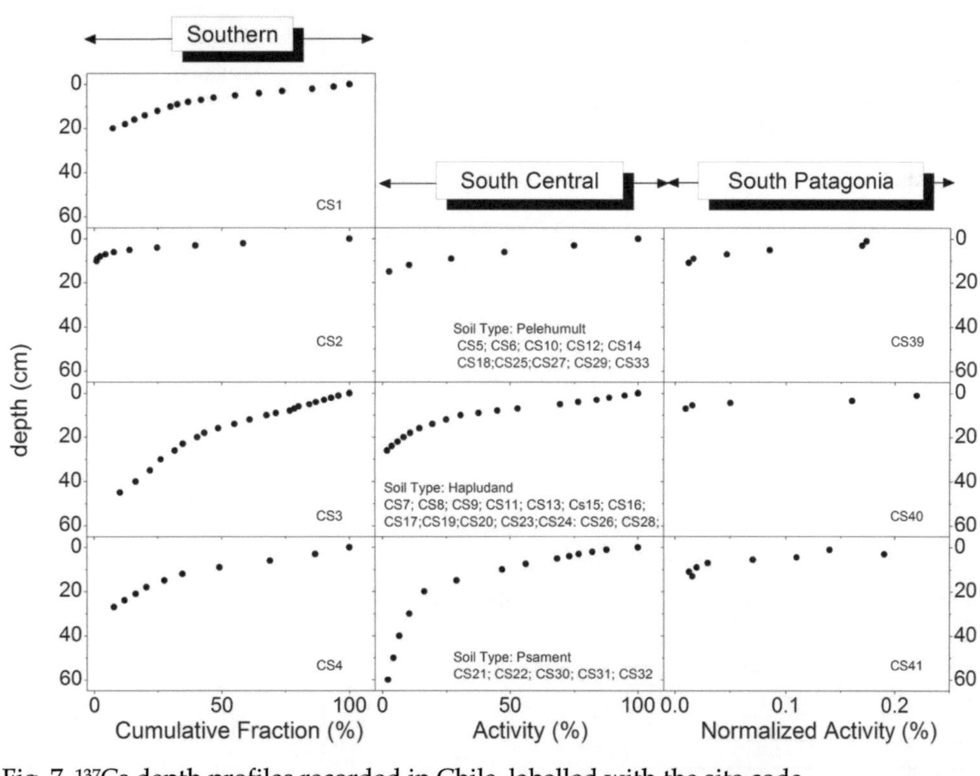

Fig. 7. ^{137}Cs depth profiles recorded in Chile, labelled with the site code.

3.4 Cs inventories analysis

The cumulated annual deposition of ^{90}Sr was compiled by UNSCEAR [UNSCEAR 2000; UNSCEAR 2008]. It is worth to mention that data of ^{137}Cs are not available due to the technological limitations on the detection of gamma emitters of the survey period. However, there is experimental evidence that the $^{90}Sr/^{137}Cs$ activity release ratio is constant and equal to 1.5, allowing using the global determination of ^{90}Sr to estimate the ^{137}Cs one. Through it should be considered that once the nuclides are incorporated in the soil, the migration rates are different due to the dissimilar soil-nuclide interaction process. Since the wind circulation is presented in latitudinal bands, this is the assumption used to evaluate the transport and deposition of nuclides [UNSCEAR, 1982; UNSCEAR, 2000; UNSCEAR, 2008].

In order to compare the inventory data with the UNSCEAR predictions [UNSCEAR, 2000; UNSCEAR, 2008], in the Table 5 are presented the inventory of ^{137}Cs data corrected by nuclide decay using the time of determination (a single input at 1965 is considered). It is observed that the experimental data do not follow the UNSCEAR prediction. The Fig. 8 shows the inventories of ^{137}Cs, the average annual precipitation vs. latitude and the inventory vs. precipitation. Globally, it seems that the inventory depends on the annual precipitation and the Andes Cordillera plays a very important role on the inventory due to the generation of higher annual precipitations, hence more ^{137}Cs deposition. It is also observed that mountains act as barrier for Argentina.

Code	*^{137}Cs inventory (Bq/m^2)*	*Latitudinal band (degree)*	*Integrated deposit (Bq/m^2)*
BS6	945±110	0-10	720
BS7	0.8±0.1		
BS8	5.15±0.07		
BS9	99±13		
BS10	83±11		
BS11	188±27		
BS12	558±71	10-20	630
BS13	654±212		
BS14	16±2		
BS17	1120±76	20-30	1050
BS16	596±143		
BS15	1691±216		
BS4	621±46		
BS20	3494±477		
BS18	1333±79		
BS19	479±61		
BS5	594±37		
BS2	771±34		
BS3	614±77		

Code	*^{137}Cs inventory (Bq/m²)*	*Latitudinal band (degree)*	*Integrated deposit (Bq/m²)*
BS22	1355±173		
BS21	1407±180		
BS23	1375±176		
BS24	320±41		
BS25	3629±463		
BS26	1344±172	30-40	1140
AS20	1877		
AS21	1285		
AS22	848		
AS16	1311		
AS17	1645		
AS18	1427		
AS19	745		
AS12	2512		
AS13	2041		
AS14	2350		
AS15	2050		
AS4	689±36		
AS3	849±38		
AS1	950±34		
AS2	1366±38		
CS2	1438		
CS3	2821		
CS1	2202		
CS5	1329		
CS6	910		
CS7	874		
CS9	110		
CS13	1565		
CS16	1511		
CS24	2111		
CS25	2020		
CS26	2584		
CS27	2894		
CS28	1893		

Code	*^{137}Cs inventory (Bq/m^2)*	*Latitudinal band (degree)*	*Integrated deposit (Bq/m^2)*
CS29	3913		
CS4	1420	40-50	1335
CS8	819		
CS10	1128		
CS11	1292		
CS12	1492		
CS14	1165		
CS15	1438		
CS17	892		
CS18	2002		
CS19	1238		
CS20	2038		
CS21	3130		
CS22	1711		
CS23	1674		
CS30	2621		
CS31	9045		
CS32	4641		
CS33	9846		
CS34	1162±89	50-60	705
CS35	1464±113		
CS36	1457±112		
CS37	1334±103		
CS45	796±61		
CS39	647±50		
CS38	630±48		
CS43	511±39		
CS40	527±41		
CS44	564±43		
CS46	991±76		
CS41	433±33		
CS42	546±42		
CS47	1673±129		

Table 5. Determined ^{137}Cs inventory together with the UNSCEAR predictions [UNSCEAR, 2008], ordered by latitudinal band.

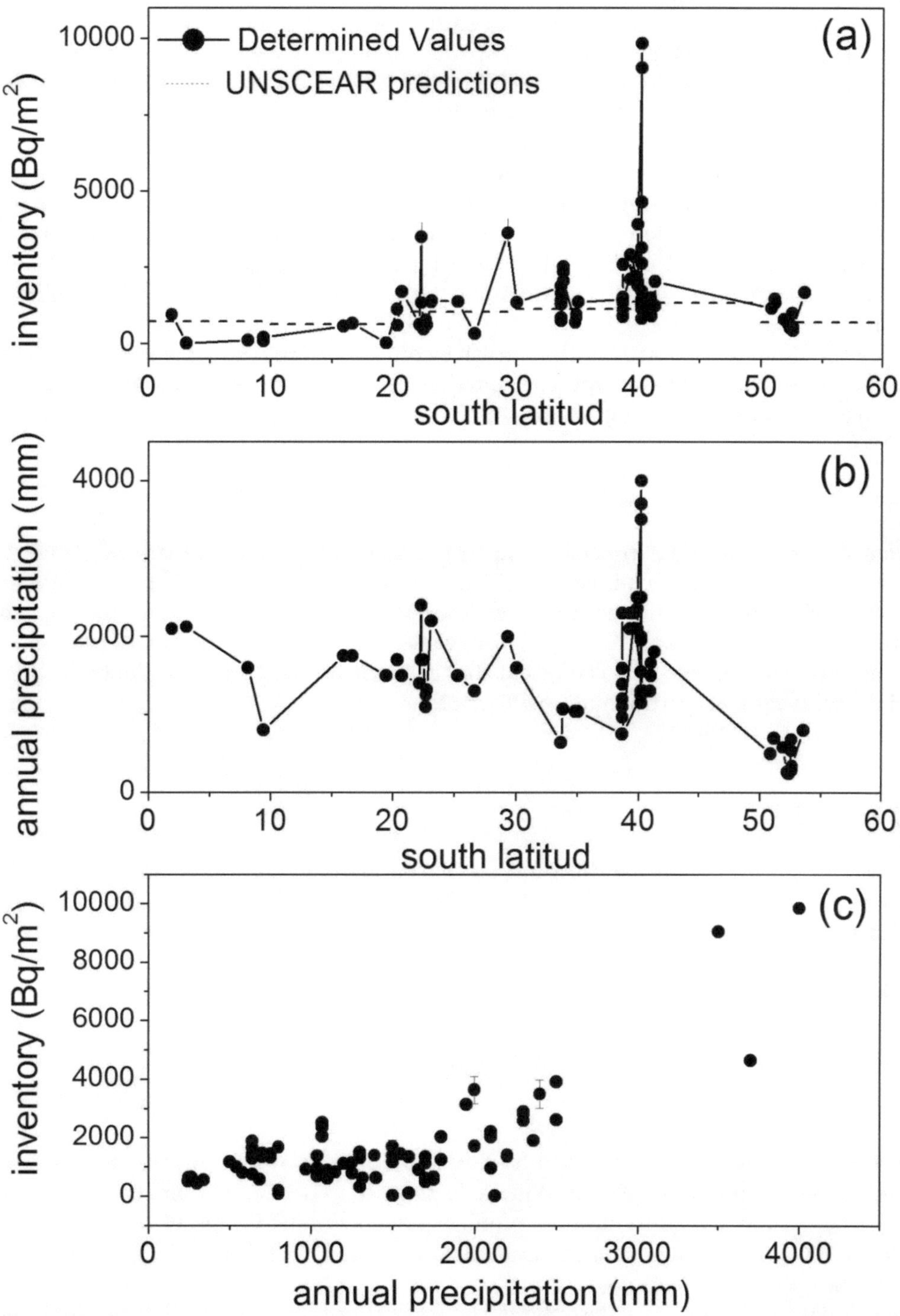

Fig. 8. a) ^{137}Cs inventory vs. annual precipitation; b) Annual precipitation vs. south latitude and c) ^{137}Cs inventory vs. annual precipitation.

3.5 Soil profile analysis

The basic processes controlling mobility of anthropogenic nuclides in soil include convective transport by water, dispersion caused by spatial variations of convection velocities, diffusive movement within the fluid, and physicochemical interaction with soil matrix. Because the slow migration velocities of Cs in soils, generally the models do not take into account the soil moisture changes in the unsaturated zone but assume mean constant water content. The spatial uniformity of the deposition rates is also considered which may be defensible for the weapon tests and, at least, on a local scale in nuclear accidents. In this frame, the two trendy for modelling the migration of radionuclides in soils are the one dimensional convection-dispersion equation (ODCDE) with constant parameters [Likar et al., 2001; Bossen & Kirchner, 2004] and the serial compartmental approach (CA) [Kirchner, 1998; Schuller et al., 1997].

The ODCDE model is based on the diffusive-convective transport, the mass conservation and the Cs-soil matrix interaction. The equation describing the migration process is usually known as the Fokker Planck equation:

$$\frac{\partial C(x,t)}{\partial t} = D_e \frac{\partial^2 C(x,t)}{\partial^2 x} - v_e \frac{\partial C(x,t)}{\partial x} - \lambda C(x,t) \tag{2}$$

where $C(x, t)$ is the ^{137}Cs concentration in the soil (mobile and sorbed), λ is the decay constant, D_e is the effective diffusion coefficient of caesium in soil, v_e is a convective velocity, x is the soil depth with respect to the soil surface and t is the time from the deposition. Assuming that all sorbed Cs is exchangeable and that the exchange process is in equilibrium, D_e is an effective coefficient that depends on the porosity ε, the bulk density ρ and the solid aqueous partitioning coefficient K_d:

$$D_e = \frac{D_w}{\left(1 + \rho K_d / \varepsilon\right)} \qquad v_e = \frac{v_w}{\left(1 + \rho K_d / \varepsilon\right)} \tag{3}$$

and where D_w *and* v_w are the diffusion coefficient of Cs in soil water and the convection velocity of water in pores of soil, respectively. The factor between parentheses in eqs. 3 is called the retardation factor. As boundary conditions, a half-infinite space-time is assumed and the considered initial condition is a pulse-like deposit at t=0 with deposition density J_0. With all these assumptions, it is obtained the well known solution:

$$.\, C(x,t) = J_0 e^{-\lambda t} \left[\frac{1}{\sqrt{\pi D_e t}} e^{-(x - v_e t)^2 / 4 D_e t} - \frac{v_e}{2 D_e} e^{-v_e / D_e} erfc\left(\frac{v_e}{2} \sqrt{\frac{t}{D_e}} + \frac{x}{2\sqrt{D_e t}}\right) \right]. \tag{4}$$

The irreversible fixation of Cs to soil has been also accounted for, however in this case, no analytical solution of the transport equations are able to obtain, so numerical methods are needed to obtain the concentration profiles [Antonopoulus-Domis et al., 1995; Toso & Velasco, 2001]. The fitting of experimental data with the ODCDE model allows determining D_e and v_e.

The CA has been used to analyse the depth layered profiles without detailed information about the site-specific processes that influence radionuclide´s mobility. Usually the soil profile is split into a series of horizontal layers (compartments) which are connected by radionuclide downward transport rates and the migration dynamic is described by a system of lineal first-order differential equations with constant coefficients. This model is applicable only if the transport of the radionuclide is dominated by convection [Kirchner et al., 2009]. It is worth to mention that neither the presence of micro-organisms nor the root intake is considered by the models.

In the following, data compilation of the transport parameters of soils of South America is presented in Fig. 9. To facilitate the comparison, the worldwide average values corresponding to weapons fallout have been also included [IAEA, 2010]. In the La Plata city area of the Buenos Aires Province, the two profiles (AS1 and AS3) that clearly have a Gaussian shape were fitted using the ODCDE [Montes et al., 2010a; Montes et al., 2010b]. For the AS1, the D_e and v_e resulted equal to 0.728 cm^2/y and 0.23 cm/y, respectively. while for AS3 the values were 0.5 cm^2/y and 0.22 cm/y, correspondingly. Since in the case of profile of AS2 site, the Gaussian-shape was established down to 7 cm in depth, the data of the top layer were disregarded in the analysis leading to diffusion coefficient and convection velocity values of 0.39 cm^2/y and 0.34 cm/y, respectively. These set of transport parameter values agrees well with the South American values [Juri Ayub et al., 2007; Juri Ayub et al., 2008; Schuller et al., 1997; Schuller et al., 2004], as observed in Fig. 6, but is slightly larger than the average values reported by IAEA [IAEA, 2010].

In the semi-natural and natural central area of Argentina (AS16-AS24), the diffusion coefficients obtained using the ODCDE varied from 0.43 cm^2/y to 2.27 cm^2/y, and the convection velocity varied from 0.13 cm/y to 0.39 cm/y. The D_e values were in the range reported in the bibliography for some Chilean and European soils, while the v_e values were one order of magnitude higher than those reported for Chilean soils and of the same order of magnitude than the European sandy ones [IAEA, 2010]. The great penetration in these soils was ascribed to the high sand and low fine materials content, i.e., high porosity facilitating water passage to deeper layers.

The Chilean Southern soil profiles of these soils were analysed using both, the CA and the ODCDE. The results of the fits were not good. In the case of the CA, the variation of the migration rates did not improve appreciably the fits. Moreover, the CS3 profile cannot be reproduced with this model. The determined migration rates resulted always low, between 0.1 cm/y and 0.3 cm/y, in the lower range of the reported data obtained for nuclear weapons and Chernobyl fallout [Schuller et al., 1997]. On the other side, the analysis of the data using the ODCDE indicated that the transport was dominated by diffusion process in agreement with the high silt and clay content of the studied soils. The agreement of D_e and v_e with reported data was better than in the CA case. However some misfit was observed at large depth, probably due to preferential transport through macropores, migration of suspended particles, spatial variability or agricultural land disturbance. Least-square fits of semi-natural and natural South Patagonia profiles using the ODCDE with constant parameters and improved by assuming a logarithmic distribution of D_e and v_e or depth dependence of both parameters were also tried. In all attempts, the mean obtained parameters were the same. The convection velocity was found to be negative (upward migration) in the CS43 site, because of the yearly flooded lowland, and at CS45 soil, where the upper soil layer was possibly disturbed by animal hoof prints. In the CS43 site, the D_e values were considerably larger than the determined for the other profiles. In the other sites, the determined median convection velocity and the diffusion coefficient values were 0.056 cm/y and 0.048 cm^2/y, respectively. The convection velocities resulted rather higher when compared with the data of temperate regions from Chile [Schuller et al., 1997], while the diffusion coefficient was close to those obtained in the Antarctic region [Schuller et al., 2002].

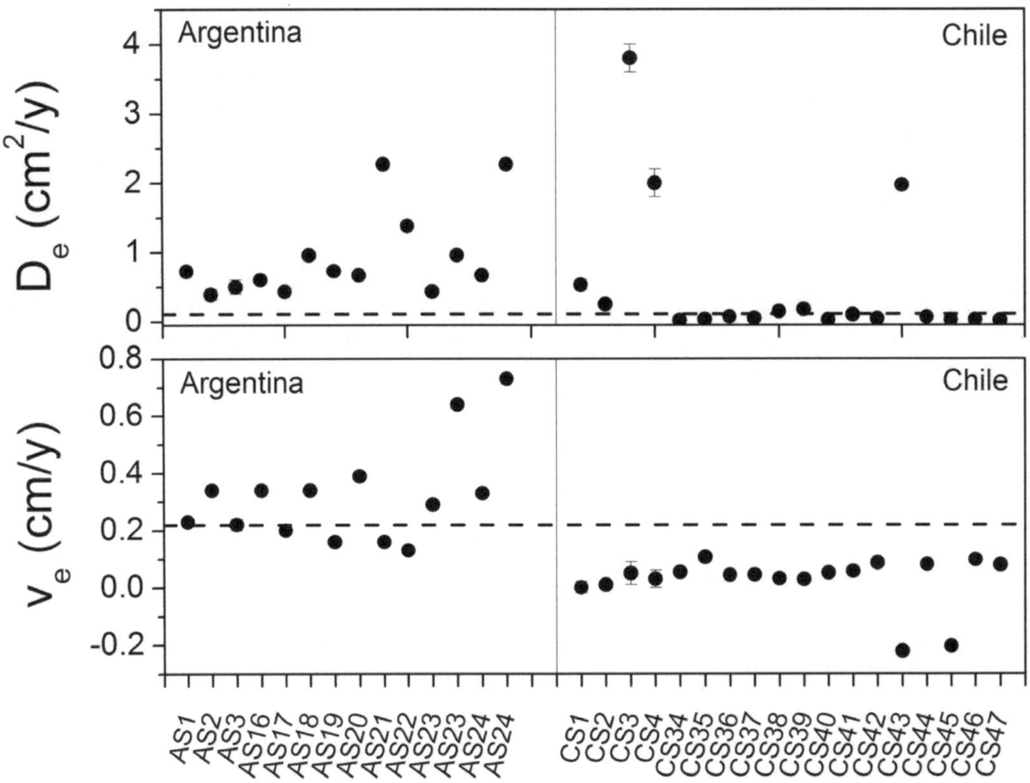

Fig. 9. Diffusion coefficient (D_e) and convection velocity (v_e) parameters together with the IAEA (dash line) [IAEA, 2010].

The present set of transport parameters of ^{137}Cs in soils is presented in Fig. 9. A scatter is observed in the data. Most of Argentinean diffusion coefficient data are slightly larger than the average values reported by IAEA for weapon test fallout [IAEA, 2010] while the Chilean data fit quite well with the average values with the exception of CS3, CS4 and CS43. Concerning the convection velocity parameter, the determined values from Provincia de Buenos Aires-Argentina soils resulted close to the worldwide average ones [IAEA, 2010], while those determined in the San Luis Province-Argentina are higher than the IAEA data [IAEA, 2010]. In Chile, the values were lower than the worldwide average [IAEA, 2010], being some of them negative, probably related to the periodic flooded lowland. Several mechanisms, such as bioturbation, horizontal transport, transport through macropores, migration of suspended particles, etc., have been used to explain the deviations from the convection-dispersion predictions, putting the ODCDE model under consideration and suggesting that the model is an oversimplification of such a complex process. However the ODCDE model is very useful to estimate the transport parameters.

4. Conclusions

A systematic compilation of radionuclide activity data in soil of South America has been completed. Radionuclide activity data concern to the natural ^{40}K, ^{238}U and ^{232}Th chains, and to the anthropogenic ^{137}Cs nuclides.

The surface activity concentrations for ^{238}U for Brazil and Argentina are higher that the worldwide mean values. The ^{232}Th activity data of Argentina are closer to the worldwide values while the Brazilian ones are quite higher than the worldwide values. In the case of Uruguay, it is not possible to extract conclusions yet due to the insufficiency and dispersion of data. The ^{40}K data are higher than the mean values in most of the cases, and fit into the worldwide range with some exceptions. The annual committed effective terrestrial exposure dose for infants, children and adults have been calculated, resulting the values slightly higher than the UNSCEAR value in the case of adults. The analysis of the ^{137}Cs inventories allows concluding that the experimental data do not follow the latitudinal band deposition predictions proposed by UNSCEAR. It is worth to mention that the analysis of the whole set of information in South America allows to establish a correlation between the inventory and the annual precipitations. Different shape type profiles have been determined for Argentina, Brazil and Chile. In several cases it was possible to reproduce the ^{137}Cs profiles with models accounting diffusion and convection process. The transport parameters agree well with the average worldwide values due to nuclear weapon test fallout. Some discrepancies were detected when bioturbation and floodedland are present, indicating that efforts to include these processes should be done to fully reproduce the caesium profiles, hence to be able to make predictions of migration in case of possible pollution.

The present set of data contributes to the establishment of regional baselines as well as help in the development of local regulations concerning to permitted activity limits to people health protection.

5. Acknowledgements

Research grants PIP 0230 from Consejo Nacional de Investigaciones Científicas y Técnicas (CONICET, Argentina) and PICT 38047-Préstamo BID from Agencia de Promoción Científica (ANCYT, Argentina) are gratefully recognized.

6. References

Antonopoulus-Domis, M.; Clouvas, A.; Hiladakis, A. & Kadi, S. (1995). Radiocesium distribution in undisturbed soil: Measurements and diffusion-advection model. *Health Physics*, Vol.69, pp. 949-953

Bellenber, J. P. & Stauton, S. (2008). Adsorption and desorption of ^{85}Sr and ^{137}Cs on reference minerals, with and without inorganic and organic surface coatings. *Journal of Environmental Radioactivity*, Vol. 99, pp. 831–840, ISSN 0265 931X

Bossen, P. & Kirchner, G. (2004). Modeling the vertical distribution of radionuclides in soils. Part I: the convection-dispersion equation revisited. *Journal of Environmental Radioactivity*, Vol. 73, pp. 127-150, ISSN 0265 931X

Bujan, A.; Santanatoglia, O. J.; Chagas, C.; Massobrio, M.; Castiglioni, M.; Yañez, M.; Ciallella, H. & Fernandez, J. (2000). Preliminary study on the use of the ^{137}Cs

method for soil erosion investigation in the pampean region of Argentina. *Acta Geologica Hispanica,* Vol. 35, pp. 271-277, ISSN 1695 6133

Bujan, A.; Santanatoglia, O. J.; Chagas, C.; Massobrio, M.; Castiglioni, M.; Yañez, M.; Ciallella, H. & Fernandez, J. (2003). Soil erosion evaluation in a small basing though the use of ^{137}Cs technique. *Soil & Tillage Research,* Vol. 69, pp. 127–137, ISSN 0167 1987

Catallette, H.; Dumonceau, J. & Ollar, P. (1998). Sorption of cesium, barium and europium on magnetite. *Journal of Contaminant Hydrology,* Vol. 35, pp. 151-159, ISNN 0169-7722

Cooper, J. R.; Randle & K.; Sokhi, R. S. (2003). In: *Radioactive Releases in the Environment,* John Wiley & Sons Ltd. ISBN 0 471 89924 0, The Atrium, Southern Gate, Chichester, West Sussex, England

Cornell, R. M. (1993). Adsorption of cesium on minerals: a review. *Journal of Radioanalytical and Nuclear Chemistry,* Vol. 171, pp. 483 - 500. ISNN 0236 5731

Correchel, V.; Oliveira Santos Bacchi, O.; Reichardt, K. & Cereci de Maria I. (2005). Random and systematic spatial variability of ^{137}Cs inventories at references sites in south-central Brazil. *Scientia Agricola,* Vol. 62, pp. 173-178, ISSN 0103 9016

Handl, J.; Sachase, R.; Jakob, D.; Michel, R.; Evangelista, H.; Gonçalves, A. C. & de Freitas, A. C. (2008). Accumulation of ^{137}Cs in Brazilian soils and its transfer to plants under different climatic conditions. *Journal of Environmental Radioactivity,* Vol. 99, pp. 271-287, ISSN 0265 931X

IAEA. (2010). Handbook of Parameter Values For The Prediction Of Radionuclide Transfer In Terrestrial And Freshwater Environments. *Technical Reports Series No. 472,* Vienna.

Imbellone, P. A.; Guichon, B. A. & Giménez, J. E. (2009). Hydromorphic soils of the River Plate coastal plain, Argentina. *Latin american Journal of sedimentology and basin analysis,* Vol. 16, pp. 3-18, ISNN 1669-73616

Juri Ayub, J.; Rizzotto, M.; Toso, J. & Velasco, H. (2007). ^{137}Cs deposition and vertical migration in soils from Argentina. *Proc. International. Conf. on Environ. Radioctivity: From Measurements and Assessments to Regulation,* Vienna, Austria, www-pub.iaea.org/mtcd/meetings/announcements.asp?confid=145

Juri Ayub, J.; Velasco, R.H.; Rizzotto, M.; Quintana E. & Aguiar J. (2008). ^{40}K, ^{137}Cs and ^{226}Ra soil and plant content in semi-natural grasslands of Central Argentina. *The Natural Radiation Environment – 8th International Symposium,* edited by A. S. Paschoa, American Institute of Physics, ISBN 978 0 7354 0559

Kirchner, G. (1998). Applicability of compartmental models for simulating the transport of radionuclides in soil. *Journal of Environmental Radioactivity,* Vol. 38, pp. 339-352, ISSN 0265 931X

Kirchner, G.; Strebl, F.; Bossew, P.; Ehlken, S. & Gerzabek, H. (2009). Vertical migration of radionuclides in undisturbed grassland soils. *Journal of Environmental Radioactivity,* Vol. 100, pp. 716-720, ISSN 0265 931X

Malanca, A.; Gaidolfi, L.; Pessina, V.& Dallara G. (1996). Distribution of ^{226}Ra, ^{232}Th, and ^{40}K in soils of Rio Grande do Norte (Brazil). *Journal of Environmental Radioactivity,* Vol. 30, pp. 55-67, ISSN 0265 931X

Marmier, N. & Fromage, F. (2000). Sorption of Cs(I) on Magnetite in the Presence of Silicates. *Journal of Colloid and Interface Sciencie*, Vol. 223, pp. 83-88

Montes, M. L.; Taylor, M. A.; Mercader, R. C.; Sives, F. R. & Desimoni, J. (2010a). Hyperfine and radiological characterization of soils of the province of Buenos Aires. Argentina. *Journal of Physics: Conference Series,* Vol. 217, pp. 012058 012062, ISSN 1742 6596

Montes, M. L.; Mercader, R. C.; Taylor, M. A.; Runco, J.; Ibellone, P.A.; Rivas, P. C. & Desimoni, J. (2010b). Radiological and Hyperfine Characterization of soils from the Northeastern region of the Province of Buenos Aires, Argentina. *Hyperfine Interactions*. Submitted, ISSN 0304 3834

Montes, M. L. (2011). Private communication

Odino Moure, M. R. (2010). Environmental Radioactivity Monitoring Plan in Uruguay. *Technical Meeting on in-situ methods for characterization of contaminated sites*. Vienna, Austria.
http://www-pub.iaea.org/mtcd/meetings/Announcements.asp?ConfID=38924

Sajó-Bohus, L.; Pálfalvi, J.; Urbani, F.; Castro, D.; Greaves, E.D.& Liendo, J.A. (1999). Environmental gamma and radon dosimetry in Venezuela. *Radiation Measurements*, Vol. 31, pp. 283-286, ISSN 1350 4487

Salbu, B. (2009). Fractionation of radionuclide species in the environment. *Journal of Environmental Radioactivity*, Vol. 100, pp. 283-289, ISSN 0265 931X

Sawhney, B. L. (1972). Selective sorption and fixation of cations by clay minerals: A review. *Clays and Clay Minerals*, Vol. 20, pp. 93-100, ISSN 0009-8604

Schuller, P.; Ellies, A. & Kirchner, G. (1997). Vertical migration of fallout ^{137}Cs in agricultural soils from southern Chile. *Science of the Total Environment,* Vol. 193, pp. 197–205, ISSN: 0048 9697

Schuller, P.; Voigt, G.; Handl, J.; Ellies, A. & Oliva, L. (2002). Global weapons fallout ^{137}Cs in soils and transfer to vegetation in south-central Chile. *Journal of Environmental Radioactivity,* Vol. 62, pp. 181-193, ISSN 0265 931X

Schuller, P.; Bunzl, K.; Voigt, G.; Ellies, A. & Castillo, A. (2004). Global fallout ^{137} Cs accumulation and vertical migration in selected soils from south Patagonia, *Journal of Environmental Radioactivity,* Vol. 71, pp. 43-60, ISSN 0265 931X

Singh, B. K.; Jain, A.; Kumar, S.; Tomar, B. S.; Tomar, R.; Manchanda, V. K. & Ramanathan, S. (2009). Role of magnetite and humic acid in radionuclide migration in the environment, *Journal of Contaminant Hydrology,* Vol. 106, pp. 144-149. ISNN 0169-7722

Staunton, S.; Dumat, C.; & Zsolnay, A. (2002). Possible role of organic matter in radiocaesium adsorption in soils. *Journal of Environmental Radioactivity,* Vol. 58, pp. 163-173, ISSN 0265 931X

Tomazini da Conceic, F. & Bonotto, D.M. (2006). Radionuclides, heavy metals and fluorine incidence at Tapira phosphate rocks, Brazil, and their industrial products. *Environmental Pollution*, Vol. 139, pp. 232-243, ISSN 0269-7491

Toso, J. P. &Velasco, R.H. (2001). Describing the observed vertical transport of radiocesium in specific soils with three time-dependent models. *Journal of Environmental Radioactivity*, Vol. 53, pp. 133-144, ISSN 0265 931X

UNSCEAR, (1982). In: *Ionizing Radiation: Sources and Biological effects*, Report of the General Assembly with Scientific Annexes, Vol. 1, Annex E, New York,

UNSCEAR, (2000). In: *Sources and effects of ionizing radiation*, Report of the General Assembly with Scientific Annexes, Vol. 1, New York

UNSCEAR, (2008). In: *Sources and effects of ionizing radiation*, Report of the General Assembly with Scientific Annexes, Vol. 1, New York

Valdés, M. E.; Blanco, M. V.; Taylor, M. A.; Sives, F.R.;Runco, J. & Desimoni J. (2011). Determinación de las actividades de ^{60}Co, ^{137}Cs y ^{235}U en muestras de suelo, sedimentos y agua provenientes de la zona aledaña a una instalación nuclear. *Ciencia Forense Latinoamericana*, in press.

Vandenberghe, R. E. (1991). In: *Mössbauer Spectroscopy and Applications in Geology*, International Training Centre for Post-graduate soil Scientists, Geological Institute, Faculteit der Wetenschappen, Faculty of Science, Belgium.

Velasco, H. R.; Jury. Ayub, J.; Belli, M.; & Sansone, U. 2006. Interaction matrices as a first step toward a general model of radionuclide cycling: Application to the ^{137}Cs behaviour in a grassland ecosystem. *Journal of Radioanalytical and Nuclear Chemistry*, Vol. 268, pp. 503-509. ISSN 0236 5731

Valkovic, V. (2000). In: Radioactivity in the Environment, Elsevier, ISBN-13: 9780444829542.

12

Radioactivity in Marine Salts and Sediments

Manuel Navarrete, José Golzarri, Guillermo Espinosa, Graciela Müller, Miguel Angel Zúñiga and Michelle Camacho
National University of Mexico/Faculty of Chemistry/Institute of Physics
Mexico

1. Introduction

Radioactivity is a natural phenomenon always taking place in our planet and in the whole universe. In the very beginning of matter, which it is evolving till now, some radioactive isotopes were created, among others, to form either in a mixture or as a single one, the ninety material units known as elements, which combined in a huge number of ways represent what is called matter, nature and universe. This sort of radioisotopes are, for example, ^{40}K, ^{50}V and ^{87}Rb, as well as every radioisotope found from bismuth to uranium, all of them radioactive, classified by Mendeleieff according their atomic number and weight in the Periodic Chart. These natural radioisotopes are called Primordial and are shown in Table 1.

Radioisotope	Half Life (years)	Isotopic Abundance (%)
^{40}K	$1.3x10^{9}$	0.0118
^{50}V	$6x10^{15}$	0.24
^{87}Rb	$4.7x10^{10}$	27.85
^{113}Cd	$9x10^{10}$	12.3
^{115}In	$6x10^{14}$	95.72
^{123}Te	$1.24x10^{13}$	0.87
^{138}La	$1.3x10^{11}$	0.089
^{144}Nd	$2.1x10^{15}$	23.85
^{147}Sm	$1.1x10^{11}$	15
^{148}Sm	$7x10^{15}$	11.2
^{152}Gd	$1.1x10^{14}$	0.20
^{156}Dy	$2x10^{14}$	0.06
^{176}Lu	$3x10^{10}$	2.6
^{174}Hf	$2x10^{15}$	0.18
^{187}Re	$5x10^{10}$	62.6
^{186}Os	$2x10^{15}$	1.6
^{190}Pt	$6x10^{11}$	0.0127
^{209}Bi	$> 2x10^{18}$	100

Table 1. Radioisotopes in the isotopic mixture of elements from K to Bi (Primordial) (Choppin a, 1980)

Two vacancies are shown in the Periodic Chart: Tc and Pm, elements not present in nature, because when they are produced by nuclear reactions, only short half life radioisotopes are produced, and so, if they have existed some time, they were quickly transformed into their neighbour elements. But nuclear reactions are taking place continuously in the earthly atmosphere by the interaction between light elements in gaseous state and nuclear particles such as α particles, fast neutrons, protons and deuterons coming from stratosphere. The products of these nuclear reactions are also radioisotopes, which are pulled down to the planet mainly by rain water and wind with no interruption. Radioisotopes of this sort are, for example: ^{3}H, ^{10}Be and ^{14}C, which in spite of their short half lives, compared with the age of solar system, reach an equilibrium state between their rates of production and decaying. These natural radioisotopes are called Cosmogenic and are shown in Table 2.

Radiosotope	Half Life	Production rate in the atmosphere (nucleus/ m^{2}-s)
^{3}H	12.35 years	2500
^{7}Be	53.4 days	81
^{10}Be	1.6x10^{6} years	360
^{14}C	5715 years	22000
^{22}Na	2.6 years	0.6
^{26}Al	7.16x10^{6} years	1.7
^{32}Si	280 years	
^{32}P	14.3 days	
^{33}P	25.3 days	
^{35}S	87.5 days	14
^{36}Cl	3x10^{5} years	11
^{39}Ar	269 years	

Table 2. Some radioisotopes found in rain water (Cosmogenic) (Choppin b, 1980)

This is a very general and rather schematic description of natural radioactivity, always existent and main indicator of earth and universe evolution, since intensity of every radioactive source is always decreasing as time goes by, that is to say, the number of nucleus decaying by unit time when emitting nuclear radiations is inversely proportional to half life, and directly proportional to mass of every radioisotope, either natural or by human creation. But over the unavoidable and omnipresent natural radioactivity, it has been added that created by man. First radioisotopes of short half life were created, such as ^{13}N and ^{30}P with half lives of 9.9 and 2.5 minutes respectively, by the irradiation of B and Al with α particles emitted by Po. This discovery was made by Frederic Joliot and his wife Irene Curie in 1934. Since then, more than 2,000 artificial radioisotopes have been created, either as a research field itself or by a huge number of technological applications.

1.1 The Oklo phenomenon, a nuclear reactor in nature

In 1972, one mine of uranium minerals called Oklo, situated in the young country of Gabon, in Western Africa, was being fully exploited. Its minerals were sent to Pierrelate Centre for industrial uranium enrichment in France. Surprisingly, some samples showed a lower ^{235}U concentration than elsewhere in the world, that is to say 0.7%, which in some cases decreased as much as 0.4%. An explanation of that anomaly was that some mass of ^{235}U had suffered fission for some time in the past. Residues of fission products with longer half-life were thus looked for in the site, and were surprisingly found. The minimum concentration of highly fissionable ^{235}U to have the critical mass for fission chain reaction is 1%. As this radioactive isotope half-life is 700 million years, while that of ^{238}U is 4.5 billion years, the necessary time span to get that minimum concentration finished 400 million years ago. But age of fission products' residues found in the field were also coincident within a much larger time span, around 2 billion years ago, which is also a common order of magnitude for some other minerals with some radioactive isotope in its composition, such as ^{40}K and ^{87}Rb. At that time, ^{235}U concentration in minerals should have been much greater, and thus very likely to make fission possible. Geographical conditions are favourable as well as rain may have washed out uranium minerals found in the surrounding hills, which could have then concentrated at the bottom of a lake. This lake could have then dried out as a result of a change in the rain cycle, or possibly as a consequence of fission heat, from which sediments can be found at Oklo mine. Therefore, the Oklo phenomenon is a fact that supports the idea of radioactivity as a natural component of material reality, and should by no means cause major concerns if the phenomenon is adequately managed, as happens with fire, explosives, acids, fuels, speed, pressure, electricity and so on. As Chang, the great chemist says: "humans are not necessarily the innovators, but merely the imitators of nature" (Chang, 2005).

Finally, some natural radioisotopes with comparable half life to planet age, such as heavy ^{232}Th, ^{235}U and ^{238}U are decaying into radioisotopes which linking one to another make a radioactive chain, each link created by decaying of the previous one and evolving to next one by its own decaying, to finish with a stable Pb isotope. These sort of natural radioisotopes are called Radiogenic and are shown in Tables 3, 4 and 5. So, they have as a link, for example, ^{215}At and ^{218}At, radioisotopes with extremely short half lives, but in spite of it always present in nature because they are continuously created in the ^{235}U and ^{238}U radioactive chains. Pu radioisotopes are formed by ^{238}U irradiated with thermal neutrons and successive beta decay. Among them, ^{239}Pu ($t_{1/2}$ = 24,400 years) and ^{241}Pu ($t_{1/2}$ = 13.2 years) are the most important, because they have a great cross section for fission with thermal neutrons, and so they are the origin of the so called breeding reactors, where calorific energy is obtained at same time that a new fissionable, nuclear fuel is produced.

^{232}Th radioactive chain is called (4n) because the mass number of every link is a multiple of 4. In the same way, as radioactive chains of ^{238}U and ^{235}U show links whose mass numbers are reproduced by algebraic expressions (4n+2) and (4n+3), where n is an entire number, they are called in this manner. While ^{241}Pu radioactive chain, which is not natural, but produced in modern enriched uranium nuclear reactors, is called (4n+1) by same reason. It is noticeable from Tables 3, 4 and 5 the presence of links with Ra, Rn and Po isotopes. Ra and Po were the first radioisotopes isolated from pechblenda minerals by Pierre and Marie Curie, while Rn radioisotopes are also found there, all of them with different half lives and radiation energies. These radioisotopes of heaviest noble gas have been always a radioactive component of earth atmosphere everywhere, specially concentrated in those indoor places where their α and γ radiations are now detected. Therefore, emissions produced by natural radioisotopes have always been in air, earth and sea, but quite a different matter is the environmental contamination produced today by ^{235}U and maybe tomorrow by ^{239}Pu and ^{241}Pu fission products.

Radioisotope	Half Life	Historical Name	Type of radioactive decay
^{232}Th	1.4×10^{10} years	Thorium	α
^{228}Ra	6.7 years	Mesothorium I	β^-
^{228}Ac	6.13 hours	Mesothorium II	β^-
^{228}Th	1.9 years	Radiothorium	α , γ
^{224}Ra	3.64 days	Thorium X	α , γ
^{220}Rn	55 seconds	Toron (emanation)	α , γ
^{216}Po	0.15 seconds	Thorium A	α
^{212}Pb	10.6 hours	Thorium B	β^- , γ
^{212}Bi	60.6 minutes	Thorium C	α , β^- , γ
^{212}Po (64%)	304 nanoseconds	Thorium C′	α
^{208}Tl (36%)	3.1 minutes	Thorium C′′	β^- , γ
^{208}Pb	stable	Thorium D	—

Table 3. ^{232}Th radioactive chain (4n)

Radioisotope	Half Life	Historical Name	Type of radioactive decay
^{238}U	4.5×10^{9} years	Uranium I	α
^{234}Th	24.1 days	Uranium X_1	β^- , γ
^{234m}Pa	1.17 minutes	Uranium X_2	β^- , γ

Radioisotope	Half Life	Historical Name	Type of radioactive decay
^{234}Pa ↓	6.75 hours	Uranium Z	β^-, γ
^{234}U ↓	2.5x10^5 years	Uranium II	α, γ
^{230}Th ↓	8x10^4 years	Ionium	α, γ
^{226}Ra ↓	1602 years	Radium	α, γ
^{222}Rn ↓	3.8 days	Radon (emanation)	α, γ
^{218}Po ↓	3.05 minutes	Radium A	α, β^-
^{214}Pb (99.98%) ↓	26.8 microseconds	Radium B	β^-, γ
^{218}At (0.02%) ↓	2 seconds	Astatine	α
^{214}Bi ↓	19.7 minutes	Radium C	α, β^-, γ
^{214}Po (99.98%) ↓	164 microseconds	Radium C´	α, β^-
^{210}T1(0.02%) ↓	1.3 minutes	Radium C´´	β^-, γ
^{210}Pb ↓	21 years	Radium D	β^-, γ
^{210}Bi ↓	5.01 years	Radium E	α, β^-
^{210}Po (100%) ↓	138.4 days	Radium F	α
^{206}T1 (0.00013%) ↓	4.19 minutes	Radium E´	β^-
^{206}Pb	stable	Radium G	—

Table 4. ^{238}U radioactive chain (4n + 2)

Radioisotope	Half Life	Historical Name	Type of radioactive decay
^{235}U ↓	7.1x10^{8} years	Actinouranium	α , γ
^{231}Th ↓	25.2 hours	Uranium Y	β^- , γ
^{231}Pa ↓	3.25x10^{4} years	Protoactinium	α , γ
^{227}Ac ↓	21.6 years	Actinium	α , β^- , γ
^{227}Th(98.6%) ↓	18.2 days	Radioactinium	α , γ
^{223}Fr(1.4%) ↓	22 minutes	Actinium K	β^- , γ
^{223}Ra ↓	11.43 days	Actinium X	α , γ
^{219}Rn ↓	4 seconds	Actinium (emanation)	α , γ
^{215}Po ↓	1.8 milliseconds	Actinium A	α , β^-
^{211}Pb (100%) ↓	36.1 minutes	Actinium B	β^- , γ
^{215}At(0.00023%) ↓	0.1 millisecond	Astatine	α
^{211}Bi ↓	2.15 minutes	Actinium C	α , β^- , γ
^{211}Po(0.28%) ↓	0.52 seconds	Actinium C´	α , γ
$^{207}T1$ (99.7%) ↓	4.79 minutes	Actinium C´´	β^- , γ
^{207}Pb	stable	Actinium D	—

Table 5. ^{235}U radioactive chain (4n + 3)

Radioisotope	Half Life	Name	Type of radioactive decay
^{241}Pu	13.2 years	Plutonium	α , β⁻ , γ
^{241}Am(100%)	458 years	Americium	α , γ
^{237}U(0.0023%)	6.75 days	Uranium	β⁻ , γ
^{237}Np	$2.14x10^6$ years	Neptunium	α , γ
^{233}Pa	27 days	Protactinium	β⁻ , γ
^{233}U	$1.6x10^5$ years	Uranium	α , γ
^{229}Th	7340 years	Thorium	α , γ
^{225}Ra	14.8 days	Radium	β⁻ , γ
^{225}Ac	10 days	Actinium	α , γ
^{221}Fr	4.8 minutes	Francium	α , γ
^{217}At	0.032 seconds	Astatine	α
^{213}Bi	47 minutes	Bismuth	α , β⁻ , γ
^{213}Po (97.8%)	4.2 microseconds	Polonium	α
$^{209}T1$(2.2%)	2.2 minutes	Thallium	β⁻ , γ
^{209}Pb	3.3 hours	Lead	β⁻ , γ
^{209}Bi	> $2x10^{18}$ years	Bismuth	α ?

Table 6. ^{241}Pu radioactive chain (4n + 1)

2. Radioactive contamination

Radioactive contamination started on the planet in 1945, when the first nuclear test was performed in Alamo Gordo, New Mexico, followed by the war actions in Hiroshima and Nagasaki. Since then, radioactive contamination at global level has been variable, depending on repeated nuclear tests, few accidents such as Three Mile Island and Chernobyl, and minor failures in nuclear power plants. These contaminants are produced mainly by fission products from ^{235}U, which according their fission yielding and half lives, they remain radioactive during a time span from seconds to a great number of eons (1 eon = 1 x 10^9 years). But certainly, burned nuclear fuels which are under control and stored accordingly the safest techniques to guarantee they will always be confined and never disseminated in the environment, same case that residues of artificially produced radioisotopes used in medicine, industry or any other purpose, they should not be considered as radioactive contaminants, as much as they are under safe enough surveillance. So, approximately 30-40% all of known radioisotopes are fission products, which when they come into environment by deliberate nuclear explosion, severe accident or failure in nuclear plant, they represent the so called radioactive contamination. From this perspective, it seems that radioactive contamination has been growing up from its beginning, with rather short equilibrium periods. Also, if it is considered that sea water represents approximately 80% of planet surface, plus the action of wind, rain and rivers current, the main repository of radioactive contamination should be the sea. However, radioactive contamination is only added to natural radioactivity. From the first elements in the Periodic Table: ^{3}H, ^{10}Be and ^{14}C, natural radioisotopes are either continuously produced by nuclear reactions in the earthly atmosphere, or they were created at same time that non radioactive ones, in the mixture of isotopes forming elements such as ^{40}K, ^{50}V and ^{87}Rb. And then from Bi to beyond uranium elements, every isotope is radioactive with no exception. Therefore, it seems that to properly quantify the importance at planet level of any radioactive contamination, it should be done on the basis of radioactivity already present since the planet birth, whose decaying becomes the most evident sign of earth evolution and it is still taking place. In this way, 0.0118% isotopic abundance, 1.28 x 10^9 years half life, ^{40}K is the natural radioisotope most abundant in the earth crust and also in the numerous salts dissolved in sea water. So, the radioactivity due to ^{40}K might be the most suitable measurement, in order to have one basis of natural radioactivity to be compared with that of any artificial radioisotope. Among these, the fission product ^{137}Cs presents the highest yielding in the fission of ^{235}U , and it is the most common radioactive pollutant found in nuclear accidents due to its half life equal to 30.07 years, and γ rays easy to detect with higher efficiency due to a low energy equal to 662 Kev. Figure 1 represents the fission products yielding from ^{235}U vs. mass number (A) and Fig. 2 represents percentage of elements on earth vs atomic number (Z).

3. Experimental

3.1 Sampling and samples conditioning

Therefore, according with the idea to consider radioactivity as a quite natural phenomenon, supported by the existence of Primordial, Cosmogenic and Radiogenic radioisotopes, as well as the Oklo phenomenon, it is proposed to identify the natural radioactivity by Primordial radioisotope ^{40}K, based on the fact that it is present in one of more abundant elements on earth, as it can be seen in Fig. 2, and as a consequence is found in the

Log % ^{235}U Fission Products Yielding

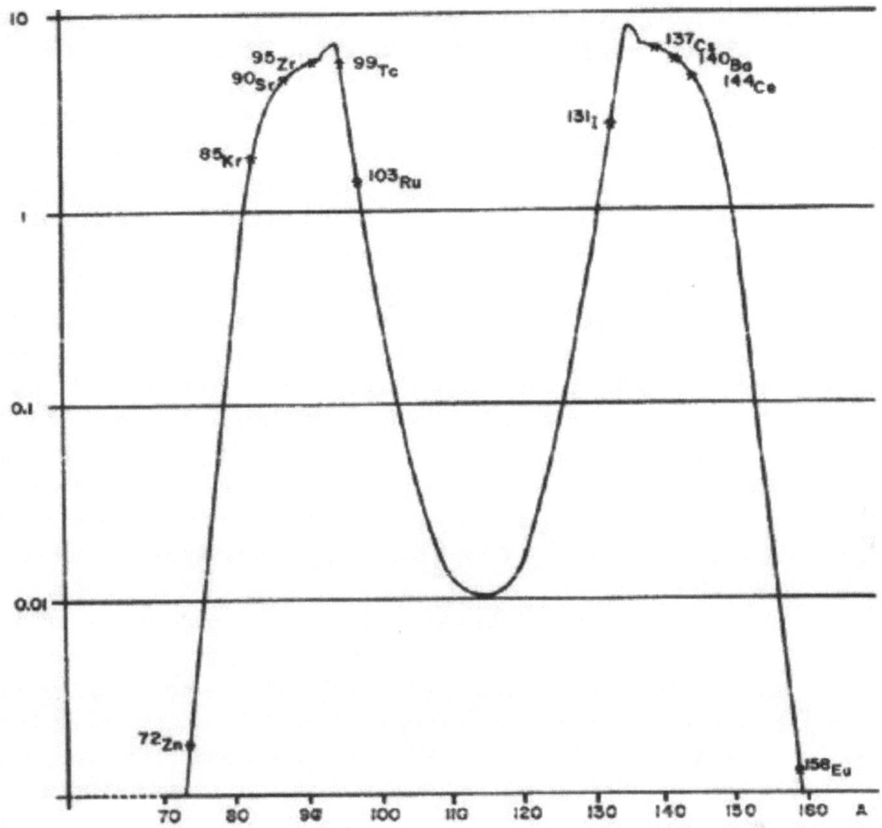

Mass Number

Fig. 1. ^{235}U Fission Products Yielding vs. Mass Number (A)

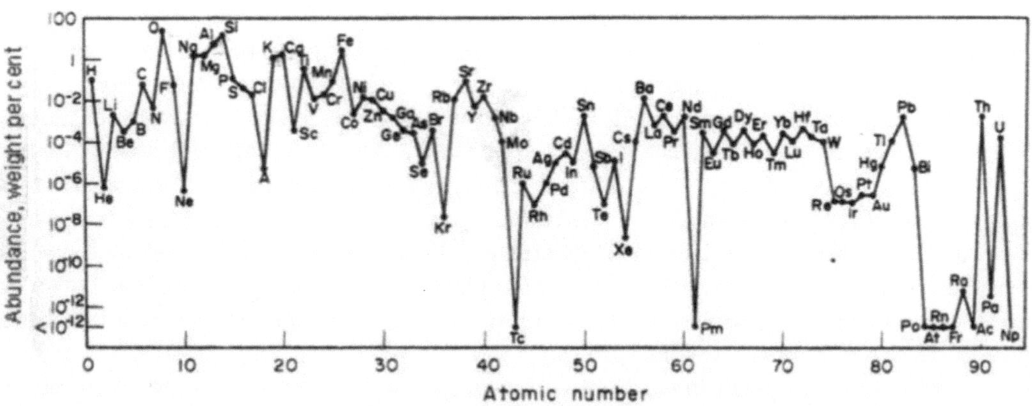

Fig. 2. Abundance of elements in earth (%) vs. Atomic Number (Choppin c, 1980)

radioactive background all over the world, while the present radioactive contamination can be easily represented by ^{137}Cs, fission product of ^{235}U. Besides, both radioisotopes are electromagnetic radiation emitters with suitable energies to be easily detected, and so one way to measure the intensity of present radioactive contamination should be to obtain a radioactive contamination factor (RCF), by dividing specific radioactivity of ^{40}K by that of ^{137}Cs in solid samples, that is to say disintegrations per time and weight units measured in both radioisotopes. This present radioactive contamination background, even when proceeds from limited portions on earth surface, where it has remained for long time as a well located radioactive source which must be left away by population and conveniently shielded, it has been unavoidable that a fraction of it spreads out to atmosphere in the gas and dust form, which can travel long distances to be finally carried down mainly by rain water on earth surface as either solutions or suspensions. But as sea represents the much larger proportion of planet surface, about 80%, and it is also the main factor of rain cycle, out of control radioactive pollutants produced anywhere in considerable amounts reach always the sea water in concentrations which can be easily measured by γ rays detection. Therefore, it seems that it is in sea water and marine sediments where global radioactive contamination should be searched and evaluated, because it is there where planet radioactive contamination has mainly created a growing deposit since the last world war. However, if it is assumed the sea water volume approximately as 1.4×10^{18} m^3, then it might be considered as an enormous natural radioactive source, not at all by contamination, but because it contains in solution an important concentration of K salts and its natural radioisotope ^{40}K (β^- and γ rays emitter after electronic capture, half life 1.28×10^9 years, 0.0118% isotopic abundance), which represents the main source of natural radioactivity as much in solid minerals (excepting those of heavy metals from Pb on), as in sea water and marine sediments. In this way, in order to asses the importance of any present or future radioactive contamination at planet scale, it might be compared by some radioactive contamination factor or some other way with natural radioactivity, which has been increased at certain extent by radioactive contamination. We are talking here about radioactivity spread out to environment from a local point, which must be immediately attended in situ, whereas that diluted in environment and reaching far away places usually produces great panic, even when it has never before been compared with natural, already existent radioactivity since the beginning of solar system. On the other hand, ^{40}K radioactivity as well as K concentration salts in sea water increases with ocean depth till a maximum value, and then decreases before reaching the bottom till a value usually lower than that at surface, as it happens with every mineral salt dissolved in sea water (Vázquez, 2001). So, it is quite possible to characterize superficial sea water in different coasts in terms of ^{40}K specific radioactivity, by sampling at about one kilometre from the coast, where it keeps constant for parallel much longer distances on the littoral, and obviously is easier to do it that in high sea, useful figure to calculate the concentration of elementary K in that particular sea zone. The way to do it is quite simple: 6-8 litres of sea water must be boiled, in order to get a suitable volume of sea salt to fill up a Marinelli container, usually about half a litre, necessary to perform low background radioactive detection. Once the dry salt sample is weighed and conditioned in the Marinelli container, it is ready to measure its natural as well as polluting radioactivity, by making use first of one heavily shielded scintillation set (NaI, Tl activated), and then one equally shielded hyper-pure Ge detector(HPGe), during 12-

24 hours detection time. Also, sediment marine samples have been picked up from 40-60 meters depth in three zones: Gulf of Mexico, to south east of Veracruz port and Laguna Verde Nuclear Power Plant, around Grijalva and Usumacinta delta rivers, as well as north, near the border with territorial USA sea water, and in Pacific Ocean between Cortés sea and Mazatlán port. Samples were taken by two ships: Puma in the Gulf and Justo Sierra in the Pacific Ocean, both at service of Sea Science and Limnology Institute, from National University of Mexico. Figure 3 presents the Puma ship. Figure 4 the Justo Sierra ship. These ships work in Oceanography research, for Institute of Sea Science and Limnology, in the National University of Mexico. Figure 5 presents one sediment sample conditioned in the Marinelli container. Figure 6 presents the low background scintillation set and Figure 7 presents the low background semi-conductor set.

Fig. 3. Ship Puma, samples collector in Pacific Ocean

Fig. 4. Ship Justo Sierra, samples collector in Gulf of Mexico

Fig. 5. Marinelli container with sediments

Fig. 6. Scintillation Detection set

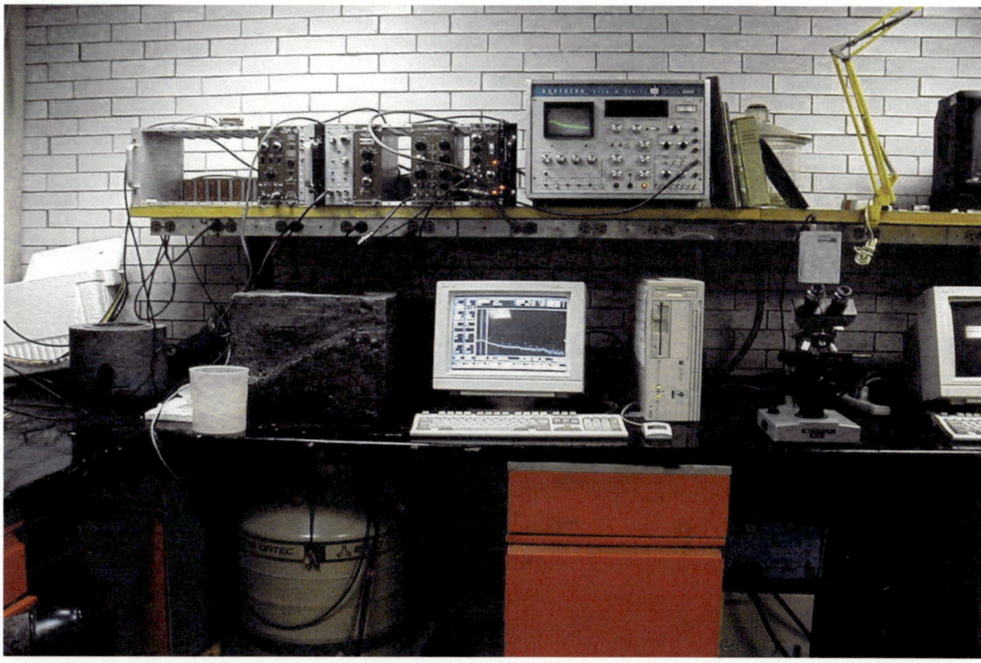

Fig. 7. HPGe Semiconductor Detetection set

3.2 Radioactive detection

In order to obtain our results either of natural or contaminant radioactivity in Bq per gram of sea salts and marine sediments, we must calculate the detection efficiency of both, scintillation and HPGe detector systems. It is easier and more precise to use one ^{40}K calibrated source formed by a known weight of KCl, and by separate one ^{137}Cs calibrated source. Detection efficiency for the 1461 Kev γ rays peak emitted by ^{40}K was determined by a standard made out by filling a Marinelli container with a weighed mass of KCl salt, AR grade. Detection time of 10-20 minutes was enough to get ±1% as statistical error. Then, the counts accumulated in the peak expressed as counts per second (cps), when divided by the specific activity expressed as disintegrations per second per gram (dps/g = Bq/g) of either KCl or elementary K, and multiplied by 100, is obtained detection efficiency for scintillation and semiconductor systems in the same way. Equations 1 and 2 show the calculation to get the specific activity of KCl and elementary K respectively, due to 11% of ^{40}K decaying nucleus by electron capture to ^{40}Ar and emitting γ rays with an energy of 1461Kev. Equation 3 show the calculation to get the total specific activity of elementary K, due to 0.0118% isotopic abundance of ^{40}K (β^- emitter 89%, EC and γ rays emitter 11%), constant value that will be used to characterize sea salts.

$$\begin{aligned} Bq^{40}K \rightarrow^{40} Ar/gKCl &= 0.693x6.02x10^{23}x0.0118\ x11\ /\ 1.28x10^{9}x365x24x60x60x100x100x74.5 \\ &= 1.8\ Bq^{40}K \rightarrow^{40} Ar\ /\ g\ KCl \end{aligned} \quad (1)$$

$$\begin{aligned} Bq^{40}K \rightarrow^{40} Ar/gK &= 0.693x6.02x10^{23}x0.0118x11\ /\ 1.28x10^{9}x365x24x60x60x100x100x39.1 \\ &= 3.4\ Bq^{40}K \rightarrow^{40} Ar/\ gK \end{aligned} \quad (2)$$

$$\begin{aligned} Bq^{40}K/gK &= 0.693x6.02x10^{23}x0.0118/1.28x10^{9}x365x24x365x60x60x100x39.1 \\ &= 31.19\ Bq/gK \end{aligned} \quad (3)$$

Where:

$Ln\ 2 = 0.693$

Avogadro's number = $6.02x10^{23}$

Isotopic abundance of $^{40}K = 0.0118/100$

Decay yielding of $^{40}K \rightarrow ^{40}Ar = 11/100$

Half life of $^{40}K = 1.28x10^{9}$ years $= 1.28x10^{9}x365x24x60x60$ seconds

KCl molecular weight = 74.5

K atomic weight = 39.1

Therefore, detection efficiency for counts accumulated in either scintillation or semiconductor detector, produced by γ rays with energy 1461 Kev, emitted by ^{40}K, is given alternatively by equations 4 and 5.

$$\text{Det. Eff.}\ (\text{electromagnetic radiation})\ (\%) = cpsx100/1.8xW_{s1} \quad (4)$$

$$\text{Det. Eff.}\ (\text{electromagnetic radiation})\ (\%) = cpsx100/3.4xW_{s2} \quad (5)$$

Where:

cps = counts accumulated per second

1.8 = specific activity of $^{40}K \rightarrow ^{40}Ar$ by EC, γ rays emission per gram of KCl $\left(Bq^{40}K \rightarrow ^{40}Ar\ /\ g\ KCl\right)$

3.4 = specific activity of $^{40}K \rightarrow ^{40}Ar$ by EC, γ rays emission per gram of elementary K $\left(Bq^{40}K \rightarrow ^{40}Ar\ /\ g\ K\right)$

W_{s1} = Weight of KCl in the Marinelli container

W_{s2} = Weight of K in the Marinelli container (52.48% of KCl)

In order to obtain the detection efficiency for gamma rays (662 Kev) emitted by radioactive contaminant ^{137}Cs, it has been used a calibrated multinuclide standard source in an identical Marinelli container to that used with KCl. In this case, calculation is only to divide counts per second accumulated in the corresponding peak (662 Kev), multiply by 100 and divide by the ^{137}Cs certificate activity in Bq at a given date and corrected to present time by decaying factor. To calculate detection efficiency by separate of γ rays from ^{40}K (1461 Kev) and γ rays from ^{137}Cs (662Kev), it is easier and more precise in our project, that to find that corresponding to ^{40}K from a graph efficiency versus energy, plotted with data obtained from the calibrated multinuclide source, because in this later case Compton distribution is much higher than in natural samples, such as KCl, marine salts and sediments. So, background correction in both detections has revealed as almost irrelevant when detections efficiencies are obtained, while on the other hand it is extremely important when marine salts and sediments are detected during much longer time periods, from 20 to 24 hours, but with similar dead time in detectors to that produced by KCl source.

If samples from Oklo uranium mine were considered as marine sediments, in order to evaluate the radiation danger they represent, it is very likely that radioactivity from natural radioisotopes of heavy metals such as ^{232}Th, ^{235}U and ^{238}U, origin of radioactive chains with several short half life radioisotopes in their links, were substantially higher than that from ^{40}K, natural radioisotope present almost everywhere, and by sure in Oklo minerals too. Since also in marine sediments have been found radioactive heavy metals, similarity between these two mineral samples becomes more understandable, besides the hypotheses that Oklo mine was a huge lake, probably of salted water in its origin. So, even when radioactive contamination by ^{137}Cs is not possible to confirm in Oklo due to its relatively short half life, it should be very easily detected in marine salts in the case of recent contamination, such as that in Fukushima, Japan, which at present should be in the mixture of natural marine salts, and in the near future will be in marine sediments, accompanying heavy metals and of course ^{40}K.

3.3 Characterization of marine salts and sediments through natural and pollutant radioactivity

Samples were taken in two points of Gulf of Mexico. One is to the south east of Laguna Verde Nuclear Plant, between delta of Usumacinta and Grijalva rivers, and the other to the north east of the Gulf, near the line with territorial USA waters. In the Pacific Ocean, samples were taken from Cortés Sea to Mazatlán port. In order to characterize sea waters by its K concentration, 5-6 litres of water samples were boiled to obtain about half a Kilogram of salt to fill up one Marinelli container. The weight of salt obtained and divided by the number of litres evaporated gives us one first figure equal to g/L, which means salinity. When counts accumulated during 20-24 hours in a low background detection system, either scintillation or HPGe, are expressed as counts per second, corrected for background in same units (cps) and divided by salt sample weight, detection efficiency for 1461 Kev γ rays (2.8% in our scintillation system and 0.22% in our HPGe detector) and the fraction of ^{40}K nucleus decaying to ^{40}Ar by EC and γ rays emission (11/100), total specific activity of ^{40}K expressed as Bq/g of salt is obtained, according the equation 6:

$$\text{Bq/g salt} = (\text{cps[Sample]} - \text{cps[Background]}) \,/\, \text{Ws x Det. Eff. x } 0.11 \tag{6}$$

Where:

Bq/g salt = Specific activity of sea salt due to ^{40}K total decaying $\left(\beta^{-}[89\%], \gamma \text{ rays } [11\%]\right)$

$\left(\text{cps[sample]} - \text{cps[Background]}\right)$ = counts accumulated per second by sample and corrected by background

Ws = Salt sample weight expressed in grams

Det. Eff = Detection efficiency for 1461 Kev γ rays emitted by ^{40}K in our detection systems, expressed as fractions $\left(\text{Scintillation}\left[2.8\text{x}10^{-2}\right], \text{HPGe}\left[0.22\text{x}10^{-2}\right]\right)$

0.11 = Fraction of ^{40}K nucleus decaying to ^{40}Ar by EC and γ rays emission (11%)

In this way, when salinity is multiplied by specific activity of sea salt, activity per litre of sea water is obtained. Also, when specific activity of sea salt is divided by specific activity of elementary K and multiplied by 100, concentration of K in sea salt is obtained as percentage, according the equations 7 and 8:

$$Bq/L = g/L \text{ x } Bq/g \text{ salt} \quad (7)$$

$$\%K = Bq/g \text{ salt x } 100 \ / \ 31.19 \ Bq/g \ K \quad (8)$$

Where:

Bq / L = Activity per litre of sea water due to ^{40}K total decaying $(\beta^{-}[89\%],\ \gamma \text{ rays } [11\%])$

g / L = Salinity of sea water expressed in grams per litre of sea water

Bq / g salt = Specific activity of sea salt due to ^{40}K total decaying $(\beta^{-}[89\%],\ \gamma \text{ rays } [11\%])$

$\%K$ = K concentration of sea salt expressed as percentage

$31.19\ Bq^{40}K\ /\ g\ K$ = Specific activity of elementary K due to ^{40}K total decaying $(\beta^{-}[89\%],\ \gamma \text{ rays } [11\%])$

So, when these figures are experimentally obtained, a great portion of sea water may be characterized from the ^{40}K natural decaying of its salt, data which should be very useful to detect and evaluate any recent contamination, such as that occurred in Fukushima, Japan, at present, and in the past those of Three Miles Island in USA, and Chernobyl in Russia, even when the nuclear accident or failure might have occurred at a large distance from the sea site. In any case, radioactive contamination should be represented by some fission product, most probably ^{137}Cs, due to its high fission yielding and easy detection of 662 Kev γ rays emission.
Nevertheless, and even when ^{137}Cs has not been detected in Mexican marine salts till now, it has been detected in every marine sediment tested in samples picked up at 60-80 meters deep. This fact maybe becomes enough evidence that it does already exists a radioactive contamination at sea bottom, creating one background from now on, which should be very important to evaluate in order to compare how it is growing up or maybe decaying when time goes by, and with no doubt nuclear power will have a great development all over the world. The main origin of this radioactive background at sea bottom, should be the test nuclear explosions at Alamo Gordo and Bikini, as well as the war actions in Hiroshima and Nagasaki, followed by nuclear test explosions performed by several countries since then, and only in a minor proportion by accidents and failure events of nuclear plants, considering that from 1945 to present day only 2.2 time spans of 30.07 years (half life of ^{137}Cs) have passed away. ^{137}Cs has not been detected so far in sea salt samples taken up from Mexican littorals, neither Pacific Ocean nor Gulf of Mexico. On the contrary, every sediment picked up from 60-80 meters depth, seems to have accumulated a small amount of ^{137}Cs, creating a certain pollutant radioactivity over the natural radioactive background present at sea bottom, which is represented mainly by ^{40}K and ^{232}Th, ^{235}U and ^{238}U radioactive chains. So, fission product ^{137}Cs should have been first dissolved in sea water, among a great diversity of ions in there, and then settled down on sediments as time goes by, because it is a rather heavy ion. In this way, ^{137}Cs present in sea salts should be indicating some recent pollution, while in marine sediments should be one of the main contributors to increase its natural background. Therefore, the proportion expressed as percentage of specific pollutant radioactivity $Bq^{137}Cs/g$ multiplied by 100 and divided by specific natural radioactivity ($Bq^{40}K/g$), should be as useful

in sea salts as in marine sediments, to have a reliable and easy to understand figure to evaluate the magnitude of recent pollution as well as to size up the possible growing or decreasing rate in already existing radioactive pollution in marine sediments.

4. Results

Figures 8 and 9 show the background and electromagnetic radiation (γ rays) of marine sediments picked up at Gulf of Mexico North, obtained with a low background scintillation detector, NaI(Tl), 3X3", coupled to a PC charged with Maestro Program.

Figures 10 and 11 show the background and electromagnetic radiation (γ rays) of marine sediments picked up at Gulf of Mexico North, obtained with a low background semiconductor detector, HPGe, coupled to a PC charged with Maestro Program II.

Table 7 shows the results obtained from sea salts samples taken up in Pacific Ocean North, between Cortes sea and Mazatlan port, and Gulf of Mexico North and South East, as well as sediments pollution measured by RCF (Radioactive Contamination Factor), where RFC = Bq ^{137}Cs x 100/g / Bq ^{40}K/g .

These results have been obtained within statistical variations given by Maestro Program I and II, maximum ± 15 % to minimum ± 1% of counts accumulated in both detection systems during detection times from 3.96 X 10^4 to 8 x 10^4 seconds or 11 and 22.2 hours. So, when subtracting background and dividing activity due to ^{137}Cs by that due to ^{40}K , statistical variations were always below ± 15%.

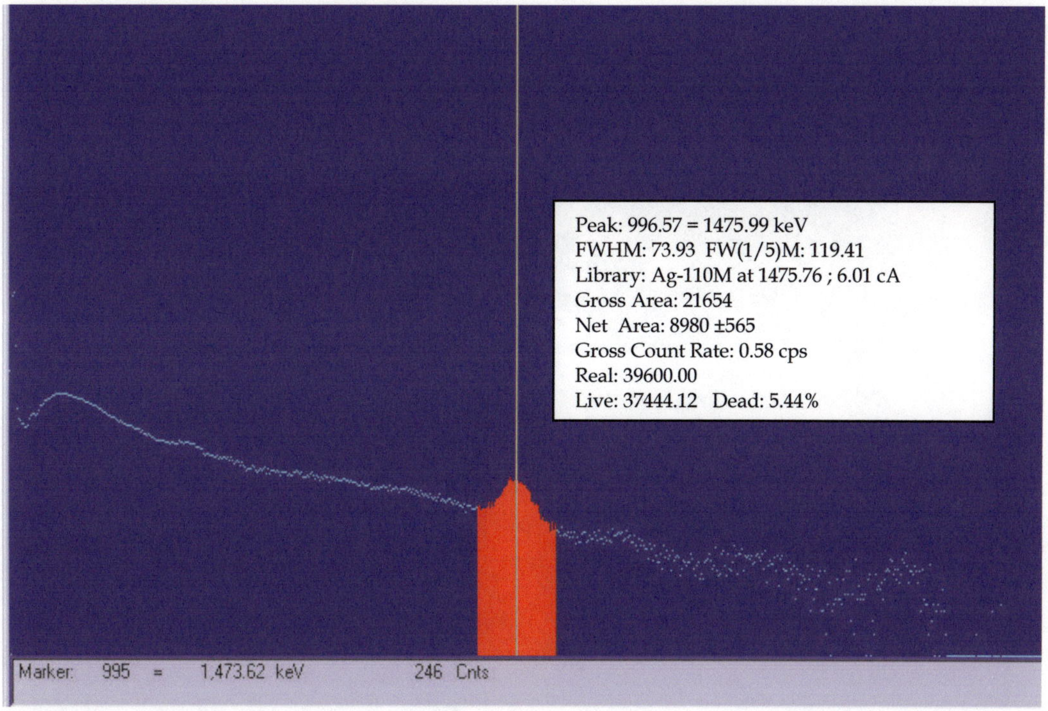

Fig. 8. Background spectrum in Scintillation Detection System

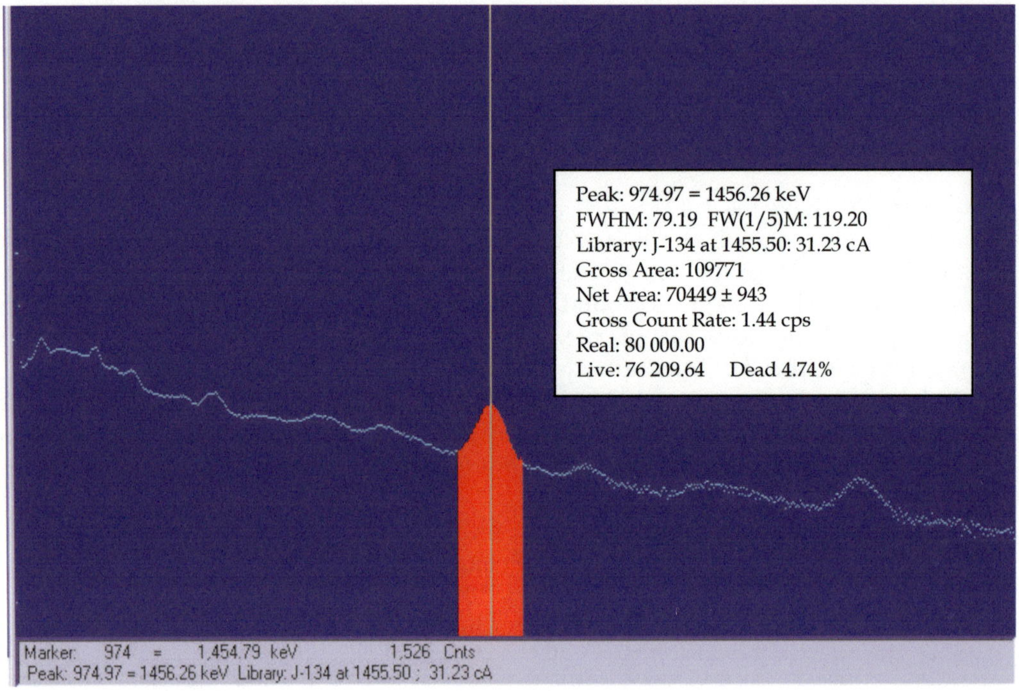

Fig. 9. Gulf of Mexico North East, sea salt spectrum sample, Scintillation Detection System

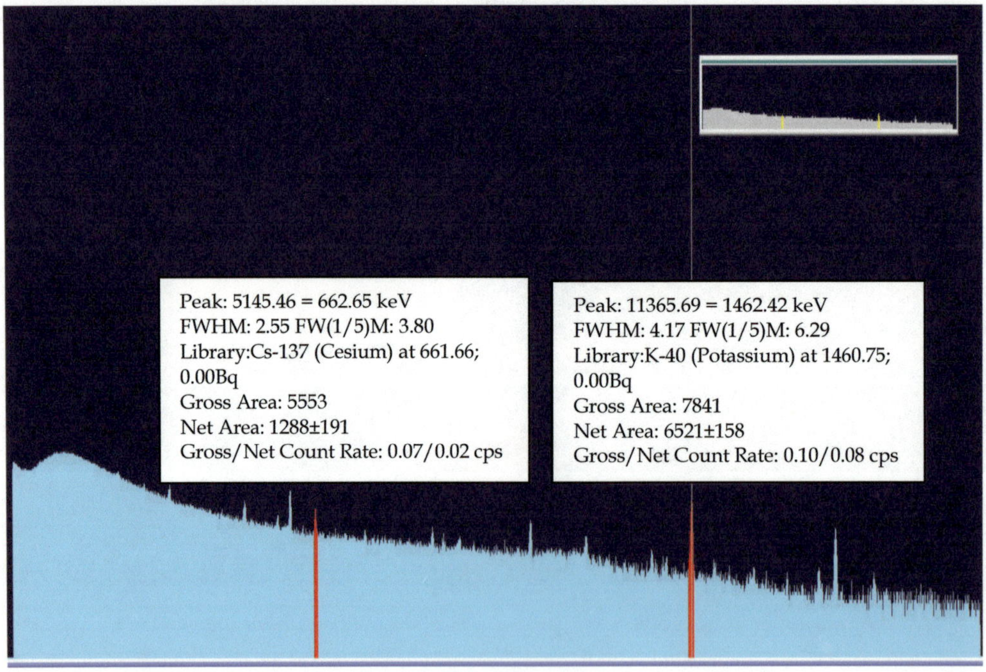

Fig. 10. Background spectrum in HPGe Detection System

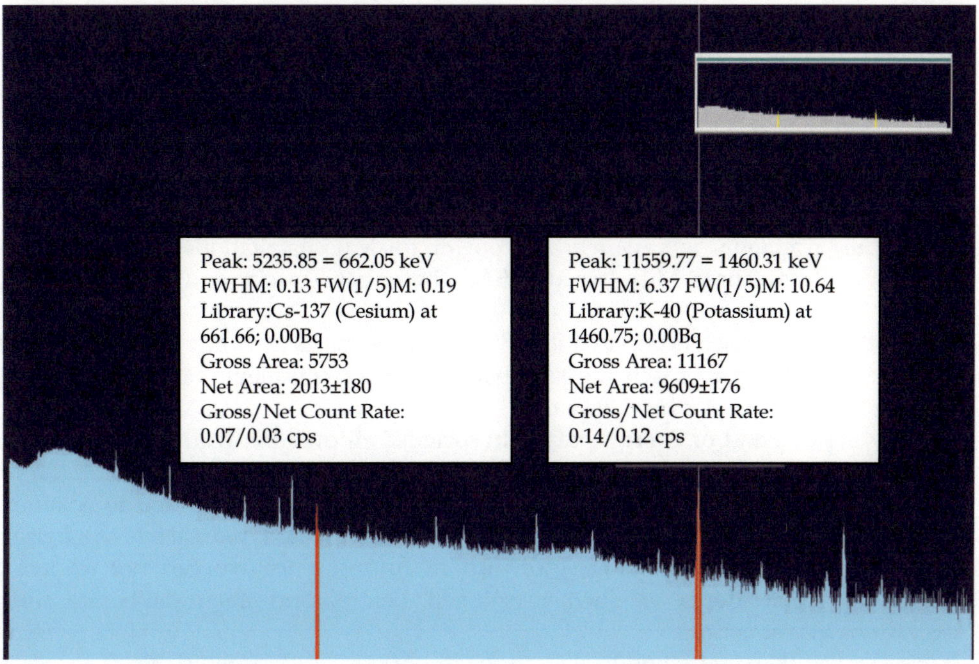

Fig. 11. Gulf of Mexico North East sediment spectrum sample, HPGe Detection System

Sea Salt Samples					Marine Sediment Samples
	Bq ^{40}K/g salt	Bq ^{40}K/L sea water	g salt/L sea water	%K in sea salt	$\%RCF = \frac{Bq\,^{137}Cs/g}{Bq\,^{40}K/g} x100$
Gulf of Mexico South East	0.276	10.1	36.7	0.88	0.89
Pacific Ocean North	0.073	2.5	34.8	0.23	0.58
Gulf of Mexico North East	0.173	7.3	42.5	0.55	0.93

Table 7. Results of natural radioactivity (^{40}K) in sea salt samples and %RCF in marine sediment samples

5. Conclusion

Conclusion of research results is based in several points, however reduced in samples number and extent too, when referring to very large littorals at Mexico.

a. It seems that radioactive pollution started on the planet at 1945, when first world war was finishing, with the first test of nuclear explosion in Alamo Gordo, followed by war actions in Hiroshima and Nagasaki, and few years later a second test in Bikini atoll.
b. Since then, a certain number of the so called industrialised countries have performed several tests in different regions of earth, including underground and submarine nuclear explosions.
c. Also, some accidents in research and power nuclear installations have taken place, notably those in Three Mile Island, USA, Chernobyl, Russia, and lately Fukushima, Japan.
d. Due to the fact that sea occupies about 80% of planet surface, every pollutant event has a larger probability to reach the sea than any other continental or insular region, starting from the point it has happened.
e. As growing demand of energy started in societies all over the world in XVIII century, when vapour machine was invented, and today nuclear energy seems to be the most powerful and suitable option to fill up energy demand, closely related to economical development, it looks like already existing, man created radioactive background, presents a strong tendency to grow up in future, since we can not neglect the possibility of accidents as such mentioned before, and even deliberate nuclear explosions as war actions.
f. It is proposed then, a method to size up the importance and growing rate of radioactive pollution all over the world, by comparing the artificial radioactivity of fission product ^{137}Cs, with that of natural radioisotope ^{40}K, both present in marine sediments at 60-80 meters depth on a great portion of sea bottom.
g. This procedure seems to be much more general than that to detect just ^{137}Cs in some vegetables such as lichens, which concentrate selectively elementary Cs, and it might be a suitable complement to it.
h. In this context, already existing radioactive pollution , seems quite possible to detect as a background in marine sediments, since ^{137}Cs half life is 30.07 years, and so it has decayed a little more than 2 half lives, about one fourth of the initial polluting radioactivity disseminated in 1945, plus the following nuclear tests and accidents.
i. Even when mathematical studies about dispersal of polluting radioisotopes have been successfully applied for limited conditions at a very small fraction of the huge sea (Periañez a, 2004), (Periañeza b, 2004), (Periañez c, 2010), it seems that this matter must be verified and treated in a quite empirical way, since natural and polluting radioactivity are facts concerning the whole planet.
j. In our samples appeared also some other peaks, such as that corresponding to ^{208}Tl (2614 Kev), with very poor resolution in the scintillation counter. Nevertheless, it is indicating the presence of other natural radioisotopes, because it is the last link of the ^{232}Th radioactive chain, in secular equilibrium with its parent and about 11 ancestors decaying at the same rate, before its own decaying to stable ^{208}Pb, with half life of just 3.1 minutes. Then, as a previous link in the chain, it is ^{228}Ac, γ rays emitter with 1459 Kev, and in consequence with possible contribution to ^{40}K peak (energy 1461 Kev) (Lavi, 2004). But as the difference of activity between these two peaks results so large in our samples ($^{40}K/^{208}Tl > 10$), then the possible contribution of ^{228} Ac peak (1% branching ratio) to that of ^{40}K (11% branching ratio) results negligible compared with our rather large calculated statistical variation.

k. Then, and based on previous points, we can say that every large sea portion might be suitably characterized by the percentage of K present in their salts. This can be made very easily in any sea of the planet, by picking up samples from the water surface near the coast. If polluting ^{137}Cs radioactivity (γ rays 662 Kev) is found out accompanying natural radioactivity from ^{40}K (γ rays 1461 Kev), the symptom is present of a rather recent polluting event, whose importance or extent might be evaluated at once by means of the ratio of specific activity per gram of sea salt, or litre of water, of polluting, divided by natural radioactivity and multiplying by 100 in order to have a percentage (Bq.^{137}Csx100/Bq^{40}K). This figure should be concernedly in the measure it approaches to 100%, which means same polluting radioactivity than natural one, and probably it might be useful to avoid the social panic. While same calculation applied to marine sediments 60-80 metres depth, should be useful to measure the already existing background polluting radioactivity, the rate of growing and the real possibility to keep it between tolerable limits.

6. Acknowledgements

Authors want to express their most sincere appreciation to Dr. Maria Leticia Rosales Hoz, Director of Sea Sciences and Limnology Institute, from the National University of Mexico, as well as to Dr. Vivianne Solís, Researcher in same Institute, for their invaluable help to make possible this effort to understand what radioactive pollution really means.

7. References

Chang, R., (2005). Nature's own Fission Reactor, In: *Chemistry,* McGrawHill Higher Education, 8th Ed., 962, ISBN 0-07-111317-7, Boston, United States of America

Choppin G. & Rydberg J., (a) (1980). Naturally occurring Radioactive Elements, In: *Nuclear Chemistry, Theory and Applications*, Pergamon Press 1st Ed., 225, ISBN 0-08-023823-8, Oxford, Great Britain

Choppin G. & Rydberg J. , (b) (1980). Naturally occurring Radioactive Elements, In: *Nuclear Chemistry, Theory and Applications*, Pergamon Press 1st Ed., 222, ISBN 0-08-023823-8, Oxford, Great Britain

Choppin G. & Rydberg J., (c) (1980). Thermonuclear Reactions and Nucleogenesis, In: *Nuclear Chemistry, Theory and Applications*, Pergamon Press, 1st Ed., 197, ISBN 0-08-023823-8, Oxford, Great Britain

Lavi N., Groppi F., Alfassi Z., (2004). On the measurement of ^{40}K in natural and synthetic materials by the method of high resolution gamma-ray spectrometry, *Radiation Measurements,* Vol. 38, (2004) pp. 139-143

Periañez R. (a) (2004). Testing the behaviour of different kinetic models for uptake/release of radionuclides between water and sediments when implanted in a marine dispersion model, *Journal of Environmental Radioactivity,* Vol. 71 (2004), pp. 243-259

Periañez R. (b) (2004). On the sensitivity of a marine dispersion model to parameters describing the transfers of radionuclides between the liquid and solid phases, *Journal of Environmental Radioactivity,* Vol. 73, (2004), pp. 101-115.

Periañez R. (c) (2010). Modelling Radioactivity Dispersion in Coastal Waters, in *Radioactive Contamination Research Developments*, Nova Science Publishers, Inc. Ed. 209-267, (2010), ISBN 978-1-60741-174-1, New York, United States of America

Vázquez A., (2001), Vertical profile determination of gamma emitting radionuclides with major concentration in Caribbean Sea and Gulf of Mexico, M. Sc. Thesis, *Environmental Engineering,* Veracruz University, Mexico, 2001, pp 23-32

13

Utilizing Radioisotopes for Trace Metal Speciation Measurements in Seawater

Croot, P.L.[1,2], Heller, M.I.[1], Schlosser C.[1] and Wuttig, K.[1]
[1]*FB2: Marine Biogeochemistry, IFM-GEOMAR, Kiel,*
[2]*Plymouth Marine Laboratory, Plymouth,*
[1]*Germany*
[2]*United Kingdom*

1. Introduction

The chemical speciation of trace metals in seawater is of critical importance to studies in marine biogeochemistry; as such information is essential for interpreting and understanding the chemical reactivity of trace metals in the environment. Foremost in this respect are studies into the role that chemical speciation plays in determining the biological availability (bioavailability) or toxicity of metals to organisms. Research on this topic over the last 30 years has clearly shown that open ocean productivity can be directly limited by iron. Other studies have revealed more subtle effects, such as co-limitation or limitation/toxicity affecting only some phytoplankton species, can occur with other trace metals and lead to controls on the composition of the phytoplankton community. Thus studies addressing chemical speciation in seawater are of relevance to the entire marine ecosystem.

Work on chemical speciation draws on skills and expertise from a diverse range of fields including; analytical chemistry, environmental chemistry, toxicology, geochemistry, genomics, proteomics, biological oceanography, physical oceanography and chemical oceanography. A tool common to all of these fields is the use of radioisotopes to examine the transfer or exchange between chemical species at environmentally relevant concentrations, which would be impossible with conventional analytical techniques. In this role radiotracers have been invaluable in the development of several key discoveries in Chemical and Biological Oceanography:

- ^{14}C measurements of primary productivity
- The development of the Free Ion Association Model (FIAM) and Biotic Ligand Model (BLM) for metal uptake kinetics by phytoplankton
- Iron limitation in the ocean and its impact on primary productivity
- The biological utilization of Cadmium by phytoplankton
- Quantification of the exchange kinetics between different metal species in solution

Oceanographic field research requires the ability to work on a moving ship in the ocean and if this was not difficult enough, work on trace metals necessitates the use of ultraclean techniques to avoid the ubiquitous contamination from the ship itself. Combining this with the normal precautions and safe working environment needed for using radio-isotopes can present researchers with a formidable challenge. However despite these problems radiotracers have always been a useful tool for marine scientists, both on land

and sea, as they allow the direct quantification of rates or fluxes and the identification of transformation pathways and mechanisms related to biogeochemical processes in the ocean. In recent years the development of extremely sensitive analytical techniques to determine the concentration of stable elements in seawater (ICP-MS) and phytoplankton (e.g. Nanosims, Synchtron XRF), coupled with the problems of using radioisotopes, has seen a general decline in their use compared to the genesis of trace metal marine biogeochemistry in the 1970's and 1980's. However in recent years a number of new questions have emerged where radioisotopes once again can provide crucial data and this has seen a mini-renaissance in their use.
The aim of this article is to provide a short overview on the previous use of radioisotopes in marine biogeochemistry (Section 2), where they have been applied directly in studies where chemical speciation is directly addressed. For the purposes of this article we only consider studies that utilize radioactive isotopes to directly assess the chemical speciation of trace metals in seawater or the use of chemical speciation techniques to determine the kinetics of exchange between different chemical species and their uptake by phytoplankton.
Research in marine biogeochemistry continues to develop rapidly and radioisotopes will continue to have a role as tools for enhancing our understanding of key processes involving trace metals in the ocean. New areas of research include the impacts of global warming, ocean acidification and ocean deoxygenation, all of which will impact on trace metal speciation and bioavailability. For this work radiotracers still have an important role to play and will continue to be utilised in major ocean-going international programs such as GEOTRACES and SOLAS and in laboratory work. As despite recent developments in analytical techniques, some important processes are still more easily followed by the use of radiotracers. In this context, here we also provide data from new applications (Section 3) of radioisotopes to current problems in this growing field.

1.1 Trace metals in seawater

In the following section we provide a short overview of trace metal chemistry in seawater and for more information we refer the reader to other review articles on this topic (Bruland and Lohan, 2003; Donat and Bruland, 1995). Other review articles examine the role of trace metals in biological cycles in the ocean (Bruland et al., 1991; Hunter et al., 1997; Morel et al., 2003).
Note on the units used in this section: It is common practice for oceanographers to report concentrations on both the molality (moles per kg of seawater) and molarity (moles per liter of seawater) scales, conversion between the two scales is easily accomplished when the density of the seawater is known (easily calculated from the salinity and temperature). In this text we report concentrations on the molarity scale (symbol M, units of mol L^{-1}) using the following standard SI abbreviations: μM (1x 10^{-6} M), nM (1x 10^{-9} M), pM (1x 10^{-12} M), fM (1x 10^{-15} M) and aM (1x 10^{-18} M). Radiochemical activities are given in Becquerels (Bq).

1.1.1 Concentrations and distributions of trace metals in seawater

Trace metals are found in seawater over a wide range of concentrations stretching from μmol kg^{-1} to amol kg^{-1} (Bruland and Lohan, 2003) and can exist in a variety of physical and chemical forms (Section 1.2). Table 1 (below) lists the typical concentrations in seawater found for the bio-active trace metals considered in this work.

Element	Concentration	Inorganic Speciation	Distribution
Al	0.3 - 40 nM	$Al(OH)_4^-$, $Al(OH)_3$	Scavenged
Ti	6 - 250 pM	$TiO(OH)_2$	Scavenged
V	30 - 36 nM	HVO_4^-	Conservative
Cr	3 - 5 nM	CrO_4^{2-}, Cr^{3+}	Nutrient
Mn	0.08 - 5 nM	Mn^{2+}	Scavenged
Fe	0.02 - 2 nM	$Fe(OH)_2^+$, $Fe(OH)_3$	Nutrient
Co	4 - 300 pM	Co^{2+},Co^{3+}	Nutrient
Ni	2 - 12 nM	Ni^{2+}	Nutrient
Cu	0.5 - 4.5 nM	Cu^{2+}	Nutrient
Zn	0.05 - 9 nM	Zn^{2+}	Nutrient
Se	0.5 - 2.3 nM	SeO_4^{2-}, SeO_3^{2-}	Nutrient
Mo	105 nM	MoO_4^{2-}	Conservative
Cd	1 - 1000 pM	$CdCl_2$	Nutrient

Table 1. Dissolved concentrations of bio-active trace metals in seawater (Bruland and Lohan, 2003; Donat and Bruland, 1995; Nozaki, 1997).

Traditionally chemical oceanographers have made a simple distinction between particulate and dissolved forms by separation via filtration (0.2 μm or 0.4 μm). More recently with the application of ultrafiltration techniques the dissolved fraction has been further divided into soluble (passes through a 1-200 kDa ultrafilter) and colloidal (difference between dissolved and soluble). Particulate forms include metals located intracellularly, or adsorbed extracellularly to biogenic particles or metals that form the matrix of minerals or are adsorbed to them.

An important concept in the development of the field of chemical oceanography was that of "Oceanographic consistency" (Boyle and Edmond, 1975), by which data for dissolved metals had to meet the following criteria:

1. Form smooth vertical profiles.
2. Have correlations with other elements that share the same controlling mechanisms.

Application of this approach has resulted in a reliable test for analytical data and led to the determination of vertical profiles for all the natural elements of the periodic table (Nozaki, 1997). Based on the shape of the vertical profile of each trace metal they can be grouped into three distinct groups reflecting their chemical behaviour in seawater:

1. *Conservative type distribution* - Metals showing this behaviour have concentrations that maintain a relatively constant ratio to salinity, have long oceanic residence times (> 10^5 years).
2. *Nutrient type distribution* - Concentrations are lowest in surface waters and increase with depth and are often strongly correlated to the distribution of the macronutrients (N, P and Si), indicating that these metals are assimilated by plankton in the euphotic zone and remineralized at depth.
3. *Scavenged type distribution* - Typical of trace metals that are adsorbed to particles (scavenged) and have oceanic residence times (~100-1000 y) less than the mixing time of the ocean. Highest concentrations are found nearest the sources of these elements to the ocean.

1.2 Trace metal speciation in seawater

Studies on trace metal speciation in seawater are concerned with determining the concentrations and processes affecting individual chemical species. Operationally this typically requires the application of specific techniques for the determination of analytically distinguishable chemical species or the application of thermodynamic or kinetic models to predict the behaviour of different species in seawater. Over the years the term 'speciation' began to be used for a number of different uses so to avoid confusion the International Union for Pure and Applied Chemistry (IUPAC) has published guidelines or recommendations for the definition of speciation analysis (Templeton et al., 2000):

i. *Chemical species.* Chemical elements: specific form of an element defined as to isotopic composition, electronic or oxidation state, and/or complex or molecular structure
ii. *Speciation analysis.* Analytical chemistry: analytical activities of identifying and/or measuring the quantities of one or more individual chemical species in a sample
iii. *Speciation of an element; speciation.* Distribution of an element amongst defined chemical species in a system
iv. *Fractionation.* Process of classification of an analyte or a group of analytes from a certain sample according to physical (e.g., size, solubility) or chemical (e.g., bonding, reactivity) properties.

1.2.1 Inorganic speciation

The inorganic speciation of trace metals in seawater (Table 1) is reasonably well described due to the extensive work performed by physical chemists in simple salt solutions. For the more complex media that is seawater, the use of Pitzer equations (Pitzer, 1973) is required but for many species in seawater this data is still missing. The reader is referred to a number of review chapters that cover the inorganic speciation of trace metals in more detail (Byrne et al., 1988; Turner et al., 1981). In particular recent reviews (Byrne, 2010; Millero et al., 2009) have focused on those elements whose inorganic speciation is dominated by hydroxide and/or carbonate species which are sensitive to decreases in pH and increasing CO_2 concentrations due to anthropogenic inputs.

1.2.2 Organic speciation

Many trace metals have been found to be strongly complexed by organic ligands in seawater, most notably iron (Gledhill and van den Berg, 1994) and copper (Coale and Bruland, 1988). However very little is known about these metal-organic complexes though it appears that they are produced by organisms in response to metal stress (Croot et al., 2000). Only a few of these ligands have been isolated and the chemical structures determined; iron complexing siderophores (Martinez et al., 2001) and heavy metal sequestering thiol complexes such as phytochelatins (Ahner et al., 1994). For a general overview of organic speciation in seawater see Hirose (2006). A recent paper by Vraspir and Butler (2009) provides a summary of the current information on trace metal binding ligands that have been isolated and identified in seawater.

1.2.3 Redox speciation – Importance of kinetics

For many trace metals there are major differences in the reactivity, bioavailability and toxicity between redox species. A critical factor here is the role of kinetics and/or oxygen concentrations in maintaining thermodynamically unstable redox species in solution where rapid reduction rates and slower oxidation rates leads to significant concentrations of the lower oxidation states of some metals in ambient seawater. For more on this and the impact of sulfide on metal speciation see general reviews on this subject (Cutter, 1992; Emerson and Huested, 1991; Morse et al., 1987).

1.3 Commonly used radioisotopes for trace metal seawater speciation studies

There are a number of trace metal radioisotopes that are commonly used for speciation work and they are listed in Table2 below.

Isotope	Half-life	Mode of decay	Detection Method (keV)
^{48}V	15.98 d	EC to ^{48}Ti	γ (983, 1312)
^{51}Cr	27.7 d	EC to ^{51}V	γ (320)
^{54}Mn	312.2 d	EC to ^{54}Cr β^- to ^{54}Fe	γ (835)
^{55}Fe	2.73 y	EC to ^{55}Mn	LSC (5.9)
^{57}Co	271.8 d	EC to ^{57}Fe	γ (122.1)
^{58}Co	70.88 d	EC to ^{58}Fe	γ (810.8)
^{59}Fe	44.51 d	β^- to ^{59}Co	γ (1099,1292)
^{60}Co	5.271 y	β^- to ^{60}Ni	β(318.7), γ (1173, 1332)
^{63}Ni	100 y	β^- to ^{63}Cu	LSC (66.9)
^{64}Cu	12.7 h	EC to ^{64}Ni β^- to ^{64}Zn	γ (511)
^{65}Zn	243.8 d	EC to ^{65}Cu	γ (1115)
^{67}Cu	2.58 d	β^- to ^{67}Zn	γ (185)
^{75}Se	119.78 d	EC to ^{75}As	γ (136,265)
^{99}Mo	2.7476 d	β^- to ^{99m}Tc	LSC (739, 778)
^{109}Cd	462 d	EC to ^{109}Ag	γ (22)

Table 2. Commonly used radioisotopes for trace metal speciation work in seawater (Data complied from sources mentioned in the text). Abbreviations used: d – days, h – hours, y – years, EC – electron capture, β^- beta decay, γ - gamma counting and LSC – liquid scintillation counting.

Most of the radioisotopes listed above are routinely available commercially and many can be obtained as high specific activity carrier free solutions. See the later sections for more details regarding experiments involving the individual metals.

1.4 Typical applications of radioisotopes to trace metal speciation in seawater

Typically speciation work in marine biogeochemistry has utilized radioisotopes for two types of experiment: (i) Biological uptake under conditions of chemical equilibrium. (ii) Kinetics of transformation of a known species in seawater.

1.4.1 Uptake of trace metals by phytoplankton

Radioisotopes have been extremely important in improving our understanding of the links between chemical speciation and bioavailability of trace metals to phytoplankton and bacteria in the ocean. The genesis of this field began with the application of trace metal buffers utilizing aminocarboxylate ligands; Nitrilotriacetic acid (NTA), Ethylenediaminetetraacetic acid (EDTA) and Diethylenetriaminepentaacetic acid (DTPA). A well characterized seawater media, AQUIL, was developed for use in trace metal uptake experiments (Price et al., 1989). New analytical tools were also required to determine the intracellular metal content from that simply adsorbed (Hudson and Morel, 1989) and this allowed the determination of metal quotas for cells (metal to carbon ratio, or metal per cell). Theoretical developments occurred simultaneously with new important paradigms and hypotheses that could be tested based on thermodynamic equilibrium between species; The Free ion association model (FIAM), see review by Campbell (1995) and later the Biotic Ligand Model (BLM) (Di Toro et al., 2001). The recognition that for some metals the system is not in equilibrium, due to slow exchange reactions (Hering and Morel, 1989; Hering and Morel, 1990), saw the use of pulse-chase experiments where a radio-isotope is added as a known species and its uptake followed over time. These new approaches led to important concepts with regard to the kinetic limitations (Hudson and Morel, 1993) on uptake by phytoplankton and how this can impact phytoplankton physiology (e.g. cell size, number of transport ligands).

Applications of the FIAM and BLM to experiments with natural seawater and phytoplankton communities are more complex as the chemical species which are bioavailable are mostly unknown. However if the added radio-isotope is in isotopic equilibrium natural uptake rates and metal quotas can be determined.

1.4.2 Kinetics of exchange between trace metal species

Experiments investigating the kinetics of exchange between chemical species in seawater have also been applied using radioisotopes. This has typically been done in a pulse-chase fashion utilizing an analytical detection method that was capable of determining the chemical species of interest. These experiments have not always been at the lab bench scale, as past experiments have been performed under controlled conditions in mesocosms, including sediments, and using multiple tracers (Li et al., 1984; Santschi et al., 1987; Santschi et al., 1980).

2. Present state of the art

In the following sections we review the current and previous use of radioisotopes in seawater speciation studies for bio relevant trace metals.

2.1 Iron (Fe)

Our understanding of the marine biogeochemistry of iron (Fe) has developed rapidly over the last 30 years. The thermodynamically favoured redox form of Fe in seawater, Fe(III), is only weakly soluble in seawater (Millero, 1998). The reduced form, Fe(II), is found in oxic waters as a transient species, primarily generated by photochemical processes (Croot et al., 2008), and existing at extremely low concentrations (picomolar or less) because of rapid oxidation by O_2 and H_2O_2 in warm surface waters. The oxidation of Fe(II) to the less soluble Fe(III) species, leads to the formation of colloidal oxyhydroxide (Kuma et al., 1996) species which coagulate and form particulate iron (Johnson et al., 1997). Dissolved iron is strongly organically complexed throughout the water column (Boye et al., 2001). Iron is an essential element for all life and is a limiting nutrient in many parts of the global ocean as has been so clearly demonstrated in the mesoscale iron enrichment experiments (de Baar et al., 2005).

Work on iron biogeochemistry has greatly benefited from the easy availability of both ^{55}Fe and ^{59}Fe for tracer studies and no other trace metal has been so widely studied.

2.1.1 Solubility of iron in seawater

The solubility of iron in seawater is a controlling factor in its distribution in the ocean and information on this topic has been achieved predominantly through the use of radioisotope experiments. Initial work (Kuma et al., 1992) focused on the determination of the rate of dissolution as a function of pH, as measured using a dialysis tube (1kDa), of amorphous ferric oxides formed upon addition of $^{59}Fe(III)$ to seawater. This approach was then adapted to determine solubility directly in seawater samples by simple syringe filtration with a 0.025 µm filter (Millipore MF) of a seawater solution that had been amended with 100 nM of radiolabelled Fe (Kuma et al., 1996). This technique has subsequently been applied to a range of oceanic environments; coastal Japan (Kuma et al., 1998b), the Pacific Ocean (Kuma et al., 1998a) and the Indian Ocean (Kuma et al., 1996). Liu and Millero (2002) used the same approach but employed a 0.02 µm Anotop filter to measure iron solubility in UV irradiated seawater as a function of temperature and salinity. Field studies using the Anotop filter and ^{55}Fe have been reported from the Mauritanian upwelling (Schlosser and Croot, 2009). Ultrafiltration (Vivaflow 50) has also been applied to studies of the effects of different ligands on iron solubility in seawater (Schlosser and Croot, 2008). The kinetics of iron hydroxide formation was determined using ^{55}Fe and ion-pair solvent extraction of chelated iron (Pham et al., 2006).

2.1.2 Kinetics of exchange between different iron species

Iron radioisotopes have proven extremely useful for examining the exchange kinetics between different iron species in seawater. Hudson et al. (1992) utilised ^{59}Fe in combination with ion-pair solvent extraction of iron chelated by sulfoxine (8-hydroxyquinoline-5-sulfonate). Using this approach they measured the rate at which the inorganic Fe(III) hydroxide species at seawater pH (referred to as Fe′) are complexed by EDTA and the natural terrestrial siderophore desferrioxamine B (DFO-B). Another approach to measuring Fe′ in UV irradiated seawater was developed by Sunda and Huntsman (2003) using solid phase extraction with EDTA on C18 Sep-Paks, by where the Fe′ was retained on the column. The phenomena of colloidal pumping, where iron initially in the colloidal size range is transformed into particles has also been investigated in seawater using ^{59}Fe (Honeyman and Santschi, 1991).

2.1.3 Iron uptake by phytoplankton and regeneration by zooplankton grazing

The use of radioisotopes to determine the rate of iron uptake by phytoplankton in trace metal buffered media is the best example there is for the advantages that this approach has over stable isotopes. The literature abounds with several key studies from Morel's group that shaped the direction of marine research on iron; the availability of Fe(II) and Fe(III) to diatoms (Anderson and Morel, 1980), iron colloids (Rich and Morel, 1990), the ability to separate intracellular from extracellular iron (Hudson and Morel, 1989) and the importance of kinetics (Hudson and Morel, 1990). Later work by Sunda and colleagues showed the differences in iron requirements between coastal and oceanic species (Sunda and Huntsman, 1995; Sunda et al., 1991) and the relationship between iron, light and cell size (Sunda and Huntsman, 1997).

There have also been a number of studies examining the role of zooplankton grazing in transforming iron contained in phytoplankton back into the dissolved phase (Hutchins and Bruland, 1994). A dual tracer (^{55}Fe and ^{59}Fe) approach has also been used to study the fate of intracellular and extracellular iron in diatoms when grazed by copepods (Hutchins et al., 1999). The direct remineralisation of colloidal iron by protozoan grazers has also been observed (Barbeau et al., 1996). Other trophic transfer mechanisms investigated include the transfer of bacterial iron to the dissolved phase by ciliates (Vogel and Fisher, 2009) and remineralisation via viral lysis (Poorvin et al., 2004).

2.1.4 Iron redox speciation

Somewhat surprisingly there have been very few studies examining iron redox processes in seawater using radioisotopes. Though in part this is most likely due to the short life-time of this species in ambient seawater and the application of chemiluminescence techniques to detect pM Fe(II) (Croot and Laan, 2002). The photoreduction of ^{59}Fe-EDTA has been used as a model system by both Hudson et al. (1992) and Sunda and Huntsman (2003). Photoreduction of natural iron complexes in the Southern Ocean as been shown to be strongly related to UV-B (Rijkenberg et al., 2005). The biological reduction of iron by phytoplankton has also been investigated by Shaked et al. (2004) who used Ferrozine as an Fe(II) chelator and then retained the complex on C18 Sep-Paks.

2.2 Manganese (Mn)

Manganese (Mn) is a redox sensitive element which is important to phytoplankton due to its involvement in photosynthesis through photosystem II in converting H_2O to O_2 (Falkowski and Raven, 1997). Mn is also utilized in superoxide dismutases (Peers and Price, 2004). Mn has a scavenged type profile (Landing and Bruland, 1987) and a secondary Mn maximum occurs in the oxygen minimum zone (Johnson et al., 1996). Mn(IV) is the thermodynamically favoured form in seawater but is strongly hydrolysed forming particulate MnO_2. Mn(II) is weakly hydrolysed in seawater and does not form strong organic complexes in seawater and is slowly oxidized to particulate Mn(III) and Mn(IV) under seawater conditions (von Langen et al., 1997).

2.2.1 Mn uptake by phytoplankton

The uptake of Mn by phytoplankton has been investigated for a number of species by using ^{54}Mn. In a series of now classical laboratory studies by Sunda and Huntsman (1983; 1996) the interactive effects between Mn and Cu, Zn and Cd in phytoplankton were investigated and showed clearly the competition for uptake by these elements for the same transport ligands in the diatom species tested. Other work has shown that the ^{54}Mn taken up by phytoplankton can be recycled back into the dissolved phase through the action of zooplankton grazing by copepods (Hutchins and Bruland, 1994).

2.2.2 Mn oxidation

The oxidation of dissolved Mn(II) in seawater to particulate manganese oxides has been studied extensively in the field via the use of ^{54}Mn (Emerson et al., 1982) and taking advantage of the differences in the solubilities of the different Mn redox states. Initial studies focused on the role of oxygen in the bacterially mediated oxidation of Mn(II) in sub-oxic zones (Tebo and Emerson, 1985). Work in oxygenated surface waters by Sunda and

Huntsman (1988) found that in the Sargasso Sea that Mn oxidation was inhibited by sunlight, consistent with photoinhibition of manganese oxidizing bacteria. Moffett (1997) confirmed this for the Sargasso Sea but found in the Equatorial Pacific that phytoplankton uptake of Mn may be more important. Similar Mn oxidation studies have been performed in the Eastern Caribbean (Waite and Szymczak, 1993) and in hydrothermal plumes (Mandernack and Tebo, 1993). A number of field studies by Moffett and co-workers have sought to link bacterial Mn oxidation to the oxidation of Co (Moffett and Ho, 1996) and Ce (Moffett, 1994).

2.2.3 Mn photoreduction

The dissolution of $^{54}MnO_2$ in seawater has been extensively investigated and found to be strongly related to the presence of H_2O_2 formed by the photoreduction of O_2 by dissolved organic matter (Sunda et al., 1983). Photoreduction of MnO_2 in shallow sediments has also been observed (McCubbin and Leonard, 1996). Laboratory studies have also investigated the impact of humic acids on the photoreduction of MnO_2 (Spokes and Liss, 1995).

2.3 Copper (Cu)

The speciation of Copper (Cu) in seawater is dominated by organic complexation (Coale and Bruland, 1988) by ligands which are believed to be produced by phytoplankton in response to Cu stress (Croot et al., 2000). While Cu(II) is the thermodynamically favoured redox state in oxygenated seawater there is growing evidence that Cu(I) may also be significant. Radiotracer studies into Cu chemistry however are limited by the short half-lives of the two isotopes available ^{64}Cu ($t_{½}$ = 12.7 hours) and ^{67}Cu ($t_{½}$ = 2.58 days).
Initial studies on Cu uptake by phytoplankton used ^{64}Cu and were focused on pulse chase experiments with NTA buffers and lipophilic ^{64}Cu complexes that could pass directly through the phytoplankton cell wall (Croot et al., 1999). Later work showed the existence of an efflux system for Cu from the Cu stressed cells of the cyanobacteria *Synechococcus* (Croot et al., 2003). Recent works on the uptake of Cu by phytoplankton have utilized the longer lived isotope ^{67}Cu to obtain important information on the uptake kinetics of Cu by diatoms (Guo et al., 2010), determined cellular Cu quotas for different phytoplankton types (Quigg et al., 2006) and showed the dependence of Cu on Fe uptake (Maldonado et al., 2006) and in turn the role of Fe in determining the cellular quota for Cu (Annett et al., 2008). However the most exciting application so far has been the first reported use of ^{67}Cu for work performed using natural phytoplankton assemblages from the North Pacific (Semeniuk et al., 2009).

2.4 Zinc (Zn)

Zinc (Zn) is a required metal for bacteria and phytoplankton in the ocean as it serves as a metal cofactor for many important processes (Vallee and Auld, 1993). Most notably Zn is utilized for both nucleic acid transcription and repair proteins (Anton et al., 2007) in the enzyme alkaline phosphatise (Shaked et al., 2006) and for the uptake of CO_2 via the enzyme Carbonic Anhydrase (CA) (Morel et al., 1994). The strong requirement for Zn by phytoplankton results in low concentrations in surface waters and a nutrient like profile in the ocean (Table 1). In most surface waters Zn is strongly organically complexed (Bruland, 1989), however in deep waters and in surface waters of the Southern Ocean inorganic complexes can dominate (Baars and Croot, 2011).
The use of ^{65}Zn was central to the first speciation studies of Zn uptake by phytoplankton performed on cyanobacteria (Fisher, 1985) and diatoms (Sunda and Huntsman, 1992).

Studies into ^{65}Zn uptake by bacteria (Vogel and Fisher, 2010) found a much lower uptake of Zn than Cd. ^{65}Zn has also been used in assessing the release of Zn from phytoplankton (Hutchins and Bruland, 1994) and bacteria (Vogel and Fisher, 2009) during zooplankton grazing.

2.5 Cobalt (Co)

Cobalt (Co) is present in seawater at very low concentrations (< 100 pM) and can exist as either inert Co(III) complexes or more labile Co(II) organic species (Saito and Moffett, 2002). Despite the range of Co isotopes available (Table 2) there have been relatively few studies examining the seawater speciation of Co. Work by Nolan et al. (1992) utilised a dual tracer approach where the uptake of ^{57}Co-cobalamine was compared to that of ^{60}Co-Co(II) and found that the cobalamine was taken up significantly faster and retained for longer in phytoplankton. ^{57}Co was also used to show that Co could replace Zn in the enzyme carbonic anhydrase in some phytoplankton (Yee and Morel, 1996). An early finding with ^{57}Co was that the oxidation of Co(II) to Co(III) in solution (Lee and Fisher, 1993) may be mediated by the same bacteria responsible for Mn oxidation (see section 2.3.2). Though new data (Murray et al., 2007) suggests no Co(II) oxidation occurs in the complete absence of Mn(II) and that the mechanism by which bacteria oxidize Co(II) is through the production of the reactive nano-particulate Mn oxide. Co has also been found to be released back to the dissolved phase from grazed and decomposing phytoplankton (Lee and Fisher, 1994).

2.6 Speciation studies with other trace metals

Studies into the biogeochemical cycles of other elements in seawater that are strongly hydrolysed and thus analogous to iron are limited by the lack of suitable radiotracers. There are no Aluminium (Al) radioisotopes suitable for use in trace metal speciation studies, as the majority of them have half-lives shorter than 10 minutes. The long lived isotope ^{26}Al (710,000 y) has found application in paleo applications (Lal et al., 2006). Similarly there are also no seawater studies on Titanium (Ti) biogeochemistry with radioisotopes due to the short half-lives of ^{45}Ti (3.08 h), ^{51}Ti (5.76 min) and ^{52}Ti (1.7 min). The longest lived Ti isotope, ^{44}Ti (43.96 y) is difficult to produce and it is not yet available commercially.

Vanadium (V) exists in oxygenated seawater as the inorganic vanadate (VO_4^{3-}) species and while a useable radio-isotope exists, ^{48}V (Table 2), it has so far been only applied to a few studies in marine systems, most notably examining the uptake of vanadate by ascidians who accumulate high concentrations of vanadium in their blood (Michibata et al., 1991). Chromium (Cr) exists in seawater in two redox states as either chromate (CrO_4^{2-}) or the reduced form Cr(III). There have only been a limited number of studies using either ^{48}Cr or ^{51}Cr (Table 2) and most have focused on the uptake of CrO_4^{2-} (Wang and Dei, 2001b) or colloidal Cr (Wang and Guo, 2000) by phytoplankton.

2.6.1 Nickel (Ni)

In seawater Nickel (Ni) shows a nutrient like behaviour and is present in surface waters at nM concentrations (Table 1). While some studies have shown organic complexation of Ni in seawater such work is complicated by the slow exchange kinetics for Ni(II) in seawater. A number of important advances in our understanding of Ni biogeochemistry in the ocean have come about through the use of ^{63}Ni. Firstly Price and Morel (1991) observed that Ni was required for growth on urea, a Ni containing enzyme, by the diatom *Thalassiosira weissflogii*. Later Dupont and co-workers investigated the uptake of ^{63}Ni by the globally important cyanobacteria *Synechococcus* in a laboratory study (Dupont et al., 2008) and made field measurements of ^{63}Ni uptake by natural phytoplankton assemblages (Dupont et al., 2010). They showed that Ni was a required element for many strains of *Synechococcus* and by comparison to the available genomic data most likely all strains of *Procholorococcus*. This was due to the use of a Ni containing superoxide dismutase and in the enzyme for urease uptake. Importantly they also observed that isotopic equilibrium was not established between the added radiotracer and the natural pools of Ni within 24 hours indicating the slow exchange kinetics of Ni in seawater (Hudson and Morel, 1993).

2.6.2 Selenium (Se)

Selenium (Se) is found in very low concentrations in seawater (< 1 nM) and its chemistry is under kinetic redox control (Cutter, 1992) with the oxyanions Selenate (SeO_4^{2-}) (thermodynamically favoured in oxygenated seawater and selenite (SeO_3^{2-}), both showing nutrient like profiles in the ocean. In surface waters enhanced concentrations of organic selenide (operationally defined) is typically present (Cutter and Bruland, 1984).

The radioisotope ^{75}Se has been used in a number of studies to elucidate the biogeochemistry of Se in phytoplankton cells. A key early finding was the identification of the pathway of uptake of selenite into the diatom *Thalassiosira pseudonana* and its conversion into the Se containing enzyme glutathione peroxidase (Price and Harrison, 1988). Later studies using the coccolithophorid *Emiliania huxleyi* as a model organism (Obata et al., 2004) , have shown in more detail the steps involved into the uptake of selenite by the cells, and identified a pool of low molecular weight compounds which are used to store Se before incorporation into specific seleno-proteins in the cell. The interspecies differences in selenite uptake and accumulation have also been assessed (Vandermeulen and Foda, 1988), and the release of Se contained in phytoplankton cells by zooplankton grazing or phytoplankton decomposition (Lee and Fisher, 1992).

2.6.3 Molybdenum (Mo)

Molybdenum (Mo) is almost conservative (105 nM) in oxygenated seawater (Collier, 1985) where it is present as the oxyanion molybdate (MoO_4^{2-}), under reducing conditions Mo is reduced to form Mo-sulfides (Erickson and Helz, 2000) and is rapidly precipitated from the water column (Helz et al., 2004). Molybdenum is an essential element for phytoplankton which is utilized as a co-factor in a number of different enzymes (Kisker et al., 1997) and in particular in nitrogenase and nitrate reductase which catalyze the reduction of N_2 and nitrate to bioavailable N in the ocean (Mendel, 2005). While Mo has been shown to be a limiting nutrient for freshwater organisms the evidence of Mo limitation in marine organisms is unclear. The use of ^{99}Mo in seawater studies requires special sample handling so that the daughter product ^{99}Tc does not interfere for this reason it has been almost used exclusively in nitrogen fixation studies where the focus has been on the potential for sulfate inhibition of Mo uptake leading to Mo limitation of nitrogen fixation (Marino et al., 2003).

2.6.4 Cadmium (Cd)

The marine biogeochemistry of Cadmium (Cd) has already yielded a number of surprises as it was long thought to be simply only toxic to organisms making its nutrient like profile in seawater and tight coupling with phosphate difficult to understand. However the finding that Cd could replace Zn in the enzyme carbonic anhydrase showed for the first time a biological function for Cd (Lane and Morel, 2000) and changed biogeochemists view of this element. Cd appears to be weakly complexed by organic ligands in surface waters with inorganic species dominating in deeper waters (Bruland, 1992). The radio-isotope ^{109}Cd is frequently used for studies investigating the uptake of Cd by phytoplankton (Sunda and Huntsman, 1996). A particular focus has been with regard to bio-dilution effects and the role that growth limitation may have on their Cd content, either through iron limitation (Sunda and Huntsman, 2000), macronutrient limitation (Wang and Dei, 2001a) or temperature and irradiance uptake (Miao and Wang, 2004). Other studies have investigated the release of Cd from phytoplankton by cellular efflux mechanisms (Lee et al., 1995) or from zooplankton grazing/phytoplankton decomposition (Xu et al., 2001). The uptake of ^{109}Cd has also been followed in natural phytoplankton communities in the English Channel (Dixon et al., 2006).

3. Potential new applications for iron speciation in seawater

In this section we outline some new applications using ^{55}Fe for determining thermodynamic and kinetic information on iron speciation in seawater. Each of the outlined methods has been evaluated during shipboard trials at sea.

3.1 Organic speciation of iron in seawater

Iron organic species are important to the biogeochemical cycling of iron in the ocean as they may determine the bioavailability of iron to phytoplankton and increase the solubility of iron in seawater. Currently most techniques to determine Fe speciation in seawater use voltammetry(Croot and Johansson, 2000; Gledhill and van den Berg, 1994; Rue and Bruland, 1995). In the following we have adapted the chemistry of an existing voltammetric method (Croot and Johansson, 2000) for use with a radiotracer.

3.1.1 Theory: Competitive Ligand Exchange (CLE) – ^{55}Fe TAC

The theory behind the CLE approach was introduced by Ruzic (1982), van den Berg (1982). A brief outline of the theory for determining dissolved iron speciation is presented below, a fully treatment can be found in Croot and Johansson (2000).

For dissolved iron in ambient seawater, a mass balance can be constructed:

$$[Fe_T] = [Fe'] + [FeL_i] \quad (1)$$

where [Fe′] represents the sum of all the inorganic species (predominantly $Fe(OH)_x^{(3-x)+}$) and $[FeL_i]$ represents the organically bound iron with L_i being classes of natural organic ligands. The speciation of Fe(II) is not considered, as in most cases the long equilibration times used in these experiments should have seen the oxidation of any Fe(II) present. Reactions between one class of the natural ligands and Fe′ can be expressed as:

$$Fe' + L' \rightarrow FeL \quad (2a)$$

$$FeL \rightarrow Fe' + L' \quad (2b)$$

where L′ is the Fe-binding ligand not already bound to Fe(III). The equilibrium expression is then:

$$K'_{Fe'L} = \frac{[FeL]}{[Fe'][L']} \tag{3}$$

where $K'_{Fe'L}$ is the conditional stability constant with respect to Fe′ under these specific conditions (in this case pH 8.0 seawater). To convert $K'_{Fe'L}$ to K'_{FeL}, the conditional stability constant for FeL with respect to free Fe^{3+} , the relationship between Fe′ and Fe^{3+} , $\alpha_{Fe'} = [Fe']/[Fe^{3+}]$, can be used (e.g. $K'_{FeL} = \alpha_{Fe'} K'_{Fe'L}$).
Upon addition of the competing ligand TAC, a new equilibrium is established between TAC, the natural organic ligands and iron:

$$[Fe_T] = [Fe'] + [FeL_i] + [Fe(TAC)_2] \tag{4}$$

The complexation of Fe′ by TAC can be described as:

$$\beta_{Fe(TAC)2} = \frac{[Fe(TAC)_2]}{[Fe'][TAC']^2} \tag{5}$$

The side reaction coefficient for $Fe(TAC)_2$ with respect to Fe′ is then denoted by:

$$\alpha_{Fe'(TAC)2} = \frac{[Fe(TAC)_2]}{[Fe']} = \beta'_{Fe(TAC)2}\,[TAC']^2 \tag{6}$$

As $[TAC'] >> [Fe_T]$ for this method, the assumption $[TAC'] = [TAC_T]$ can be used.
Titrations performed using CLE-ACSV yield the fraction of Fe complexed by TAC at different Fe concentrations. This fraction is related to the side reaction coefficient by the following relationship (all relative to Fe′):

$$\frac{[Fe(TAC)_2\,]}{[Fe_T]} = \frac{\beta'_2[TAC]^2}{1 + \Sigma K'_i\, L_i + \beta'_2[TAC]^2} = \frac{\alpha_{Fe'(TAC)2}}{1 + \alpha_o + \alpha_{Fe'(TAC)2}} \tag{7}$$

K_i is the conditional stability constant, L_i the concentration of the ith natural ligand, α_o is the side-reaction coefficient for the naturally occurring ligands, and $\beta'_2[TAC]^2$ the side-reaction coefficient for TAC complexes, which was determined previously (Croot and Johansson, 2000). The side-reaction coefficient of Fe (αFe) for all naturally occurring ligands (including inorganic ligands) is related to the concentration of [Fe′] by the relationship:

$$\frac{[Fe']}{[Fe_T] - [Fe(TAC)_2]} = \frac{1}{1 + \Sigma K_i\, L'_i} \tag{8}$$

Data in this study were analyzed with a single ligand model that was a nonlinear fit to a Langmuir adsorption isotherm (Gerringa et al., 1995). The single ligand model is derived from equation 3, where $[L_T] = [L'] + [FeL]$. Rearranging Eq. 3, 4 and 8 yields a reciprocal Langmuir isotherm:

$$\frac{[FeL\,]}{[Fe']} = \frac{K[L_T]}{1 + K[Fe']} \tag{9}$$

We solved Eq. 9 for K and [L] by nonlinear regression analysis (Levenberg, 1944; Marquardt, 1963) with Fe′ as the independent variable and [FeL]/[Fe′] as the dependent variable using a purpose built program running Labview™ . For the case of multiple ligands a more correct form of the equation is:

$$\frac{\Sigma[\mathrm{FeL}_i]}{[\mathrm{Fe}']} = \Sigma K'_{\mathrm{i}}\, \mathrm{L}_{i(i>1)} + \frac{[\mathrm{L}_1]K'_1}{1 + K'_1[\mathrm{Fe}']} \tag{10}$$

$\Sigma\ K_i\ L_{i(i>1)}$ is the side-reaction coefficient for the weaker ligands, and K_1 and L_1 represent K and L in Eq. 7.

3.1.2 Methodology: Competitive Ligand Exchange (CLE) – ^{55}Fe TAC

The seawater samples analyzed here were collected by the snorkel sampling system on the Polarstern (Schüßler and Kremling, 1993) from the Atlantic sector of the Southern Ocean during Polarstern expedition ANTXXIII-9 ((1) March 9, 2007 at 61° 55.67′ S, 72° 43.23′ E (2) March 19, 2007 at 61° 58.51′ S, 82° 50.00′ E).

A 0.01 M stock solution of TAC 2-(2-Thiazolylazo)-p-cresol; Aldrich was prepared in HPLC grade methanol, when not in use the stock solution is kept refrigerated. A 1.0 M stock buffer of EPPS (N-(2-hydroxyethyl)piperazine-N′-2-propanesulfonic acid; pKa 8.00; SigmaUltra) was prepared in 1 M NH_4OH (Fluka, TraceSelect). Ultrapure (R > 18 MΩ cm^{-1}) deionized water (denoted UP water) was produced using a combined systems consisting of a Millipore Elix 3 and Synergy point of use system. All equipment used was trace metal clean and performed under a class 100 laminar airflow bench (AirClean Ssystems). Waters C18 Sep-Pak cartridges (holdup volume 1 mL) were pre-cleaned using 5 mL of quartz-distilled methanol (Q-MeOH) and 5 mL of UP water.

In this work the ^{55}Fe (Hartmann Analytics, Braunschweig, Germany) had a specific activity of 157.6 MBq/mg Fe, a total activity of 75 MBq and was dissolved in 0.51 mL of 0.1 M HCl. The ^{55}Fe stock solution was diluted to form working stock solutions with UP water and acidified with quartz-distilled HCl (Q-HCl) to a pH < 2 to prevent precipitation of the iron.

Subsamples (20 mL) of seawater were pipetted into a series of 12 Teflon bottles (60 mL) and 100 µL of 1M EPPS added. Iron (^{55}Fe) was added to all but two of the bottles, yielding concentrations from 0 to 12 nM. The added Fe was allowed to equilibrate with the natural ligands for one hour at laboratory temperature. At the end of this equilibration period, 20 µL of 10 mM TAC was added and then left to equilibrate for 24 hours. At the end of this time the complete sample was pumped through the C18 column, followed by 5 mL of UP water and the filtrate (including UP water rinse) collected for counting. The ^{55}Fe TAC complex retained on the column was recovered by a 5 mL rinse with Q-MeOH.

The activity of the samples was quantified using a liquid scintillation counter (Packard Tri-Carb 2900TR) with the scintillation cocktail Lumagel Plus (Lumac LSC). The efficiency of the instrument was obtained by quench curve calibration measurements. Separate quench curves were obtained for samples with seawater or methanol/TAC.

3.1.3 Example: Competitive Ligand Exchange (CLE) – ^{55}Fe TAC

The results of the ligand titration are shown in Figure 1 below and are analogous to similar titrations using electrochemical detection (Croot and Johansson, 2000). The ^{55}Fe-TAC complex is efficiently retained by the C18 column (Baliza et al., 2009). Comparison of the FeL concentration determined by difference between the measured ^{55}Fe-TAC concentration and that which directly passed through the C18 column indicates that either an appreciable amount of the FeL was hydrophobic and retained on the C18 column after the methanol rinse (see section 3.3). Non-linear fitting of the data (Figure 2) to equation 9 gave log K = 21.23 ± 0.08 and L_T = 1.50 ± 0.07, values consistent with other data from the Southern Ocean using electrochemical techniques (Croot et al., 2004). The data shown here provides initial confirmation that this approach can be applied to measuring iron speciation in seawater and could potentially be less labour intensive and time consuming than the current electrochemical method.

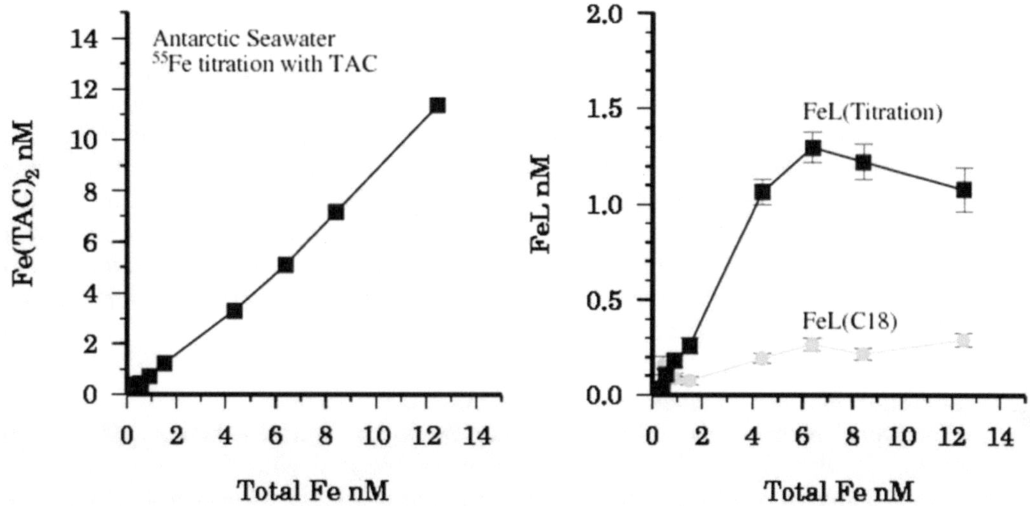

Fig. 1. (left) Recovery of $Fe(TAC)_2$ as a function of the total iron in solution (natural iron plus added radiotracer). (right) The concentration of organic iron (FeL) measured in the samples as a function of the total iron in solution (natural iron plus added radiotracer). FeL(Titration) refers to the FeL determined by difference from the measured ^{55}Fe-TAC concentrations and FeL(C18) is the directly measured concentration of the seawater filtrate that passed through the C18 Sep-Pak.

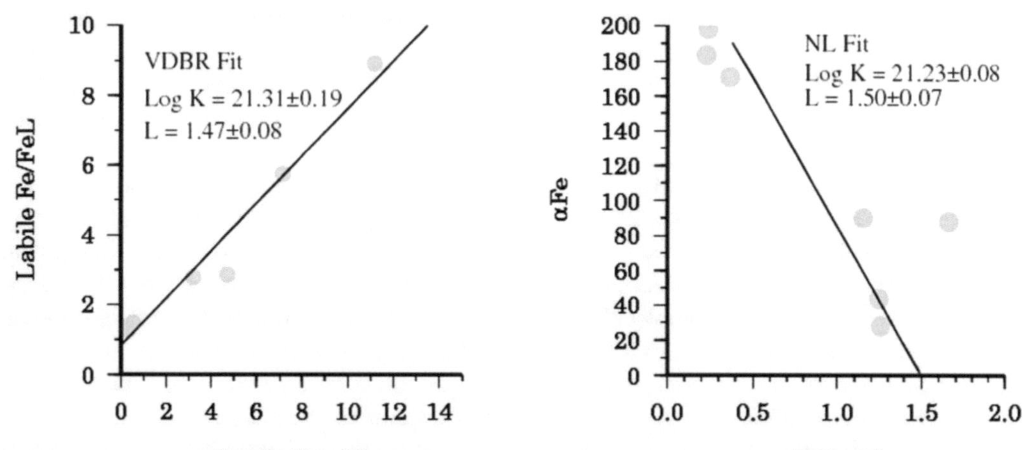

Fig. 2. (left) Van den berg/Ruzic fit to the data shown in Figure 1. (right) Non-linear fit (equation 9) to the data shown in Figure 1.

3.1.4 Issues: Competitive Ligand Exchange (CLE) – ^{55}Fe TAC

If this method is to be used more routinely there are a number of issues that would need to be addressed further. A critical factor in the interpretation of the data is whether the C18 column also retains hydrophobic organic Fe in addition to the ^{55}Fe TAC complex as this could result in the retention of ^{55}Fe not bound to TAC and lead to an underestimation of L_T if these complexes are removed by the MeOH rinse used to elute the ^{55}Fe TAC complex. We examine the issue of natural Fe hydrophobic organic complexes in more detail in section 3.3. Additionally this method relies on an accurate measurement of the dissolved iron concentration in the seawater and this needs to be taken into account for the addition of ^{55}Fe, as the ratio of ^{55}Fe to stable iron increases with each subsequent addition of ^{55}Fe. This variation in the overall specific activity of the solution could have an impact on the time required to establish isotopic equilibrium between radiotracer and ambient iron. In the present case it is assumed that the 24 hour equilibration time used was sufficient given that TAC most likely reacts with natural iron ligands via an adjunctive mechanism (Hering and Morel, 1990), though this is not yet confirmed. Finally the use of a high specific activity ^{55}Fe source is essential if low level (< nM) work is performed.

3.2 Dissociation kinetics of weak iron binding complexes

The earlier work on iron solubility in seawater by Kuma and co-workers (1992, 1993) assumed that the decrease in the concentration of soluble iron with time was due to the aging of meta-stable iron colloids and reduction in their solubility. However an alternative explanation is also possible as subsequent research has suggested that much of the added iron is initially complexed by weaker iron binding ligands (Gerringa et al., 2007) that slowly dissociate over time resulting in the loss of soluble iron from solution. In the following we adapt an existing radiotracer protocol for iron solubility measurements to determine the kinetics of the dissociation of the weak iron binding complexes in seawater.

3.2.1 Filtration and size exclusion

Recently we became aware of a potential problem when comparing between different filters. In comparing an ultrafiltration system (Vivaflow 50) and the Anotop (Whatman) syringe filters (Figure 3) we found that the Anotop retained far more ^{55}Fe than the ultrafilters. Further measurements comparing the 0.02 μm Anotop filters with another type of ultrafiltration membranes (5, 10, 30, and 100 kDa) also found that the Anotop filters have a much smaller molecular weight cut off (< 5 kDa) than 20 nm (C. Schlosser, unpublished data). It seems likely then that the aluminium oxide matrix of the Anotop filter may also interact and adsorb some inorganic and organically complexed Fe species. Our finding agrees with an earlier study by Chen et al. (2004) which reported that they had observed that the Anotop filters were considerably different from its rated pore size of 0.02 μm (or ~2000 kDa) as they found by using fluorescein tagged macromolecular compounds that it had an actual cutoff of ~3 kDa. Our own initial work with 0.025 μm Millipore MF filters suggest that these have a cutoff more in keeping with their stated poresize based on comparison with ultrafiltration. These results contrast with an ICP-MS study reporting on the existence of colloidal Fe in the ocean (Wu et al., 2001) which found apparently good agreement between the Millipore MF 0.025 μm filters and the Anotop 0.02 μm syringe filters. More work is needed urgently to address and understand the differences between ultrafiltration systems and how this effects our interpretation of the natural system being investigated.

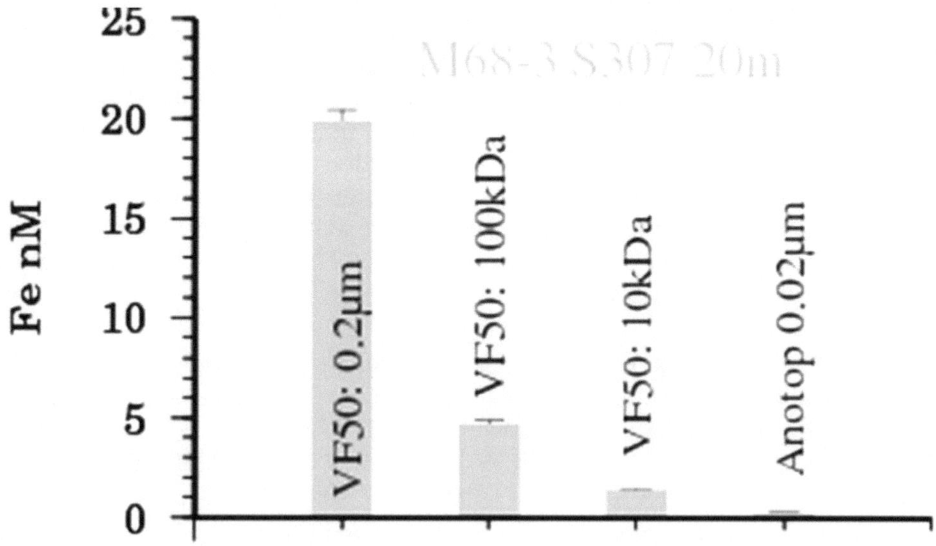

Fig. 3. Comparison of ultrafiltration methods using water collected by GO-FLO and amended with 20 nM ^{55}Fe

3.2.2 Dissociation kinetics of weak complexes

The following approach is based on the assumption that the observed decrease in soluble iron with time is due to the exchange of Fe between the weak organic ligands and the colloidal phase which does not pass through the filter. Support for this assumption lies partly in the findings that inorganic iron colloids formed from oversaturation of the solution will be formed very rapidly (Nowostawska et al., 2008) and be considerably larger (Hove et al., 2007) than the cutoff of the Millipore MF filter (25 nm) or Anotop (20 nm - though see 3.2.1 above). An earlier study by Okumura confirms that in the absence of a strong chelator over 95% of the Fe is found in the > 0.025 µm fraction (Okumura et al., 2004).

Thus the formation and dissociation of Fe complexes can be described by equations 2a and 2b from section 3.1.1. We now further assume that the ligands can be divided into two groups; a strong ligand (L_S) that is practically inert to dissociation and a weaker ligand (L_W) that at equilibrium is not able to keep iron in solution. Thus the soluble Fe concentration can be described by the following equation as a function of time, assuming that the formation of both weak and strong complexes is equally rapid.

$$Fe_{sol} = FeL_S + FeL_W(e^{-kt}) \qquad (11)$$

where Fe_{sol} is the measured solubility of iron, FeL_S is the concentration of the strong ligand and FeL_W is the concentration of the weaker ligands which at thermodynamic equilibrium do not prevent the precipitation of iron from solution, k is the observed dissociation rate of the weaker iron organic complexes.

3.2.3 Methodology – Dissociation kinetics of weak iron-organic complexes

Filtered (0.2 µm) seawater samples were obtained from throughout the water column using GO-FLO sampling bottles on a trace metal clean line at two stations in the Tropical Atlantic

during the Polarstern expedition ANTXXVI-4. All sample handling was performed in a clean room container. In this work the ^{55}Fe (Perkin Elmer) had a specific activity of 1985.42 MBq/mg Fe, a total activity of 75 MBq and a concentration of 1466.79 MBq/mL. The ^{55}Fe stock solution was diluted as described in section 3.1.2.
Seawater (200 mL) from different depths was transferred into Teflon FEP bottles (1 L) and an aliquot of ^{55}Fe was added to the bottles to give an addition of 21 nM. Subsamples (20 mL) for filtration were taken after 3, 6, 24 and 48 hours and were filtered through 47 mm 0.025 µm Millipore MF filters using an all Teflon filtration unit (Savillex), the filtrate was collected in a Teflon vial. All experiments were performed at 23° C. The activity of the samples was quantified as described in section 3.1.2.

3.2.4 Example – Dissociation kinetics of weak iron-organic complexes

Samples for the kinetic experiments were obtained from vertical profiles at two stations in the Tropical Atlantic; (i) S283 - April 28, 2010 at 01° 46.62′ N, 23° 00.18′ W in the Equatorial Atlantic and (ii) S287 - May 4, 2010 at 17° 34.98′ N, 24° 15.18′ W, the TENATSO time series site (for more information on TENATSO see Heller and Croot (2011)). Figure 4 below shows example results for the time dependent decrease in soluble iron from two different depths from the TENATSO site.

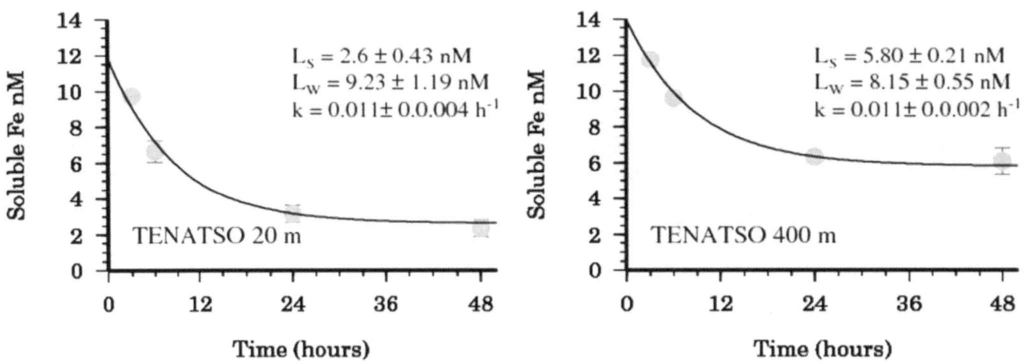

Fig. 4. Change in the concentration of Soluble Fe (0.025 µm Millipore MF) over 48 hours after the addition of 21 nM ^{55}Fe to water samples from the TENATSO station in the Eastern Tropical Atlantic. (left) 20 m depth. (right) 400 m depth. Error estimates are the result of duplicate measurements and correspond to the 95% confidence interval. The least squares fit to equation 11 are also shown (solid line).

In all samples soluble Fe was initially high (10-13 nM) and declined rapidly over the first 24 hours with only small or no changes in the concentration over a further 24 hours. This indicates that the equilibrium value for L_S was reached typically within 24 hours and that the weaker L_W ligands had dissociated within the same timeframe. Previous studies have indicated that this equilibrium is established over timescales ranging from 1-2 weeks for studies using 0.45 µm pre-filtration (Kuma et al., 1996; Liu and Millero, 2002) to less than 24 hours when 0.2 µm pre-filtration was used (Chen et al., 2004). This highlights the role of colloidal matter/ligands in the time it takes to reach equilibrium. All data was fit to equation 11 using least squares regression.

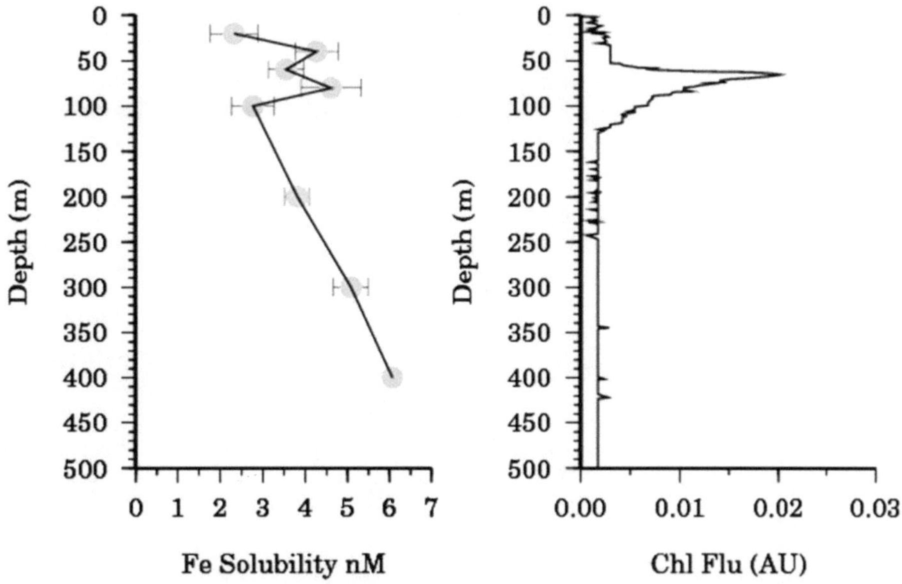

Fig. 5. (left) Vertical profile of iron solubility after 24 hours at TENATSO (ANTXXVI-4, S287). (right) Chlorophyll fluorescence (arbitary units) at TENATSO.

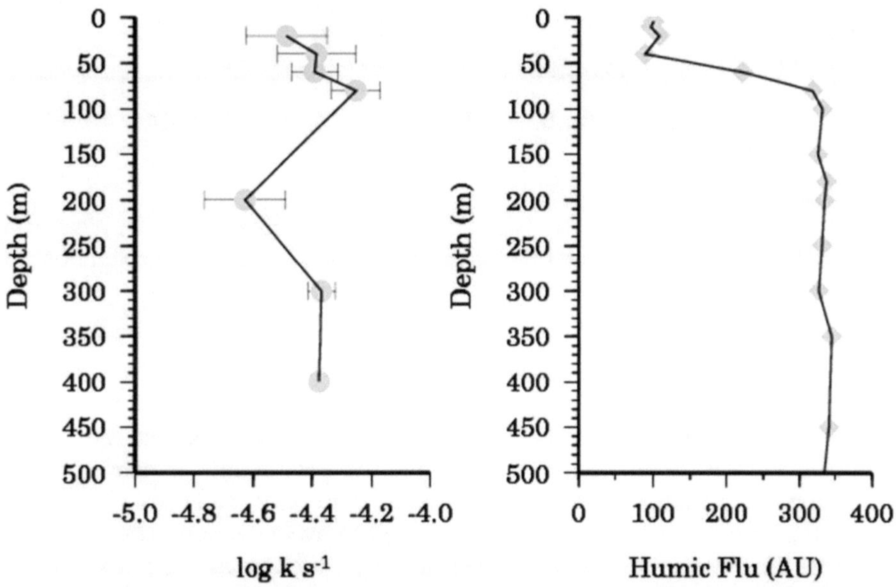

Fig. 6. (left) Vertical profile of dissociation rate (*k*) for FeL_W at TENATSO. (right) Vertical profile of marine humic fluorescence (arbitrary units, 320 nm excitation, 420 nm emission).

In earlier studies in the Pacific, iron solubility in intermediate and deep waters has been found to be highly correlated to the fluorescence of marine humic substances (Tani et al., 2003). The humic fluorescence profile at the TENATSO station is shown in figure 6 and is clearly poorly correlated with L_S at this location. However at station (S283) in the Equatorial

Atlantic (Figure 7) we did observe a strong correlation between L_S and humic fluorescence (Figure 8).
The vertical distribution of iron solubility after 24 hours at TENATSO is shown in figure 5 and shows a generally increasing trend with depth with a small local maximum in the surface waters in the vicinity of the chlorophyll maximum consistent with a biological source for L_S. Values for L_W showed no systematic variation in the water column and ranged from 8-10 nM. The estimated dissociation rate, *k*, for L_W ranged from 2×10^{-5} to 8×10^{-5} s^{-1} similar to voltammetric observations of weak iron binding ligands (Gerringa et al., 2007).

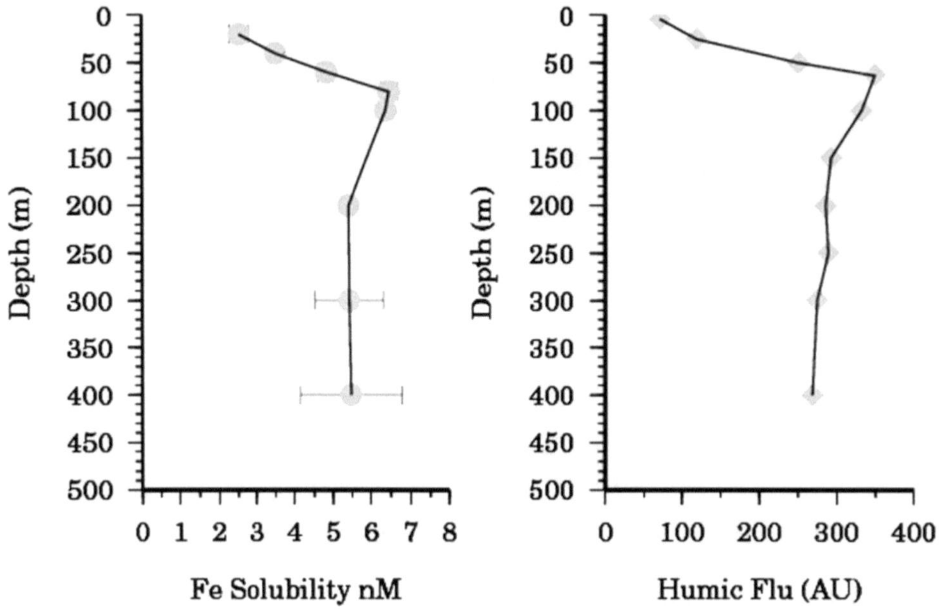

Fig. 7. (left) Vertical profile of Iron solubility after 24 hours at S283 in the Tropical Atlantic (ANTXXVI-4). (right) Marine Humic fluorescence (arbitrary units, 320 nm excitation, 420 nm emission).at the same location.

At S283 L_W showed no consistent pattern over the depth range examined with values from 6-10 nM. Estimated rates for the dissociation of the iron complexes from the weak ligands, *k*, ranged from 0.4 - 4.4×10^{-5} and also showed no discernable pattern with depth. The correlation between humic fluorescence and L_S suggests that in this case the ligands were mostly derived by the same process inferred for the production of marine humics, the remineralisation of organic matter by microbial action. It furthermore suggests the photochemical destruction of the ligands in near surface waters at both S283 and TENATSO. The differences in the profiles between S283 and TENATSO may be related to a greater production of iron binding ligands by phytoplankton or bacteria at TENATSO. This may be in response to the greater dust flux this site receives as it is lies directly under the path of the Saharan dust plume (Heller and Croot, 2011).
Our approach here clearly provides important information with regard to the kinetics of processes relevant to dust deposition to the ocean (Baker and Croot, 2010) and highlights the role that weaker ligands may play in solubilising iron from aerosols and allowing phytoplankton a critical few extra hours were it is still soluble and potentially bioavailable.

More work is clearly needed on this subject and the method outlined here should be a key contribution to this.

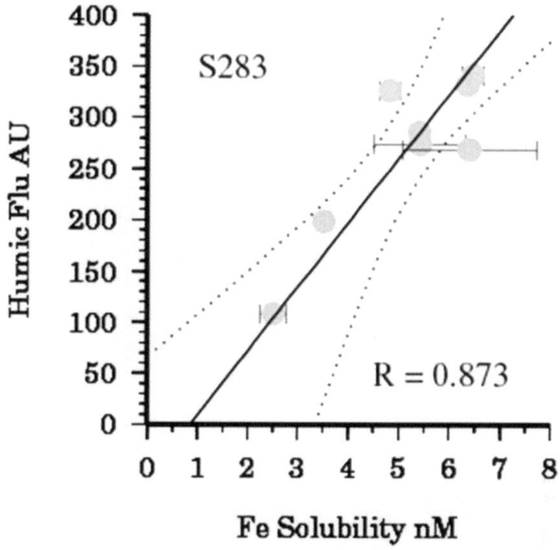

Fig. 8. Correlation between humic fluorescence and Fe solubility (after 24 hours) for samples from Station 283 in the Tropical Atlantic (Polarstern ANTXXVI-4).

3.3 Hydrophobic organic Fe complexes

As noted early in section 3.1 information on hydrophobic Fe complexes is important for the interpretation of methods using C18 columns to recovery the Fe from solution. Such information is also useful for assessing the scavenging behaviour of iron organic complexes to particles in seawater. Hydrophobic Fe complexes are known to exist as many siderophores possess a hydrophobic tail which facilitates the uptake of iron by the phytoplankton (Martinez et al., 2000). A number of siderophore complexes are quantitatively retained by C18 columns including the terrestrial siderophore desferrioxamine B and its Fe chelate, ferrioxiamine B (Gower et al., 1989). This has lead to the development of extraction techniques for siderophores from seawater using C18 solid phase extraction (Freeman and Boyer, 1992). Other dissolved organic matter is also retained by this approach (Mopper et al., 2007) including marine humic complexes, though recoveries are highest when the sample is acidified (Amador et al., 1990).
There have been a number of studies that have utilized C18 or similar substrates to trap organic complexes using solid phase extraction techniques (Mackey, 1983). Previous work combining C18 solid phase extraction with radioisotopes in seawater has utilized ^{64}Cu, finding that there is a significant but variable concentration of hydrophobic Cu complexes (Croot et al., 2003).

3.3.1 Methodology – Hydrophobic organic Fe complexes

The description of the seawater sampling, sample handling and ^{55}Fe standard preparation are the same as described in section 3.2.3. In these experiments, 20 mL of the seawater samples described in section 3.2.3 were pumped through Waters C18 Sep-Paks (cleaned as described in section 3.1.2), rinsed with 5 mL UP water and then the ^{55}Fe retained on the C18 column was eluted with 5 mL Q-MeOH. The activity of the MeOH samples was quantified as described in section 3.1.2. Samples were taken after 3, 6, 24 and 48 hours after the addition of the ^{55}Fe.

3.3.2 Example – Hydrophobic organic Fe complexes

Seawater for this experiment was obtained from 4 depths at S283 (see description in section 3.2.4) and run as described above. The results from two of the kinetic runs are shown in Figure 9 below. In the sample from 40 m there was a clear decrease with time in the concentration of the hydrophobic Fe trapped by the C18 column. This was similar to the decrease in Fe solubility for the same sample (data not shown) suggesting that for this sample a significant portion of the weak organic ligands (section 3.2) were hydrophobic in nature. Contrastingly samples from deeper in the water column showed little variation with time (Figure 9) indicating that the bulk of the hydrophobic component here were stronger iron binding ligands.

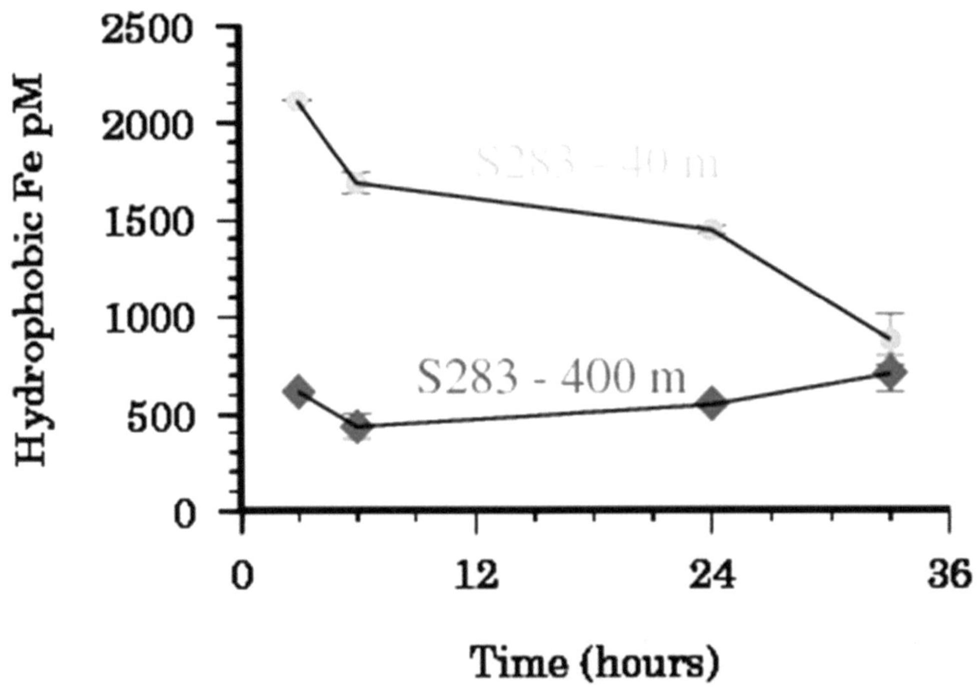

Fig. 9. Hydrophobic organic Fe complexes at S283 in the Equatorial Atlantic. Samples were obtained from 40 m depth (circles) and 400 m depth (triangles). The 95% confidence intervals for the data are represented as error bars.

The vertical distribution of hydrophobic Fe complexes is shown in Figure 10 and indicates a maximum in near surface waters with lower concentrations in deep waters suggesting a biological source. Comparison with the iron solubility data from section 3.2 indicates that the percentage of hydrophobic Fe that was in the soluble phase was high in surface waters and decreased rapidly to be only ~10 % below 200 m (Figure 10).

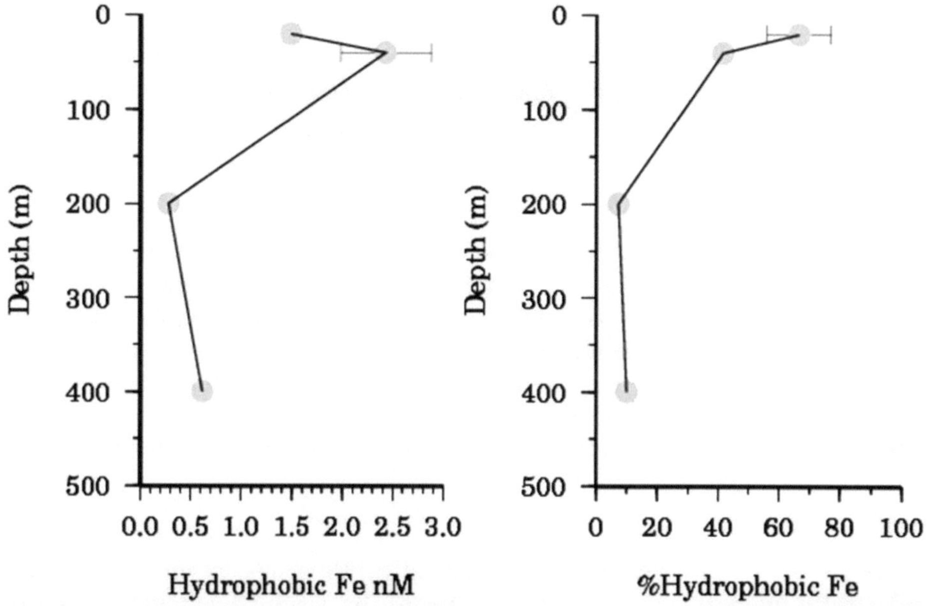

Fig. 10. Data from Station 283 (Polarstern expedition ANTXXVI-4. (left) Hydrophobic iron recovered by C18 Sep-Pak (t=24 hours). (right) percentage of iron passing through 0.025 μm filter that is hydrophobic.

Thus at S283 it suggests that in waters below 200 m the Fe complexation was dominated by hydrophilic humic complexes while nearer to the surface complexation was by ligands that were more hydrophobic in nature. This is the first data on iron complexation in the ocean to show information on the hydrophobic nature of complexes in the water column. This data also points to the role that hydrophobic Fe complexes may play in the biogeochemical cycling of iron in the ocean. The chemistry of these complexes is a rapidly developing field as shown in a recent review article on metallosurfactants (Owen and Butler, 2011) and their role in bioinorganic processes.

3.3.3 Issues: Hydrophobic organic Fe complexes

The data we present above is for Fe retained on a C18 column and eluted with methanol. Previous work has shown that the retention of the Fe complex under similar conditions to those employed here varies between different iron species and that the methanol elution may not remove all the Fe that was trapped (Freeman and Boyer, 1992). Other studies have used nitric acid instead of methanol to recover the Fe retained on the C18 column (Abbasse et al., 2002). In the present study we did not attempt a nitric acid rinse due to potential problems with this acid in the scintillation cocktail. Further work is needed to develop a more complete protocol that includes a complete mass balance.

Similarly colloidal inorganic iron has been found to be almost quantitatively trapped on C18 Sep-Paks (Sunda and Huntsman, 2003), in the present case it is unlikely that there is significant concentrations of this form of iron as outlined early in section 3.2. It would be even more unlikely that colloidal inorganic iron was formed in the ^{55}Fe TAC experiment as this solution would be under saturated with respect to iron precipitation and the TAC is likely to dissolve any amorphous iron colloids quite rapidly (Croot and Johansson, 2000).

4. Conclusions and future prospects

Radioisotopes are a vital tool for trace metal marine biogeochemists as they allow pulse chase experiments for rapid assessment of the kinetics of processes under natural conditions. The applications that radioisotopes can be applied to in seawater speciation studies is strongly linked to, and drives, new analytical developments in techniques to distinguish between individual chemical species. In the present work we have outlined new approaches for examining Fe organic speciation in seawater, but these methods could easily be applied to other trace metal radioisotopes also. New research is needed urgently for assessing the current status and the potential for change in oceanic systems due to global warming, ocean acidification and ocean deoxygenation. In this context we encourage future research into the use of trace metal radioisotopes for determining the changes in speciation and kinetic reactivity for oceanic redox processes in oxygen minimum zones. As this is a critical area of research that needs to be developed over the next decade in order to better assess the impact ocean deoxygenation may play on trace metal redox cycles in the ocean.

5. Acknowledgments

This work is a contribution of the Collaborative Research Centre 754 "Climate - Biogeochemistry Interactions in the Tropical Ocean" (www.sfb754.de), which is supported by the German Research Association. Financial support for this work has come from the DFG through SFB754 "Climate – Biogeochemistry Interactions in the Tropical Ocean" and research grants to PLC (CR145/5, CR145/9 and CR145/18). The technical support of Uwe Rabsch, Kerstin Nachtigall and Peter Streu is gratefully acknowledged.

6. References

Abbasse, G., Ouddane, B. and Fischer, J.C., 2002. Determination of total and labile fraction of metals in seawater using solid phase extraction and inductively coupled plasma atomic emission spectrometry (ICP-AES). Journal of Analytical Atomic Spectrometry, 17(10): 1354-1358.

Ahner, B.A., Price, N.M. and Morel, F.M.M., 1994. Phytochelatin production by marine phytoplankton at low free metal ion concentrations: Laboratory studies and filed data from Massachusetts Bay. Proceedings of the National Academy of Sciences, 91: 8433-8436.

Amador, J., Milne, P.J., Moore, C.A. and Zika, R.G., 1990. Extraction of chromophoric humic substances from seawater. Marine Chemistry, 29: 1-17.

Anderson, M.A. and Morel, F.M., 1980. Uptake of Fe(II) by a diatom in oxic culture medium. Marine Biology Letters, 1: 263-268.

Annett, A.L., Lapi, S., Ruth, T.J. and Maldonado, M.T., 2008. The effects of Cu and Fe availability on the growth and Cu:C ratios of marine diatoms. Limnology and Oceanography, 53(6): 2451-2461.

Anton, M. et al., 2007. Identification and comparitive Genomic Analysis of Signalling and Regulatory Components in the diatom *Thalassiosira Psuedonana*. Journal of Phycology, 43(3): 585-604.

Baars, O. and Croot, P.L., 2011. The speciation of dissolved zinc in the Atlantic sector of the Southern Ocean. Deep Sea Research Part II: Topical Studies in Oceanography, In Press, Corrected Proof.

Baker, A.R. and Croot, P.L., 2010. Atmospheric and marine controls on aerosol iron solubility in seawater. Marine Chemistry, 120: 4-13.

Baliza, P.X., Ferreira, S.L.C. and Teixeira, L.S.G., 2009. Application of pyridylazo and thiazolylazo reagents in flow injection preconcentration systems for determination of metals. Talanta, 79(1): 2-9.

Barbeau, K., Moffett, J.W., Caron, D.A., Croot, P.L. and Erdner, D.L., 1996. Role of protozoan grazing in relieving iron limitation of phytoplankton. Nature, 380: 61-64.

Boye, M. et al., 2001. Organic complexation of iron in the Southern Ocean. Deep Sea Research, 48: 1477-1497.

Boyle, E. and Edmond, J.M., 1975. Copper in surface waters south of New Zealand. Nature, 253: 107-109.

Bruland, K.W., 1989. Complexation of zinc by natural organic ligands in the central North Pacific. Limnology and Oceanography, 34: 269-285.

Bruland, K.W., 1992. Complexation of cadmium by natural organic ligands in the central North Pacific. Limnology and Oceanography, 37: 1008-1017.

Bruland, K.W., Donat, J.R. and Hutchins, D.A., 1991. Interactive influences of bioactive trace metals on biological production in oceanic waters. Limnology and Oceanography, 36: 1555-1577.

Bruland, K.W. and Lohan, M.C., 2003. Controls of Trace Metals in Seawater. In: D.H. Heinrich and K.T. Karl (Editors), Treatise on Geochemistry. Pergamon, Oxford, pp. 23-47.

Byrne, R.H., 2010. Comparative carbonate and hydroxide complexation of cations in seawater. Geochimica et Cosmochimica Acta, 74(15): 4312-4321.

Byrne, R.H., Kump, L.R. and Cantrell, K.J., 1988. The Influence of Temperature and pH on Trace Metal Speciation in Seawater. Marine Chemistry, 25: 163-181.

Campbell, P.G.C., 1995. Interactions between Trace Metals and Aquatic Organisms: A Critique of the Free-ion Activity Model. In: A. Tessier and D.R. Turner (Editors), Metal Speciation and Bioavailability in Aquatic Systems. IUPAC Series on Analytical and Physical Chemistry of Environmental Systems. John Wiley & Sons, Chichester, pp. 45-102.

Chen, M., Wang, W.-X. and Guo, L., 2004. Phase partitioning and solubility of iron in natural seawater controlled by dissolved organic matter. Global Biogeochemical Cycles, 18: doi:10.1029/2003GB002160.

Coale, K.H. and Bruland, K.W., 1988. Copper complexation in the northeast Pacific. Limnology and Oceanography, 33: 1084-1101.

Collier, R.W., 1985. Molybdenum in the Northeast Pacific Ocean. Limnology and Oceanography, 30: 1351-1354.

Croot, P.L., Andersson, K., Öztürk, M. and Turner, D., 2004. The Distribution and Speciation of Iron along 6° E, in the Southern Ocean. Deep-Sea Research II, 51(22-24): 2857-2879.

Croot, P.L. et al., 2008. Regeneration of Fe(II) during EIFeX and SOFeX. Geophysical Research Letters, 35(19): L19606,doi:10.1029/2008GL035063.

Croot, P.L. and Johansson, M., 2000. Determination of iron speciation by cathodic stripping voltammetry in seawater using the competing ligand 2-(2-Thiazolylazo)-p-cresol (TAC). Electroanalysis, 12(8): 565-576.

Croot, P.L., Karlson, B., van Elteren, J.T. and Kroon, J.J., 1999. Uptake of 64Cu-Oxine by Marine Phytoplankton. Environmental Science and Technology, 33(20): 3615-3621.

Croot, P.L., Karlson, B., van Elteren, J.T. and Kroon, J.J., 2003. Uptake and efflux of ^{64}Cu by the marine cyanobacterium *Synechococcus* (WH7803). Limnology and Oceanography, 48: 179-188.

Croot, P.L. and Laan, P., 2002. Continuous shipboard determination of Fe(II) in Polar waters using flow injection analysis with chemiluminescence detection. Analytica Chimica Acta, 466: 261-273.

Croot, P.L., Moffett, J.W. and Brand, L., 2000. Production of extracellular Cu complexing ligands by eucaryotic phytoplankton in response to Cu stress. Limnology and Oceanography, 45: 619-627.

Cutter, G.A., 1992. Kinetic controls on metalloid speciation in seawater. Marine Chemistry, 40(1-2): 65-80.

Cutter, G.A. and Bruland, K.W., 1984. The marine biogeochemistry of selenium: A re-evaluation. Limnology and Ocenography, 29: 1179-1192.

de Baar, H.J.W. et al., 2005. Synthesis of 8 Iron Fertilization Experiments: From the Iron Age to the Age of Enlightenment. Journal of Geophysical Research, 110: C09S16, doi:10.1029/2004JC002601.

Di Toro, D.M. et al., 2001. Biotic ligand model of the acute toxicity of metals. 1. Technical Basis. Environmental Toxicology and Chemistry, 20(10): 2383-2396.

Dixon, J.L. et al., 2006. Cadmium uptake by marine micro-organisms in the English Channel and Celtic Sea. Aquatic Microbial Ecology, 44(1): 31-43.

Donat, J.R. and Bruland, K.W., 1995. Trace Elements in the Oceans, pp. 247-281.

Dupont, C.L., Barbeau, K. and Palenik, B., 2008. Ni uptake and limitation in marine Synechococcus strains. Applied And Environmental Microbiology, 74(1): 23-31.

Dupont, C.L., Buck, K.N., Palenik, B. and Barbeau, K., 2010. Nickel utilization in phytoplankton assemblages from contrasting oceanic regimes. Deep Sea Research Part I: Oceanographic Research Papers, 57(4): 553-566.

Emerson, S. et al., 1982. Environmental oxidation rate of manganese(II): bacterial catalysis. Geochimica et Cosmochimica Acta, 46: 1073-1079.

Emerson, S.R. and Huested, S.S., 1991. Ocean anoxia and the concentrations of molybdenum and vanadium in seawater. Marine Chemistry, 34(3-4): 177-196.

Erickson, B.E. and Helz, G.R., 2000. Molybdenum(VI) speciation in sulfidic waters: Stability and lability of thiomolybdates. Geochimica Et Cosmochimica Acta, 64(7): 1149-1158.

Falkowski, P.G. and Raven, J.A., 1997. Aquatic Photosynthesis. Blackwell Scientific.

Fisher, N., 1985. Accumulation of metals by marine picoplankton. Marine Biology, 87: 137-142.

Freeman, R.A. and Boyer, G.L., 1992. Solid phase extraction techniques for the isolation of siderophores from aquatic environments. Journal of Plant Nutrition, 15(10): 2263 - 2276.

Gerringa, L.J.A., Herman, P.M.J. and Poortvliet, T.C.W., 1995. Comparison of the linear van den Berg/Ruzic transformation and a non-linear fit of the Langmuir isotherm applied to Cu speciation data in the estuarine environment. Marine Chemistry, 48: 131-142.

Gerringa, L.J.A. et al., 2007. Kinetic study reveals weak Fe-binding ligand, which affects the solubility of Fe in the Scheldt estuary. Marine Chemistry, 103: 30-45.

Gledhill, M. and van den Berg, C.M.G., 1994. Determination of complexation of iron(III) with natural organic complexing ligands in seawater using cathodic stripping voltammetry. Marine Chemistry, 47: 41-54.

Gower, J.D., Healing, G. and Green, C.J., 1989. Determination of desferrioxamine-available iron in biological tissues by high-pressure liquid chromatography. Analytical Biochemistry, 180(1): 126-130.

Guo, J. et al., 2010. Copper-uptake kinetics of coastal and oceanic diatoms. Journal of Phycology, 46(6): 1218-1228.

Heller, M.I. and Croot, P.L., 2011. Superoxide decay as a probe for speciation changes during dust dissolution in Tropical Atlantic surface waters near Cape Verde. Marine Chemistry, In Press, Corrected Proof. doi:10.1016/j.marchem.2011.03.006.

Helz, G.R., Vorlicek, T.P. and Kahn, M.D., 2004. Molybdenum scavenging by iron monosulfide. Environmental Science & Technology, 38(16): 4263-4268.

Hering, J.G. and Morel, F.M.M., 1989. Slow coordination reactions in seawater. Geochimica et Cosmochimica Acta, 53: 611-618.

Hering, J.G. and Morel, F.M.M., 1990. Kinetics of trace metal complexation: ligand exchange reactions. Environmental Science and Technology, 24: 242-252.

Hirose, K., 2006. Chemical speciation of trace metals in seawater: a review. Analytical Sciences, 22(8): 1055-1063.

Honeyman, B.D. and Santschi, P.H., 1991. Coupling Adsorption and Particle Aggregation: Laboratory Studies of "Colloidal Pumping" Using 59Fe-Labeled Hematite. Environmental Science and Technology, 25: 1739-1747.

Hove, M., van Hille, R.P. and Lewis, A.E., 2007. Iron solids formed from oxidation precipitation of ferrous sulfate solutions. AIChE Journal, 53(10): 2569-2577.

Hudson, R.J.M., Covault, D.T. and Morel, F.M.M., 1992. Investigations of iron coordination and redox reactions in seawater using 59Fe radiometry and ion-pair solvent extraction of amphiphilic iron complexes. Marine Chemistry, 38: 209-235.

Hudson, R.J.M. and Morel, F.M.M., 1989. Distinguishing between extra- and intracellular iron in marine phytoplankton. Limnology and Oceanography, 34: 1113-1120.

Hudson, R.J.M. and Morel, F.M.M., 1990. Iron transport in marine phytoplankton: Kinetics of cellular and medium coordination reactions. Limnology and Oceanography, 35: 1002-1020.

Hudson, R.J.M. and Morel, F.M.M., 1993. Trace metal transport by marine microorganisms: implications of metal coordination kinetics. Deep-Sea Research, 40: 129-150.

Hunter, K.A., Kim, J.P. and Croot , P.L., 1997. Biological roles of trace metals in natural waters. Environmental Monitoring and Assessment, 44: 103-147.

Hutchins, D.A. and Bruland, K.W., 1994. Grazer-mediated regeneration and assimilation of Fe, Zn and Mn from planktonic prey. Marine Ecology Progress Series, 110: 259-269.

Hutchins, D.A., Wang, W.-X., Schmidt, M.A. and Fisher, N.S., 1999. Dual-labeling techniques for trace metal biogeochemical investigations in aquatic plankton communities. Aquatic Microbial Ecology, 19: 129-138.

Johnson, K.S., Coale, K.H., Berelson, W.M. and Gordon, R.M., 1996. On the formation of the manganese maximum in the oxygen minimum. Geochimica Et Cosmochimica Acta, 60(8): 1291-1299.

Johnson, K.S., Gordon, R.M. and Coale, K.H., 1997. What controls dissolved iron concentrations in the world ocean? Marine Chemistry, 57: 137-161.

Kisker, C., Schindelin, H. and Rees, D.C., 1997. Molybdenum-cofactor-containing enzymes: Structure and mechanism. Annual Review of Biochemistry, 66: 233-267.

Kuma, K., Katsumoto, A., Kawakami, H., Takatori, F. and Matsunaga, K., 1998a. Spatial variability of Fe(III) hydroxide solubility in the water column of the northern North Pacific Ocean. Deep-Sea Research, 45: 91-113.

Kuma, K., Katsumoto, A., Nishioka, J. and Matsunaga, K., 1998b. Size-fractionated iron concentrations and Fe(III) hydroxide solubilities in various coastal waters. Estuarine Coastal and Shelf Science, 47(3): 275-283.

Kuma, K., Nakabayashi, S., Suzuki, Y. and Matsunaga, K., 1992. Dissolution rate and solubility of colloidal hydrous ferric oxide in seawater. Marine Chemistry, 38: 133-143.

Kuma, K., Nishioka, J. and Matsunaga, K., 1996. Controls on iron(III) hydroxide solubility in seawater: The influence of pH and natural organic chelators. Limnology and Oceanography, 41: 396-407.

Lal, D. et al., 2006. Paleo-ocean chemistry records in marine opal: Implications for fluxes of trace elements, cosmogenic nuclides (Be-10 and Al-26), and biological productivity. Geochimica Et Cosmochimica Acta, 70(13): 3275-3289.

Landing, W.M. and Bruland, K.W., 1987. The contrasting biogeochemistry of iron and manganese in the Pacific Ocean. Geochimica et Cosmochimica Acta, 51: 29-43.

Lane, T.W. and Morel, F.M.M., 2000. A biological function for cadmium in marine diatoms. Proceedings of the National Academy of Sciences of the United States of America, 97(9): 4627-4631.

Lee, B.-G. and Fisher, N.S., 1994. Effects of sinking and zooplankton grazing on the release of elements from planktonic debris. Marine Ecology Progress Series, 110: 272-281.

Lee, B.G. and Fisher, N.S., 1992. Decomposition and release of elements from zooplankton debris. Marine Ecology Progress Series, 88(2-3): 117-128.

Lee, B.G. and Fisher, N.S., 1993. Microbially Mediated Cobalt Oxidation In Seawater Revealed By Radiotracer Experiments. Limnology And Oceanography, 38(8): 1593-1602.

Lee, J.G., Roberts, S.B. and Morel, F.M.M., 1995. Cadmium: A nutrient for the marine diatom Thalassiosira weissflogii. Limnology and Oceanography, 40: 1056-1063.

Levenberg, K., 1944. A Method for the Solution of Certain Non-Linear Problems in Least Squares. The Quarterly of Applied Mathematics, 2: 164-168.

Li, Y.-H., Burkhardt, L., Buchholtz, M., O'Hara, P. and Santschi, P.H., 1984. Partition of radiotracers between suspended particles and seawater. Geochimica et Cosmochimica Acta, 48(10): 2011-2019.

Liu, X. and Millero, F.J., 2002. The solubility of iron in seawater. Marine Chemistry, 77: 43-54.

Mackey, D.J., 1983. Metal-Organic Complexes in Seawater - an Investigation of Naturally Occurring Complexes of Cu, Zn, Fe, Mg, Ni, Cr, Mn and Cd using High-Performance Liquid Chromatography with Atomic Fluorescence Detection. Marine Chemistry, 13: 169-180.

Maldonado, M.T. et al., 2006. Copper-dependent iron transport in coastal and oceanic diatoms. Limnology And Oceanography, 51(4): 1729-1743.

Mandernack, K.W. and Tebo, B.M., 1993. Manganese scavenging and oxidation at hydrothermal vents and in vent plumes. Geochimica et Cosmochimica Acta, 57(16): 3907.

Marino, R., Howarth, R.W., Chan, F., Cole, J.J. and Likens, G.E., 2003. Sulfate inhibition of molybdenum-dependent nitrogen fixation by planktonic cyanobacteria under seawater conditions: a non-reversible effect. Hydrobiologia, 500(1): 277-293.

Marquardt, D.W., 1963. An Algorithm for Least-Squares Estimation of Nonlinear Parameters. SIAM Journal on Applied Mathematics, 11(2): 431-441.

Martinez, J.S., Haygood, M.G. and Butler, A., 2001. Identification of a natural desferrioxamine siderophore produced by a marine bacterium. Limnology and Oceanography, 46: 420-424.

Martinez, J.S. et al., 2000. Self-Assembling Amphiphilic Siderophores from Marine Bacteria. Science, 287: 1245-1247.

McCubbin, D. and Leonard, K.S., 1996. Photochemical dissolution of radionuclides from marine sediment. Marine Chemistry, 55(3-4): 399-408.

Mendel, R.R., 2005. Molybdenum: biological activity and metabolism. Dalton Transactions(21): 3404-3409.

Miao, A.J. and Wang, W.X., 2004. Relationships between cell-specific growth rate and uptake rate of cadmium and zinc by a coastal diatom. Marine Ecology-Progress Series, 275: 103-113.

Michibata, H. et al., 1991. Uptake of V-48 Labeled Vanadium by Subpopulations of Blood-Cells in the Ascidian, Ascidia-Gemmata. Zoological Science, 8(3): 447-452.

Millero, F.J., 1998. Solubility of Fe(III) in seawater. Earth and Planetary Science Letters, 154: 323-329.

Millero, F.J., Woosley, R., Ditrolio, B. and Waters, J., 2009. Effect of Ocean Acidification on the Speciation of Metals in Seawater. Oceanography, 22(4): 72-85.

Moffett, J.W., 1994. The relationship between cerium and manganese oxidation in the marine environment. Limnology and Oceanography, 39: 1309-1318.

Moffett, J.W., 1997. The importance of microbial Mn oxidation in the upper ocean: a comparison of the Sargasso Sea and equatorial Pacific. Deep-Sea Research I, 44: 1277-1291.

Moffett, J.W. and Ho, J., 1996. Oxidation of cobalt and manganese in seawater via a common microbially catalysed pathway. Geochimica et Cosmochimica Acta, 60: 3415-3424.

Mopper, K., Stubbins, A., Ritchie, J.D., Bialk, H.M. and Hatcher, P.G., 2007. Advanced Instrumental Approaches for Characterization of Marine Dissolved Organic Matter: Extraction Techniques, Mass Spectrometry, and Nuclear Magnetic Resonance Spectroscopy. Chemical Reviews, 107(2): 419-442.

Morel, F.M.M., Milligan, A.J. and Saito, M.A., 2003. Marine Bioinorganic Chemistry: The Role of Trace Metals in the Oceanic Cycles of Major Nutrients. In: D.H. Heinrich and K.T. Karl (Editors), Treatise on Geochemistry. Pergamon, Oxford, pp. 113-143.

Morel, F.M.M. et al., 1994. Zinc and carbon co-limitation of marine phytoplankton. Nature, 369: 740-742.

Morse, J.W., Millero, F.J., Cornwell, J.C. and Rickard, D., 1987. The chemistry of the hydrogen sulfide and iron sulfide systems in natural waters. Earth-Science Reviews, 24(1): 1-42.

Murray, K.J., Webb, S.M., Bargar, J.R. and Tebo, B.M., 2007. Indirect Oxidation of Co(II) in the Presence of the Marine Mn(II)-Oxidizing Bacterium Bacillus sp. Strain SG-1 Appl Environ Microbiol., 73(21): 6905-6909.

Nolan, C.V., Fowler, S.W. and Teyssie, J.-L., 1992. Cobalt speciation and bioavailabilty in marine organisms. Marine Ecology Progress Series, 88: 105-116.

Nowostawska, U., Kim, J.P. and Hunter, K.A., 2008. Aggregation of riverine colloidal iron in estuaries: A new kinetic study using stopped-flow mixing. Marine Chemistry, 110(3-4): 205.

Nozaki, Y., 1997. A Fresh Look at Element Distribution in the North Pacific. EOS, 78(21): 221-223.

Obata, T., Araie, H. and Shiraiwa, Y., 2004. Bioconcentration Mechanism of Selenium by a Coccolithophorid, Emiliania huxleyi. Plant and Cell Physiology, 45(10): 1434-1441.

Okumura, C., Hasegawa, H., Mizumoto, H., Maki, T. and Ueda, K., 2004. Size fractionation of iron compounds in phytoplankton cultures in the presence of chelating ligands. Bunseki Kagaku, 53(11): 1215-1221.

Owen, T. and Butler, A., 2011. Metallosurfactants of bioinorganic interest: Coordination-induced self assembly. Coordination Chemistry Reviews, 255(7-8): 678-687.

Peers, G. and Price, N.M., 2004. A role for manganese in superoxide dismutases and growth of iron-deficient diatoms. Limnology and Oceanography, 49: 1774-1783.

Pham, A.N., Rose, A.L., Feitz, A.J. and Waite, T.D., 2006. Kinetics of Fe(III) precipitation in aqueous solutions at pH 6.0-9.5 and 25 degrees C. Geochimica Et Cosmochimica Acta, 70(3): 640-650.

Pitzer, K.S., 1973. Thermodynamics of Electrolytes. I. Theoretical Basis and General Equations. The Journal of Physical Chemistry, 77: 268-277.

Poorvin, L., Rinta-Kanto, J.M., Hutchins, D.A. and Wilhelm, S.W., 2004. Viral release of iron and its bioavailabiltiy to marine plankton. Limnology and Oceanography, 49: 1734-1741.

Price, N.M. et al., 1989. Preparation and Chemistry of the Artificial Algal Culture Medium Aquil. Biological Oceanography, 6: 443-461.

Price, N.M. and Harrison, P.J., 1988. Specific Selenium-Containing Macromolecules in the Marine Diatom Thalassiosira pseudonana. Plant Physiology, 86: 192-199.

Price, N.M. and Morel, F.M.M., 1991. Colimitation of phytoplankton growth by nickel and nitrogen. Limnology and Oceanography, 36: 1071-1077.

Quigg, A., Reinfelder, J.R. and Fisher, N.S., 2006. Copper uptake kinetics in diverse marine phytoplankton. Limnology and Oceanography, 51(2): 893-899.

Rich, H.W. and Morel, F.M.M., 1990. Availability of well-defined iron colloids to the marine diatom Thalassiosira weissflogii. Limnology and Oceanography, 35: 652-662.

Rijkenberg, M.J.A. et al., 2005. The influence of UV irradiation on the photoreduction of iron in the Southern Ocean. Marine Chemistry 93: 119-129.

Rue, E.L. and Bruland, K.W., 1995. Complexation of Iron(III) by Natural Organic Ligands in the Central North Pacific as Determined by a New Competitive Ligand Equilibration/Adsorptive Cathodic Stripping Voltammetric Method. Marine Chemistry, 50: 117-138.

Ruzic, I., 1982. Theoretical Aspects of the Direct Titration of Natural Waters and its Information Yield for Trace Metal Speciation. Analytica Chimica Acta, 140: 99-113.

Saito, M.A. and Moffett, J.W., 2002. Temporal and spatial variability of cobalt in the Atlantic Ocean. Geochimica et Cosmochimica Acta, 66: 1943-1953.

Santschi, P.H. et al., 1987. Relative mobility of radioactive trace elements across the sediment-water interface in the MERL model ecosystems of Narragansett Bay. Journal Of Marine Research, 45(4): 1007-1048.

Santschi, P.H., Li, Y.H. and Carson, S.R., 1980. The fate of trace metals in Narragansett Bay, Rhode Island: Radiotracer experiments in microcosms. Estuarine and Coastal Marine Science, 10(6): 635-654.

Schlosser, C. and Croot, P., 2009. Controls on seawater Fe(III) solubility in the Mauritanian upwelling zone. Geophys. Res. Lett., 36: L18606, doi:10.1029/2009GL038963.

Schlosser, C. and Croot, P.L., 2008. Application of cross-flow filtration for determining the solubility of iron species in open ocean seawater. Limnology and Oceanography: Methods, 6: 630-642.

Schüßler, U. and Kremling, K., 1993. A pumping system for underway sampling of dissolved and particulate trace elements in near-surface waters. Deep Sea Research, 40: 257-266.

Semeniuk, D.M. et al., 2009. Plankton copper requirements and uptake in the subarctic Northeast Pacific Ocean. Deep-Sea Research Part I-Oceanographic Research Papers, 56(7): 1130-1142.

Shaked, Y., Kustka, A.B., Morel, F.M.M. and Erel, Y., 2004. Simultaneous determination of iron reduction and uptake by phytoplankton. Limnology and Oceanography: Methods, 2: 137-145.

Shaked, Y., Xu, Y., Leblanc, K. and Morel, F.M.M., 2006. Zinc availability and alkaline phosphatase activity in *Emiliania huxleyi*: Implications for Zn-P co-limitation in the ocean. Limnology and Oceanography, 51(1): 299-309.

Spokes, L.J. and Liss, P.S., 1995. Photochemically induced redox reactions in seawater, I. Cations. Marine Chemistry, 49: 201-213.

Sunda, W. and Huntsman, S., 2003. Effect of pH, light, and temperature on Fe-EDTA chelation and Fe hydrolysis in seawater. Marine Chemistry, 84(3-4): 35-47.

Sunda, W.G. and Huntsman, S.A., 1983. Effect of competitive interactions between manganese and copper on cellular manganese and growth in estuarine and oceanic species of the diatom Thalassiosira. Limnology and Oceanography, 28: 924-934.

Sunda, W.G. and Huntsman, S.A., 1988. Effect Of Sunlight On Redox Cycles Of Manganese In The Southwestern Sargasso Sea. Deep-Sea Research Part A-Oceanographic Research Papers, 35(8): 1297-1317.

Sunda, W.G. and Huntsman, S.A., 1992. Feedback interactions between zinc and phytoplankton in seawater. Limnology and Oceanography, 37: 25-40.

Sunda, W.G. and Huntsman, S.A., 1995. Iron uptake and growth limitation in oceanic and coastal phytoplankton. Marine Chemistry, 50: 189-206.

Sunda, W.G. and Huntsman, S.A., 1996. Antagonisms between cadmium and zinc toxicity and manganese limitation in a coastal diatom. Limnology and Oceanography, 41: 373-387.

Sunda, W.G. and Huntsman, S.A., 1997. Interrelated influence of iron, light and cell size on marine phytoplankton growth. Nature, 390: 389-392.

Sunda, W.G. and Huntsman, S.A., 2000. Effect of Zn, Mn, and Fe on Cd accumulation in phytoplankton: Implications for oceanic Cd cycling. Limnology and Oceanography, 45(7): 1501-1516.

Sunda, W.G., Huntsman, S.A. and Harvey, G.R., 1983. Photoreduction of manganese oxides in seawater and its geochemical and biological implications. Nature, 301(5897): 234-236.

Sunda, W.G., Swift, D.G. and Huntsman, S.A., 1991. Low iron requirement for growth in oceanic phytoplankton. Nature, 351: 55-57.

Tani, H. et al., 2003. Iron(III) hydroxide solubility and humic-type fluorescent organic matter in the deep water column of the Okhotsk Sea and the northwestern North Pacific Ocean. Deep-Sea Research, 50: 1063-1078.

Tebo, B.M. and Emerson, S., 1985. Effect of Oxygen Tension, Mn(II) Concentration, and Temperature on the Microbially Catalyzed Mn(II) Oxidation Rate in a Marine Fjord. Appl. Environ. Microbiol., 50(5): 1268-1273.

Templeton, D.M. et al., 2000. Guidelines for terms related to chemical speciation and fractionation of elements. Definitions, structural aspects, and methodological approaches (IUPAC Recommendations 2000). Pure and Applied Chemistry, 72(8): 1453-1470.

Turner, D.R., Whitfield, M. and Dickson, A.G., 1981. The equilibrium speciation of dissolved components in freshwater and seawater at 25C and 1 atm pressure. Geochimica et Cosmochimca Acta, 45: 855-881.

Vallee, B.L. and Auld, D.S., 1993. Zinc: biological functions and coordination motifs. Accounts of Chemical Research, 26(10): 543-551.

van den Berg, C.M.G., 1982. Determination of Copper Complexation with Natural Organic Ligands in Seawater by Equilibration with MnO2 I. Theory. Marine Chemistry, 11: 307-322.

Vandermeulen, J.H. and Foda, A., 1988. Cycling of selenite and selenate in marine phytoplankton. Marine Biology, 98(1): 115-123.

Vogel, C. and Fisher, N.S., 2009. Trophic transfer of Fe, Zn and Am from marine bacteria to a planktonic ciliate. Marine Ecology-Progress Series, 384: 61-68.

Vogel, C. and Fisher, N.S., 2010. Metal accumulation by heterotrophic marine bacterioplankton. Limnology And Oceanography, 55(2): 519-528.

von Langen, P.J., Johnson, K.S., Coale, K.H. and Elrod, V.A., 1997. Oxidation kinetics of manganese (II) in seawater at nanomolar concentrations. Geochimica Et Cosmochimica Acta, 61(23): 4945-4954.

Vraspir, J.M. and Butler, A., 2009. Chemistry of Marine Ligands and Siderophores. Annual Review Of Marine Science, 1: 43-63.

Waite, T.D. and Szymczak, R., 1993. Manganese Dynamics in Surface Waters of the Eastern Caribbean. Journal of Geophysical Research-Oceans, 98(C2): 2361-2369.

Wang, W.-X. and Dei, R.C.H., 2001a. Effects of major nutrient additions on metal uptake in phytoplankton. Environmental Pollution, 111(2): 233-240.

Wang, W.-X. and Dei, R.C.H., 2001b. Influences of phosphate and silicate on Cr(VI) and Se(IV) accumulation in marine phytoplankton. Aquatic Toxicology, 52(1): 39-47.

Wang, W.-X. and Guo, L., 2000. Bioavailability of colloid-bound Cd, Cr, and Zn to marine plankton. Marine Ecology Progress Series, 202: 41-49.

Wu, J., Boyle, E., Sunda, W. and Wen, L.-S., 2001. Soluble and Colloidal Iron in the Oligotrophic North Atlantic and North Pacific. Science, 293: 847-849.

Xu, Y., Wang, W.-X. and Hsieh, D.P.H., 2001. Influences of metal concentration in phytoplankton and seawater on metal assimilation and elimination in marine copepods. Environmental Toxicology and Chemistry, 20(5): 1067-1077.

Yee, D. and Morel, F.M.M., 1996. In vivo substitution of zinc by cobalt in carbonic anhydrase of a marine diatom. Limnology and Oceanography, 41: 573-577.

14

Cesium (^{137}Cs and ^{133}Cs), Potassium and Rubidium in Macromycete Fungi and *Sphagnum* Plants

Mykhailo Vinichuk[1, 3], Anders Dahlberg[2] and Klas Rosén[1]
[1]*Department of Soil and Environment, Swedish University of Agricultural Sciences,*
[2]*Department of Forest Mycology and Pathology, Swedish University of Agricultural Sciences,*
[3]*Department of Ecology, Zhytomyr State Technological University,*
[1,2]*Sweden*
[3]*Ukraine*

1. Introduction

1.1 Cesium (^{137}Cs and ^{133}Cs), potassium and rubidium in macromycete fungi

Radiocesium (^{137}Cs) released in the environment as result of nuclear weapons tests in the 1950s and 1960s, and later due to the Chernobyl accident in 1986, is still a critical fission product because of its long half-life of 30 years and its high fission yield. The study of the cesium radioisotope ^{137}Cs is important, as production and emission rates are much higher than other radioisotopes. This chapter comprises results obtained in several experiments in Swedish forest ecosystems and aims to discuss the behavior of cesium isotopes (^{137}Cs and ^{133}Cs) and their counterparts potassium (K) and rubidium (Rb) in the "soil-fungi-plants transfer" system. The chapter consists of two parts: one mainly dealing with ^{137}Cs, ^{133}Cs, K and Rb in forest soil and macromycete fungi, and the other with the same isotopes in separate segments of *Sphagnum* plants.

The bioavailability of radionuclides controls the ultimate exposure of living organisms and the ambient environment to these contaminants. Consequently, conceptually and methodologically, the understanding of bioavailability of radionuclides is a key issue in the field of radioecology. Soil-fungi-plants transfer is the first step by which ^{137}Cs enters food chains.

1.1.1 The role of fungi in ^{137}Cs transfer in the forest

The availability of radionuclides (^{137}Cs in particular) in soils of different ecosystems is to a large extent regulated by various vascular plants and fungal species. Thus, the behavior of ^{137}Cs in forest ecosystems differs substantially from other ecosystems, foremost due to the abundance of fungal mycelia in soil, which contribute to the persistence of the Chernobyl radiocesium in the upper horizons of forest soils (Vinichuk & Johanson, 2003). Both saprotrophic and mycorrhizal fungi have key roles in nutrient and carbon cycling processes in forest soils. The mycelium of soil fungi has a central role in breaking down organic matter and in the uptake of nutrients from soil into plants via the formation of symbiotic mycorrhizal associations (Read & Perez-Moreno, 2003). The fungi facilitate nutrient uptake into the host plant, both as a consequence of the physical geometry of the mycelium and by the ability of the fungi to mobilize nutrients from organic substrates through the action of extracellular catabolic enzymes (Leake & Read, 1997). In addition to acquiring essential macronutrients, mycorrhizal fungi are efficient at taking-up and accumulating microelements (Smith & Read, 1997), this ability results in the accumulation of non-essential elements and radionuclides, particularly ^{137}Cs and can have important consequences for the retention, mobility and availability of these elements in forest ecosystems (Steiner et al., 2002).

Although fungal biomass, in comparison to plant biomass, is relatively low in forest soil (Dighton et al., 1991; Tanesaka et al., 1993), many fungal species accumulate more ^{137}Cs than vascular plants do and ^{137}Cs activity concentrations in many fungi are 10 to 100 times higher than in plants (Rosén et al., 2011). Fungi (particularly sporocarps) accumulate ^{137}Cs against a background of low ^{137}Cs activity concentrations, thus, the contribution of fungi to ^{137}Cs cycling in forest systems is substantial.

Fungi are important in radiocesium migration in nutrient poor and organic rich soils of forest systems (Rafferty et al., 1997). In organic matter, the presence of single strains of saprotrophic fungi considerably enhances the retention of Cs in organic systems (Parekh et al., 2008): $\approx$ 70% of the Cs spike is strongly (irreversibly) bound (remains non-extractable) compared to only $\approx$ 10% in abiotic (sterilized) systems. Fungal mycelium may act as a sink for radiocesium (Dighton et al., 1991; Olsen et al., 1990), as it contains 20–30% ^{137}Cs in soil inventories, and as much as 40% of radiocesium can leached from irradiated samples compared to control samples (Guillitte et al., 1994). Mycelium in upper organic soil layers may contain up to 50% of the total ^{137}Cs located within the upper 0-10 cm layers of Swedish and Ukrainian forest soils (Vinichuk & Johanson 2003). In terms of the total radiocesium within a forest ecosystem, fungal sporocarps contain a small part of activity and may only account for about 0.5 % (McGee et al., 2000) or even less – 0.01 to 0.1% (Nikolova et al., 1997) of the total radiocesium deposited within a forest ecosystem. However, these estimates are based on the assumption radionuclide concentration in fungal sporocarps is similar to that of the fungal parts of mycorrhizae (Nikolova et al., 1997). The activity concentration in sporocarps is probably higher than in the mycelium (Vinichuk & Johanson, 2003, 2004) and sporocarps constitute only about 1% of the total mycelia biomass in a forest ecosystem. Due to the high levels of ^{137}Cs in sporocarps, their contribution to the internal dose in man may be high through consumption of edible mushrooms (Kalač, 2001). Consequently, the consumption of sporocarps of edible fungi (Skuterud et al., 1997) or of game animals that consumed large quantities of fungi with high ^{137}Cs contents (Johanson & Bergström, 1994) represents an important pathway by which ^{137}Cs enters the human food system.

The ^{137}Cs activity concentration in edible fungi species has not decreased over the last 20 years (*Suillus variegatus*) or significantly increased (*Cantharellus* spp.) (Mascanzoni, 2009; Rosén et al., 2011).

1.1.2 ^{137}Cs, ^{133}Cs and alkali metals in fungi

Although fungi are important for ^{137}Cs uptake and migration in forest systems and since the Chernobyl accident, fungal species may contain high concentrations of radiocesium, the reasons and mechanisms for the magnitude higher concentration of radiocesium in fungi than in plants remains unclear (Kuwahara et al., 1998; Bystrzejewska-Piotrowska & Bazala, 2008). In addition to radiocesium, fungi effectively accumulate potassium (K), rubidium (Rb) and stable cesium (^{133}Cs) (Gaso et al., 2000) and the concentrations of ^{137}Cs, ^{133}Cs and Rb in fungal sporocarps can be one order of magnitude higher than in plants growing in the same forest (Vinichuk et al., 2010b).

The chemical behavior of the alkali metals, K, Rb and ^{133}Cs, can be expected to be similar to ^{137}Cs, due to similarities in their physicochemical properties, e.g. valence and ion diameter (Enghag, 2000). Potassium is a macronutrient and an obligatory component of living cells, which depend on K+ uptake and K+ flux to grow and maintain life. In radioecology cesium is assumed to behave similarly to potassium. At the cellular level, K is accumulated within cells and is the most important ion for creating membrane potential and excitability. Myttenaere et al. (1993) summarize the relationship between radiocesium and K in forests and suggest the possible use of K as an analogue for predicting radiocesium behavior.

Generally, ^{137}Cs is positively associated with K concentration across plant species in an undisturbed forest ecosystem, which suggests ^{137}Cs, stable ^{133}Cs and K are assimilated in a similar way and the elements pass through the biological cycle together (Chao et al., 2008). Cs influx into cells and its use of K transporters is reviewed by White & Broadley (2000) and potassium transport in fungi is reviewed by Rodríguez-Navarro (2000).

Rubidium is another rarely studied alkali metal, which may be an essential trace element for organisms, including fungi. However, there is scarce information on the concentrations and distribution of Rb in fungi and its behavior in food webs originating in the forest. Rubidium is often used in studies on K uptake and appears to emulate K to a high degree (Marschner, 1995): both K and Rb have the same uptake kinetics and compete for transport along concentration gradients in different compartments of soil and organisms (Rodríguez-Navarro, 2000). The concentrations of K, Rb and ^{133}Cs have been analyzed in fungal sporocarps (Baeza et al., 2005; Vinichuk et al., 2010b; 2011) and a relation between the uptake of Cs and K has been found (Bystrzejewska-Piotrowska & Bazal, 2008). Cesium uptake in fungi is affected by the presence of K and Rb and the presence of ^{133}Cs (Gyuricza et al., 2010; Terada et al., 1998). Although in fungal sporocarps, the relationships between these alkali metals and ^{137}Cs when taken up by fungi and their underlying mechanisms are insufficiently understood, as Cs does not always have high correlation with K and it is suggested there is an alternative pathway for Cs uptake into fungal cells (Yoshida & Muramatsu, 1998).

The correlations between ^{137}Cs and these alkali metals suggest the mechanism of fungal uptake of ^{133}Cs and ^{137}Cs is different from K and that Rb has an intermediate behavior between K and ^{133}Cs (Yoshida & Muramatsu, 1998). However, this interpretation is based on a few sporocarp analyses from each species, and comprised different ectomycorrhizal and saprotrophic fungal species. Although fungal accumulation of ^{133}Cs is reported as species-dependent, there are few detailed studies of individual species (Gillet & Crout, 2000). The variation in ^{137}Cs levels within the same genotype of fungal sporocarps can be as large as the variation among different genotypes (Dahlberg et al., 1997).

Another way to interpret and understand the uptake and relations between ^{137}Cs, ^{133}Cs, K and Rb in fungi is to use the isotopic (atom) ratio ^{137}Cs/^{133}Cs. Chemically, ^{133}Cs and ^{137}Cs are the same, but the atom abundance and isotopic disequilibrium differ. Among other factors, uptake of ^{133}Cs and ^{137}Cs by fungi depends on whether equilibrium between the two isotopes is achieved. An attainment of equilibrium between stable ^{133}Cs and ^{137}Cs in the bioavailable fraction of soils within forest ecosystems is reported Karadeniz & Yaprak (2007) but in cultivated soils, equilibrium between fallout ^{137}Cs and stable ^{133}Cs among exchangeable, organic bound and strongly bound fractions has not reached, even though most ^{137}Cs was deposited on the soils more than 20 years before (Tsukada, 2006).

The important roles fungi play in nutrient uptake in forest soils, in particular its role in ^{137}Cs transfer between soil and fungi, requires better understanding of the mechanisms involved. Although transfer of radioactive cesium from soils to plants through fungi is well researched, there is still limited knowledge on natural stable ^{133}Cs and other alkali metals (K and Rb) and the potential role as a predictor for radiocesium behavior, and less is known about the relationships between ^{133}Cs and other alkali metals (K and Rb) during uptake by fungi.

To explore mechanisms governing the uptake of radionuclides (^{137}Cs) data on uptake of stable isotopes of alkali metals (K, Rb, ^{133}Cs) by fungal species, and the behavior of the three alkali metals K, Rb and ^{133}Cs in bulk soil, fungal mycelium and sporocarps are required. Therefore, an attempt was made to quantify the uptake and distribution of the alkali metals in the soil–mycelium–sporocarp compartments and to study the relationships between K, Rb and ^{133}Cs in the various transfer steps. Additionally, the sporocarps of ectomycorrhizal fungi *Suillus variegatus* were analyzed to determine whether i) Cs (^{133}Cs and ^{137}Cs) uptake was correlated with K uptake; ii) intraspecific correlation of these alkali metals and ^{137}Cs activity concentrations in sporocarps was higher within, rather than among different fungal species; and, iii) the genotypic origin of sporocarps affected uptake and correlation.

Substantial research in this area has been conducted in Sweden after the fallout from nuclear weapons tests and the Chernobyl accident. Some results are published in a series of several articles in collaboration with Profs K.J. Johanson, H. Rydin and Dr. A. Taylor (Vinichuk et al., 2004; 2010a; 2010b; 2011).

This chapter aims to summarize the acquired knowledge from studies in Sweden and to place them in a larger context. The results are summarized and discussed and address the issues of K, Rb and ^{133}Cs concentrations in soil fractions and fungal compartments (Section 1.3.); concentration ratios of K, Rb and ^{133}Cs in soil fractions and fungi (Section 1.4); relationships between K, Rb and ^{133}Cs in soil and fungi (Section 1.5); the isotopic (atom) ratios $^{137}Cs/K$, $^{137}Cs/Rb$ and $^{137}Cs/^{133}Cs$ in fungal species (Section 1.6); K, Rb and Cs (^{137}Cs and ^{133}Cs) in sporocarps of a single species (Section 1.7); mechanisms of ^{137}Cs and alkali metal uptake by fungi (Section 1.8); Cs (^{137}Cs and ^{133}Cs), K and Rb in *Sphagnum* plants (Section 2); distribution of Cs (^{137}Cs and ^{133}Cs), K and Rb within *Sphagnum* plants (Section 2.3); mass concentration and isotopic (atom) ratios between ^{137}Cs, K, Rb and ^{133}Cs in segments of *Sphagnum* plants (Section 2.4); relationships between ^{137}Cs, K, Rb and ^{133}Cs, in segments of *Sphagnum* plants (Section 2.5); mechanisms of ^{137}Cs and alkali metal uptake by *Sphagnum* plants (Section 2.6); and conclusions from the Swedish studies (Section 3). Before presenting and discussing results a short description of study area, study design and methods used is presented (section 1.2).

1.2 Study area, study design and methods for results presented

1.2.1 Study area

The K, Rb and ^{133}Cs concentrations in soil fractions and fungal compartments were studied in an area located in a forest ecosystem on the east coast of central Sweden (60°22'N, 18°13'E). The soil was a sandy or clayey till and the humus mainly occurred in the form of mull. A more detailed description of the study area is presented by Vinichuk et al. (2010b).

Sporocarps of ectomycorrhizal fungi *Suillus variegatus* was studied in an area located about 40 km north-west of Uppsala in central Sweden (N 60°08'; E 17°10'). The forest is located on moraine and is dominated by Scots pine (*Pinus sylvestris*) and Norway spruce (*Picea abies*), with inserts of deciduous trees, primarily birch (*Betula pendula* and *Betula pubescens*). The field layer consisted mainly of the dwarf shrubs bilberry (*Vaccinium myrtillus* L.), lingonberry (*Vaccinium vitis-idaea* L.) and heather *Calluna vulgaris* L.): for details about the area and sampling see Dahlberg et al. (1997).

1.2.2 Study design

For studies of K, Rb and ^{133}Cs concentrations in soil fractions and fungal compartments, samples of soil and fungal sporocarps were collected from 10 sampling plots during September to November 2003. Four replicate soil samples were taken, with a cylindrical steel tube with a diameter of 5.7 cm, from around and directly underneath the fungal sporocarps (an area of about 0.5 m^2) and within each 10 m^2 area to a depth of 10 cm. Soil cores were divided horizontally into two 5-cm thick layers. Sporocarps of 12 different fungal species were collected and identified to species level, and the ^{137}Cs activity concentration in fresh material was determined. The sporocarps were dried at 35°C to constant weight and concentrations of ^{133}Cs, K and Rb were determined.
A selection of dried sporocarps of *S. variegatus* (n=51), retained from a study by Dahlberg et al. (1997) on the relationship between ^{137}Cs activity concentrations and genotype identification, was used. The sporocarps were collected once a week during sporocarp season (end of August through September) in 1994 and were taken from five sampling sites (100 to 1600 m^2 in size) within an area of about 1 km^2. Eight genotypes with 2 to 8 sporocarps each were tested (in total 32 sporocarps) and are referred to here as individual genotypes. Sporocarps within genotypes were spatially separated by up to 10-12m. All genotypes were used for the estimation of correlation coefficients, but only genotypes with at least four sporocarps were included in the alkali metal analysis. In addition, 19 individual sporocarps with unknown genotype (i.e. not tested for genotype identity) were included: these sporocarps consisted of both the same and different genotypes. The combined set of sporocarps refers to all sporocarps: for further details about the sampling and identification of genotypes see Dahlberg et al. (1997). The ^{137}Cs activity concentration values corrected to sampling date and expressed as kBq kg^{-1} dry weight (DW) for each sporocarp, as reported by Dahlberg et al. (1997), were used.

1.2.3 Methods

For the studies of K, Rb and ^{133}Cs concentrations in soil fractions and fungal compartments, fungal mycelia were separated from the soil samples (30-50 g, 0-5 cm layer depth) under a dissection microscope (magnification X64) with forceps and by adding small amounts of distilled water to disperse the soil. The prepared fraction of mycelium (30−60 mg DW g^{-1} soil) was not identified to determine of the mycelia extracted from the soil samples and the sporocarps belonged to the same species, as it assumed a majority of the prepared mycelia belonged to the same species as the nearby sporocarps. The method for mycelium preparation is described in Vinichuk & Johanson (2003). Mycelium samples were dried at 35°C to constant weight for determination of K, Rb and ^{133}Cs.

The soil samples (0–5 cm layer) were partitioned by the method described in Gorban & Clegg (1996). First, soil was gently sieved through a 2 mm mesh giving a bulk soil fraction. The remaining soil aggregates containing roots were further crumbled and gently squeezed between the fingers: this was called the rhizosphere fraction. The residue (finest roots with adhering soil particles) was called the soil–root interface fraction. Nine samples of bulk soil fraction and mycelium, 12 samples of fungal sporocarps, and six samples of rhizosphere and soil–root interface fraction were analyzed for K, Rb and ^{133}Cs.

The ^{137}Cs activity concentrations in the bulk soil samples and sporocarps were determined with calibrated HP-Ge detectors, corrected to sampling date and expressed as Bq kg^{-1} DW. The measuring time employed provided a statistical error ranging between 5 and 10%. For element analyses, a 2.5 g portion of each sample was analyzed by inductively coupled plasma in the laboratories of ALS Scandinavia (Luleå, Sweden) with recoveries 97–101% for K; 97.5–99.4% for Rb, and 93.7–102.5% for, ^{133}Cs. For soil, CRM SO-2 (heavy metals in soil) was used which had no certified values for K, Rb or ^{133}Cs. Element concentrations in the analyzed fractions are reported as mg kg^{-1} DW.

For element analyses (K, Rb and ^{133}Cs) of *S. variegatus* sporocarps, aliquots of about 0.3 g of each sample were analyzed by the same technique. Element concentrations are reported as mg kg^{-1} DW and the isotopic ratio of ^{137}Cs/^{133}Cs was calculated with Equations 1 and 2 (Chao et al., 2008):

$$\frac{^{137}Cs}{^{133}Cs} = \frac{A}{C} \times \frac{\alpha}{\lambda \times N} \times 10^3 \tag{1}$$

where: A is the ^{137}Cs radioactivity (Bq kg^{-1}); λ is the disintegration rate of ^{137}Cs 7.25 x10^{-10} s^{-1}; a is the atomic weight of cesium (132.9); N is the Avogadro number, which is 6.02 x10^{23}; and, ^{133}C and C are the ^{133}Cs concentration (mg g^{-1}). Eq. (1) can be simplified to Eq. 2:

$$\frac{^{137}Cs}{^{133}Cs} = 3.05 \times 10^{-10} \times \frac{A}{C} \tag{2}$$

where: A is the ^{137}Cs activity concentration in Bq kg^{-1} and ^{133}C is the ^{133}Cs concentration in mg kg^{-1}. Thus, the units of the isotope ratio are dimensionless.

Relationships between K, Rb, ^{133}Cs and ^{137}Cs concentrations in different fractions and sporocarps of *S. variegatus* were identified by Pearson correlation coefficients. Correlation coefficients were analyzed in five separate sets of samples: in four sets, all samples had known genotype identity, and in the last set, there was a combined set of samples containing both genotypes that had been tested by somatic incompatibility sporocarps and genotypes that had not been tested. Correlation analyses for genotypes with three or less sporocarps were omitted. All statistical analyses were run with Minitab® 15.1.1.0. (© 2007 Minitab Inc.) software, with level of significance of 5% (0.05), 1% (0.01) and 0.1% (0.001).

1.3 K, Rb and ^{133}Cs concentrations in soil fractions and fungal compartments

K, Rb and ^{133}Cs concentrations values in soil fractions and fungal compartments are necessary for calculating the concentration ratio at each step of its transfer in the soil-fungi system, differences in the uptake between elements and the relationships. This in turn will be the main reason for the different K, Rb and ^{133}Cs concentrations observed in sporocarps of various fungal species. Concentrations of K, Rb and ^{133}Cs in bulk soil were not significantly different from those in the rhizosphere, although the values for all three elements were slightly higher in the rhizosphere fraction (Table 1).

Element	Bulk soil	Rhizosphere	Soil root-interface	Fungal mycelium	Fruit bodies
K	642.6 (214.6)a	899.3 (301.4)a	3215 (842.8)b	2 867(727.5)b	43 415 (20 436)b
Rb	3.9 (2.7)a	5.4 (4.4)a	6.8 (1.7)a	13.8 (6.9)b	253.9 (273.6)b
^{133}Cs	0.3 (0.2)a	0.4 (0.3)a	0.2 (0.05)a	0.8 (0.8) a	5.65 (7.1)b

[1]Means within rows with different letters (a or b) are significantly different ($p < 0.001$).

Table 1. Mean concentrations of K, Rb and ^{133}Cs (mg kg^{-1} DW (standard deviation)) in soil fractions and fungi[1].

Potassium concentrations were higher in both the soil-root interface and fungal mycelium fractions than in the bulk soil and rhizosphere fraction. A comparison of K, Rb and ^{133}Cs concentrations revealed fungal sporocarps accumulated much greater amounts of these elements than mycelium. For example, K concentrations in fungal sporocarps collected from the same plots where soil samples and mycelium were extracted were about 15 times higher than K concentrations found in mycelium. The concentrations of Rb in fungal sporocarps were about 18-fold higher than in corresponding fungal mycelium, and those of ^{133}Cs were about 7-fold higher (Table 1).

Thus, potassium concentration increased in the order bulk soil<rhizosphere<fungal mycelium<soil-root interface<fungal sporocarps and was higher in the soil-root interface fraction and fungi than in bulk soil. The high concentrations of K in fungal sporocarps may reflect a demand for this element as a major cation in osmoregulation and that K is an important element in regulating the productivity of sporophore formation in fungi (Tyler, 1982).

Rb in mycelium was 3.5-fold higher than in bulk soil and 2.5-fold higher than in rhizosphere, and concentrations increased in the order bulk soil<rhizosphere<soil-root interface<fungal mycelium<fungal sporocarps. The concentrations of Rb were slightly higher in the soil-root interface fraction than in bulk soil; thus, fungi appeared to have high preference for this element, as the accumulation of Rb by fungi, and especially fungal sporocarps, was pronounced. Rubidium concentrations in sporocarps were more than one order of magnitude higher than those in mycelium extracted from soil of the same plots where fungal sporocarps were sampled. The ability of fungi to accumulate Rb is documented: mushrooms accumulate at least one order of magnitude higher concentrations of Rb than plants growing in the same forest (Yoshida & Muramatsu, 1998).

Concentrations of stable cesium varied considerably among samples but no significant differences were found among the different fractions analyzed. Cesium concentrations increased in the order soil-root interface<bulk soil<rhizosphere<fungal mycelium<fungal sporocarps, and were only significantly higher in fungal sporocarps, compared with bulk soil. Stable ^{133}Cs was generally evenly distributed within bulk soil, rhizosphere and soil-root interface fractions, indicating no ^{133}Cs enrichment in those forest compartments. However, ^{133}Cs concentrations in sporocarps were nearly one order of magnitude higher than those found in soil mycelium.

Radioactive ^{137}Cs presented similar to ^{133}Cs behavior, where ^{137}Cs activity increased in the order soil<mycelium<fungal sporocarps (Vinichuk & Johanson, 2003; Vinichuk et al., 2004). The differences between fungal species in their preferences for uptake of ^{137}Cs or stable ^{133}Cs appear to reflect the location of the fungal mycelium relative to that of cesium within the soil profile (Rühm et al., 1997). Unlike ^{137}Cs, stable ^{133}Cs originates from soil; therefore, the

amount of unavailable ^{133}Cs, compared to the total amount of ^{133}Cs, in soil presumably higher than that of ^{137}Cs. As a result, stable ^{133}Cs is considered less available for uptake as it is contained in mineral compounds and is difficult for fungi or plants to access: the concentration ratio of stable ^{133}Cs in mushrooms is lower than for ^{137}Cs (Yoshida & Muramatsu, 1998). The differing behavior of the natural and radioactive forms of ^{133}Cs may derive from their disequilibrium in the ecosystem (Horyna & Řanad, 1988).

1.4 Concentration ratios of K, Rb and ^{133}Cs in soil fractions and fungi

The concept of concentration ratios (CR, defined as concentration of the element (mg kg^{-1} DW) in a specific fraction or fungi divided by concentration of the element (mg kg^{-1} DW) in bulk soil) is widely used to quantify the transfer of radionuclides from soil to plants/fungi. This approach allows the estimation of differences in uptake of elements. The elements concentration ratio data followed a similar pattern, but the enrichment of all three elements in fungal material was more evident, particularly in the sporocarps (Table 2).

Element	Rhizosphere	Soil root-interface	Fungal mycelium	Fruit bodies
K	1.7 (0.4)	6.1 (1.9)	5.1 (1.4)	68.9 (23.1)
Rb	1.3 (0.4)	2.7 (1.1)	3.9 (1.1)	121.7 (172.2)
Cs	1.1 (0.5)	0.8 (0.3)	2.1 (0.9)	39.7 (67.6)

Table 2. Concentration ratios CR (defined as concentration of the element (mg kg^{-1} DW) in the specific fraction divided by concentration of the element (mg kg^{-1} DW) in bulk soil) (mean values (standard deviation)).

Thus, for all three alkali metals studied, the levels of K, Rb, ^{133}Cs and ^{137}Cs in sporocarps were at least one order of magnitude higher than those in fungal mycelium (Table 2). The concentration ratios for each element varied considerably between the species sampled. The saprotrophic fungus *Hypholoma capnoides* had the lowest values and the mycorrhizal fungus *Sarcodon imbricatus* had the highest. Sporocarp:bulk soil concentration ratios are presented in Table 3.

Sarcodon imbricatus accumulates nearly 100 000 Bq kg^{-1} of ^{137}Cs, giving TF values (defined as ^{137}Cs activity concentration (Bq kg^{-1} DW) in fungi divided by ^{137}Cs deposition (kBq m^{-2})) about 22 (Vinichuk & Johanson, 2003). The sporocarps of *Sarcodon imbricatus* had distinctively high concentration ratios of Rb and ^{133}Cs than other species analyzed. The mycorrhizal fungus *Cantharellus tubaeformis,* is another species showing relatively high concentration ratios, particularly for K and Rb. *Cantharellus tubaeformis* accumulates several tens of thousands Bq kg^{-1} of ^{137}Cs (Kammerer et al., 1994). Among those with moderate concentration ratios for each element are *Boletus edulis, Tricholoma equestre, Lactarius scrobiculatus* and *Cortinarius* spp.

Thus, the levels of K, Rb, ^{133}Cs and ^{137}Cs in sporocarps were at least one order of magnitude higher than those in fungal mycelium indicating biomagnification through the food web in forest ecosystems.

		Concentration ratios		
Plot	Species	K	Rb	^{133}Cs
4	*Boletus edulis*	62.7	77.4	37.4
6	*Cantharellus tubaeformis*	104.7	109.7	15.5
7	*Cortinarius armeniacus*	67.5	69.6	19.2
5	*C. odorifer*	71.8	70.9	34.7
8	*C. spp.*	90.9	157.2	14.8
8-10	*Hypholoma capnoides*[1]	26.6	13.1	6.9
1	*Lactarius deterrimus*	29.9	17.2	2.6
3	*L. scrobiculatus*	67.8	26.2	3.7
6	*L. trivialis*	77.5	126.9	52.2
5-7	*Sarcodon imbricatus*	101.7	675.7	258.8
2	*Suillus granulates*	58.6	41.4	14.7
10-11	*Tricholoma equestre*	66.6	75.4	15.4

[1]Saprophyte, all other analyzed fungal species are ectomycorrhizal

Table 3. Element concentration ratios (mg kg^{-1} DW in fungi)/(mg kg^{-1} DW in bulk soil) in fungi for fungal sporocarps.

1.5 Relationships between K, Rb and ^{133}Cs in soil and fungi

Although correlation analysis may be not definitive, it is a useful approach for elucidating similarities or differences in uptake mechanisms of cesium (^{137}Cs and ^{133}Cs), K and Rb: close correlation between elements indicates similarities in their uptake mechanisms. No significant correlations between K in soil and in either mycelium (r=0.452, ns) or in sporocarps (r=0.338, ns) has been identified and sporocarp Rb and ^{133}Cs concentrations were unrelated to soil concentrations, however, in mycelium both elements were correlated with soil concentrations (Rb: r=0.856, p=0.003; Cs: r=0.804, p=0.009). There was a close positive correlation (r=0.946, p=0.001) between the K:Rb ratio in soil and in fungal mycelium (Figure 1b) and this relationship was also apparent between soil and sporocarps, but was weak and not significant (r=0.602, ns: Figure1b).

The K:^{133}Cs ratio in soil and fungal components had a different pattern: the K:Cs ratio in mycelium was closely positively correlated (r=0.883, p=0.01) to the K:^{133}Cs ratio in soil (Figure 1a), but was relatively weakly and non-significantly correlated to soil in fungal sporocarps. No significant correlations were found between the concentrations of the three elements in fungi, soil pH or soil organic matter content (data not shown).

The competition between K, Rb and ^{133}Cs in the various transfer steps was investigated in an attempt to estimate the relationships between the concentrations of these three elements in soil, mycelia and fungal sporocarps. The lack of a significant correlation between K in soil and in either mycelium or sporocarps indicated a demand for essential K in fungi, regardless of the concentration of this element in soil. Regardless of fungal species, K concentration in fungi appears to be controlled within a narrow range, (Yoshida & Muramatsu, 1998), and supports the claim K uptake by fungi is self-regulated by the internal nutritional requirements of the fungus (Baeza et al., 2004).

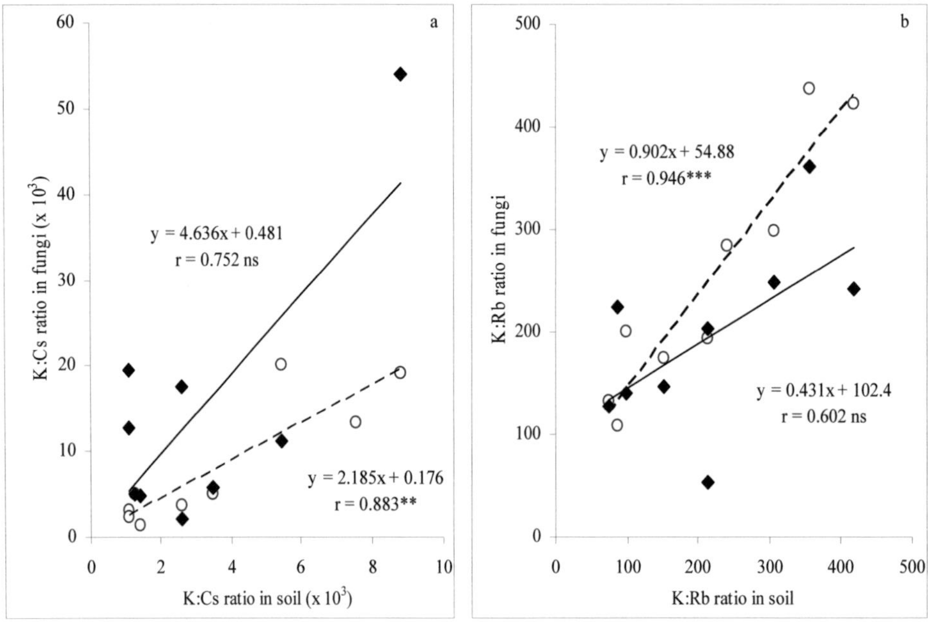

Fig. 1. Ratio of (a) K: ^{133}Cs and (b) K:Rb in fungal sporocarps (♦, solid line) and soil mycelium (○, dotted line) in relation to the soil in which they were growing. ** p=0.01, *** p=0.001

The relationships observed between K:Rb and K:^{133}Cs ratios in fungal sporocarps and soil mycelia, with respect to the soil in which they were growing (Figure 1), also indicated differences in uptake of these alkali metals by fungi. Although correlation analyses is not the best tool for analyzing the uptake mechanism, the closest positive correlations between K:Rb ratios in fungal mycelium and in soil indicated similarities in the uptake mechanism of these two elements by fungi, although the relationships between K: ^{133}Cs ratios in soil mycelium and in soil were less pronounced. These findings were in good agreement with the suggestion by Yoshida & Muramatsu (1998) that there might be an alternative pathway for ^{133}Cs uptake into cells and the mechanism of ^{133}Cs uptake by fungi could be similar to that for Rb, as ^{133}Cs does not show a good correlation with K. The high efficiency of Rb uptake by fungi indicates Rb, but not ^{133}Cs, eventually replaces essential K due to K limitation (Brown & Cummings, 2001) and Rb has the capacity to partially replace K, but ^{133}Cs does not (Wallace, 1970 and references therein). Forest plants apparently discriminate between K+ and Rb+ in soils and a shortage of K+ favors the uptake of the closely related Rb+ ion (Nyholm & Tyler, 2000), whereas, increasing K+ availability in the system decreases Rb+ uptake (Drobner & Tyler, 1998). These results provided new insights into the use of transfer factors or concentration ratios.

1.6 The isotopic (atom) ratios ^{137}Cs/K, ^{137}Cs/Rb and ^{137}Cs/^{133}Cs in fungal species

The isotopic ratios of ^{137}Cs/K, ^{137}Cs/Rb and ^{137}Cs/^{133}Cs in the fungal sporocarps belonging to different species were used to interpret the distribution of ^{137}Cs and the alkali metals in fungi and to provide better understanding of its uptake mechanisms. Measurements of trace levels of stable ^{133}Cs could be another way of obtaining information about the biological behavior of ^{137}Cs. To obtain better estimates, the isotopic ratios for fungal sporocarps in this study (Vinichuk et al., 2010b) were calculated and compared with estimates calculated in similar studies by Yoshida & Muramatsu (1998). Mean values of isotopic ratios of $^{137}Cs/K$, $^{137}Cs/Rb$ and $^{137}Cs/^{133}Cs$ in the fungal sporocarps, and range and correlation coefficients between concentration ratios $^{137}Cs/^{133}Cs$ and K, Rb and ^{133}Cs are presented in Table 4.

Data set	n	Isotopic ratios		
		$^{137}Cs/K$	$^{137}Cs/Rb$	$^{137}Cs/^{133}Cs$
Vinichuk et al. (2010b), Sweden	12	14.4(1.54−45.4)x10^{-13}	7.8(0.55−30.9)x10^{-10}	4.9(0.30−15.1)x10^{-8}
Yoshida & Muramatsu (1998), Japan	29	5.2(0.15−23.0)x10^{-16}	3.4(0.14−18.2)x10^{-13}	4.1(1.53−5.94)x10^{-9}
		Correlation coefficients		
		$^{137}Cs/^{133}Cs$:K	$^{137}Cs/^{133}Cs$:Rb	$^{137}Cs/^{133}Cs:^{133}Cs$
Vinichuk et al. (2010b), Sweden	12	0.25	−0.35	−0.31
Yoshida & Muramatsu (1998), Japan	29	0.12	0.39	0.26

Table 4. Isotopic (atom) ratios of $^{137}Cs/K$, $^{137}Cs/Rb$, $^{137}Cs/^{133}Cs$, correlation coefficients between isotopic ratios $^{137}Cs/^{133}Cs$ and mass concentrations of K, Rb and ^{133}Cs in fungal sporocarps (n = number of sporocarps analyzed).

The activity concentrations of ^{137}Cs in fungal sporocarps were about 13 to 16 orders of magnitude lower than mass concentrations of K, 10 to 13 orders of magnitude lower than mass concentrations for Rb, and 8 to 9 orders of magnitude lower than mass concentrations for ^{133}Cs. Isotopic (atom) ratios in the fungal sporocarps collected in Sweden were two-three orders of magnitude narrower than those collected in Japan, which reflected the level of ^{137}Cs concentrations in mushrooms: the median value for all fungi species was 4151 Bq kg^{-1} DW in Swedish forests and 135 Bq kg^{-1} DW in Japanese forests. Isotopic (atom) ratios of $^{137}Cs/K$, $^{137}Cs/Rb$, $^{137}Cs/^{133}Cs$ were variable in both datasets and appeared independent of specific species of fungi. These ratios might reflect the isotopic ratios of soil horizons from which radiocesium is predominantly taken up and be a possible source of the variability in isotopic ratios in fungal fruit bodies. Rühm et al. (1997) used the isotopic ratio $^{134}Cs/^{137}Cs$ to localize mycelia of fungal species *in situ;* alternatively, the isotopic (atom) ratio $^{137}Cs/^{133}Cs$ can be used to localize fungal mycelia *in situ*. However, this approach is only appropriate for organic soil layers, which contain virtually no or very little clay mineral to which cesium can bind. The isotopic ratios $^{137}Cs/^{133}Cs$ in fruit bodies of fungi were similar to those found in organic soil layers of forest soil (Rühm et al., 1997; Karadeniz & Yaprak, 2007).

The relationships observed between the concentration ratios $^{137}Cs/^{133}Cs$ and K, Rb and ^{133}Cs in fungal sporocarps also varied widely and were inconsistent (Table 4). The concentration of K, Rb and ^{133}Cs in sporocarps appeared independent of the $^{137}Cs/^{133}Cs$ isotopic ratio, suggesting differences in uptake of these alkali metals by fungi and complex interactions between fungi, their host and the environment.

1.7 K, Rb and Cs (^{137}Cs and ^{133}Cs) in sporocarps of a single species

Most results presented in this Chapter are already published (Vinichuk et al., 2011), and are based on sporocarp analysis of different ectomycorrhizal and saprotrophic fungal species. Fungal accumulation of ^{137}Cs is suggested to be species-dependent, thus, ^{137}Cs activity concentration and mass concentration of K, Rb and ^{133}Cs in fungal sporocarps belonging to the mycorrhizal fungus *Suillus variegatus* were analyzed. *S. variegatus* form *mycorrhiza* with Scots pine and predominantly occur in sandy, acidic soils and have a marked ability to accumulate radiocesium (Dahlberg et al., 1997): as this is an edible mushroom, high radiocesium contents present some concern with regard to human consumption.

The concentrations of K (range 22.2-52.1 g kg^{-1}) and Rb (range 0.22-0.65 g kg^{-1}) in sporocarps of *S. variegatus* varied in relatively narrow ranges, whereas, the mass concentration of ^{133}Cs had a range of 2.16 to 21.5 mg kg^{-1} and the activity concentration of ^{137}Cs ranged from 15.8 to 150.9 kBq kg^{-1}. Both ^{133}Cs and ^{137}Cs had wider ranges than K or Rb within sporocarps from the same genotype or across the combined set of sporocarps (Table 5). The mean of the ^{137}Cs/^{133}Cs isotopic ratio in the combined set of sporocarps was 2.5 x 10^{-7} (range 8.3 x 10^{-8} and 4.4 x 10^{-7}). The ^{137}Cs/Cs isotopic ratios from identified genotypes were site-genotype dependent: the ratio values of genotypes at site 4 were about two-times higher than the ratios of genotypes at site 2 (Table 6).

Site-genotype[1]	n	K			Rb			^{133}Cs			^{137}Cs		
		g kg^{-1}		%	g kg^{-1}		%	mg kg^{-1}		%	kBq kg^{-1}		%
		M	SD	CV	M	SD	CV	M	SD	CV	M	SD	CV
Sporocarps with identified genotypes													
2-1	8	30.6	8.06	26.4	0.47	0.12	24.7	12.1	4.23	35.1	67.3	35.1	52.2
2-2	6	28.0	6.99	25.0	0.50	0.07	13.8	16.6	2.19	13.2	75.9	23.2	30.6
4-3	4	28.5	2.13	7.5	0.39	0.16	4.0	6.6	0.44	6.7	68.9	11.7	17.0
4-4	3	33.6	8.60	-	0.30	0.04	-	3.0	0.60	-	39.1	9.38	-
4-5	2	38.9	2.40	-	0.36	0.02	-	3.8	0.04	-	35.7	28.2	-
4-6	2	35.2	8.84	-	0.37	0.11	-	3.7	2.16	-	26.8	8.54	-
7-7	5	33.7	5.79	17.2	0.34	0.06	17.9	6.7	0.80	12.0	71.4	9.30	13.0
6-8	2	25.4	1.34	-	0.31	0.03	-	8.7	2.16	-	63.3	18.3	-
Sporocarps with unknown genotypes													
	19	33.4	6.69	20.0	0.38	0.08	20.3	7.7	1.97	25.5	66.0	21.3	32.3
Combined set of sporocarps (identified and unknown genotypes)													
	51	31.9	6.79	21.3	0.40	0.09	23.6	8.7	4.36	50.1	63.7	24.2	38.0

[1]Site numbering according to Dahlberg et al. (1997), the second figure is a running number of the study's different genotypes.

Table 5. Potassium, rubidium and cesium (^{133}Cs) mass concentrations and ^{137}Cs activity concentrations in sporocarps of *S. variegatus* (DW) from identified and unknown genotypes, where n = number of sporocarps of each genotype analyzed, M = mean, SD = standard deviation, CV = coefficient of variation.

Site-genotype[1]	Identified genotypes								Unidentified genotypes	Combined set of sporocarps
	2-1	2-2	4-3	4-4	4-5	4-6	7-7	6-8		
M	1.67	1.43	3.16	3.95	2.86	2.43	3.27	2.24	2.62	2.50
CV (%)	97.1	36.4	10.4	5.1	78.1	29.5	9.2	3.9	20.0	34.6

[1]Site numbering according to Dahlberg et al. (1997), the second figure is a running number of the study's different genotypes.

Table 6. ^{137}Cs/^{133}Cs isotopic (atom) ratios in sporocarps of *S. variegatus* from identified genotypes, with unknown genetic belonging, and the two combined groupings, x10^{-7}. M = mean, CV = coefficient of variation.

Similarly, in results obtained from a previous study (Vinichuk et al. 2004) the concentrations of K in sporocarps of *S. variegatus* were not related to the concentrations of ^{137}Cs (r=0.103) or ^{133}Cs (r=−0.066) in the combined data set (Figure 2: c, b). In contrast, the concentrations of K and Rb were significantly correlated in the combined dataset (r=0.505, Figure 2: a).
Rubidium was strongly correlated with stable ^{133}Cs (r=0.746) and moderately correlated with ^{137}Cs (r=0.440) and K (r=0.505: Figure 2: d, e, a). Both ^{133}Cs and ^{137}Cs were significantly correlated in the combined dataset (Figure 2: f).
The ^{137}Cs/^{133}Cs isotopic ratio in the combined dataset was not correlated to K concentration, but correlated moderately and negatively with both ^{133}Cs (r=−0.636) and Rb (r=−0.500) concentrations (Figure 3: a, c, b).
Thus, the study of *S. variegatus* revealed no significant correlations between ^{133}Cs mass concentration or ^{137}Cs activity concentration and the concentration of K in sporocarps, either within the whole population or among the genotypes.
Potassium, ^{133}Cs and ^{137}Cs within the four genotypes were also not correlated, with one genotype exception (Table 7). However, the exception was conditional due to a one single value. Three of four analyzed sporocarp genotypes had high correlation between K and Rb: the forth was only moderately correlated (Table 7).
However, the correlations between ^{137}Cs and K and Rb and ^{133}Cs in the four genotypes were inconsistent (Table 3). Potassium, Rb, ^{133}Cs and ^{137}Cs were correlated in genotype 2-1 (due to one single value), whereas, no or negative correlations were found between the same elements/isotopes for the other three genotypes. In two of four genotypes, the ^{137}Cs/^{133}Cs isotopic ratio was not correlated with ^{133}Cs, K or Rb; however, there was a negative correlation with Rb in one genotype (2-2) and positive correlation with ^{133}Cs in another (4-3) (Table 7).
Data obtained for *S. variegatus* supported results from earlier studies (Ismail, 1994; Yoshida & Muramatsu, 1998) on different species of fungi, suggesting cesium (^{137}Cs and ^{133}Cs) and K are not correlated in mushrooms. Thus, correlation analysis may be a useful, although not definitive, approach for elucidating similarities or differences in uptake mechanisms of cesium (^{137}Cs and ^{133}Cs) and K.

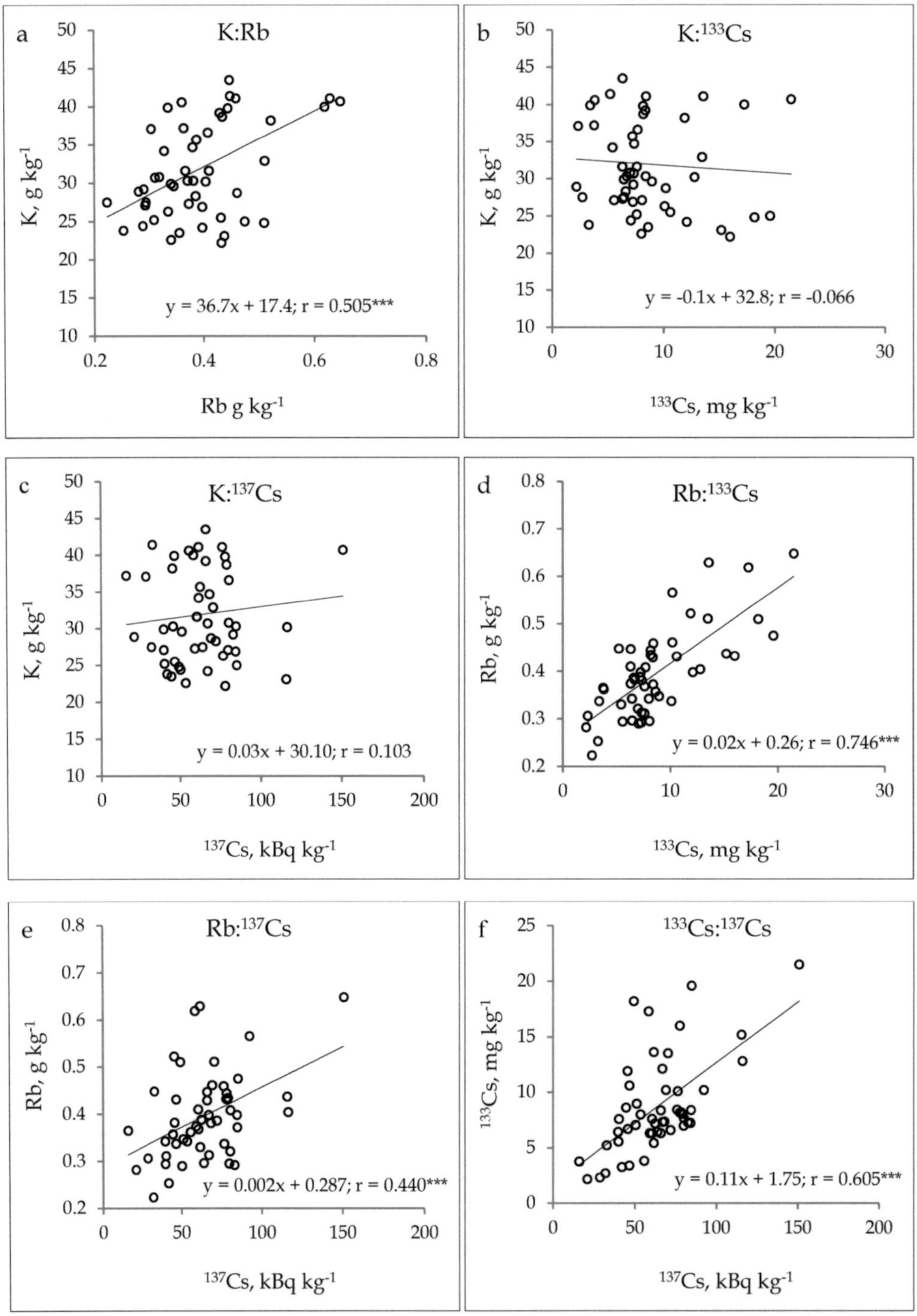

Fig. 2. Relationship between ^{137}Cs and K, Rb and ^{133}Cs concentrations in sporocarps in the combined set of all *S. variegatus* sporocarps (a-f). K:Rb (a); K:^{133}Cs (b); K:^{137}Cs (c); Rb:^{133}Cs (d); Rb:^{137}Cs (e); and, ^{133}Cs:^{137}Cs (f). *** p=0.001

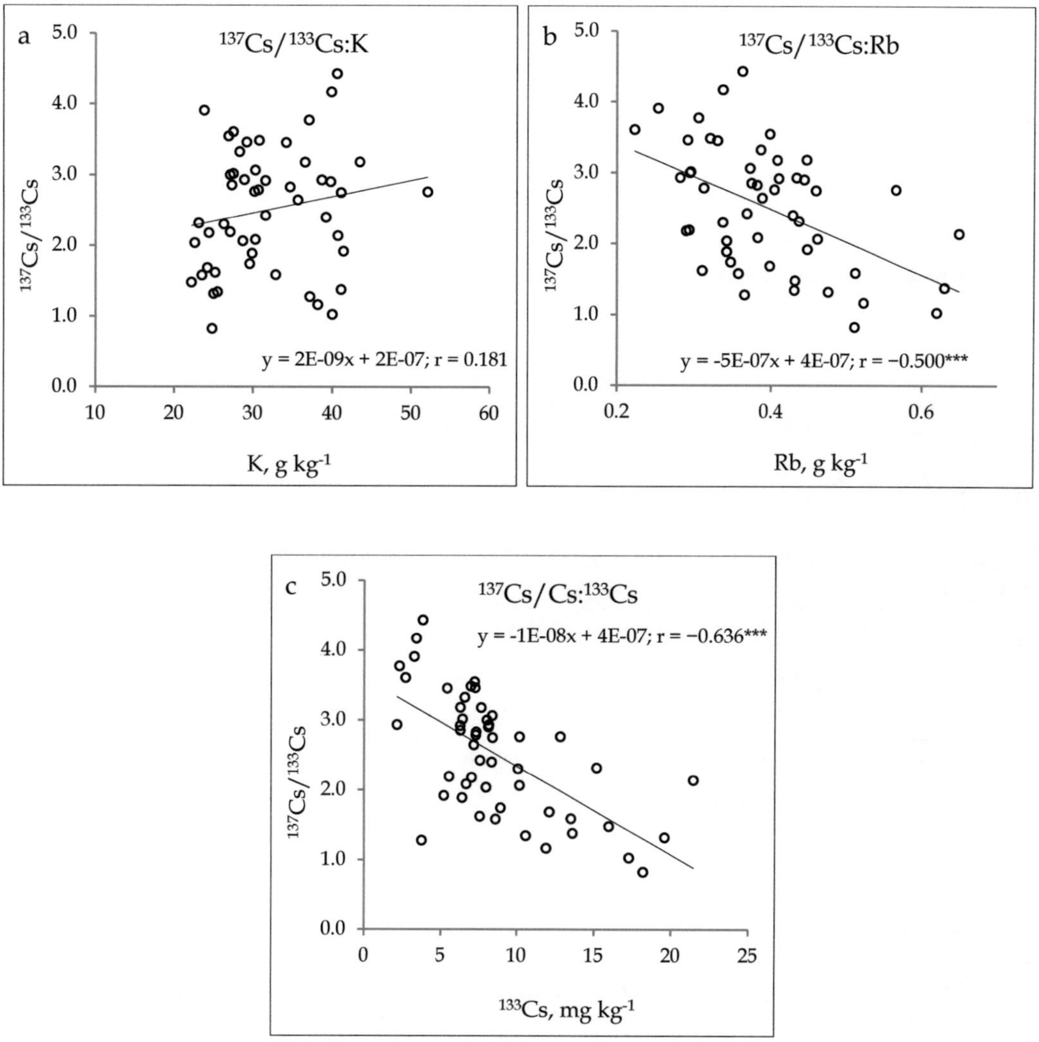

Fig. 3. Relationship between the ^{137}Cs/^{133}Cs isotopic (atom) ratios (x10^{-7}) and K, Rb and ^{133}Cs mass concentrations in the combined set of *S. variegatus* sporocarps, (a) ^{137}Cs/^{133}Cs:K; (b) ^{137}Cs/^{133}Cs:Rb; and, (c) ^{137}Cs/^{133}Cs:^{133}Cs. *** p=0.001

	^{137}Cs	K	Rb	^{133}Cs
Genotype 2-1 (8 sporocarps)				
K	0.502			
Rb	0.626*	0.966***		
^{133}Cs	0.908**	0.745*	0.837**	
$^{137}Cs/^{133}Cs$		−0.172	−0.058	0.240
Genotype 2-2 (6 sporocarps)				
K	−0.472			
Rb	−0.658	0.928**		
^{133}Cs	−0.263	−0.138	0.159	
$^{137}Cs/^{133}Cs$		−0.352	−0.608	−0.586
Genotype 4-3 (4 sporocarps)				
K	−0.531			
Rb	0.177	0.696		
^{133}Cs	0.979*	−0.569	0.182	
$^{137}Cs/^{133}Cs$		−0.488	0.163	0.930
Genotype 7-7(5 sporocarps)				
K	−0.562			
Rb	−0.472	0.987**		
^{133}Cs	0.699	−0.528	−0.404	
$^{137}Cs/^{133}Cs$		−0.115	−0.155	−0.345

[1]* p=0.05; ** p=0.01; *** p=0.001

Table 7. Correlation coefficients between concentrations of potassium, rubidium and cesium (^{133}Cs and ^{137}Cs) in genotypes of *S. variegatus* with more than four sporocarps analyzed[1].

The concentration of K in sporocarps appeared independent of the $^{137}Cs/^{133}Cs$ isotopic ratio in both the whole population (Figure 3) and among the genotypes, with one exception (Table 7). The absence of correlation between ^{137}C (or ^{133}Cs) and K in fungi may be due to the incorporation of K being self-regulated by the nutritional requirements of the fungus, whereas, incorporation of ^{137}Cs is not self-regulated by the fungus (Baeza et al., 2004). Although K and cesium (^{133}Cs and ^{137}Cs) concentrations did not correlate within *S. variegatus*, both K^+ and Cs^+ ions may compete for uptake by fungi. In experiments under controlled conditions and with sterile medium (Bystrzejewska-Piotrowska & Bazala, 2008), the competition between Cs^+ and K^+ depends on Cs^+ concentration in the growth medium and on the path of Cs^+ uptake. In studies of Cs uptake by hyphae of basidiomycete *Hebeloma vinosophyllum* when grown on a simulated medium (Ban-Nai et al., 2005), the addition of monovalent cations of K^+, Rb^+, and NH_4^+ reduced uptake of Cs. In addition, radiocesium transport by arbuscular mycorrhizal (AM) fungi decreases if K concentration increases in a compartment accessible only to AM (Gyuricza et al., 2010), and a higher Cs:K ratio in the nutrient solution increases uptake of Cs by ectomycorrhizal seedlings (Brunner et al., 1996). A noticeable (20-60%) and long-lasting (at least 17 years) reduction in ^{133}Cs activity concentration in fungal sporocarps *in situ* due to a single K fertilization of 100 kg ha^{-1} in a Scots pine forest is reported by Rosén et al., (2011).

The relation between ^{137}Cs and K, and Rb and ^{133}Cs within S. *variegatus* (Figure 2) was similar to an earlier report on different species of fungi (Yoshida & Muramatsu, 1998). Rubidium concentration in sporocarps was positively correlated with ^{133}Cs and ^{137}Cs, but generally negatively correlated with ^{137}Cs/^{133}Cs isotopic ratio, i.e. a narrower ^{137}Cs/^{133}Cs ratio in sporocarps resulted in higher Rb uptake by fungi. This ratio may reflect the soil layers explored by the mycelia (Rühm et al., 1997), as fungi have a higher affinity for Rb than for K and cesium (Ban-Nai et al., 2005; Yoshida & Muramatsu, 1998), and Rb concentrations in sporocarps can be more than one order of magnitude greater than in mycelium extracted as fungal sporocarps from soil of the same plots (Vinichuk et al., 2011). Soil mycelia always consist of numerous fungal species and the intraspecific relationships between soil mycelia and sporocarps has not yet been estimated; however, the development of molecular methods with the ability to mass sequence environmental samples in combination with quantitative PCR may now enable such analysis to be conducted.
Mass concentration of ^{133}Cs and activity concentration of ^{137}Cs have different relations in fungal sporocarps: in three of four genotypes, there was a high correlation, two of which were significant (r=0.908** and r=979*), and there was no correlation in the fourth genotype (r=−0.263, Table 7), whereas, correlation between ^{137}Cs and ^{133}Cs within the whole population was only moderate (r=0.605*** Figure 2). In terms of ^{133}Cs and ^{137}Cs behavior, there would be no biochemical differentiation, but there could be differences in atom abundance and isotopic disequilibrium within the system. Fungi have large spatiotemporal variation in ^{133}Cs and ^{137}Cs content in sporocarps of the same species and different species (de Meijer et al., 1988), and the variation in K, Rb, ^{133}Cs and ^{137}Cs concentrations within a single genotype appeared similar, or lower, than the variation within all genotypes. The results for ^{137}Cs and alkali elements in a set of samples of *S. variegatus,* collected during the same season and consisting of sporocarps from both different and the same genotype, indicated the variability in concentrations was similar to different fungal species collected in Japan over three years (Yoshida & Muramatsu, 1998).
The relatively narrow range in K and Rb variation and the higher ^{133}Cs and ^{137}Cs variations might be due to different mechanisms being involved. The differences in correlation coefficients between ^{137}Cs and the alkali metals varied among and within the genotypes of *S. variegatus,* suggesting both interspecific and intrapopulation variation in the uptake of K, Rb, stable ^{133}Cs and, ^{137}Cs and, their relationships could be explained by factors other than genotype identity. The variability in ^{137}Cs transfer depends on the sampling location of fungal sporocarps (Gillett & Crout, 2000), for *S. variegatus,* these interaction factors might include the spatial pattern of soil chemical parameters, heterogeneity of ^{137}Cs fallout, mycelia location, and heterogeneity due to abiotic and biotic interactions increasing over time (Dahlberg et al., 1997).
Within the combined set of sporocarps the concentration of Rb and ^{137}Cs activity concentration in *S. variegatus* sporocarps were normally distributed but the frequency distribution of ^{133}Cs and K was not: asymmetry of ^{137}Cs frequency distributions is reported in other fungal species (Baeza et al., 2004; Gaso et al., 1998; Ismail, 1994). According to Gillett & Crout (2000), the frequency distribution of ^{137}Cs appears species dependent: high accumulating species tend to be normally distributed and low accumulating species tend to be log-normally distributed. However, lognormal distribution is almost the default for concentration of radionuclides and is unlikely to be a species-specific phenomenon, as it also occurs in soil concentrations, which implies normal distribution would not be expected, even if large set of samples were analyzed.

1.8 Mechanisms of ^{137}Cs and alkali metal uptake by fungi

Generally, little is known about the mechanisms involved in the uptake and retention of radionuclides by fungi. Studies of uptake mechanisms and affinity for alkali metals in fungi are scarce, but some results are reviewed by Rodríguez-Navarro (2000). Compared to plants, fungal fruit bodies can be characterized by high ^{137}Cs, ^{133}Cs and Rb concentrations and low calcium (Ca) and strontium (Sr) concentrations. In a laboratory experiment with the wood-inhabiting mushroom *Pleurotus ostreatus* (Fr.) Kummer Y-1 (Terada et al., 1998), ^{137}Cs uptake by mycelia decreased with increasing of ^{133}Cs, K or Rb concentration in the media, and K uptake by mycelia decreased with increasing of ^{133}Cs concentration. In an experiment with pure cultures of mycorrhizal fungi (Olsen et al., 1990) some species had preference for Cs over K and in the experiments with yeast (Conway & Duggan, 1958), K had preference over Cs and the affinity for alkali metal uptake decreased in the order $K^+ < Rb^+ < Cs^+$ followed by Na^+ and Li^+, with a relative ratio of 100:42:7:4:0.5. Fungi (mycelium and sporocarps) have a higher affinity for uptake of Rb and K to Cs, and based on the CR values for fungal sporocarps (Table 3), alkali metal can be ranked in the order $Rb^+ > K^+ > Cs^+$, with a relative ratio of 100:57:32, which is within the range of 100:88:50 derived by Yoshida & Muramatsu (1998).

The affinity for an alkali metal depends on the nutritional status of the organism, which at least partly explains differences reported between field experiments and laboratory experiments with a good nutrient supply. The mycorrhizal species *Sarcodon imbricatus* was found to be the most efficient in accumulating K, Rb and Cs, which was in agreement with results obtained by Tyler (1982), where a mean CR for Rb in litter decomposing fungus *Collybia peronata* was reported to be 41, and the mean CR for Rb in *Amanita rubescens*, which is mycorrhizal with several tree species, was above 100. However, lower ^{40}K content for mycorrhizal species is reported by Römmelt et al. (1990), which means mycorrhizal species do not necessarily accumulate alkali metals more efficiently than saprotrophic ones.

Accumulation of stable and radioactive cesium by fungi is apparently species-dependent but is affected by local environmental conditions. According to de Meijer et al. (1988), the variation in concentrations of stable and radioactive cesium in fungi of the same species is generally larger than the variation between different species and the variation in ^{137}Cs levels within the same genet of *S. varegatus* is as large as within non-genet populations of the species (Dahlberg et al., 1997), suggesting both interspecific and intrapopulation variation in the uptake of K, Rb, stable ^{133}Cs and ^{137}Cs, and that their relationships can be explained by factors other than genotype identity (Vinichuk et al., 2011). There is about two orders of magnitude variation in Cs uptake, with the highest CR value in e.g. *S. imbricatus* (256) and the lowest in *Lactarius deterrimus* (2.6), although other studies (Seeger & Schweinshaut, 1981) report the highest accumulation of stable Cs is in *Cortinarius* sp.

2. Cs (^{137}Cs and ^{133}Cs), K and Rb in *Sphagnum* plants

2.1 Introduction

Peatlands are areas where remains of plant litter have accumulated under water-logging as a result of anoxic conditions and low decomposability of the plant material. They are generally nutrient-poor habitats, particularly temperate and boreal bogs in the northern hemisphere, in which peat formation builds a dome isolating the vegetation from the surrounding groundwater. Hence, bogs are ombrotrophic, i.e. all water and nutrient supply to the vegetation is from aerial dust and precipitation, resulting in an extremely nutrient-poor ecosystem often formed and dominated by peat mosses (*Sphagnum*). *Sphagnum*-dominated peatlands with some groundwater inflow (i.e. weakly minerotrophic 'poor fens') are almost as nutrient poor and acid as true bogs. *Sphagnum* plants absorb and retain substantial amounts of fallout-derived radiocesium, and some attention has been given to the transfer of the radioactive cesium isotope ^{137}Cs within raised bogs (Bunzl & Kracke, 1989; Rosén et al., 2009), and relatively high ^{137}Cs bioavailability to bog vegetation and mosses in particular are found (Bunzl & Kracke, 1989).

The transfer of ^{137}Cs within a peatland ecosystem is different from that in forest or on agricultural land. In soils with high clay content, there is low bioavailability and low vertical migration rate of radiocesium due to binding to some clay minerals (Cornell, 1993). In nutrient-poor but organic-matter-rich forest soils, the vertical migration rate of ^{137}Cs is also low, but bioavailability is often high, particularly for mycorrhizal fungi (Olsen et al., 1990; Vinichuk & Johansson, 2003; Vinichuk et al., 2004; 2005). In forests and pastures, extensive fungal mycelium counteracts the downward transport of ^{137}Cs by an upward translocation flux (Rafferty et al., 2000); this results in a slow net downward transport of ^{137}Cs in the soil profile.

In peatlands, ^{137}Cs appears to move through advection in peat water (review by Turetsky et al., 2004). Small amounts of clay mineral in the peat reduce Cs mobility (MacKenzie et al., 1997), but most *Sphagnum* peat is virtually clay mineral free organic matter. In wet parts of open peatlands that lack fungal mycelium, the downward migration of ^{137}Cs in the *Sphagnum* layers is expected to be faster than in forest soil and Cs is continuously translocated towards the growing apex of the *Sphagnum* shoots, where it is accumulated. Some attempts have been made to investigate whether ^{137}C is associated with essential biomacromolecules in mosses and to determine the ^{137}Cs distribution among intracellular moss compartments (Dragović et al., 2004).

The chemical behavior of radiocesium is expected to be similar to that of stable ^{133}Cs and other alkali metals, i.e. K, Rb, which have similar physicochemical properties. Moreover, stable ^{133}Cs usually provides a useful analogy for observing long-term variation and transfer parameters of ^{137}Cs in a specific environment, particularly in peatlands that are cut off from an input of stable Cs from the mineral soil. As the relationship between K and Rb in fungi is not clearly understood, whether Cs follows the same pathways as K in *Sphagnum* is also unclear.

Thus, the ^{137}Cs activity concentration and mass concentration of K, Rb and ^{133}Cs was analyzed within individual *Sphagnum* plants (down to 20 cm depth) growing on a peatland in eastern central Sweden and its distribution in the uppermost capitulum and subapical segments of *Sphagnum* mosses were compared to determine the possible mechanisms involved in radiocesium uptake and retention within *Sphagnum* plants.

Additionally, the isotopic (atom) ratios of ^{137}Cs/K, ^{137}Cs/Rb and ^{137}Cs/^{133}Cs within individual *Sphagnum* plants were recorded for determining the distribution of ^{137}Cs and alkali metal, and to obtain a better understanding of the uptake mechanisms and the biological behavior of ^{137}Cs in nutrient-poor *Sphagnum* dominated ecosystem. There are few studies on the influence of alkali metals (K, Rb, ^{133}Cs) on ^{137}Cs distribution and cycling processes in peatlands.

Plant species growing on peat have varying degree capacities for influencing uptake and binding of radionuclides, but no systematic study has covered all the dominant species of *Sphagnum* peatlands their competition for radionuclides and nutrients. The important role of *Sphagnum* mosses in mineral nutrient turnover in nutrient-poor ecosystems, in particular their role in ^{137}Cs uptake and binding, necessitates a clear understanding of the mechanisms involved.

The general aim was to gain better insight into mechanisms governing the uptake of both radionuclides (^{137}Cs) and stable isotopes of alkali metals (K, Rb, ^{133}Cs) by *Sphagnum* mosses. The specific aim was to compare the distribution of ^{137}Cs, K, Rb and ^{133}Cs in the uppermost capitulum and subapical segments of *Sphagnum* mosses to be able to discuss the possible mechanisms involved in radiocesium uptake and retention within *Sphagnum* plants. Most results obtained in this study are published in collaboration with Prof. H. Rydin (Vinichuk et al., 2010a).

2.2 Study area and methods

2.2.1 Study area

The study area was a small peatland (Palsjömossen) within a coniferous forest in eastern central Sweden, about 35 km NW of Uppsala (60°03'40"N, 17°07'47"E): the peatland area sampled was open and *Sphagnum*-dominated (Figure 4). A weak minerotrophic influence was indicated by the dominance of *Sphagnum papillosum*, and the presence of *Carex rostrata*, *Carex pauciflora* and *Menyanthes trifoliata* (fen indicators in the region). The area had scattered hummocks, mostly built by *Sphagnum fuscum*, and was dominated by dwarf-shrubs such as *Andromeda polifolia*, *Calluna vulgaris*, *Empetrum nigrum* and *Vaccinium oxycoccos*. Sampling was within a 25 m^{-2} low, flat 'lawn community' (Rydin & Jeglum, 2006) totally covered by *S. papillosum*, *S. angustifolium* and *S. magellanicum* with an abundant cover of *Eriophorum vaginatum*. The water table was generally less than 15 cm below the surface: surface water was pH 3.9–4.4 (June 2009).

2.2.2 Methods

Samples of individual *Sphagnum* shoots that held together down to 20 cm were randomly collected in 2007 (May and September) and 2008 (July, August and September). Thirteen samples of *Sphagnum* plants were collected and analyzed; three in 2007 and 10 sets in 2008. Each sample consisted of approximately 20–60 individual *Sphagnum* plants (mostly *S. papillosum*, in a few cases *S. angustifolium* or *S. magellanicum*). In the laboratory, the fresh, individual, erect and tightly interwoven *Sphagnum* plants were sectioned into 1 cm (0–10) or 2 cm (10–20 cm) long segments down to 20 cm from the growing apex. The ^{137}Cs activity concentrations were measured in fresh *Sphagnum* segments. Thereafter, the samples were dried at 40°C to constant weight and analyzed for K, Rb and ^{133}Cs.

The activity concentration (Bq kg^{-1}) of ^{137}Cs in plant samples was determined by calibrated HP Ge detectors. Statistical error due to the random process of decay ranged between 5 and 10%. Plant material was measured in different geometries filled up, except a few samples that contained about 1 g of dry material. All ^{137}Cs activity concentrations were recalculated to the sampling date and expressed on a dry mass basis. The analysis of *Sphagnum* segments for K, Rb and Cs was by a combination of ICP-AES and ICP-SFMS techniques at ALS Scandinavia AB. For K concentration determination, ICP-AES was used and for ^{133}Cs and Rb, ICP-SFMS was used. The detection limits were 200 mg kg^{-1} for K, 0.04 mg kg^{-1} for ^{133}Cs and 0.008 mg kg^{-1} for Rb. The isotopic (atom) ratio of $^{137}Cs/^{133}Cs$ was calculated with Equations 1 and 2 (Chao et al., 2008). Relationships between K, Rb and ^{133}Cs concentrations in different *Sphagnum* segments were determined by Pearson correlation coefficients. All statistical analyses were with Minitab (© 2007 Minitab Inc.) software.

Fig. 4. The study area of peatland, Palsjömossen: *Sphagnum*-dominated bog.

2.3 Distribution of Cs (^{137}Cs and ^{133}Cs), K and Rb within *Sphagnum* plants

Concentration values of Cs (^{137}Cs and ^{133}Cs) and neighboring alkali counterparts K and Rb in different segments of plant provide information on differences in their uptake, distribution and relationships. The averaged ^{137}Cs activity concentrations in *Sphagnum* segments are presented in Figure 5a. Within the upper 10 cm from the capitulum, ^{137}Cs activity concentration in *Sphagnum* plants was about 3350 Bq kg^{-1}, with relatively small variations. Below 10-12 cm, the activity gradually declined with depth and in the lowest segments of *Sphagnum*, ^{137}Cs activity concentrations was about 1370 Bq kg^{-1}.

For individual samples, K concentrations ranged between 508 and 4970 mg kg^{-1} (mean 3096); Rb ranged between 2.4 and 31.4 mg kg^{-1} (mean 18.9) and ^{133}Cs ranged between 0.046 and 0.363 mg kg^{-1} (mean 0.204): averaged concentrations of K, Rb and ^{133}Cs in *Sphagnum* segments are presented in Figure 5b. Concentrations of Rb and ^{133}Cs were constant in the upper 0-10 cm segments of *Sphagnum* moss and gradually declined in the lower parts of the plant length; whereas, the concentration of K decreased with increasing depth below 5 cm. Generally, the distribution of all three alkali metals was similar to ^{137}Cs, but with a weaker increase of Rb towards the surface. The ^{137}Cs activity concentrations had the highest coefficient of variation (standard deviation divided by the mean) in *Sphagnum* (43%). The coefficients of variation were 35% for K, 35% for Rb and 37% for ^{133}Cs concentrations.

Two important features should be mentioned when discussing distributions of K, Rb, ^{133}Cs and ^{137}Cs in a *Sphagnum*-dominated peatland. Firstly, this type of peatland is extremely nutrient-poor, where only a few plant and fungal species producing small fruit bodies can grow and no mycorrhiza, except ericoid mycorrhiza, exists. Secondly, the upper part of the stratigraphy is composed of living *Sphagnum* cells that selectively absorb mineral ions from the surrounding water, and the binding of K, Rb and ^{133}Cs can be at exchange sites both outside and inside the cell.

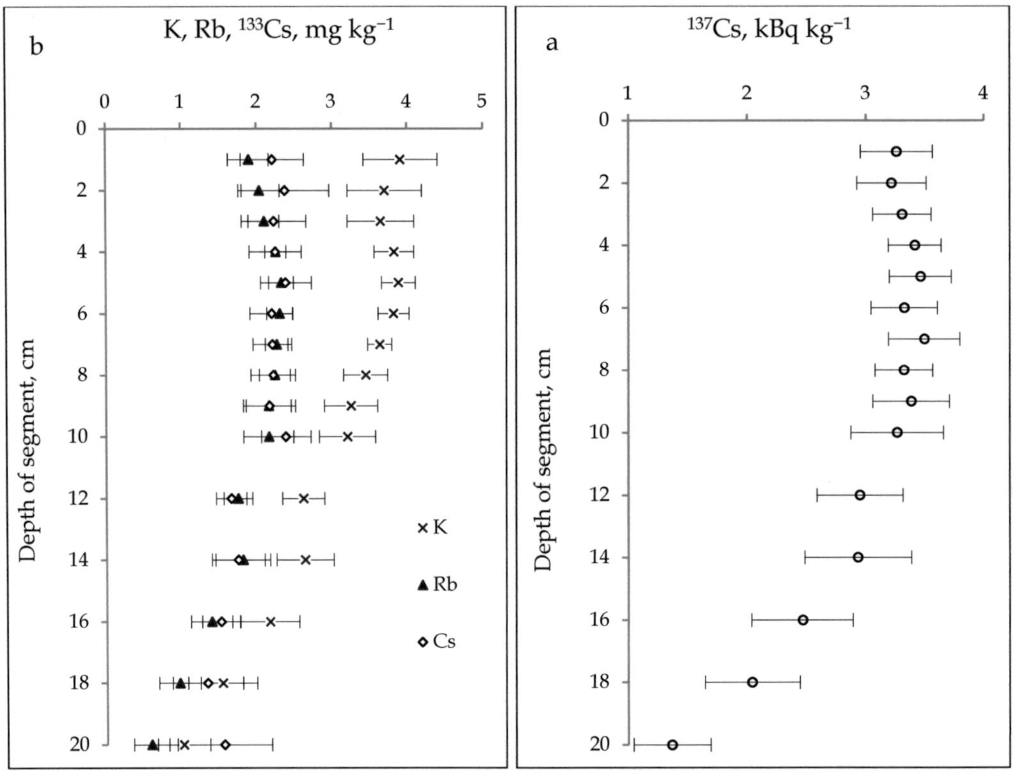

Fig. 5. ^{137}Cs and alkali metals in Sphagnum: (a) average ^{137}Cs activity concentration (kBq kg^{-1}) in *Sphagnum* segments (+/− SE, n = 13); (b) average concentrations of K (scale values should be multiplied by 10^3), Rb ($x10^1$) and ^{133}Cs ($x10^{-1}$) (mg kg^{-1}) in *Sphagnum* segments (+/− SE, n=4).

The distribution of ^{137}Cs within *Sphagnum* plants was similar to stable K, Rb and ^{133}Cs. The ^{137}Cs activity concentrations and K, Rb and ^{133}Cs concentrations were always highest in the uppermost 0-10 cm segments of *Sphagnum* (in the capitula and the subapical segments) and gradually decreased in older parts of plant. Such distribution could be interpreted as dependent on the living cells of capitula and living green segments in the upper part of *Sphagnum*. Similar patterns of K distribution within *Sphagnum* plants are reported (Hájek, 2008). ^{137}Cs is taken up and relocated by *Sphagnum* plants in similar ways to the stable alkali metals, as the ratios between K, Rb, Cs and ^{137}Cs in *Sphagnum* segments (Figure 6) were much the same down to about 16 cm, and displayed a slightly different pattern in the lower part of the plant.

2.4 Mass concentration and isotopic (atom) ratios between ^{133}Cs, K, Rb and ^{133}Cs, in segments of *Sphagnum* plants

Ratios between mass concentrations of all three alkali metals and ^{137}Cs activity concentrations, i.e. ^{133}Cs:^{137}Cs; K:^{137}Cs, Rb:^{137}Cs and ^{133}Cs:^{137}Cs, were constant through the upper part (0-16 cm) of *Sphagnum* plants (Figure 6). The ratio K/Rb was higher in the uppermost (0-2 cm) and the lowest (18-20 cm) parts of the plant (Figure 6).

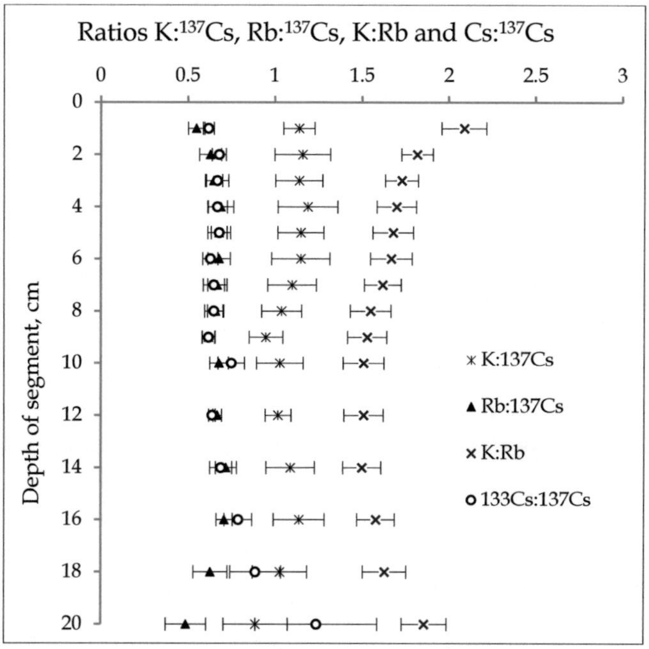

Fig. 6. Ratios between K:^{137}Cs, Rb:^{137}Cs (scale values should be multiplied by 10^{-2}), K:Rb (x10^{2}) and ^{133}Cs:^{137}Cs (x10^{-4}) in *Sphagnum* segments. Calculations based on concentrations in mg kg^{-1} for stable isotopes and Bq kg^{-1} for ^{137}Cs (+/− SE, n=13 for ^{137}Cs; n=4 for each of K, Rb and ^{133}Cs).

However, the isotopic (atom) ratios between ^{137}Cs activity concentrations and mass concentrations of alkali metals, i.e. $^{137}Cs/K$, $^{137}Cs/Rb$ and $^{137}Cs/^{133}Cs$, had distinctively different pattern of distribution through the upper part (0-20 cm) of *Sphagnum* plants (Figure 7). The $^{137}Cs/K$ ratio was relatively narrow through the upper part (0-16 cm) of *Sphagnum* plants and wider with increasing depth, whereas, the $^{137}Cs/^{133}Cs$ ratio was fairly constant through the upper part (0-12 cm) of *Sphagnum* plants and becomes narrower in the lower (14-20 cm) parts. The $^{137}Cs/Rb$ ratio was constant through the middle part (4-16 cm) of *Sphagnum* plants and somewhat narrower in the uppermost (0-4 cm) and lowest (16-20 cm) parts (Figure 7).

The distribution of the isotopic (atom) ratios between ^{137}Cs activity concentrations and mass concentrations of alkali metals K and Rb through the upper part (0-20 cm) of *Sphagnum* plants are probably conditioned by at least three processes: physical decay of ^{137}Cs atoms with time; attainment of equilibrium between stable ^{133}Cs and ^{137}Cs in the bioavailable fraction of peat soil; and, relation between cesium (^{133}Cs and ^{137}Cs), K and Rb when taken up by the *Sphagnum* plant.

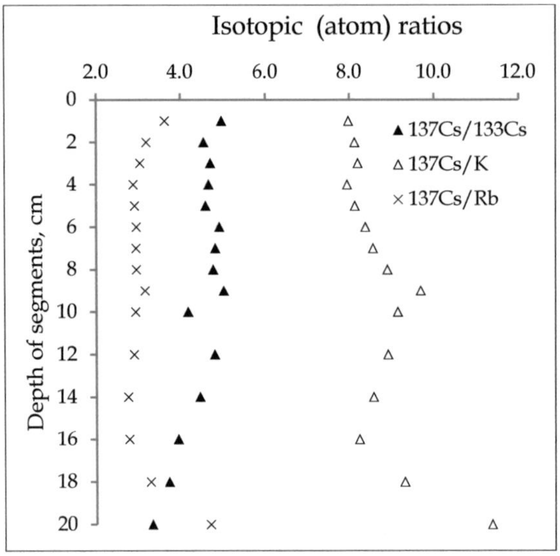

Fig. 7. Isotopic (atom) ratios ^{137}Cs/K (scale values should be multiplied by 10^{-12}), ^{137}Cs/Rb (x 10^{-09}), and ^{137}Cs/^{133}Cs (x10^{-07}) in *Sphagnum* segments. Calculations based on ^{137}Cs activity concentrations and mass concentrations of K, Rb ^{133}Cs (Eq. 2) (mean values, n=4 for each of ^{137}Cs, K, Rb and ^{133}Cs).

2.5 Relationships between ^{133}Cs, K, Rb and ^{133}Cs in segments of *Sphagnum* plants

Relationships between ^{133}Cs, K, Rb and ^{133}Cs in separate segments of *Sphagnum* plants is a tool allowing future investigate its uptake mechanism. There were close positive correlations between K, Rb and ^{133}Cs mass concentrations and ^{137}Cs activity concentrations in *Sphagnum* segments (Table 8). Correlation between ^{137}Cs activity concentrations and Rb mass concentrations (r=0.950; p=0.001) and correlation between K and Rb mass concentrations (r=0.952; p=0.001) in 10-20 cm length of *Sphagnum* plants were highest, but ^{137}Cs and K had a weaker correlation only when the upper 0-10 cm part of *Sphagnum* plants were analyzed (r=0.562; p=0.001). ^{137}Cs/^{133}Cs isotope (atom) ratios and mass concentrations of alkali metals (K, Rb and ^{133}Cs) were not or negatively correlated (Table 8).

The marked decrease in ^{137}Cs activity concentration below 14 cm (Figure 5a) raises the question as to at what depth the 1986 Chernobyl horizon was when the sampling was done. A peat core was sampled in May 2003 at Åkerlänna Römosse, an open bog about 14 km SW of Pålsjömossen, by van der Linden et al. (2008). Detailed dating by ^{14}C wiggle-matching indicated the Chernobyl horizon was then at a depth of 17 cm. Depth-age data estimated a linear annual peat increment of 1.3 cm yr^{-1} over the last decade (R^2=0.998), indicating the Chernobyl horizon would be at about 23 cm deep when the ^{137}Cs sampling was done in 2007-08. Even if there are uncertainties in applying data from different peatlands, the Chernobyl horizon should be at, or below, the lowest segments sampled. Thus, an upward migration of ^{137}Cs was obvious, but no downward migration could be tested in the study.

The relatively unchanged ^{137}Cs/K, ^{137}Cs/Rb and ^{137}Cs/^{133}Cs isotopic (atom) ratios in the upper 0-14 cm part of Sphagnum plant and the noticeable widening below 14-16 cm supported this assumption. An upward migration of ^{137}Cs has been observed in earlier studies (Rosén et al., 2009); similarly, most ^{137}Cs from the nuclear bomb tests from 1963 was retained in the top few cm of *Sphagnum* peat 20 years after, but there was also a lower peak at the level where the 1963 peat was laid down (Clymo, 1983): *Cladonia* lichens also retain high activity concentrations in the shoot apices.

	^{137}Cs	K	Rb	^{133}Cs
0-10 cm length				
K	0.562***			
Rb	0.893***	0.632***		
^{133}Cs	0.840***	0.792***	0.802***	
^{137}Cs/^{133}Cs	–	−0.262	0.270	−0.157
10-20 cm length				
K	0.856***			
Rb	0.950***	0.952***		
^{133}Cs	0.645***	0.651***	0.664***	
^{137}Cs/^{133}Cs	–	0.122	0.219	−0.401

Table 8. Correlation coefficients between concentrations of potassium, rubidium and cesium (^{133}Cs and ^{137}Cs) in *Sphagnum* segments (*** p=0.001).

2.6 Mechanisms of ^{137}Cs and alkali metal uptake by *Sphagnum* plants

Presumably, ^{137}Cs is bound within capitula, living green segments and dead brown segments of *Sphagnum* plants. According to Gstoettner and Fisher (1997), the uptake of some metals (Cd, Cr, and Zn) in *Sphagnum papillosum* is a passive process as they living and dead moss accumulate metal equally. For a wide range of bryophytes, Dragović et al. (2004) found ^{137}Cs was primarily bound by cation exchange, with only a few percent occurring in biomolecules. *Sphagnum* mosses have remarkably high cation exchange capacity (Clymo, 1963), and according to Russell (1988), a high surface activity of *Sphagnum* is related to its high cation exchange capacity, which ranges between 90-140 meq/100 g. In a water saturated peat moss layer, water washes (1 L de-ionised water added to a column of about 1.4 L volume) removed about 60% of K from *Sphagnum* (Porter B. Orr, 1975), indicating this element was held on cation exchange sites. In turn, the desiccation of living moss usually causes cation leakage from cell cytoplasm, during which most of the effused K^+ is retained on the exchange sites and reutilized during recovery after rewetting (Brown & Brümelis, 1996; Bates, 1997).
However, this is not necessarily the case for ^{137}Cs, as ^{137}Cs has a weaker correlation with K, especially in the uppermost parts of the plant, which means ^{137}Cs uptake might be somewhat different from that of K. Even within the same segments of the plant, ^{137}Cs activity concentrations has higher variation than K concentration. An even stronger decoupling between ^{137}Cs and K is observed in the forest moss *Pleurozium schreberi*, in which ^{137}Cs is retained to a higher degree in senescent parts (Mattsson & Lidén, 1975). However, close correlations, were found between Rb and ^{137}Cs, which suggests similarities in their

uptake and relocation: these observations complied with results reported for fungi (Vinichuk et al., 2010b; 2011).

Some lower parts of *Sphagnum* plants are still alive and able to create new shoots (Högström, 1997), however, although still connected to the capitulum, much of lower stem is dead. Thus, the decrease of ^{137}Cs activity concentration in plant segments below 10 cm indicates a release of the radionuclide from the dying lower part of *Sphagnum* and internal translocation to the capitulum.

The mechanism of radiocesium and alkali metal relocation within *Sphagnum* is probably the same active translocation as described for metabolites by Rydin & Clymo (1989). Although external buoyancy-driven transport (Rappoldt et al., 2003) could redistribute ^{137}Cs, field evidence suggests buoyancy creates a downward migration of K (Adema et al., 2006); thus, this mechanism appears unlikely. Likewise, a passive downwash and upwash (Clymo & Mackay, 1987) cannot explain accumulation towards the surface.

3. Conclusions from the Swedish studies

The concentrations of the three stable alkali elements K, Rb and ^{133}Cs and the activity concentration of ^{137}Cs were determined in various components of Swedish forests – bulk soil, rhizosphere, soil–root interface fraction, fungal mycelium and fungal sporocarps. The soil–root interface fraction was distinctly enriched with K and Rb, compared with bulk soil. Potassium concentration increased in the order bulk soil < rhizosphere < fungal mycelium < soil–root interface < fungal sporocarps, whereas, Rb concentration increased in the order bulk soil < rhizosphere < soil–root interface < fungal mycelium < fungal sporocarps.

Cesium was generally evenly distributed within bulk soil, rhizosphere and soil–root interface fractions, indicating no ^{133}Cs enrichment in these forest compartments.

The uptake of K, Rb and ^{133}Cs during the entire transfer process between soil and sporocarps occurred against a concentration gradient. For all three alkali metals, the levels of K, Rb and ^{133}Cs were at least one order of magnitude higher in sporocarps than in fungal mycelium.

Potassium uptake appeared to be regulated by fungal nutritional demands for this element and fungi had a higher preference for uptake of Rb and K than for Cs. According to their efficiency of uptake by fungi, the three elements may be ranked in the order $Rb^+ > K^+ > Cs^+$, with a relative ratio 100:57:32. Although the mechanism of Cs uptake by fungi could be similar to that of Rb, uptake mechanism for K appeared to be different. The variability in isotopic (atom) ratios of $^{137}Cs/K$, $^{137}Cs/Rb$ and $^{137}Cs/^{133}Cs$ in the fungal sporocarps suggested they were independent on specific species of fungi. The relationships observed between concentration ratios $^{137}Cs/^{133}Cs$ and K, Rb and ^{133}Cs in fungal sporocarps also varied widely and were inconsistent. The concentration of K, Rb and ^{133}Cs in sporocarps appeared independent of the $^{137}Cs/^{133}Cs$ isotopic ratio.

The study of *S. variegatus* sporocarps sampled within 1 km^2 forest area with high ^{137}Cs fallout from the Chernobyl accident confirmed ^{133}Cs and ^{137}Cs uptake is not correlated with uptake of K; whereas, the uptake of Rb is closely related to the uptake of ^{133}Cs. Furthermore, the variability in ^{137}Cs and alkali metals (K, Rb and ^{133}Cs) among genotypes in local populations of *S. variegatus* is high and the variation appears to be in the same range as found in species collected at different localities. The variations in concentrations of K, Rb and ^{133}Cs and ^{137}Cs activity concentration in sporocarps of *S. variegatus* appear to be influenced more by local environmental factors than by genetic differences among fungal genotypes.

For *Sphagnum* the distribution of ^{137}Cs can be driven by several processes: cation exchange is important and gives similar patterns for monovalent cations; uptake/retention in living cells; and downwash and upwash by water outside the plants. However, the most important mechanism is internal translocation to active tissue and the apex, which can explain the accumulation in the top layer of the mosses.

4. Acknowledgements

The authors gratefully acknowledge the Swedish University of Agricultural Sciences (SLU), Sweden, for supporting the project. We would like to express our thanks to Dr. I. Nikolova for her assistance with the experiments and to the staff of the Analytica Laboratory, Luleå, Sweden, for ICP-AES *and* ICP-SFMS analyses. The project was financially supported by SKB (Swedish Nuclear Fuel and Waste Management Co).

5. References

Adema, E.; Baaijens, G.; van Belle, J.; Rappoldt, A.; Grootjans, A. & Smolders, A. (2006). Field evidence for buoyancy-driven water flow in a *Sphagnum* dominated peat bog. *Journal of Hydrology,* Vol.327, pp. 226- 234, ISSN 0022-1694

Baeza, A.; Hernández, S.; Guillén, F.; Moreno, J.; Manjón, J.L. & Pascual, R. (2004). Radiocaesium and natural gamma emitters in mushrooms collected in Spain. *Science of the Total Environment,* Vol.318, No.1-3, pp. 59-71, ISSN 0048-9697

Baeza, A.; Guillén, J.; Hernández, S.; Salas, A.; Bernedo, M.; Manjón, J. & Moreno, G. (2005). Influence of the nutritional mechanism of fungi (mycorrhize/saprophyte) on the uptake of radionuclides by mycelium. *Radiochimica Acta,* Vol.93, No.4, pp. 233-238, ISSN 0033-8230

Ban-nai T.; Yoshida S.; Muramatsu Y. & Suzuki A. (2005). Uptake of Radiocesium by Hypha of Basidiomycetes - Radiotracer Experiments. *Journal of Nuclear and Radiochemical Sciences,* Vol. 6, No.1, pp. 111-113, ISSN 1345-4749

Bates, J. (1997). Effects of intermittent desiccation on nutrient economy and growth of two ecologically contrasted mosses. *Annals of Botany,* Vol.79, pp.299–309, ISSN 0305-7364

Brown, D. & Brümelis, G. (1996). A biomonitoring method using the cellular distribution of metals in moss. *Science of the Total Environment,* Vol.187, pp. 153–161, ISSN 0048-9697

Brown, G. & Cummings, S. (2001). Potassium uptake and retention by *Oceanomonas baumannii* at low water activity in the presence of phenol. *FEMS Microbiology Letters,* Vol.205, No.1, pp. 37–41, ISSN 0378-1097

Brunner, I.; Frey, B. & Riesen, T. (1996). Influence of ectomycorrhization and cesium/potassium ratio on uptake and localization of cesium in Norway spruce seedlings. *Tree Physiology,* Vol.16, pp. 705-711, ISSN 0829-318X

Bunzl, K. & Kracke, W. (1989). Seasonal variation of soil-to-plant transfer of K and fallout ^{134,137}Cs in peatland vegetation. *Health Physics,* Vol.57, pp. 593-600, ISSN 0017-9078

Bystrzejewska-Piotrowska, G. & Bazala, M. (2008). A study of mechanisms responsible for incorporation of cesium and radiocaesium into fruitbodies of king oyster mushroom (*Pleurotus eryngii*). *Journal of Environmental Radioactivity,* Vol.99, pp. 1185-1191, ISSN 0265-931X

Chao J.; Chiu, C. & Lee, H. (2008). Distribution and uptake of ^{137}Cs in relation to alkali metals in a perhumid montane forest ecosystem. *Applied Radiation and Isotopes,* Vol.66, pp. 1287-1294, ISSN 0969-8043

Cornell, R. (1993). Adsorption of cesium on minerals: A review. *Journal of Radioanalytical and Nuclear Chemistry,* Vol.171, pp. 483-500, ISSN 0236-5731

Clymo, R. (1963). Ion exchange in Sphagnum and its relation to bog ecology. *Annals of Botany,* (Lond.) Vol.27, pp. 309-324, ISSN 0305-7364

Clymo, R. (1983). Peat. In: Gore, A.J.P. (Ed.), Ecosystems of the world. 4A. Mires: swamp, bog, fen and moor. General studies, 159-224, ISBN 0-444-42003-7, Elsevier, Amsterdam.

Clymo, R.; & Mackay, D. (1987). Upwash and downwash of pollen and spores in the unsaturated surface layer of *Sphagnum*-dominated peat. *New Phytologist,* Vol.105, pp. 175-183, ISSN 0028-646X

Conway, E.; & Duggan, F. (1958). A cation carrier in the yeast cell wall. *Biochemistry Journal,* Vol.69, pp. 265–274, ISSN 0006-2960

Dahlberg, A.; Nikolova, I.; Johanson, K. (1997). Intraspecific variation in ^{137}Cs activity concentration in sporocarps of *Suillus variegatus* in seven Swedish populations. *Mycological Research,* Vol.101, pp.545-551, ISSN 0953-7562

de Meijer, R.; Aldenkamp, F. & Jansen, A. (1988). Resorption of cesium radionuclides by various fungi. *Oecologia,* Vol.77, pp. 268-272, ISSN 0029-8549

Dighton, J.; Clint, G. & Poskitt J. (1991). Uptake and accumulation of ^{137}Cs by upland grassland soil fungi: a potential pool of Cs immobilisation. *Mycological Research,* Vol.95, No.9, pp. 1052–1056, ISSN 0953-7562

Dragović, S.; Nedić, O.; Stanković, S. & Bačić, G. (2004). Radiocaesium accumulation in mosses from highlands of Serbia and Montenegro: chemical and physiological aspects. *Journal of Environmental Radioactivity,* Vol.77, pp. 381-388, ISSN 0265-931X

Drobner, U. & Tyler G. (1998). Conditions controlling relative uptake of potassium and rubidium by plants from soils. *Plant and Soil,* Vol.201, pp. 285–293, ISSN 0032-079X

Enghag, P. (2000). *Jordens grundämnen och deras upptäckt.* ISBN 9789175485904, Industrrilitteratur. Stockholm.

Gaso, M.; Segovia, N.; Morton, O.; Cervantes, M.; Godinez, L.; Peña, P. & Acosta, E. (2000). ^{137}Cs and relationships with major and trace elements in edible mushrooms from Mexico. *Science of the Total Environment,* Vol.262, No.1–2, pp. 73–89, ISSN 0048-9697

Gillett, A. & Crout, N. (2000). A review of ^{137}Cs transfer to fungi and consequences for modeling environmental transfer. *Journal of Environmental Radioactivity,* Vol.48, pp. 95-121, ISSN 0265-931X

Gorban, G. & Clegg, S. (1996). A conceptual model for nutrient availability in the mineral soil-root system. *Canadian Journal of Soil Science,* Vol.76, pp. 125–131, ISSN 1918-1841

Gstoettner, E. & Fisher, N. (1997). Accumulation of cadmium, chromium, and zinc by the moss *Sphagnum papillosum* Lindle. *Water, Air, Soil Pollution,* Vol.93, pp. 321-330, ISSN 0049-6979

Guillitte, O.; Melin, J. & Wallberg, L. (1994). Biological pathways of radionuclides originating from the Chernobyl fallout in a boreal forest ecosystem. *Science of the Total Environment,* Vol.157, pp. 207–215, ISSN 0048-9697

Gyuricza, V.; Dupré de Boulois, H. & Declerck, S. (2010). Effect of potassium and phosphorus on the transport of radiocaesium by arbuscular mycorrhizal fungi. *Journal of Environmental Radioactivity*, Vol.101, pp. 482–487, ISSN 0265-931X

Hájek, T. (2008). Ecophysiological adaptations of coexisting *Sphagnum* mosses. PhD. thesis. University of South Bohemia, Faculty of Science, Czech Republic, 98 pp.

Högström, S. (1997). Habitats and increase of Sphagnum in the Baltic Sea island Gotland, Sweden. *Lindbergia*, Vol.22, pp. 69-74, ISSN 0105-0761

Horyna, J. & Řanad, Z. (1988). Uptake of radiocaesium and alkali metals by mushrooms. *Journal of Radioanalytical and Nuclear Chemistry*, Vol.127, pp. 107–120, ISSN 0236-5731

Ismail, S. (1994). Distribution of Na, K, Rb, Cs and ^{137}Cs in some Austrian higher fungi. *Biology of Trace Element Research*, Vol.43-45, No.1, pp. 707–714, ISSN 01634984

Johanson, K.J. &, Bergström, R. (1994). Radiocaesium transfer to man from moose and roe deer in Sweden. *Science of the Total Environment*, Vol.157, pp. 309-316, ISSN 0048-9697

Kalač P. (2001). A review of edible mushroom radioactivity. *Food Chemistry*, Vol.75, pp. 29–35, ISSN 0308-8146

Kammerer, L.; Hiersche, L. & Wirth, E. (1994). Uptake of radiocaesium by different species of mushrooms. *Journal of Environmental Radioactivity*, Vol.23, pp. 135-150, ISSN 0265-931X

Karadeniz, Ö. & Yaprak, G. (2007). Dynamic equilibrium of radiocesium with stable cesium within the soil–mushroom system in Turkish pine forest. *Environmental Pollution*, Vol.148, No.1, pp. 316-324, ISSN 0269-7491

Kuwahara, C.; Watanuki, T.; Matsushita, K.; Nishina, M. & Sugiyama, H. (1998). Studies on uptake of cesium by mycelium of the mushroom *(Pleurotus ostreatus)* by ^{133}Cs-NMR. *Journal of Radioanalytical and Nuclear Chemistry*, Vol.235, No.1-2, pp. 191-194, ISSN 0236-5731

Leake, J. & Read, D. (1997). Mycorrhizal fungi in terrestrial habitats. Chapter 18. In: *The Mycota, Vol.IV: Environmental and Microbial Relationship*, D.T. Wicklow & B. Söderström, (Ed.), 281-301, ISBN 978-3540-71839-0, Springer Verlag, Berlin

MacKenzie, A.; Farmer, J. & Sugden, C. (1997). Isotopic evidence of the relative retention and mobility of lead and radiocaesium in Scottish ombrotrophic peats. *Science of the Total Environment*, Vol.203, pp. 115–127, ISSN 0048-9697

Marschner, H. (1995). *Mineral nutrition of higher plants*. Academic press, 2nd ed, ISBN 0-12-473543-6, London, UK

Mascanzoni, D. (2009). Long-term transfer of ^{137}Cs from soil to mushrooms in a semi-natural environment. *Journal of Radioanalytical and Nuclear Chemistry*, Vol.282, pp. 427-431, ISSN 0236-5731

Mattsson, S. & Lidén, K. (1975). ^{137}Cs in carpets of the forest moss Pleurozium schreberi, 1961-1973. *Oikos*, Vol.26, pp. 323-327, ISSN 0030-1299

McGee, E.; Synnott, H.; Johanson, K.; Fawaris, B.; Nielsen, S. Horrill, A.D., et al., (2000). Chernobyl fallout in a Swedish spruce forest ecosystem. *Journal of Environmental Radioactivity*, Vol.48, pp. 59-78, ISSN 0265-931X.

Myttenaere, C.; Schell, W.; Thiry, Y.; Sombre, L.; Ronneau, C.; van der Stegen de Schriek, J. (1993). Modeling of Cs-137 cycling in forests: recent developments and research needed. *Science of the Total Environment*, Vol.136, pp. 77–91, ISSN 0048-9697

Nikolova, I.; Johanson, K. & Dahlberg, A. (1997). Radiocaesium in fruitbodies and mycorrhizae in ectomycorrhizal fungi. *Journal of Environmental Radioactivity*, Vol.37, pp. 115-125, ISSN 0265-931X

Nyholm, N. & Tyler G. (2000). Rubidium content of plant, fungi and animals closely reflects potassium and acidity conditions of forest soils. *Forest Ecology and Management*, Vol.134, pp. 89–96, ISSN 0378-1127

Olsen, R.; Joner, E. & Bakken. L. (1990). Soil fungi and the fate of radiocaesium in the soil ecosystem - a discussion of possible mechanisms involved in the radiocaesium accumulation in fungi, and the role of fungi as a Cs-sink in the soil. In: G. Desmet, P. Nassimbeni, & M. Belli (Eds.), *Transfer of radionuclides in natural and semi-natural environment*, pp. 657–663, ISBN 9781851665396, Luxemburg: Elsevier Applied Science

Parekh, N.; Poskitt, J.; Dodd, B.; Potter, E. & Sanchez, A. (2008). Soil microorganisms determine the sorption of radionuclides within organic soil systems. *Journal of Environmental Radioactivity*, Vol.99, pp. 841-852, ISSN 0265-931X

Porter B. Orr. 1975. Available from http://scholar.lib.vt.edu/ejournals/JARS/v30n3/v30n3-orr.htm

Rafferty, B.; Dawson, D, & Kliashtorin, A. (1997). Decomposition in two pine forests: the mobilisation of ^{137}Cs and K from forest litter. *Soil Biology and Biochemistry*, Vol.29, No.11/12, pp. 1673–1681, ISSN 0038-0717

Rafferty, B.; Brennan, M.; Dawson, D. & Dowding, D. (2000). Mechanisms of ^{137}Cs migration in coniferous forest soils. *Journal of Environmental Radioactivity*, Vol.48, pp. 131-143, ISSN 0265-931X

Rappoldt, C.; Pieters, G.; Adema, E.; Baaijens, G.; Grootjans, A. & van Duijn, C. (2003). Buoyancy-driven flow in peat bogs as a mechanism for nutrient recycling. *Proceedings of the National Academy of Sciences of the United States of America*, Vol.100, pp. 14937-14942, ISSN 1091-6490

Read, D.; Perez-Moreno, J. (2003). Mycorrhizas and nutrient cycling in ecosystems – a journey towards relevance? *New Phytologist*, Vol.157, pp. 475–492, ISSN 0028-646X

Rodríguez-Navarro, A. (2000). Potassium transport in fungi and plants. *BBA - Biochimica et Biophysica Acta*, Vol.1469, pp. 1–30, ISSN 1388-1981

Römmelt, R.; Hiersche, L.; Schaller, G. & Wirth, E. (1990). Influence of soil fungi (basidiomycetes) on the migration of ^{134}Cs+^{137}Cs and ^{90}Sr in coniferous forest soils. In: G. Desmet, P. Nassimbeni and M. Belli, Editors, *Transfer of radionuclides in natural and seminatural environments*, pp. 152–160, ISBN 9781851665396, Elsevier, London

Rosén, K.; Vinichuk, M. & Johanson, K. (2009). ^{137}Cs in a raised bog in central Sweden. Journal of Environmental Radioactivity, Vol.100, pp. 534–539, ISSN 0265-931X

Rosén, K.; Vinichuk, M.; Nikolova, I. & Johanson, K. (2011). Long-term effects of single potassium fertilization on ^{137}Cs levels in plants and fungi in a boreal forest ecosystem. *Journal of Environmental Radioactivity*, Vol.102, pp. 178-184, ISSN 0265-931X

Rühm, W.; Kammerer, L.; Hiersche, L. & Wirth, E. (1997). The ^{137}Cs/^{134}Csratio in fungi as an indicator of the major mycelium location in forest soil. *Journal of Environmental Radioactivity*, Vol.35, pp. 129–148, ISSN 0265-931X

Russell, E. (1988). *Soil conditions and plants growth,* Eleventh Edition, Longmans, ISBN 0582446775, London, England

Rydin, H. & Clymo, R. (1989). Transport of carbon and phosphorus about Sphagnum. *Proceedings of Royal Society,* Biol. Sci., Vol.237, pp. 63-84, ISSN 1471-2954, London

Rydin, H. & Jeglum, J. (2006). *The biology of peatlands,* ISBN 978-0-19-852871-5, Oxford University Press

Seeger, R. & Schweinshaut, P. (1981). Vorkommen von caesium in höheren pilzen. *Science of the Total Environment,* Vol.19, pp. 253–276, ISSN 0048-9697

Skuterud, L.; Travnicova, I.; Balonov, M.; Strand, P. & Howard, B. (1997). Contribution of fungi to radiocaesium intake by rural populations in Russia. *Science of the Total Environment,* Vol.193, No.3, pp. 237–242, ISSN 0048-9697

Smith, S. & Read, D. (1997). *Mycorrhizal symbiosis* (2nd ed.), ISBN 0-12-652840-3, London: Academic Press

Steiner, M.; Linkov, I. & Yoshida, S. (2002). The role of fungi in the transfer and cycling of radionuclides in forest ecosystems. *Journal of Environmental Radioactivity,* Vol.58, pp. 217–241, ISSN 0265-931X

Tanesaka, E.; Masuda, H. & Kinugawa K. (1993). Wood degrading ability of basidiomycetes that are wood decomposers, litter decomposers, or mycorrhizal symbionts. *Mycologia,* Vol.85, pp. 347–354, ISSN 0027-5514

Terada, H.; Shibata, H.; Kato, F. & Sugiyama, H. (1998). Influence of alkali elements on the accumulation of radiocesium by mushrooms. *Journal of Radioanalytical and Nuclear Chemistry,* Vol.235, No.1-2, pp. 195–200, ISSN 0236-5731

Tsukada, H.; Takeda, A.; Hisamatsu, S. & Inaba, J. Inequilibrium between Fallout ^{137}Cs and Stable Cs in Cultivated Soils. Radionuclides in Soils and Sediments, and their Transfer to Biota, In: *Proceedings of 18th World Congress of Soil Science,* 15 July 2006 Available from http://crops.confex.com/crops/wc2006/techprogram/P17531.HTM

Turetsky, M.; Manning, S. & Wieder, R. (2004). Dating recent peat deposits. *Wetlands,* Vol.24, pp. 324-356, ISSN 0277-5212

Tyler, G. (1982). Accumulation and exclusion of metals in *Collybia peronata* and *Amaniuta rubescens. Transactions of British Mycological Society,* Vol.79, pp. 239–241, ISSN 0007-1536

van der Linden, M.; Vickery, E.; Charman, D. & van Geel, B. (2008). Effects of human impact and climate change during the last 350 years recorded in a Swedish raised bog deposit. *Palaeogeography, Palaeoclimatology, Palaeoecology,* Vol.262, pp. 1-31, ISSN 0031-0182

Vinichuk, M. & Johanson, K. (2003). Accumulation of ^{137}Cs by fungal mycelium in forest ecosystems of Ukraine. *Journal of Environmental Radioactivity,* Vol.64, pp. 27-43, ISSN 0265-931X

Vinichuk, M.; Johanson, K. & Taylor, A. (2004). ^{137}Cs in the fungal compartment of Swedish forest soils. *Science of the Total Environment,* Vol.323, pp. 243–251, ISSN 0048-9697

Vinichuk, M.; Johanson, K.; Rosén, K. & Nilsson, I. (2005). Role of fungal mycelium in the retention of radiocaesium in forest soils. *Journal of Environmental Radioactivity,* Vol.78, pp. 77–92, ISSN 0265-931X

Vinichuk, M.; Johanson, K.; Rydin, H. & Rosén, K. (2010a). The distribution of ^{137}Cs, K, Rb and Cs in plants in a Sphagnum-dominated peatland in eastern central Sweden. *Journal of Environmental Radioactivity*, Vol.101, pp. 170–176, ISSN 0265-931X

Vinichuk, M.; Taylor, A.; Rosén, K. & Johanson, K. (2010b). Accumulation of potassium, rubidium and cesium (^{133}Cs and ^{137}Cs) in various fractions of soil and fungi in a Swedish forest. *Science of the Total Environment*, Vol.408, pp. 2543–2548, ISSN 0048-9697

Vinichuk, M.; Rosén, K.; Johanson, K. & Dahlberg, A. (2011). Correlations between potassium, rubidium and cesium (^{133}Cs and ^{137}Cs) in sporocarps of *Suillus variegatus* in a Swedish boreal forest. *Journal of Environmental Radioactivity,* Vol.102, No.4, pp. 386-392, ISSN 0265-931X

Wallace, A. (1970). Monovalent–ion carrier effects on transport of ^{86}Rb and ^{137}Cs into bush bean plants. *Plant and Soil*, Vol.32, pp. 526–520, ISSN 0032-079X

White, P. & Broadley, M. (2000). Mechanisms of caesium uptake by plants. *New Phytologist*, Vol.147, pp. 241-256, ISSN 0028-646X

Yoshida, S. & Muramatsu, Y. (1998). Concentration of alkali and alkaline earth elements in mushrooms and plants collected in a Japanese pine forest, and their relationship with ^{137}Cs. *Journal of Environmental Radioactivity*, Vol.41, No.2, pp. 183–205, ISSN 0265-931X

15

Body Composition Analyzer Based on PGNAA Method

Hamed Panjeh and Reza Izadi-Najafabadi
Ferdowsi University of Mashhad, Faculty of Science, Mashhad, Iran

1. Introduction

Determination of the elemental compositions of a human body is a useful tool for understanding general physiology relationships, diagnosing some disease and cancers. Measurements of body composition yield data about normal growth, maturity and the process of ageing.

Practically, these measurements provide standards against which departures from normality may be judged. It is necessary to define differences between genetic groups, the sexes within each group, the systematic variations with age and body size and the distribution of the seemingly random differences between individuals that remain unexplained. Knowledge of the range of normality is of value in studying trends in disease processes and monitoring the response to treatment. Body composition data may influence the choice of the most appropriate treatment of wasting illness, sepsis, trauma, renal failure and nutritional disorders. So many experimental methods employed in the measurement of the composition of the human body over the past 50 years and in the consequence a lot of techniques have been applied to determine the weight percentage of body chemical compositions.

Early methods such as hydrodensitometry and skinfold anthropometry have been superseded by dual-energy x-ray absorptiometry and bioelectrical impedance spectroscopy. Also x-ray fluorescence can give important information of clinical significance. The relatively simple, rapid and risk-free electrical methods such as multifrequency bioelectrical impedance analysis, which can be employed at the bedside, have been found to be more complicated in their interpretation. Electromagnetic methods may only measure the composition of the human body at its surface. X-ray computed tomography and magnetic resonance imaging have not yet been employed much in body composition measurements.

One of the non-destructive and the most sensitive approaches is Prompt Gamma Neutron Activation Analysis (PGNAA) method (Miri & Panjeh, 2007, Chichester, 2004, Metwally, 2004) but neutron activation facilities in practice remain available in only a few centers worldwide.

In this method the sample is excited with neutrons. When an atom in the sample captures a neutron, that atom is transformed to another nuclear state of the same element. The new atom can be radioactive. If it decays with a short half life the radioactive signal can be measured by special detectors simultaneously. So the active sample which is composed of some elements promptly releases several prompt gamma rays with various intensions and energies. The gamma rays are produced immediately and stop appearing as soon as the neutron source is removed. Ordinarily, the energy spectrum of these gamma rays is the characteristic sign of the special constituting elements.

2. Whole body counting and neutron activation analysis

In this method one of the practical measurements is total body measurement of nitrogen (TBN). A total body measurement of nitrogen provides a quantitative estimate of muscle mass and may prove of value in the assessment of patients with diseases associated with muscle wasting, for example, in malabsorption syndrome. Also the technique of in vivo neutron activation analysis has proved successful for measurement of total body calcium (TBC) (McNeill, **1973).**
Several body elements may be measured by the following prompt gamma reactions resulting from the capture of thermal neutrons. The most abundant or useful gamma energies measureable detectors such as NaI(Tl) scintillation are:

$$\begin{array}{lll} {}^{1}\mathrm{H}(\mathrm{n},\gamma)\,{}^{2}\mathrm{H} & E = 2.223\ \mathrm{MeV} & \\ {}^{14}\mathrm{N}(\mathrm{n},\gamma)\,{}^{15}\mathrm{N} & E = 10.8\ \mathrm{MeV},\ \text{other energies} & \\ {}^{35}\mathrm{Cl}(\mathrm{n},\gamma)\,{}^{36}\mathrm{Cl} & E = 6.11, 8.57\ \mathrm{MeV},\ \text{many other energies.} & \end{array}$$

$$\begin{array}{lll} {}^{16}\mathrm{O}(\mathrm{n},\mathrm{n}'\gamma)\,{}^{16}\mathrm{O} & E = 6.134\ \mathrm{MeV} & E_T = 6.6\ \mathrm{MeV} \\ {}^{12}\mathrm{C}(\mathrm{n},\mathrm{n}'\gamma)\,{}^{12}\mathrm{C} & E = 4.439\ \mathrm{MeV} & E_T = 4.9\ \mathrm{MeV}. \end{array}$$

Figure 1 shows the prompt-gamma-emission spectra from bilateral irradiation of a normal male volunteer from the shoulder to the knee (Ryde *et al.* 1989). Regions of interest for hydrogen, carbon, chlorine and nitrogen are indicated.
The abundance of gamma ray emissions from other elements necessitates the use of a high resolution semiconductor detector to determine them. The most prominent feature of the prompt-gamma emission spectrum from neutron irradiation of the human body is the full-energy peak at 2.223 MeV from hydrogen, since these nuclei are most abundant. The gamma ray emission from nitrogen at 10.8 MeV is the highest-energy emission from any body element. Consequently it is free from any interference except background noise due to random summing of lower-energy gamma rays (at high counting rates) and neutron irradiation of the detectors. This has become the standard method for the determination of TBN, and therefore total body protein (TBPr), since nitrogen comprises 16% of protein (Mernagh *et al* 1977, Beddoe *et al* 1984, Ryde *et al* 1989, Baur *et al* 1991).
Nitrogen is a direct indicator of total body protein. A high nitrogen reading indicates healthy tissue. When this measurement is combined with measurements from the total body potassium, scientists can determine total organ and muscle mass. For online information refer o the following address (http://www.bcm.edu/bodycomplab/ivnamainpage.htm).

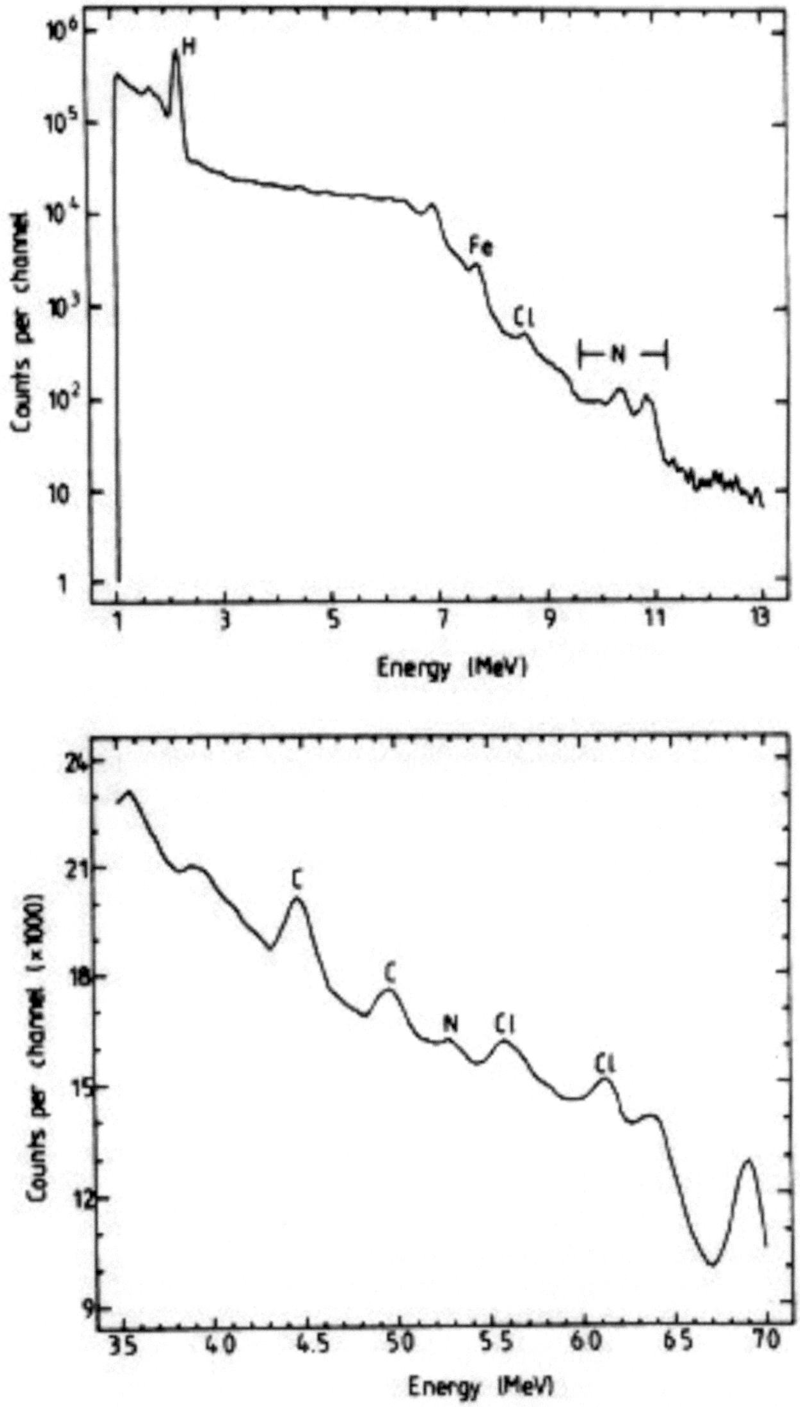

Fig. 1. The prompt gamma spectrum from bilateral irradiation of a male with a 252 Cf neutron source

3. Detectors

NaI(Tl) scintillation detectors are more suitable for this measurement than semiconductor detectors because of their greater stopping power. The most precise measurement of TBN reported is 1.6% for a neutron dose of 0.45 mSv (Ryde *et al* 1989) using a 4 GBq 252Cf fission source. A commonly employed technique in the measurement of body nitrogen is to measure the ratio of the emissions from nitrogen and hydrogen. This ratio is much less sensitive to variations in body size, neutron fluence and detector characteristics, which affect the signal from each element alone. It also permits the determination of TBN from partial-body irradiation (of the torso and thighs, thereby minimizing the radiation dose to radiosensitive tissues such as the eyes) assuming that hydrogen comprises one-tenth of body weight (Vartsky *et al* 1979). This requires correction since the proportion of body weight due to hydrogen has been estimated to vary from 9.5 to 10.8% in a large population of patients.

Chlorine may be determined from its emission at 8.57 MeV after deduction of the underlying background noise due to random summing and scattered gamma rays from nitrogen (Mitra et al. 1993). It may also be determined from its prominent emission at 6.11 MeV, but a high-resolution semiconductor detector (Ge(Li) or hyper pure Ge) must be employed to distinguish this emission from the emission from oxygen at 6.134 MeV.

4. Simulation and advantages

In according to the latest recommendations of international institutes of radioprotection, an increasing attention must be paid to the patient protection during cancer radiotherapy. Therefore one of the primary attempts should be protection of the patient from hazardous radiation and minimizing un-useful doses. Designing a Body Chemical Composition Analyzer (BCCA) in order to use for cancer therapy while having the lowest gamma and neutron dose equivalent rate in the soft tissue is desirable. The Design of the BCCA need to be modeled by Monte Carlo N-particle general code (MCNP) (Briesmeister, 2000) before the construction. By this way we can assess all the geometry and material's effects and other parameters affect the dose received by the patient and the personnel. Also if we have an improving idea we can investigate its subsequent role in simulation design before the real structure.

5. Sources

In applying this technique many kinds of neutron sources have been used and suggested. The compact and portable neutron sources such as ^{252}Cf and ^{241}Am-Be are commonly used in the PGNAA method because of their high flux and reliable neutron spectrum. Also Anderson et al. (1964) and then Cohn et al. (1972) suggested using the fast neutron reaction 14N(n, 2n)I3N. This proved unsuitable because of interferences from other reactions and because of problems in maintaining a uniform fast (>11.3 MeV) neutron flux. The Birmingham group (Harvey et al. 1973), however, have shown that a suitable nitrogen measurement can be made by using the thermal neutron capture gamma rays from the reaction 14N(n,y)15N*. 15N is stable but in this reaction is formed in the excited state, 15N*; 15% of the time de-excitation results in the release of a 10.83 MeV gamma ray.

In another works we see that viable signal/background ratio can be obtained using Pu-Be neutron sources and heavy shielding of both sources and detector. (Mernagh et al. 1977)

6. Absorbed dose quantities and attentions

Absorbed dose, D, is the energy imparted by ionizing radiation to matter per unit mass at a point given in units of J kg^{-1} (commonly called the Gray, Gy) (Alpen, 1998).

$$D = \frac{dE}{dM}$$

The effective dose, *E*, which is a summation of differing risks to organs in the human body in units of Sieverts (Sv), is given by (Clark et al., 1993).

$$E = \sum_T w_T H_T$$

Table 1 lists all the tissue weighting factor based on two reports.

Because of biological effects and absorbed dose don't always have one-to-one correspondence, so another factor called quality factor is introduced.
And H_T is the equivalent dose (in Sv) in tissue or organ, T, and is given by (Clark et al, 1993).

$$H_T = \sum_R w_R D_{T,R}$$

Where w_R is the radiation weighting factor (or quality factor) due to radiation of type R (for example neutron, alpha etc.) and $D_{T,R}$ is the absorbed dose averaged over a tissue or organ, T, due to a radiation of type R.
Radiation weighting factors (w_R) for neutrons, according to ICRP Publication 60 can be chosen from either a step function or a continuous function to avoid discontinuity. The following formula (ICRP 60,1991) is used to calculate the w_R continuous values:

$$w_R = 5.0 + 17.0e^{\frac{-[\ln(2E_n)]^2}{6}}$$

where E_n is the neutron energy in MeV. Another set of new w_R data, is also released from ICRP Publication 103 (ICRP 103, 2008). The new radiation weighting factors function was expressed as:

$$w_R = \begin{cases} 2.5 + 18.2\, e^{\frac{-[\ln(E_n)]^2}{6}} & , E_n < 1MeV \\ 5.0 + 17.0\, e^{\frac{-[\ln(2E_n)]^2}{6}} & , 1MeV \leq E_n < 50MeV \\ 2.5 + 3.25\, e^{\frac{-[\ln(0.04E_n)]^2}{6}} & , E_n > 50MeV \end{cases}$$

ICRP 1991		**ICRP 2005 DraftReport**	
Tissue or organ	Tissue weighting factor, *wT*	Tissue or organ	Tissue weighting factor, *wT*
Gonads	0.20	Gonads	0.05
Bone marrow (red)	0.12	Bone marrow (red)	0.12
Colon	0.12	Colon	0.12
Lung	0.12	Lung	0.12
Stomach	0.12	Stomach	0.12
Bladder	0.05	Bladder	0.05
Breast	0.05	Breast	0.12
Liver	0.05	Liver	0.05
Oesophagus	0.05	Oesophagus	0.05
Thyroid	0.05	Thyroid	0.05
Skin	0.01	Skin	0.01
Bone surface	0.01	Bone surface	0.01
Remainder: adrenals, brain, Lower Large Intestine, Upper Large Intestine, Kidneys, muscle, pancreas, spleen, thymus, uterus	0.05	Brain	0.01
		Kidneys	0.01
		Salivary glands	0.01
		Remainder: adipose tissue, adrenals, connective tissue, extrathoracic airways, gall bladder, heart wall, lymphatic nodes, muscle, pancreas, prostate, small intestine wall, thymus, uterus/cervix	0.10

Table 1. ICRP 60 (1991) and ICRP 2005 proposed tissue-weighting factors.

7. Technique problems

One of the disadvantages of the neutron sources is that they don't generate only neutron but also they emit high-intensive gamma-rays. When using PGNAA method for medical purposes, the sample is a human body so these gamma-rays can cause destructive effects on it.

Another major problem of this technique is thermal and epithermal neutron capture by the iodine in the detecting crystal (NaI(Tl)), plus pile-up of gamma-rays from lower energy reactions or from the source of the neutrons. The Birmingham group has largely solved this problem by the use of a pulsed neutron beam and gated circuits (Harvey *et al.* **1973).**

Note that the activation of gamma detector is only in prompt gamma technique but in the delay gamma neutron activation analysis since the detection of delayed gamma rays is after irradiation so this worry vanishes.

8. Delayed-gamma-emission neutron activation analysis

When the body is irradiated with neutrons, penetrating gamma rays are emitted both during irradiation (prompt) and for some time afterwards (delayed). These gamma rays originate from atomic nuclei which have absorbed energy from the neutrons or captured the neutrons themselves, and the energies of the gamma rays are characteristic of the nucleus which emits them. Therefore energy sensitive detectors may identify the emitting nucleus and the number of gamma rays detected at a given energy may be used to determine the abundance of the emitting nucleus in the body.

The majority of gamma rays are emitted during irradiation, but the elements sodium, chlorine, calcium, nitrogen and phosphorus may be determined after irradiation, if the subject is transferred from the irradiation facility into a whole-body counter within a short period, typically 5 min. Sodium and chlorine are extracellular ions from which the extracellular fluid space of the body may be determined. Calcium is contained almost entirely within the skeleton, comprising 34% of bone mineral. Phosphorus occurs mainly in the skeleton but is also found in lean soft tissue, in association with the energy metabolism. Nitrogen is uniquely a constituent of protein, 16% by weight, so that measurement of total body nitrogen (TBN) is used to determine total body protein (TBPr). These nuclear reactions are given as follows:

$$
\begin{array}{lll}
{}^{23}\mathrm{Na}(\mathrm{n},\gamma)\,{}^{24}\mathrm{Na} & E = 1.369, 2.754\ \mathrm{MeV} & t_{1/2} = 15\ \mathrm{h} \\
{}^{37}\mathrm{Cl}(\mathrm{n},\gamma)\,{}^{38}\mathrm{Cl} & E = 1.642, 2.168\ \mathrm{MeV} & t_{1/2} = 37.3\ \mathrm{min} \\
{}^{48}\mathrm{Ca}(\mathrm{n},\gamma)\,{}^{49}\mathrm{Ca} & E = 3.084\ \mathrm{MeV} & t_{1/2} = 8.72\ \mathrm{min} \\
{}^{40}\mathrm{Ca}(\mathrm{n},\alpha)\,{}^{37}\mathrm{Ar} & E = 2.6\ \mathrm{keV} & t_{1/2} = 35.1\ \mathrm{d} \\
{}^{14}\mathrm{N}(\mathrm{n},2\mathrm{n})\,{}^{13}\mathrm{N} & E = 0.511\ \mathrm{MeV} & t_{1/2} = 9.96\ \mathrm{min} \\
{}^{31}\mathrm{P}(\mathrm{n},\alpha)\,{}^{28}\mathrm{Al} & E = 1.779\ \mathrm{MeV} & t_{1/2} = 2.24\ \mathrm{min} \\
{}^{16}\mathrm{O}(\mathrm{n},\mathrm{p})\,{}^{16}\mathrm{N} & E = 6.134\ \mathrm{MeV} & t_{1/2} = 7.2\ \mathrm{s}
\end{array}
$$

Where E denotes the energy of the characteristic gamma rays emitted and t1/2 is the half life of the induced activity.

The minor elements magnesium, copper, iodine and iron may also be determined from the delayed emission of gamma rays.
The reaction with oxygen has been successfully employed, where the subject was transferred (within 30 s) from the irradiation facility to a whole-body counter. The reactions with nitrogen, oxygen and phosphorus only occur with fast neutrons above an energy threshold: 11 MeV for the reactions with oxygen and nitrogen and 2 MeV in the case of phosphorus. Two configurations of the delayed gamma neutron activation analysis system have been shown in Figures 2. And 3.

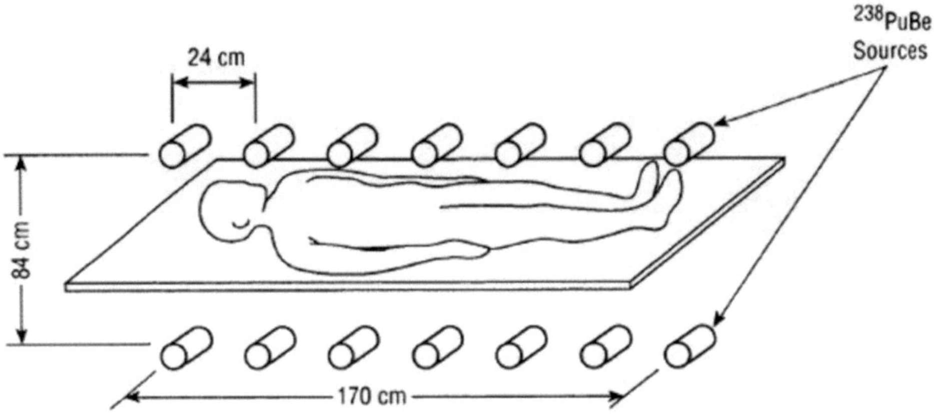

Fig. 2. Pu-Be neutron source arrangement for the Delayed Gamma Neutron Activation Analysis

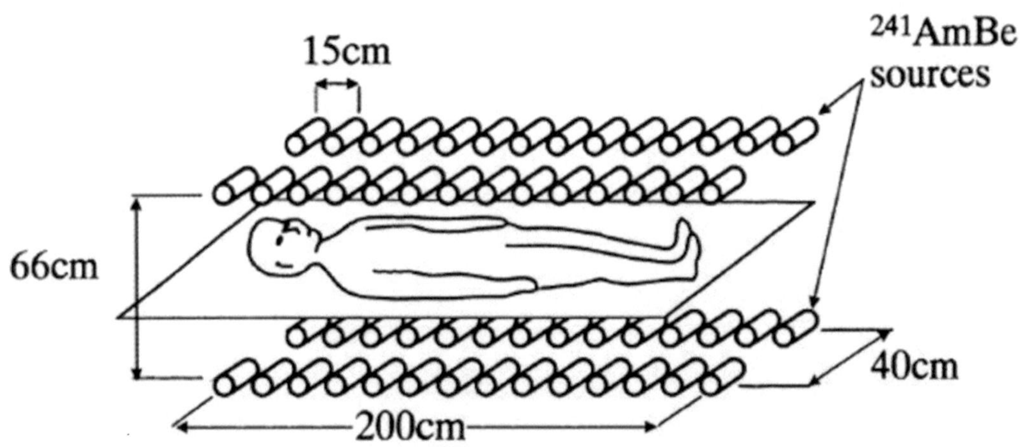

Fig. 3. 241Am-Be neutron source arrangement for the Delayed Gamma Neutron Activation Analysis

The reaction with nitrogen suffers from the disadvantage that the positron annihilation radiation (0.511 MeV) is common to many nuclear reactions, and it is not possible to distinguish this from another reaction which produces the same daughter nuclide which decays with the same half life:

$$^{16}\mathrm{O}(\mathrm{p},\alpha)\,^{13}\mathrm{N} \qquad E = 0.511\ \mathrm{MeV} \qquad t_{1/2} = 9.96\ \mathrm{min}$$
$$^{12}\mathrm{C}(\mathrm{n},2\mathrm{n})\,^{11}\mathrm{C} \qquad E = 0.511\ \mathrm{MeV} \qquad t_{1/2} = 20.3\ \mathrm{min}.$$

The protons which produce the interfering reaction with oxygen originate as the result of elastic collisions between neutrons and hydrogen nuclei, the most numerically abundant element in the human body. There are many other minor reactions which also interfere, producing positron annihilation radiation at 0.511 MeV.

9. Prompt-gamma neutron activation analysis

The vast majority of gamma rays induced by the inelastic scattering and capture of neutrons by atomic nuclei in the human body are emitted within a few microseconds. The abundance of the emission, at all energies up to 11 MeV, makes it difficult to distinguish gamma rays of similar energies from different elements unless a high energy resolution detector is employed, such as the semiconductors Ge(Li) or hyperpure Ge. Otherwise NaI(Tl) crystal scintillation detectors, with an optically coupled photomultiplier tube, are usually employed, since they have a larger sensitive volume and greater stopping power for gamma rays.

The reason for the better energy resolution of the semiconductor detectors is that it requires the deposition of only approximately 3 eV of energy from the gamma ray in the detector's depletion layer to produce an electron–hole pair, whereas it requires around 100 times as much energy to be deposited in the NaI(Tl) crystal to produce one photoelectron at the photocathode of the photomultiplier due to losses of light in the crystal. The total number of electrons released in each type of detector is a measure of the amount of energy absorbed. Although the signal is amplified many times in the photomultiplier tube attached to the NaI(Tl) crystal, the anode current reflects the fluctuations in the number of electrons emitted from the photocathode. Therefore, for the detection of gamma rays of a given energy, there is a greater statistical variation in the signal from a NaI(Tl) detector than a semiconductor, so that the latter is used for high resolution gamma spectroscopy.

If the energy resolution of a Germanium semiconductor detector is 2 keV, the corresponding energy resolution of a NaI(Tl) crystal scintillator is around 80 keV. The semiconductor detectors suffer the disadvantage of having to be cooled with liquid nitrogen when in use, and, in the case of Ge(Li) detectors, cooled continuously.

Another problem associated with prompt-gamma neutron activation analysis is neutron irradiation of the detectors themselves, which in the case of semiconductors produces dislocations in the crystal lattice, and in NaI(Tl) crystals activates both the sodium and the iodine nuclei, from which the resulting gamma rays are counted with great efficiency.

This increases the background in the gamma ray spectrum upon which the characteristic emissions of body elements are superimposed. Therefore suitable neutron shielding of the detectors is necessary.

Bismuth germinate scintillation detectors have a greater stopping power for gamma rays than sodium iodide, and therefore may improve the signal to background for nitrogen, but this advantage has not been realized in practice due to activation of germanium. Multiple small NaI(Tl) crystals were found to give a better signal to noise ratio than a few larger crystals.

A third problem associated with prompt-gamma neutron activation analysis is the high count rate encountered. Since the output pulse from a detector is of a finite length (typically with a rise time of 0.25 _s and a fall time of up to 10 _s), any radiations being detected within this interval may be added to the original event, producing a pulse of greater amplitude. This process of random summing at high count rates has the effect of increasing the background in the gamma ray spectrum further. The statistical uncertainties in the determination of the abundance of any element in the body from the number of events in the corresponding full-energy peak in the spectrum are increased by the contribution from the underlying background. It is necessary to minimize this background. One method to reduce the random summing background to nitrogen is to electronically suppress the counting of events below 5 MeV for the major part of the measurement, and only count the whole spectrum (including the 2.223 MeV peak from hydrogen) for a short interval .

This increases the nitrogen signal to background by 18%. Since many (inelastic or non-elastic scattering) reactions (e.g. with carbon, oxygen) have an energy threshold of several mega-electron-volts, the optimum signal for a given dose is achieved when the subject is irradiated with monoenergetic neutrons at 14.4 MeV from a D-T neutron generator. These neutron generators, or alternatively cyclotrons, can be temperamental to operate, so that often neutron sources, comprising an alloy of beryllium and an alpha emitting radionuclide, are preferred.

These sources (241Am/Be, 238Pu/Be) produce a 4.439 MeV gamma ray per neutron which may interfere with the determination of carbon and add significantly to the problem of random summing of gamma rays in the detectors unless the source is well shielded.

Moreover, it is possible to improve the signal to background ratio in operating a neutron generator or cyclotron in a pulsed or cycled mode by counting short-lived induced activity between pulses of neutrons, thereby reducing the lower limit of the target element that can be measured.

10. Configurations of the PGNAA facility

In the progress of using PGNAA method for medical purposes many kind of setups and configurations suggested and applied. In the following some of them have been shown

Figure 5 is a schematic of a conventional machine used to measure the body composition. Another gold design is shown in figure 6. In this setup the uniform neutron flux will meet the patient tissue so we can get good results.

Figure 7 shows a cross-sectional view of the modified BCCA. Sheets of 2 cm thickness of Lead surround the neutron moderator (here paraffin wax) to provide radiation shielding for personnel. To protect personnel from biological effects of neutrons and to reduce background counts, neutron shielding must be considered. Since high-speed neutrons are more difficult to shield, at first neutrons must be moderated by a hydrogenous material such as paraffin wax (14.86 % H, 85.14 % C). Because of Hydrogen has a great absorption Cross Section for thermal neutrons, the risk of neutrons for personnel vanishes. One of the benefits that the moderator has been covered with a 2cm layer of Pb is that the gamma-rays (2.224 MeV) produced by the H(n, γ) interaction are filtered.

A sphere of Lead has been centered at the source position to filter gamma-rays of the neutron source. Another part of this configuration is an invert, rectangular, cuneus void cast within the paraffin wax block ($40_{cm} \times 50_{cm} \times 60_{cm}$). To protect patient body from high-rate

2.224 MeV gamma-rays, the inner wall of the valley, made above the neutron source (Figure 7), was lined by Pb sheet of 2cm thickness. By this way, a rectangular neutron-beam aperture measuring 40 cm length (perpendicular to the paper sheet) and 20 cm (width) at the sample location is defined.

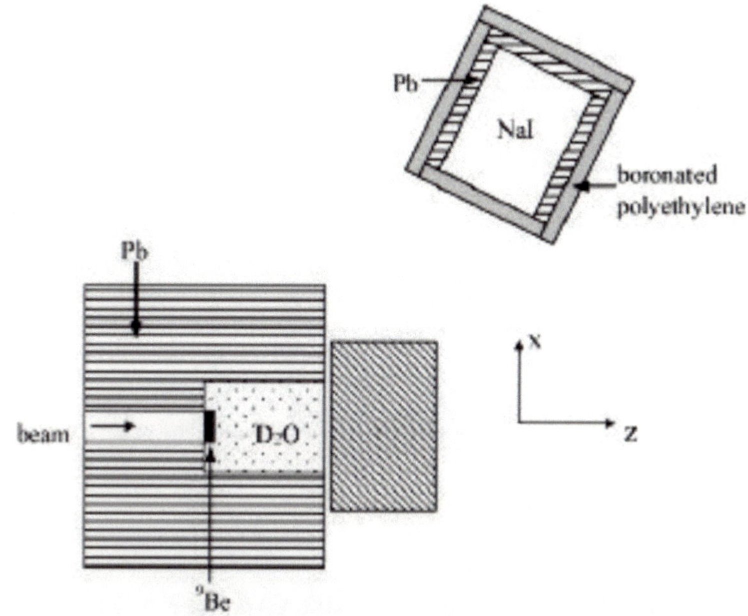

Fig. 4. A typical schematic representation of PGNAA setup based accelerator. In this setup the D2O is used as moderator.

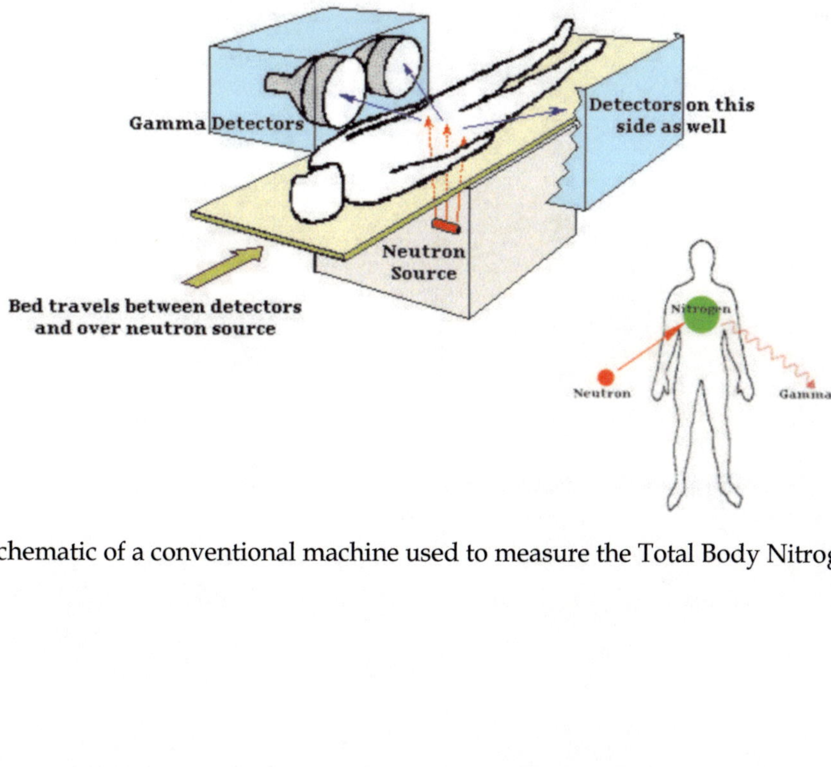

Fig. 5. Schematic of a conventional machine used to measure the Total Body Nitrogen (TBN)

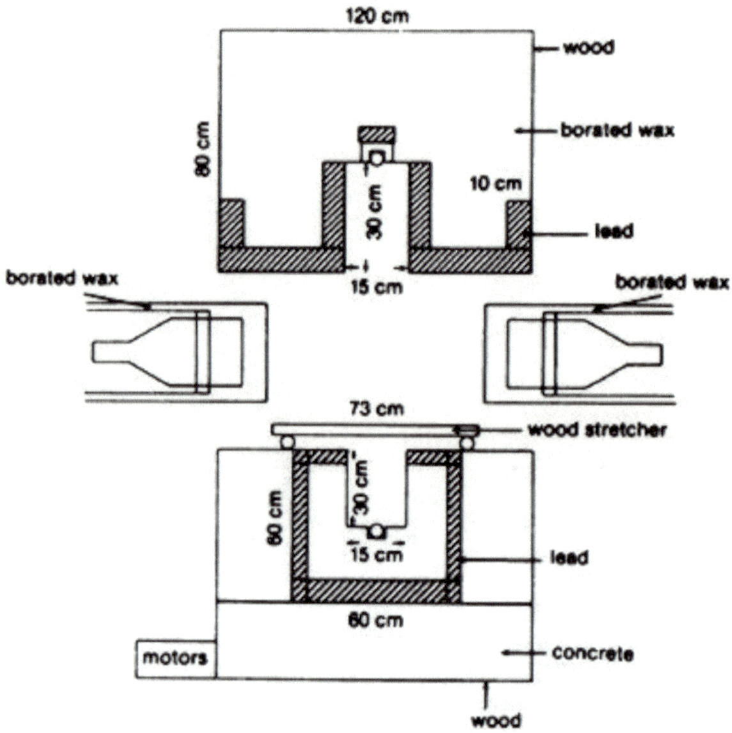

Fig. 6. A unique design of Body Chemical Composition Analyzer

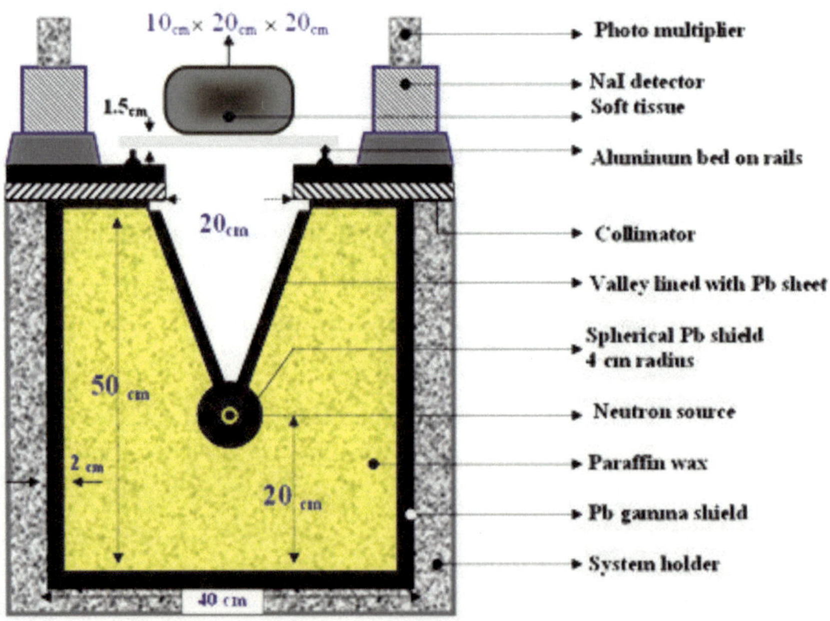

Fig. 7. Design and Geometry of a Body Chemical Composition Analyzer.

11. Future works

The authors are attempting to investigate all the aspects related to this topic and welcome any idea and proposal. Designing of a PGNAA setup for medical and industrial purposes need time and considering a lot of parameters which is under construction in Ferdowsi University of Mashhad, FUM Radiation Detection and Masurement Lab. For more information please don't hesitate to contact me (panjeh@gmail.com).

12. References

Alpen, E L. (1998). *Radiation Biophysics 2nd edition*. (Academic Press).

Anderson, J., Osborn, S B., Tomlinson, R W S., Newton, D., Rundo, J., Salmon, L., & Smith, J W. (1964), *Lancet*, Vol. 2, pp. 1201-1205.

Baur, L A., Allen, B J., Rose, A., Blagojevic, N., & Gaskin, K J. (1991) A total body nitrogen facility for paediatric use *Phys. Med. Biol.*Vol. 36, pp. 1363–1375.

Beddoe, A H., Zuidmeer, H., & Hill, G L. (1984). A prompt gamma in vivo neutron activation analysis facility formeasurement of total body nitrogen in the critically ill *Phys. Med. Biol.* Vol. 29, pp. 373–83.

Briesmeister, J F. (2000), MCNP A General Monte Carlo N-particle Transport Code, Version 4C, Los Alamos National laboratory, LA-13709-M.

Chichester, D L., & Empey, E. (2004). Measurement of nitrogen in the body using a commercial PGNAA system—phantom experiments. *Appl. Radiat. Isot.* Vol. 60, pp. 55-61.

Clark M, J., Bartlett, D T., Burgess, P H., Francis, T M., Marshall, T O., & Fry, F A. (1993). Dose quantities for protection against external radiation. *Documents of the NRPB* Vol. 4, pp. 3. (NRPB: Chilton).

Harvey T C., Dykes P W., Chen N S., Ettinger K V., Jain S., James H., Chettle D R., & Fremlin J H., (1973), *Lancet*, Vol. 2, pp. 395.

Cohn, S H., Cinque, T J., Dombrowski, C S., & Letteri J M. (1972), *J. Lab. Clin. Med.* Vol. 7, pp. 978.

ICRP 60. (1991). Recommendations of the International Commission on Radiological Protection, International Commission on Radiological Protection, Pergamon Press,

Oxford. ICRP (2005) recommendations of the International Commission on Radiological Protection. Pergamon Press, Oxford. ICRP 103. (2008). Recommendations of the ICRP, International Commission on Radiological Protection, Pergamon Press, Oxford.

Mernagh, J R., Harrison, J E., & McNeill, K G. (1977). In vivo determination of nitrogen using Pu-Be sources *Phys. Med. Biol.* Vol. 22, pp. 831–5

Mcneill, K. G. (1973). *J. Nucl. Engng*, Vol. 14, No. 3, PP. 84-86.

Mernagh, J., Harrison, E., & McXeill, K. G. (1977). In vivo Determination of Nitrogen using Pu-Be Sources. *Phys. Med. Biol.*, Vol. 22, No. 5, pp.831-835.

Metwally, W M., & Gardner, R P. (2004). Stabilization of prompt gamma-ray neutron activation analysis (PGNAA) spectra from NaI detectors. *Nucl. Instr. and Meth. A*, Vol. 525, 518-521.

Miri Hakimabad, H., Panjeh, H. & Vejdani, A. R. (2007). Evaluation the nonlinear response function of a NaI Scintillation detector for PGNAA applications. Appl. Radiat. Isot. *Applied Radiation and Isotopes* Vol. 65, pp. 918–926.

Mitra, S., Plank, L D., Knight, G S., & Hill, G L. (1993). In vivo measurement of total body chlorine using the 8.57 MeV prompt de-excitation following thermal neutron capture. *Phys. Med. Biol.* Vol. 38, pp. 161–72.

Ryde, S J S., Morgan, W D., Evans, C J., Sivyer, A., & Dutton, J. (1989). Calibration and evaluation of a 252Cf based neutron activation analysis instrument for the determination of nitrogen *Phys. Med. Biol.* Vol. 34, pp.1429–41.

Vartsky, D., Prestwich, W V., Thomas, B J., Dabek, J T., Chettle, D R., Fremlin, J H., & Stammers, K. (1979). The use of body hydrogen as an internal standard in the measurement of nitrogen in vivo by prompt neutron capture gamma ray analysis. *J. Radioanal. Chem.* Vol. 48, pp. 243–52

16

Transportation Pathway of Potassium and Phosphorous in Grape Fruit

Zhenming Niu, Yi Wang, Yanqing Lu, Xuefeng Xu and Zhenhai Han
Institute of Horticultural Plants, China Agriculture University, Beijing, China

1. Introduction

Since the balance proportion of mineral elements affects the fruit quality, reasonable fertilization is an important way to increase fruit yield and quality. With regard to N, P and K elements, K element is the largest element required by the grape. Various studies have shown that phosphorus and potassium affect the appearance of the fruit quality due to that potassium element could increase the size of grape, citrus and peach fruit, extending the shelf-life, Increasing hardness, beautiful color, and effective anti-browning (Cummings, 1980) simultaneously. P and K elements have an important role in the formation of intrinsic quality of the fruits, for instance, the soluble solid concentration in the apple fruits is in positive correlation with the potassium concentration (Tagliavini et al, 2000). Organic acid content in peach fruit is affected by the nutrition conditions of potassium and nitrogen elements, since potassium could stimulate the accumulation of acid in the fruit, while neutralizing the fruit acidity in part (Habib et al, 2000). Over the annual growth cycle of grapes, whether single application of P and K-fertilization or in coordination with nitrogen fertilizer, spraying on the surface of leaves could obtain different levels of production increase and quality improvement, As a consequence, controlling nitrogen, increasing phosphate and necessary potassium prior to the development and ripening of grapes are the essential measures to obtain superior quality grapes. However, the actual situation in China is attaching great importance on nitrogen while neglecting the application of potassium and phosphorus.which has seriously impacted the grape quality. Therefore, the rational application of P and K-fertilization has great significance in the grape production.

Ascertaining the effects of mineral elements on fruit quality, and the features of absorption, transportation, distribution of mineral elements are the premise of rational fertilization. Some scientists have applied isotope tracer technique to research the nutrient uptake and distribution, which shows that fruit is one of the centers for nutrient distribution (Hu Shi Bi et al, 1998; Huang Weidong et al, 2002; Xie Shenxi and Zhang Qiuming, 1994). Hu Shi Bi et al (1998) found that soil-applied ^{86}Rb before bloom of grape, the distribution rate at its stems, shoot tips, and leaves at the earlier stage was higher than the inflorescence; foliar application of ^{86}Rb at the same time, the absorption of ^{86}Rb by inflorescence was significantly higher than that of the soil application of ^{86}Rb; foliar ^{86}Rb application at full bloom, the largest distribution was found at the side tip and inflorescence; additionally, the absorption of ^{86}Rb in fruit at ripening period was significantly reduced compared with the previous periods. Fu Yu-man et al (1997) using ^{32}P tracer technique revealed that the ^{32}P uptake rate of fruit showed an increasing trend against time, accumulated in the developing fruits and roots. The uptake of ^{32}P by leaf surface rapidly involved in the metabolism process, synthesized into various organic phosphorus compounds from inorganic element, 80% of which incorporated with acid soluble components. The Research of Pei Xiaobo et al (2002) suggested that nitrogen, phosphorus and potassium were transported by stems and leaves at the earlier stage of cucumber growth, and then increased into the fruits after fruiting. Zhou Yurong and Chen Mingli (1996) utilized ^{32}P tracer study showed that the ^{32}P uptake sequence of citron daylily was autumn seedling > spring seedling> bolting> squaring >earlier bolting. The trend of ^{32}P uptake in citron daylily was high - low – high over its life. Most of the absorbed phosphorus was distributed in roots while the amount in the leaves was small. Slender roots were the major organs to absorb inorganic phosphorus and to turn it into acid soluble phosphorus. Currently, the transportation, distribution characteristics of phosphorus and potassium into the grape fruits are in need for further study. Based on the study of the absorption, transportation, distribution and regulation of P, K mineral elements, the aim of this experiment was to provide basis for the reasonable fertilization and high-quality cultivation of grapes, so as to explore approaches for improving fruit quality from the perspective of mineral nutrition.

2. Experimental materials and methods

2.1 Experimental materials

This experiment was carried out in the Science Park at the China Agricultural University from April to October in 2003. 3-year-old potted 'JingYou' grapes (Vitis *vinifera* L. × *V. labrusca* L.) were selected as the experimental samples, which were provided by the Institute of Forestry Fruit Trees, Beijing Academy of Agriculture and Forestry. In April 2003, the sprouting grapes were dug out from the exposed flowerpots and planted in a greenhouse at a spacing of 1. 0 × 2.0m. Each plant remained a main vine, and two fruiting branch main tips reserved an inflorescence, while the rest were cut. 8 leaves were left at the upper part of the inflorescence and pinched at early flowering stage. 50 pieces of fruits will be reserved on each cluster after petal fall, and the management was applied according to the standard recommendation in addition to other experimental treatment. While the selected experimental elements of ^{32}P, ^{86}Rb were $NaH_2{}^{32}PO_4$ and $^{86}RbCl$, respectively.The selected experimental elements of ^{86} Rib, ^{32} P were NaH $_2$ ^{32} PO $_4$ and ^{86} Rebel respectively, which were supplied by the Institute of Atomic Energy Application, the Chinese Academy of Agricultural Sciences.

2.2 Experimental methods

The processing and determination approach for the transportation pathway of P, K mineral elements from the leaves to the fruits. (1) Carpopodium micro-girdling method: gently strip 0.5cm of the phloem with stainless steel blade 2cm at the base of the spicate peduncle. (2) 25d after full bloom, the first leaves at the upper part of 3-year-old potted 'Jingyou' grape clusters with same growth potential were selected, and uniformly coated with 1. 6× 10^5Bq of $NaH_2{}^{32}PO_4$ or 0.77 × 10^5Bq of $^{86}RbCl$ by means of micro injector facilities at 9:00 of a sunny morning, 48h later, cut the grape cluster and packaged in a brown paper bag, degrading enzyme at 105°C for 10min, drying to counter weight at 80°C, and then grinding and weighing 50mg of samples. Afterwards, BH1216 low background α, β measuring device should be utilized to measure the activity of ^{86}Rb and ^{32}P; adjusting the measure time to control the measurement error within 5% or less, then its mean value will be adopted for analysis. Repeat this experiment for three times, the total weight of isotope in the fruit (cpm) = total dry weight of fruit (mg) / sample weight (50mg) × experimental sample value (cpm).

The processing and determination approach for the transportation pathway of P, K mineral elements from the roots to the fruits. 25d after full bloom, the roots of 3-year-old potted 'Jingyou' grape clusters with same growth potential were selected and placed in water for soaking, and rinsed off soil on the roots. Afterwards, 3.57×10^5Bq of $NaH_2{}^{32}PO_4$ or 2.66×10^5 Bq of $^{86}RbCl$ would be added into 2L of complete nutrient solution, and then completely immersed the roots of the grapes into the $NaH_2{}^{32}PO_4$, $^{86}RbCl$ complete nutrient medium for 48h after mixing. Finally, the grape clusters should be cut to measure the activity of ^{86}Rb, ^{32}P by virtue of the same sampling methods and testing equipment as above-mentioned.

Study on the critical period of P, K mineral elements transporting into the fruits. The first leaves at the upper part of 3-year-old potted 'Jingyou' grape clusters with same growth potential were selected, and uniformly coated with 2.36×10^5Bq of $NaH_2{}^{32}PO_4$ or 7.47×10^4Bq of $^{86}RbCl$ by means of micro-injector at the first stage of fruit development (15d after bloom), the second stage of fruit development (25d afterbloom), veraision (50d after bloom) and the third stage of fruit development (20d before harvest) respectively, as shown in Figure l. Selecting the grape clusters after processing for 2, 4, 8, 12, 24, 48, 72, 96 and 120h to measure the activity of ^{86}Rb, ^{32}P in the fruits by virtue of the same sampling methods and testing equipment as above-mentioned.

Study on the distribution of P, K mineral elements in the grape fruiting branches. The first leaves at the upper part of 3-year-old potted 'Jingyou' grape clusters with same growth potential were selected, and uniformly coated with 2.36×10^5Bq of $NaH_2{}^{32}PO_4$ or 7.47×10^4Bq of $^{86}RbCl$ by means of micro-injector at the first stage of fruit development (15d after bloom), the second stage of fruit development (25d afterbloom), veraision (50d after bloom) and the third stage of fruit development (20d before harvest) respectively at 9:00 of a sunny morning. Cut the whole fruit branches 5d after processing, divided into processing leaves, un-processing leaves, fruits, and stems. Measure the activity of ^{86}Rb, ^{32}P in those four parts by virtue of the same sampling methods and testing equipment as above-mentioned.

3. Results

3.1 The principal transportation pathway of P, K mineral elements from the leaves to the fruits

Through the method of ear stem girdling the transportation pathway of the phloem was blocked, then observe the amount of ^{86}Rb, ^{32}P absorbed by leaves to the fruits through the xylem. The results illustrated in Figure 1 shows that the amount of ^{86}Rb, ^{32}P transported to the fruits has been significantly reduced after girdling, which is only 5.44% (^{32}P) and 7.28%(^{86}Rb) of the control, there were extreme significiant differences between treatments and control. Therefore, the principal transportation pathway for P, K mineral elements absorbed by leaves to the fruits is the phloem. It also shows that, when the phloem is blocked, a small amount of P, K mineral elements can be transported through xylem. Nonetheless, whether the P, K mineral elements that transported through the xylem are directly conducted by the xylem of petiole - stem - ear stem, or through horizontal transportation to the xylem due to the blocking of ear stem phloem, requires further testing for confirmation.

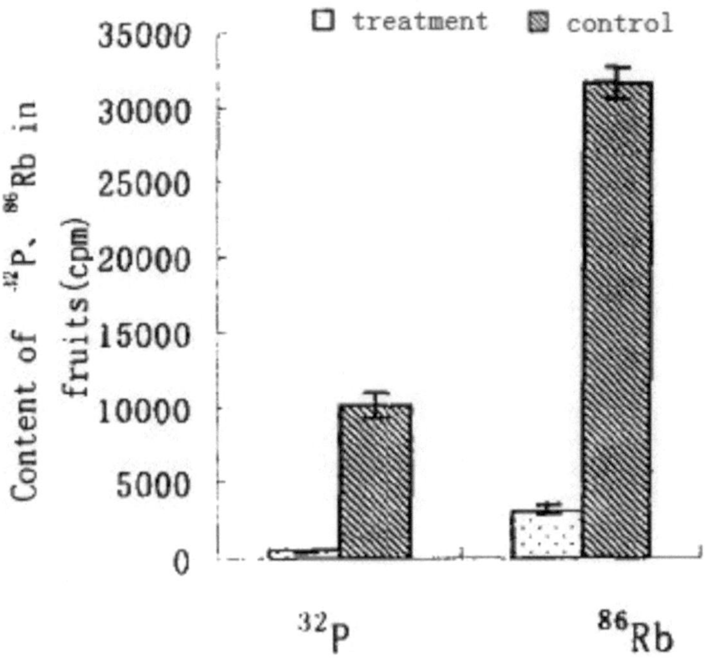

Fig. 1. Effect of fruit stalk microgirdling on influx into fruit of ^{32}P, ^{86}Rb absorbed from leaves

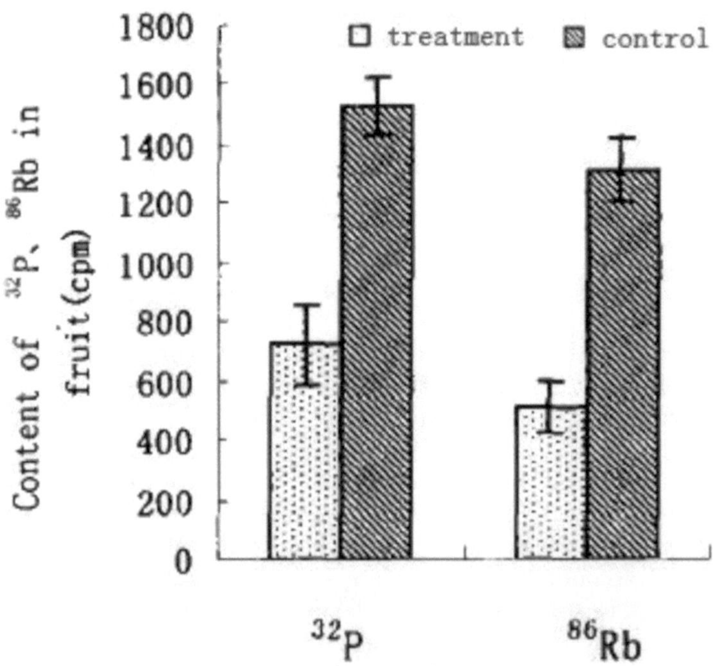

Fig. 2. Effect of fruit stalk microgirdling on influx into fruits of ^{32}P, ^{86}Rb absorbed from roots

3.2 The principal transportation Pathway of P, K mineral elements from the roots to the fruits

The principal transportation pathway for ^{86}Rb, ^{32}P absorbed by roots is ear stem phloem, and the ^{86}Rb , ^{32}P transported to the fruits will be significantly reduced ($P < 0.01$) in case of blocking of the transportation pathway at the phloem, which only accounts for 46.75%(^{32}P) and 38.67%(^{86}Rb) as compared with the control(Figure 2), it can be seen that the phloem and xylem have played an important role for the transportation of P, K mineral elements absorbed by roots to the fruits, additionally, the phloem is the principal transportation pathway, or there is more horizontal transportation between the xylem and phloem in the course of upward transportation, which is quite different from the results obtained by Zhao Jinchun (2000). Thus, it is need carry out more experimental evidence and in-depth research to further determine the principal transportation pathway of P, K mineral elements absorbed by roots to the fruits.

3.3 The P, K uptake of fruits at different developmental stages

Supplying P, K on the surface of the leaves at different growth stages of grape fruits will have a significant effect on the accumulation of P and K in the fruits (Figure 3). Foliar application of P at the first stage of fruit development (15d after bloom) will lead to the higher uptake and accumulation of P mineral element as compared with Other stages. Foliar application of P at different periods has presented a regular impact on fruits, and it shows that Supplying P at the two fast growing periods of fruit will incrase P absorption in fruit. And it has demonstrated that the accumulation of P is gradually increased over time, especially 72h later, which is the fastest period of P transported into the fruit. Therefore, the results show that the earlier stage of fruit development demands the most P elements and absorbs the fastest, which is also the critical period of application of P in the production.

The absorption of K mineral elements by fruits at the ripening and young fruit stage is similar with P, The absorption was more in the two fast growing periods of fruits. K elements presented two stages of demand and efficient absorption as compared with P, namely, the first and the third stage of fruit development, while the maximum absorption was the third stage of fruit development. Thus, it should supply with potassium accroding to the nutrient condition in the orchard.

Figure 4 shows that the absorption efficiency of P, K fertilizer for the grape fruits at different stages had significant variation. The absorption of P at the rapid growth period of young fruit displayed a respective peak at 3d, 5d, accounting for 47.15% and 23.98% of total amount of absorption in 5d; in contrast, the 5th day had the maximum absorption at slow growth stage, accounting for 73.58% of total absorption in 5d ; there was no significant difference in continuous 5d at the veraision stage of the grape fruits; supplied with P at 20d before harvest, the latest 2d showed more absorption, accounting for 36.79% and 39.80% of total absorption in 5d, respectively.

The absorption of K at the rapid growth period of young fruit displayed a peak at 3d, accounting for 47.62% of total absorption in 5d; 4d had the minimum absorption, merely accounting for 6.96% of total absorption in 5d; rebounded after the 5d, accounting for 18.74% of total absorption in 5d. In contrast, there was no significant difference for the absorption rate in the slow growth stage, the 4d had the maximum absorption at the veraision stage, accounting for 50.54% of total absorption in 5d; supplied with K at 20d before harvest,the absorbtion of K at 1d after treatment was low,while the peak absorption was found at the 4d, and the rest 3ds had no significant difference.

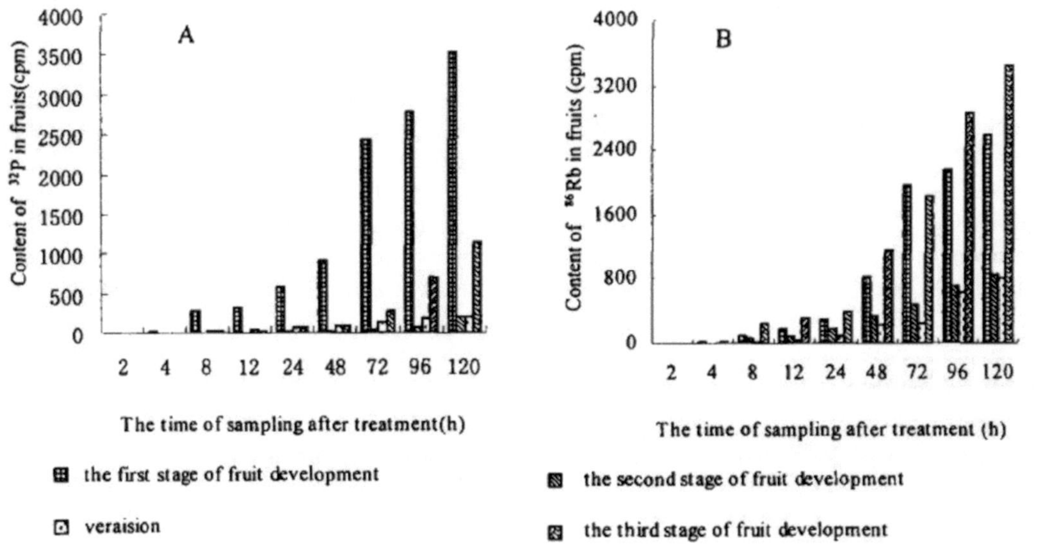

Fig. 3. Effect of different stages of fruit development on influx into fruit of ^{32}P(A), ^{86}Rb(B)

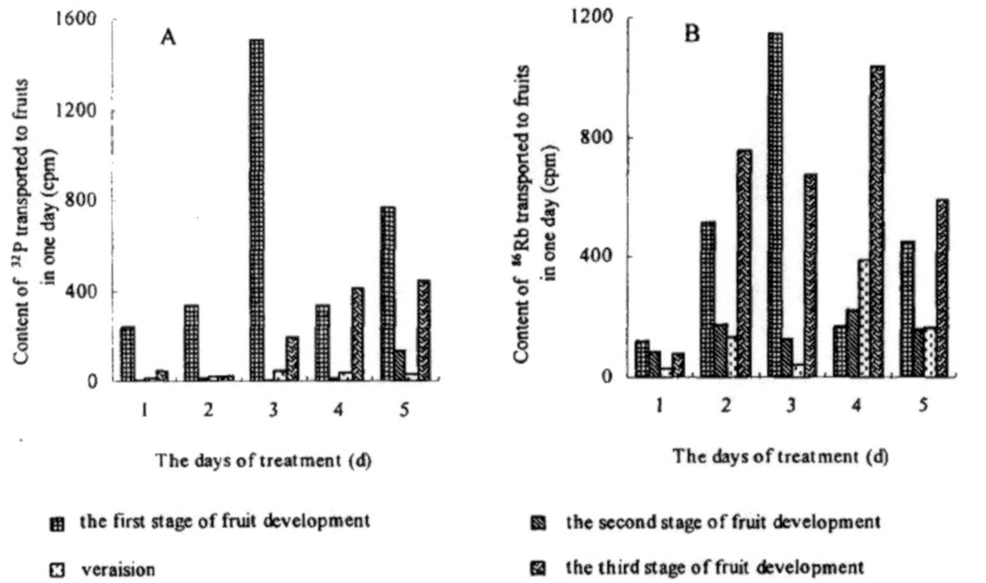

Fig. 4. Content of ^{32}P(A), ^{86}Rb(B) in fruits absorbed from leaves in one day

3.4 The distribution of P, K mineral elements in the branches of grapes

The measurement results of the 5th day after foliar application of $NaH_2{}^{32}PO_4$ showed that 82.90% -95.79% ^{32}P was detained in the leaves (Figure 5A). It indicated that the leaf growth itself required a certain amount of P, and ^{32}P had significantly different distribution in the fruit branches for foliar coating with ^{32}P at different stages, while the fruits were the major organs absorbing ^{32}P in addition to the leaves. The distribution of ^{32}P by virtue of foliar spray at different stages of fruit development presented variable proportion: the first stage of fruit development (15.13%)> the third stage of fruit development (7.18%)> veraision Stage (4.87%)> the second stage of fruit development (3.53 %). The smallest proportion was in the stem, less than 1%, while the proportion of un-treated leaves was approximately 2%.

76.94%-85.58% of ^{86}Rb was detained in the leaves after 5d of the foliar application of $^{86}RbCl$ (Figure 5B). There was significant difference in distribution of ^{86}Rb in the fruit branches after treating at different stages, while the fruits were the major organs absorbing ^{86}Rb from the leaves. After foliar application of ^{86}Rb, The distribution of ^{86}Rb in fruit at different stages of fruit development presented variable proportion: the third stage of fruit development (26.86%)> the first stage of fruit development (15.44%)> the second stage of fruit development (11.40%)> veraision Stage (9.06%). The distribution proportion in the stems was 4-7%, while the proportion of un-treated leaves was approximately 1%. Compared with P element, the distribution proportion of K was higher than that of P in fruit (Figure 5B).

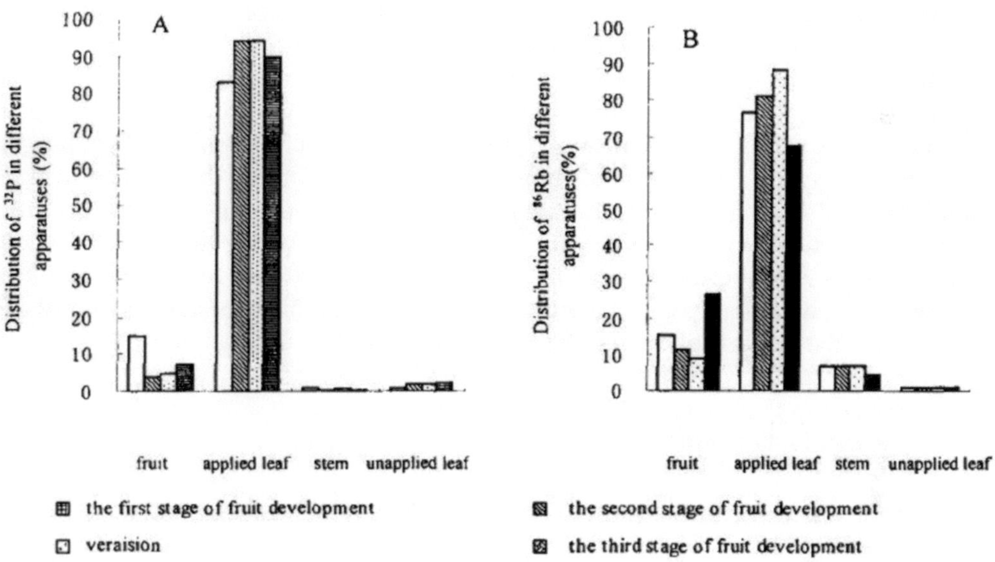

Fig. 5. Distribution of ^{32}P(A), ^{86}Rb(B) absorbed from leaves in bearing branch of grape vine

4. Discussions

4.1 The transportation pathway of P, K mineral elements in the grape fruits

At the 40s of 20th century, ^{42}K tracer technique proved that the upward channel for the transportation of inorganic nutrients is catheter, while existing active horizontal transportation from the xylem to the phloem. Circulation and redistribution process are taken place inside the plants. Literatures on the transportation pathway of nutrients showed that N, P, K mineral elements can be transported from the xylem and the phloem to apple fruits at the growing season, and the both transportation pathways came into play at the early and mid stage of apple fruit development, however, phloem sap played a major role in the fruit enlargement before harvest; with respect to peach fruits, there was still high rate of xylem sap transported to the fruits before harvest, and organic nitrogen, magnesium, potassium in the ripening leaves were transported to the fruits through the xylem

(Tagliavini et al, 2000). Zhao Jinchun (2000) applied micro-girdling stems (Han Zhenhai et al, 1995) to study the transportation pathway of apple fruits, which suggested that the absorbed K mineral element by roots was transported into the fruits mainly through the phloem in the normal development conditions of apple fruits. From the perspective of this experiment, there are two sources of P, K mineral elements for the grape fruits, namely, one is absorbed by roots and directly transported to the fruits through the xylem; the second is transported from the phloem, which may be derived from the horizontal transportation from the xylem to the phloem, and the cycling transportation from the ground leaves and so forth. However, whether P elements at the two parts are involved in the transportation to the fruit and its proportion are needed further study.

Although, it is certain that the root xylem is the important way to transport the P, K mineral elements by roots to the fruits. Taking into account that the ear stem girdling has blocked the phloem transportation pathway, it may stimulate the horizontal transportation in the plant phloem and xylem, and the transported P and K mineral elements from the xylem to the clusters in normal conditions may be less than the measured results under experimental conditions. However, according to the conclusions of this study, there is sufficient argument to deem that the transported P, K mineral elements from the xylem to the fruits cannot be ignored. Foliar application of P, K mainly transported through the phloem, also taking into account the ear stem girdling side effects, the transported P, K mineral elements from the xylem to the clusters in normal conditions may be higher than the measured results under experimental conditions. Potassium maintains a high concentration in the phloem sap, easy to juice up and down for long-distancetransportation, and gives priority to supplying for the tender leaves, meristem, fruits and other parts. According to the results of this experiment, spraying P, K fertilizers on the surface of leaves are mainly transported through the phloem, hence, foliar application of P, K can meet the P, K demands at metabolic locations in a short time, thereby providing theoretical evidences to demonstrate that foliar application of P, K is able to give rapid relief of nutrient deficiency for the production.

4.2 The critical period for the grape fruit to absorb P, K mineral elements

It should be noted that this study found that young fruit at its rapid growth stage also requires a lot of potassium, however, it have paid little attention to the importance of applying K at the earlier stage of grape growth currently, and hence it is necessary to supply K for one or two times on the surface of leaves from petal full to 15d after bloom, to meet the high demand for K nutrition of rapid growth of young fruit. This study showed that the quantity demanded for P, K mineral elements of grape fruits at different developmental stages was variable, which was consistent with the research results provided by Hu Shibi et al (1998). K mineral element plays a positive effect in the late enlargement of the fruits, and thus it should timely supply K mineral element at start coloring of the grapes and meets the high demand for K mineral element at cluster late development, so as to promote the ripening of fruits and improve fruit quality of grapes and efficiency of fertilizer utilization. In the beginning of fruit ripening and slow growth period, the fruits require small amount of P, K mineral elements, and it should be supplied appropriately according to the actual nutrition situation in the orchard.

The content of nutrient elements in the fruits has an important influence on fruit quality, and rational fertilization is one of the main measures to improve grape yield and quality.

Deficiency of phosphorus element will block the protein synthesis of fruit trees, affecting cell division, thereby resulting in fruit growth retardation, and quality decline. Vitis liking K fruit tree has a high demand for K, thus, applying appropriate amount of P and K fertilizer will improve the nutritional balance of trees, improve fruit quality, and increase fruit resistance. In conclusion, the earlier stage of fruit development, and the stage from coloring to harvest are the critical periods for the grape to absorb P, K mineral elements, and hence it should timely apply K-fertilizer and appropriate amount of P-fertilizer. In this way, it will not only meet the requirements for a large amount of P, K mineral elements by the grape fruits, but also will increase yield and fruit quality, as well as the efficiency of fertilizer utilization.

4.3 Absorption and distribution of P, K mineral elements

On the basis of preceding studies, it indicated that the demand for K by fruits will increase as the approaching of ripening period (Tagliavini et al, 2000). According to the experimental results, the distribution ratio to P, K mineral elements into fruits by virtue of spraying on the surface of leaves showed the maximum at the rapid growth stage of young fruit and approaching the ripening stage, which were relatively consistent with the demands for P, K mineral elements by grape fruits at these stages in this study indeed. During the slow growth period, the growth rate of grape fruit slowed, the sink strength weakened, and the demand for nutrition declined. In addition, the new branch tip was still in its growth peak period, requiring relatively additional competitiveness of nutrition, and hence the distribution ratio of P, K at stems and leaves was increased accordingly.

It is noteworthy that, supplying P, K on the surface of leaves will leave some P, K mineral elements in the treated leaves after 5d. As a consequence, it can be inferred that the leaves are able to store nutrient, and the variation of vacuoles ion concentration in the leaf cells within a certain range indicates that the leaves could effectively accumulate and store P, K mineral elements. Comparison of the radioactivity of P, K elements in the leaves which had been applied at different period, it shows that the amount of external output by the leaves depends on the intensity of the pool (including the fruits, leaves and other organs). As the main metabolic pool, the changes in the strength of fruit pool will direct regulate the output volume of P, K mineral elements. At the rapid growth and near fruit ripening stage, the fruits will absorb much more P, K mineral elements, and thus the isotope detained in the leaves will be reduced accordingly. This indicates that the leaves have nutrient storage function during spraying fertilizer on the surfaces of the leaves. The measurement of nutrient variable range in the leaves by virtue of different concentrations is possible to determine the appropriate concentration of leaf fertilizer. In addition, the observation of ion content dynamics in leaves after supplying mineral elements at different developmental stages can provide reference to determine reasonable intervals of leaf fertilizer for the production.

5. Acknowledgements

This work was funded by 973 project (2011CB100600), Science and Technology in Trade Project (200903044) and Key Laboratory of Beijing Municipality of Stress Physiology and Molecular Biology for Fruit Tree.

6. References

Cummings GA. K-fertilization increases yield and quality of peaches. Better Crops with Plant Food,1980,64:20-21

Tagliavini M, Zavalloni C, Rombola A D. Mineral nutrient partitioning to fruits of deciduous trees .Acta Horti.,2000,512,131-140

Habib R, Possingham J V, Neilsen G H. Modeling fruit acidity in peach trees effects of nitrogen and potassium mutriton. Acata Horti.,2000,512,141-148

Hu Shibi, Zhao Qiang, He Shoulin.The absorption,distribution,storage and redistribution of 86Rb in kyoho grapevine. Acta Horticulturae Sinica, 1998, 25 (1) : 6-10

Huang Weidong, Zhang Ping,Li Wenqing.The effects of 6-BA on the fuit development and transportation of carbon and nitrogen assimilates in grape. Acta Horticulturae Sinica, 2002, 29 (4) : 303-306

Xie Shenxi Zhang Qiuming.Absorption of 32P by leaf and peel of citrus unshiu during fruit development.Subtrop plant research commun,1994,23(2):8-13

Fu Yuman, Suo Binhua,Chen Guang, Liu Tong, Liu Zhaorong.Absorption,distribution and metabolism of phosphorus in ginseng. Journal of Jilin Agricultural University,1997, 19 (2) : 58-61

Pei Xiao bo,Zhang Fu man,Wang Liu.Effect of light and temperature on uptake and distribution of nitrogen,phosphorus and potassium of solar greenhouse cucumber. Scientia Ag ricultura Sinica, 2002, 35(12) :1510-1513

Zhou Yurong, Chen Mingli. Studies on the absorption and distribution of 32P in day-lily.Journal of Southwest Agricultural University,1996,18(5):416-420

Han Zhenhai, Wang Qian. Microgirdling-a new method for investigating the pathway of nutrition forward into fruit. The commition of china scicene and technowledge,1995

17

Intercellular Communication in Response to Radiation Induced Stress: Bystander Effects in Vitro and in Vivo and Their Possible Clinical Implications

Maria Widel
Institute of Automatics, Electronics and Informatics
Silesian University of Technology
Poland

1. Introduction

Communication between cells is important for maintaining homeostasis, the physiological regulatory processes that keep the internal environment of a system in a constant state. A disease can disturb the internal equilibrium of cells, and this can be further disrupted by various therapies. Malignances are the diseases that need to be treated by highly aggressive methods, such as radiotherapy, which affects not only tumor cells but also normal cells adjacent to the tumor and usually included in the radiation field. This treatment may interfere with normal intercellular communication. It has been a central radiobiological dogma for decades that damaging effects of ionizing radiation are the result of direct ionization of cell structures, particularly DNA, or are due to indirect damage *via* water radiolysis products. Indeed, DNA damage such as chromosomal aberrations, micronuclei, sister chromatid exchange and mutagenesis result from ionizing radiation. All of these types of damage, if unrepaired, can lead to cell death or, if misrepaired, can lead to genomic instability and carcinogenesis. Recently however, the attention was focused on the third mechanism, a phenomenon termed "radiation induced bystander effect" (RIBE). This phenomenon is a non-targeted effect where molecular signal(s) produced by directly irradiated cells elicit subsequent responses in unirradiated neighbors. These responses are manifested as decreased survival, increased sister chromatid exchanges (SCE), chromosomal aberrations (CA), micronucleus (MN) formation, gene mutations, apoptosis, genomic instability, neoplastic transformation and a variety of damage-inducible stress responses (reviewed in Morthersill and Seymour, 2001, Lorimore et al., 2003, Morgan, 2003a, 2003b, Little, 2006a,b, Chapman et al. 2008, Rzeszowska-Wolny et al., 2009a). Bystander effect accompanies very low doses of alpha particles (mGy and cGy), (Nagasawa and Little, 1992, Lorimore et al., 1998), as well as irradiation of cells with a low LET radiation (X- and gamma rays), even at conventionally used higher clinical doses (Morthersill and Seymour, 1997, 1998, 2002b, Przybyszewski et al., 2004). The mechanisms responsible for RIBE are complex and not quite well-known. Mechanisms by which bystander signals may be transmitted from irradiated to non-irradiated cells involve direct cell-to-cell contact mediated by gap junction intercellular communication (GJIC), and indirect communication by means of soluble factors secreted by irradiated cells into the surrounding medium. It is believed that molecular signaling factors released by cells irradiated and dispatched to the medium or transferred through GJIC induce various signaling pathways in neighboring cells, leading to the observed effects. The nature of these factors may be different and they have not been definitely defined. In addition to short-lived oxygen and nitrogen free radicals (Matsumoto et al., 2001, Azzam et al., 2002), long-lived radicals (Koyama et al., 1998), interleukin 8 (Narayanan et al., 1999), TGF-β (Shao et al., 2008 a, b, Massague and Chen, 2000) and other agents can be included. Potentially, bystander phenomenon could play an important role in the appearance of undesirable localized or systemic radiotherapeutic effects in tissues not included in the irradiation field. Furthermore, the effect may appear after low-dose irradiation during diagnostic radiology procedures and following application of a radioisotope for diagnosis or treatment (Prise and O'Sullivan, 2009). Factors emitted by irradiated cells may have impact on risk of genetic instability and the induction of mutation. However, the radiation-induced bystander effect may have both detrimental and potentially beneficial consequences. If cells directly hit by ionizing energy will, through their signals (secreted or transmitted through the gap junction) damage adjacent cancer cells, or will initiate differentiation of these cells, it is desirable. However, if normal cells are damaged (epithelial and endothelial cells, fibroblasts, leucocytes, etc.), then the effect may be a disadvantage that increases the unwanted effects of radiotherapy such as late complications and second primary tumors. Bystander effect can be particularly important in the case of the use of current techniques of irradiation, such as 3D conformal radiation therapy (3D-CRT) and intensively modulated radiotherapy (IMRT), the purpose of which is to reduce the irradiation dose in healthy tissues (Followill et al., 1997). Some data indicate that bystander effect also occurs *in vivo* (Koturbash et al., 2006, 2007, Ilnytskyy et al. 2009). The studies of bystander effect in *in vivo* animal models show that the post- radiation damage can appear in tissues distant from the place of irradiation, and the effect may vary depending on the type of tissue. However, recent experimental results (Mackonis et al., 2007), including our own (Widel et al. 2008, and unpublished), show that cross-talk between irradiated and un-irradiated cells may be sometimes protective and non-irradiated cells, which are in the vicinity of irradiated cells can hamper the effects caused by their irradiation. Furthermore, a radioprotective bystander effect has been observed in several studies with low-dose exposure in the form of increased cell redioresistance to subsequent higher doses (e.g. Sawant et al., 2001, Prise et al., 2006). Less known are the consequences of bystander effect in the case of dose fractionation during external irradiation. Our preliminary results from *in vitro* fractionation dose experiments, presented in this Chapter indicate that apoptosis is even more effectively induced in human melanoma radiation-targeted and bystander cells when the same dose is delivered in 3 fractions than in one single dose. A growing body of experimental *in vitro* and *in vivo* data indicate the occurrence of bystander phenomenon in radionuclide-based radiotherapy (Xue et al., 2002, Gerashchenko and Howell, 2004, Boyd et al., 2006, Mairs et al. 2007). However, studies of radionuclide-induced bystander effect demonstrate varying responses (compared to low LET radiation-induced ones), being either damaging or protective depending on dose and type of emitters. The practical consequences, as well as capacities of the bystander effect, in terms of modulating radiotherapeutic approaches, are therefore still uncertain and are the subject of intensive research. It is possible that the impact of bystander signaling on both cancer and healthy tissue responses is more relevant than it is believed at present. Below is a comprehensive review of the various aspects of radiation-induced bystander effect, based on the current knowledge and our own experimental results.

2. History of bystander effect phenomenon

First observations of the bystander effect phenomenon appeared in the nineties of the last century. Using a low-dose of alpha particles which targeted only 1% of cultured Chinese hamster ovary cells (CHO), Nagasawa and Little (1992) noticed cell damage in the form of sister chromatid exchanges (SCE) appearing in about 30% of cells. The level of damage increased with 0.3-2.5 mGy dose, but not with higher ones. Subsequent experiments showed an increase in the number of cells with overexpression of *TP53* gene after 6 mGy alpha irradiation, but not after exposure to the same dose of X-rays (Hickman et al., 1994). Very soon, it appeared that this effect also occurs in cells exposed to radiation with a low LET radiation. It was observed that the factors inducing the observed effects in non-irradiated cells are soluble and can be passed through the growth medium (Deshpande et al., 1996, Morthersill and Seymour, 1997), or by an intercellular connection slot (Azam et al., 1998). Morthersill and Seymour (1997) showed that factors present in the culture medium collected from epithelial cells exposed to gamma radiation decreased survival of clonogenic non-irradiated cancer and epithelial cells in culture; therefore for the bystander effect to occur the contact of irradiated cells with non-irradiated is not necessary. Furthermore, reduced cell survival did not occur when medium harvested from irradiated fibroblasts was used. The cytotoxic effect of irradiation-conditioned medium (ICM) has been observed in several experimental systems following both particle (Deshpande et al., 1996, Lorimore et al., 1998) and photon irradiation (Clutton et al., 1996, Matsumoto et al., 2001). It was found that the bystander effect-signaling molecules may include tumor necrosis factor beta (TGFβ) and interleukin-8 (Narayanan et al., 1999) secreted to the medium or transferred through GJIC. Closing these connections by *lindane*, an inhibitor of gap junction, lead to the inhibition of bystander effect, evidenced as the reduced expression of *TP53*, *CDKN1A* (p21) and *CDC2* genes (Azzam et al., 1998), or increased survival of clonogens (Bishayee et al., 1999). Several studies have demonstrated that the radiation-induced bystander effect triggers apoptosis (Prise et al., 1998, 2006, Morthersill and Seymour, 2001, Przybyszewski et al., 2004) and increase of micronucleus frequency, DNA double-strand breaks (DSBs) measured as histone H2AX phosphorylation (Sokolov et al., 2007, Burdak-Rothkam et al., 2007), accumulation of p53 (Tartier et al., 2007) and ATM and ATR proteins (Burdak-Rothkam et al., 2008), epigenetic changes, such as DNA hypomethylation, as well as the expression of other genes (Chaudry, 2006, Iwakawa et al., 2008, Rzeszowska-Wolny et al., 2009b). Many of these experiments showed that higher doses of radiation, including those used in conventional radiotherapy, also induce bystander effects in non-irradiated cells. They confirmed the quantitative biophysical model of Nikjoo and Kvostunov (2003, 2006) which assumes that RIBE may be a component of neighborhood responses to radiation, both at low and high doses. The results obtained in tissue explant culture (Belyakov et al., 2002, 2006, Mothersill and Seymour, 2002b), tri-dimensional cell culture, *in vivo*-like models (Bishayee et al., 1999, 2001, Belyakov et al., 2005), and in animal studies (Koturbash et al., 2006, 2007, 2008) all point out to the bystander phenomenon relevance to clinical radiotherapy. Therefore, one cannot exclude that the intensity of side effects in healthy tissues following fractionated radiotherapy may be partly related to bystander effect. It is suspected that this effect may also lead to genetic instability, the consequence of which can involve development of secondary cancers (Hendry, 2001). Not always, however, radiation induced bystander effect has a damaging action. The signals emitted to the microenvironment by irradiated cells seem to induce in cells unexposed to radiation more complex effects, inter alia their differentiation, probably as a comprehensive response in order to preserve the integrity of the tissue (Belyakov et al., 2006, Vines et al. 2009).

3. Radiation induced bystander effect, genetic instability and adaptive response

Bystander effect, genetic instability and adaptive response seem to be related. Known as the genetic instability are the delayed effects such as lethal mutation, unstable chromosome aberrations, and delayed reproductive death (DRD) in distant generations of cells previously exposed to radiation (Gorgojo et al., 1989, Mendonca et al. 1989), or arising *de novo* chromosome aberrations (Kadim et al., 1995, Marder and Morgan 1993, Weissenborn and Streffer, 1989) and gene mutations (Little et al., 1997). Delayed reproductive death (DRD), manifested as diminution of clonogenic cell survival, appears to be caused neither by apoptosis nor by necrosis. DRD is mainly observed in cells with uninterrupted mechanisms of DNA double-strand breaks repair (Little et al., 1990, Little, 1999), but is not observed in cells with impairment of these mechanisms (Chang and Little, 1992). It was demonstrated that cell clones with post-radiation genetic instability evolve through many generations of descendants, the cytotoxic factors affecting non-irradiated cells (Kadim et al., 1995) and, the effect being independent of intercellular gap junctions (Nagasawa et al., 2003). Studies of genetic instability in which only some mouse marrow stem cells were targeted by alpha particles showed higher numbers of cells with chromosome aberrations than those of irradiated cells. These lesions are transferred to the descendant cells forming colonies (Loroimore et al., 1998). In addition, the surviving fraction of clonogenic cells decreases deeper with the dose than would result from the dose absorbed, provided the damage resulted from communication of lethally-irradiated cells with non-irradiated cells. Increased mutation frequency of hypoxanthine-guanine-phosphoribosyl transferase gene (*HPRT*) in distant generations of murine hematopoietic stem cells irradiated *in vitro* with both the X-rays and neutrons was also observed (Harper et al., 1997). Furthermore, human T-lymphocytes showed chromosome aberrations transferred through generations of their progenitor cells that had been irradiated with 3Gy X-rays dose (Holmberg et al., 1995). Factors inducing the bystander effects can be passed through gap junctions (Zhou et al., 2000, Azzam et al., 2002], or secreted to the surroundings (Lyng et al., 2000, Morthersill and Seymour, 1998). Some of them are clastogenic and can induce chromosomal damage in non-irradiated cells, analogous to that in directly-hit cells. Huang et al. (2007) observed that growth medium conditioned by some chromosomally unstable RKO derivatives induced genomic instability, indicating that these cells can secrete factor(s) that elicit responses in non- irradiated cells. Furthermore, low radiation doses suppressing the induction of delayed genomic instability by a subsequent high dose, are indicative of an adaptive response for radiation-induced genomic instability. Adaptive response is a phenomenon by which cells irradiated with a sub-lethal radiation dose (mGy or cGy) may become less susceptible to subsequent high-dose (a few Gys) radiation exposure (Wolff, 1996, Marples and Skov, 1996). The mechanism of this phenomenon is not sufficiently known. Irradiation leads to disturbances of the balance between pro-oxidant and anti-oxidant signaling molecules; one of such molecules can be nitric oxide (NO) (Spitz et al., 2004). An increase of radioresistance was observed in human glioblastoma A-172 cells with functional *TP53* gene when they were co-incubated with irradiated (1-10 Gy X-rays) cells of the same line transfected with mutated *TP53* gene (A-172/mp53), or incubated in the presence of conditioned medium from irradiated cells (Matsumoto et al. 2001). The sign of radioresistance was the accumulation of HSP72 and p53 protein which had declined in the presence of nitrogen oxide scavenger or inducible nitrogen oxide synthase inhibitor. Another probable mechanism thought to be a cellular adaptive response is the low-dose enhancement of DNA repair ability and antioxidant activity, resulting in more proficient cellular responses to the subsequent challenge. Sawant et al. (2001) observed that the exposure of C3H 10T91/2 cells to single alpha particle radiation, which hit only 10% of cells, caused the death of a much larger number of cells. However, the use of 2cGy gamma rays 6 hours before exposure to the alpha particles continuously reduced the bystander effect expressed as increased surviving cell fraction. Increased resistance induced by large dose of gamma radiation was also observed in cells of the same line if they were pre-exposed to a cGy dose of 60-Co (Azzam et al., 1996), and the reduction in the percentage of micronuclei was accompanied by an increase in the repair of DNA double-strand breaks (Azzam et al. 1994). Recently, it was presented that different cell lines can show different pattern of response to low priming dose (Ryan et al. 2009). An adaptive response was detected in cell lines known to produce hypersensitive response, and was inversely correlated with the bystander effect suggesting that an adaptive response may be mutually exclusive to the bystander effect.

4. The mechanisms of radiation induced bystander effect

The ionizing radiation acts through direct ionization of organic macromolecules or through reactive oxygen species (ROS), namely, hydroxyl radical ($OH^{\bullet}$), hydrogen peroxide (H_2O_2) and superoxide radical anion ($O_2^{\bullet -}$), the effect of which is primarily oxidative DNA damage (Marnett, 2000, Matsumoto et al., 2007). Half-life of ROS is extremely short and penetration distance is expressed in micrometers. Therefore, these factors may not reach non-irradiated cells. Electron spin resonance studies have shown, however, that long-lived radicals with a period of half-lives ca. 20 hours may appear in cells after irradiation, even at room temperature (Koyama et al., 1998); if transferred to the surroundings, they may be the factors inducing DNA damage in non-irradiated cells. The long-lived secondary radicals are likely to be less active in damaging DNA than the extremely active primary radicals generated during irradiation time. Therefore, DNA damage induced by secondary radicals may not be a sufficient barrier to stop the replication of DNA and can lead to duplication of altered DNA through generations of cells, and finally to mutation and neoplastic transformation (Azzam et al., 2003, Clutton et al., 1996, Iyer and Lehnert, 2000, Lala and Chakraborty, 2001). DMSO, a radical scavenger, reduced the level of DNA damage in irradiated cells and inhibited the bystander effect which seems to confirm the role of reactive forms of oxygen in initiating signaling molecules (Hussain et al., 2003, Kashino et al., 2007). Also, the use of vitamin C as a scavenger of long-lived radicals compromised the level of micronuclei in human fibroblasts co-incubated with irradiated cells (Harada et al., 2008), as well as in K562 myelogenous lukemia cells treated with medium from irradiated cultures of the same cell line collected one hour post irradiation (Konopacka and Rzeszowska-Wolny, 2006). However, not only DNA is the target for ROS; no less important are the fatty acid molecules, in which the peroxidation chain reactions lead, through short-lived lipid radicals, to stable end-products such as malondialdehyde (MDA), 4-

hydroxynonenal (4HNE) and other with mutagenic and carcinogenic properties and which can form massive DNA adducts (Marnet, 2000, Zhong et al., 2001). The end-products of lipid peroxidation have secondary signaling molecule properties and can activate a cascade of signals leading to either DNA damage repair, or to damage stabilization or apoptosis (Hu et al., 2006). In our research we found increased MDA concentration in irradiated Me45 human melanoma cells growing in the form of megacolonies, as well as in the neighboring megacolonies growing in the same flask but protected against irradiation with a lead shield (Przybyszewski et al., 2004). At the same time, we found in both the irradiated and shielded megacolonies, decreased glutathione peroxidase (GSH-Pox) and mitochondrial superoxide dismutase (MnSOD), as well as elevated numbers of single- and double-strand DNA breaks (SSBs and DSBs), as assessed by single cell gel electrophoresis. The level of DNA breaks in non-irradiated cells was lower and appeared with several-hour delay compared to that observed in irradiated cells, which may suggest participation of long-lived radicals in the bystander effect induction (Przybyszewski et al., 2004). Time-shifted appearance of DSBs in neighboring cells estimated as the expression of phosphorylated histone H2AX (γH2AX foci) has been observed in the *in vitro* (Hu et al., 2006, Sokolov et al. 2007) as well as in *ex vivo* (Sedelnikova et al., 2007) conditions. While the phosphorylation of histone H2AX at serine 139 is a very early-stage event in cells directly exposed to radiation, the appearance of gamma-H2AX foci in cells co-cultured with irradiated ones, or treated with ICM only, may even take several hours. The gamma-H2AX foci, which indicate the presence of DNA DSBs in cells exposed to the signals transmitted by irradiated cells, co-localize with other proteins involved in the cell cycle control and DNA damage repair, such as ATM, MRE11, NBS1, Rad50 and 53BP1 (Sokolov et al., 2007). It is worth noticing that, based on ATM foci enumeration, Ojima et al. (2009) found that DSBs induced by the radiation-induced bystander effect persist for long periods (over 24 h), whereas DSBs induced by direct radiation effects are repaired relatively quickly. However, ATM foci persisted even longer (48 h) if bystander fibroblasts were co-incubated with very low (1.2 mGy) irradiated counterparts. This indicates that bystander signals coming from irradiated cells *induce* chromatin damage which differs from that induced by direct irradiation. It has been shown that not exclusively irradiation of DNA but irradiation of cytoplasm induces cytogenetic damage in both irradiated and bystander glioma cells and fibroblasts to a comparable extent (Shao et al., 2004) The bystander responses were completely eliminated when the populations were treated with nitric oxide scavenger or agent which disrupt membrane rafts. This finding shows that direct DNA damage is not required for induction of important cell-signaling mechanisms after low-dose irradiation and that, the whole cell should be considered a sensor of radiation exposure. The use of compounds that compromise the level of nitrogen oxide abolishes the bystander effect elicited as γH2AX expression. Nitric oxide (NO) seems to be an important signaling molecule transmitted by irradiated cells, which initiates the changes in cells not exposed to radiation (Matsumoto et al., 2001, 2007, Shao et al., 2008a, b). This small molecule is also a free radical which is synthesized from the L-arginine with the participation of nitric oxide synthase (NOS). It plays important, often contradictory roles in many biological processes, stimulating either the proliferation or apoptosis, which primarily depends on its concentration (Shao et al., 2008b). Nitric oxide is vasodilatator, neurotransmitter and an immunomodulatory agent, but it may also cause damage to DNA by generating peroxynitrite anion (ONOO-), which may cause oxidation or nitration of DNA (Xu et al., 2002). Shao et al. (2008a, b) demonstrated that radiation-generated NO induced in glioma cells TGFβ1, the multifunctional transcription factor involved in the transcription of proteins engaged in cell proliferation and differentiation, immunomodulation, cell-cycle control and apoptosis (Massague and Chen, 2000). The use of inducible nitric oxide synthase inhibitor, or anti-TGF antibodies which compromise micronuclei in cells directly irradiated with alpha particles and adjacent non-irradiated cells indicates a positive feedback. However, NO role as a mediator of the bystander effect has not been observed in all tested glioma cell lines (Matsumoto et al., 2001). In several types of cancer (colon, lung, throat) expression of inducible nitric oxide synthase (iNOS) was also linked to the *TP53* gene mutation (Lala and Chakraborty, 2001) indicating that the correct protein of p53 gene may negatively regulate the accumulation of iNOS. Many other factors were proposed as the bystander effect mediators, among them interleukin 8 (Narayanan et al., 1999), soluble tumor necrosis factor (TNFα) as well as Fas and TRAIL death ligands (Lucen et al. 2009). Also, multiple pathways are activated that take part in transmitting the bystander effect signals. Those induced in human fibroblasts by alpha particles (0.3-3 cGy) and transmitted through the GJIC or surrounding environment activated in adjacent cells various proteins such as MAP- kinase, NFκB, Raf-1, ERK1/2, JNK, AP-1 and others (Azzam et al. 2002, Lyng et al., 2006). Since application of SOD and catalase neutralizes the resulting oxygen radicals and hydrogen peroxide and hampers the bystander effect (reduction in the level of micronuclei, inhibition of nuclear factor κB and p38 MAPK activation), the mediators of these processes appear to be reactive oxygen and nitrogen species (Azzam et al., 2002). Targeting the nucleus or cytoplasm of HeLa cells by single helium ions induced expression of 53BP1, the protein which marks double-stand breaks in DNA (Tartier et al., 2007). The use of aminoguanidine, an inducible NO synthase inhibitor, or radical scavenger DMSO, cause inhibition of 53BP1 protein expression in both irradiated and co-incubated non-irradiated cells, pointing to the NO and ROS as the mediators of these lesions. At the same time, it was observed that antibiotic *filipin*, which damages the glycosphingolipid microdomains in cellular membrane, inhibited cellular signals from irradiated cells and led to a drastic reduction in the 53BP1 foci in neighboring cells. This reveals that transmission of bystander signals is dependent on the integrity of the cellular membranes, whereas membrane integrity was not necessary to generate the damage in irradiated cells. Also, the presence of mitochondria was necessary to generate bystander signals by irradiated cells, but was not necessary to their reception (Tartier et al., 2007). Calcium ion channels seem to play a role in the transmission of bystander signals. It was observed that biogenic amines, such as serotonin (5-hydroxytryptamine, 5-HT) and dopamine, may be the transducers of signals emitted by irradiated cells. The level of 5-HT neurotransmitter in culture medium decreased after irradiation of cells, likely due to its binding to the receptors which form the calcium channels, and leads to increased level of micronuclei (Poon et al., 2007). These effects were abolished after treatment of cells with calcium channel blockers calcicludin or rezerpin, which are the natural antagonists for serotonin (Poon et al., 2007, Shao et al., 2006). The study of transcript levels using DNA microchips may indicate signaling pathways and genes that are involved in the radiation-induced bystander effect. Gandhi et al. (2008), when examining the overall gene expression (global genome expression), after irradiation of human lung fibroblasts with alpha particles (0.5 Gy and 4-hour co-incubation with non-irradiated cells), observed that the expression of over 300 genes in both groups (hit and non-hit) was changed, and that 165 genes were common to both groups. Among them were genes mainly over-expressed in irradiated cells (CDKN1) and those that were over-expressed equally in irradiated and neighboring cells, namely NFκB-regulated PTGS2 (cyclooxygenase 2), IL8 and BCL2A1. However, Chaudhry (2006) observed that gene expression profile differs in irradiated human fibroblasts and in non-irradiated cells treated only with radiation-conditioned medium. In the former, over-expressed were the genes of early response to radiation, while in the bystander cells the over-expressed ones included genes involved in the intercellular communication. In our genome-wide microarray study, we compared transcript profile changes in Me45 human melanoma cells grown in culture medium from irradiated cells with those which occurred after irradiation and we also observed the bystander effect at the genome level (Rzeszowska-Wolny et al., 2009). Using the criterion of a greater than ±10% change, transcripts of >10,000 genes were shown to be expressed at increased or decreased levels under both conditions, and almost 90% of these were common to ICM-treated and X-rays-treated cells. Among them were genes involved in the neuronal receptor-ligand interactions, oxidative phosphorylation, cytokine–cytokine receptor interactions, proteasomes, ribosomes and cell cycle regulation. All these tests indicate a very complex mechanism of cell response to both ionizing radiation and for signals transmitted by them to communicate with the neighboring cells.

5. The role of the p53 protein in the response to bystander signals

The *TP53* gene is a tumor suppressor gene which participates in the regulation of cell cycle and apoptosis. Its main role is to prevent the transmission of genetic disorders in cells to daughter cells by extending G1 phase, which allows the cell to repair DNA damage induced by various egzo-and endogenous agents, mainly the oxidative stress. When the damage is too bulky or the repair is ineffective, *TP53* initiates apoptosis through its own product, p53 protein, which is a transcription factor for multiple genes involved in DNA repair, regulation of cell cycle and apoptosis (Chipuk and Green, 2006, Tlsty, 2002). The role of p53 protein in the bystander effect is debatable, however. Research carried out using human fibroblasts cell lines, where only a small fraction of cells was exposed to alpha particles has shown a significant increase in p53, as well as $p21^{Waf1}$ protein, not only in the targeted, but also in the non-targeted cells (Azzam et al., 1998). The effect disappears after inhibition of the gap junction intercellular communication. Similarly, expression of p53 protein was observed in the rat lung epithelial cells adjacent to alpha particle-targeted cells (Hickman et al, 1994). However, survival of clonogenic fibroblasts after 2 and 4 Gy was increased when they were exposed to the medium from fibroblasts gamma-irradiated with a dose of just 1cGy. This was accompanied by the reduction of p53 protein level in addition to the increase in intracellular pool of reactive oxygen radicals and DNA-repair protein nuclease APE (Iyer and Lehnert, 2000). The appearance of DSBs is accompanied by DNA binding protein 53BP1 which may be detected immunochemically using fluorescent-labeled antibodies. It was shown that the irradiation of cell cytoplasm with single alpha particles, induced increased numbers of 53BP1 foci not only in nuclei of irradiated cell, but also in adjacent to them non-irradiated cells (Tartier et al., 2007). The use of inhibitors targeting reactive oxygen radicals and nitric oxide prevented the formation of DNA breaks in irradiated and adjacent cells. This indicates that the bystander effect signals are transmitted not only between cells but even between cell compartments. Also, the use of membrane specific antibiotic (*filipin*) to disrupt membrane-dependent signaling has resulted in lowering the number of clusters of 53BP1 foci an important sensors of DNA double strand breaks, in cells co-incubated with irradiated ones, indicating that reception of bystander effect signaling molecules requires the integrity of the cellular membranes (Tartier et al., 2007). The tests in rats which were given 1Gy doses of X-rays, both whole-body or head-area-only, revealed expression of p53 protein in the spleen of animals, pointing to the involvement of the TP53 gene in the bystander effect *in vivo* (Koturbash et al., 2008). However, in our own research using HCT116 colon cancer cells lines differing in *TP53* status, and the transwell system of co-cultivation, we observed that *TP53 gene* is not required to uncover the bystander effect. Non-irradiated *TP53*-knockout cells (HCT116p53 -/-) were even more sensitive to apoptosis induced by signals sent by irradiated (2 Gy) cells than wild-type cells (HCT116p53+/+) (Widel et al., 2009). In the same experiments we noticed that the level of micronuclei induced in cells co-cultured with non-irradiated ones did not differ between both lines. Recently, He et al (2010) found that the bystander effect after irradiation can be modulated by the p53 status of irradiated hepatoma cells and that a p53-dependent release of cytochrome c may be involved in the RIBE. Following irradiation cytochrome c was released from mitochondria into the cytoplasm only in HepG2 (wild-type p53) cells, but not in PLC/PRF/5 (p53 mutated) or Hep3B (p53-null) cells. Only irradiated HepG2 cells induced bystander effect elicited as micronuclei (MN) formation in the neighboring Chang liver cells. In conclusion, the various criteria for assessing the role of *TP53* gene reveal differences in its response to bystander effect signals.

6. Bystander effect can function bi-directionally

Recent studies have shown an interplay between adjacent irradiated and non-irradiated cell populations. Thus, signals leading to damage in non-irradiated cells, sent by the irradiated ones, are answered by non-hit cells affecting in turn the directly-irradiated ones. Experiments performed on MM576 melanoma cells, the goal of which was to investigate the impact of modulating irradiation fields in a way to resemble the intensity-modulated radiotherapy technique (IMRT) on survival showed, that the mutual communication works in three different manners (Mackonis et al., 2007). The first type of this communication, the classic "bystander effect", occurs when irradiated cells growing in one part of the field damage the adjacent non-irradiated cells growing in another part of the field. The second type of communication, causes an increase in the survival of non-irradiated cells, when they are co-cultured with cells exposed to high doses (6-20 Gy) or even a lethal dose. One of the factors responsible for this process is, according to these authors, the eruption of "death-burst signals", which promotes proliferation of the non-irradiated cells, although the authors do not specify the chemical nature of these signals. The third type of communication causes increased survival of cells that have received a high dose of radiation, through signaling from neighboring cells exposed to low-dose in another part of the field (Mackonis et al., 2007). Also, the irradiation of human fibroblasts with low doses of alpha-particles resulted in an increased proliferation, reduction of the level of p53 and CDKN1 (p21Waf-) proteins and an increase in the level of the CDC2 kinase. The promitogenic effect was associated with an increase in the level of the TGFβ1-induced by reactive oxygen species (Iyer and Lehnert, 2002). Our recent study revealed bystander effect of the third type, similar to that described by the Mackonis, indicating the bilateral signaling of irradiated and non-irradiated cells (Widel et al., 2008, and unpublished). Using the transwell system of co-incubated irradiated mouse lung cancer cells (LLC) with non-irradiated fibroblasts (NIH3T3) growing in inserts we studied the mutual interaction of cells in terms of micronuclei and apoptosis induction. The membrane of insert bottom with 0.4 μm pores separates both types of cells but enables free circulation of medium between them. LLC cells growing in 6-well plates were irradiated with doses of 2 and 4 Gy X-rays generated by a therapeutic accelerator (Clinac 600). Immediately after irradiation the inserts with non-irradiated (bystander) fibroblasts were inserted into the wells and co-incubated for a desired time. Another set of irradiated LLC cells was incubated without cells in inserts, the latter filled with medium only. Micronuclei and apoptosis were scored in microscopic slides prepared from cells harvested at different time-points. The results show that the irradiated cells induced apoptosis and micronuclei in bystander fibroblasts. For the first time we show the radioprotective effect of normal cells on irradiated cancer cells (the opposite bystander effect); thus the percentage of micronuclei and apoptosis in irradiated LLC cells co-incubated with NIH3T3 fibroblasts was significantly decreased in comparison with analogous levels in the irradiated LLC cells incubated without fibroblasts growing in inserts (Figure 1).

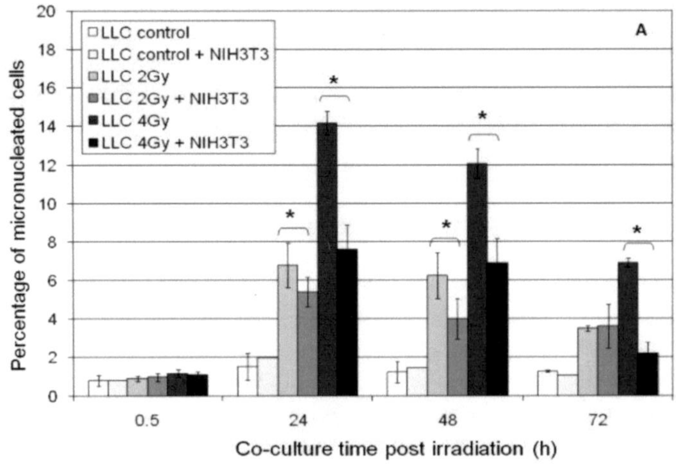

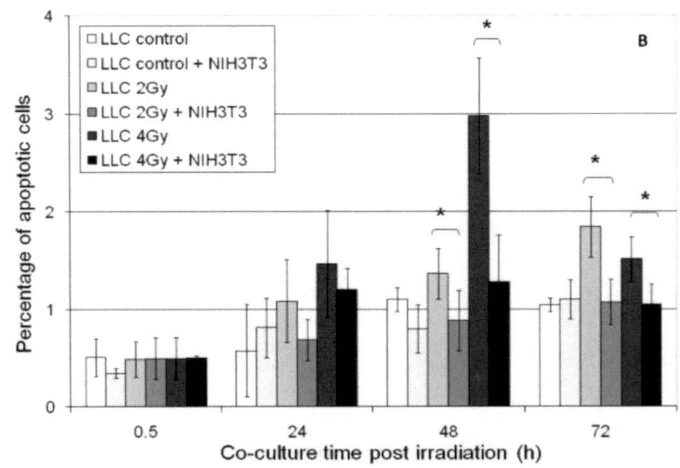

Fig. 1. Non-irradiated murine NIH3T3 fibroblasts co-cultured with irradiated Lewis lung carcinoma cells significantly diminish micronuclei (A) and apoptosis frequency (B) in irradiated (2 and 4 Gy) cancer cells compared with those irradiated and incubated without fibroblasts. Results are means ± standard deviation from three independent experiments (*$p<0.05$, Student's t-Test).

The mechanism of this phenomenon requires clarification. It seems that the radioprotective bystander effect is a feature of normal fibroblasts. Indeed, the same effect, i.e. a significant reduction in the level of micronuclei and apoptosis in irradiated human melanoma Me45 cells was observed when these were co-incubated with normal human fibroblasts (Widel et al., unpublished). The progressive increase of micronuclei and apoptosis was paralleled by an increase of ROS; however, the ROS level in irradiated melanoma cells, which were co-cultured with fibroblasts, was significantly diminished. Such a radioprotection was not observed in irradiated Me45 cells co-cultured with cells of the same line of melanoma (Widel et al, unpublished). We believe that the observed radio-protective effect of non-irradiated fibroblasts exerted on irradiated melanoma cells may result from signaling molecule(s) modifying the redox status of irradiated cells. Similar effect is likely to occur during cancer radiotherapy, causing some decrease of damage to cancer cells owing to fibroblasts present in tumor tissue.

7. Fractionated irradiation and bystander effect

Experimental data on bystander effect mostly come from single-dose application experiments *in vitro*. However, there is a lack of knowledge, which would have potential clinical implication, e. g. whether bystander effect occurs during fractionated treatment. Mothersill and Seymour (2002a) performed experiments involving repeated treatment of bystander cells with medium collected from irradiated cells as well as involving repeated dose exposure of cells producing bystander signals, as a way of mimicking fractionated exposures. The recovery factor was defined as the surviving fraction of the cells receiving two doses (direct, or ICM) separated by an interval of 2 h divided by the surviving fraction of cells receiving the same dose in one exposure. The authors observed that fractionated bystander treatments removed the effect of dose sparing that is observed after conventional fractionated regime, during which cells can repair DNA damage. Using Me45 human melanoma cell line established at the Center of Oncology in Gliwice (Kramer-Marek et al, 2006) we compared frequency of apoptosis and micronuclei formation in directly irradiated and bystander cells after single doses (1.5 - 6 Gy) and after doses divided into 3 fractions given at consecutive days (3 x 0.5 Gy – 3 x 2 Gy). We used a transwell system of co-incubation which allows co-culturing the irradiated cells growing in wells with non- irradiated cells growing in inserts. This system to some extent resembles situation *in vivo*, due to prolonged contact of non-irradiated and irradiated cells. As a source of X-rays (6 MV) Clinac 600 therapeutic accelerator was used. Non irradiated control cells were-sham exposed. After irradiation, inserts with growing non irradiated cells were placed into wells with irradiated ones and co-incubated. Before irradiation medium in both, wells and inserts, was replaced by fresh aliquots. To observe the response of hit and bystander cells after the set time of incubation (0, 24 and 48 h), we performed microscopic analysis of micronuclei induction and apoptosis. The results obtained show that both single dose irradiation and fractionation of the dose into three fractions effectively induced bystander effect in malignant Me45 melanoma cells. However, fractionated irradiation at low doses (Fig. 2) appears to be much more effective in inducing micronuclei in directly hit and bystander cells, whereas higher apoptosis induction was clearly seen in hit, and especially in bystander cells, at all doses in fractionated system (Fig. 3).

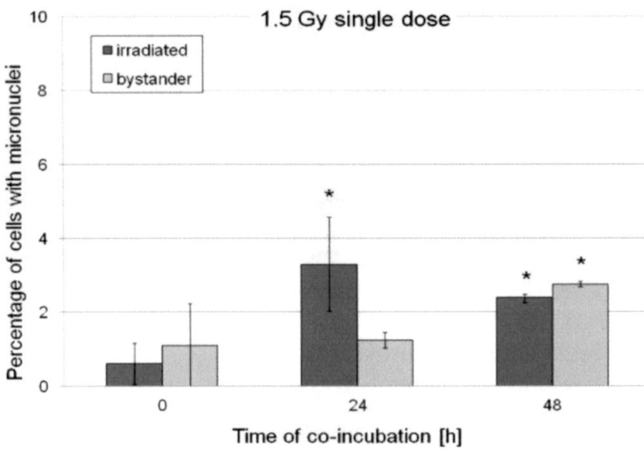
1.5 Gy single dose
irradiated
bystander
Percentage of cells with micronuclei
10
8
6
4
2
0
0
24
48
Time of co-incubation [h]

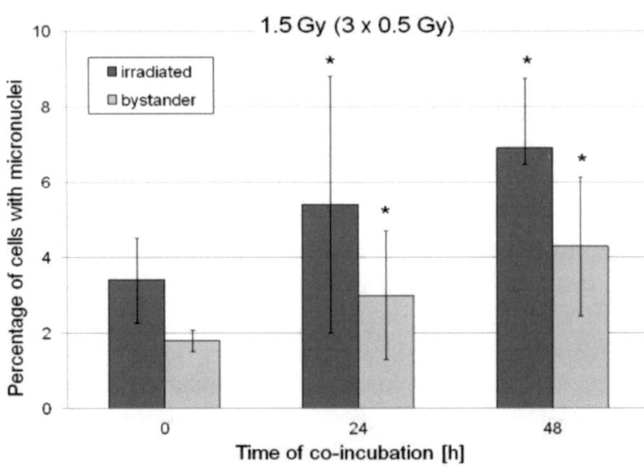
1.5 Gy (3 x 0.5 Gy)
irradiated
bystander
Percentage of cells with micronuclei
10
8
6
4
2
0
0
24
48
Time of co-incubation [h]

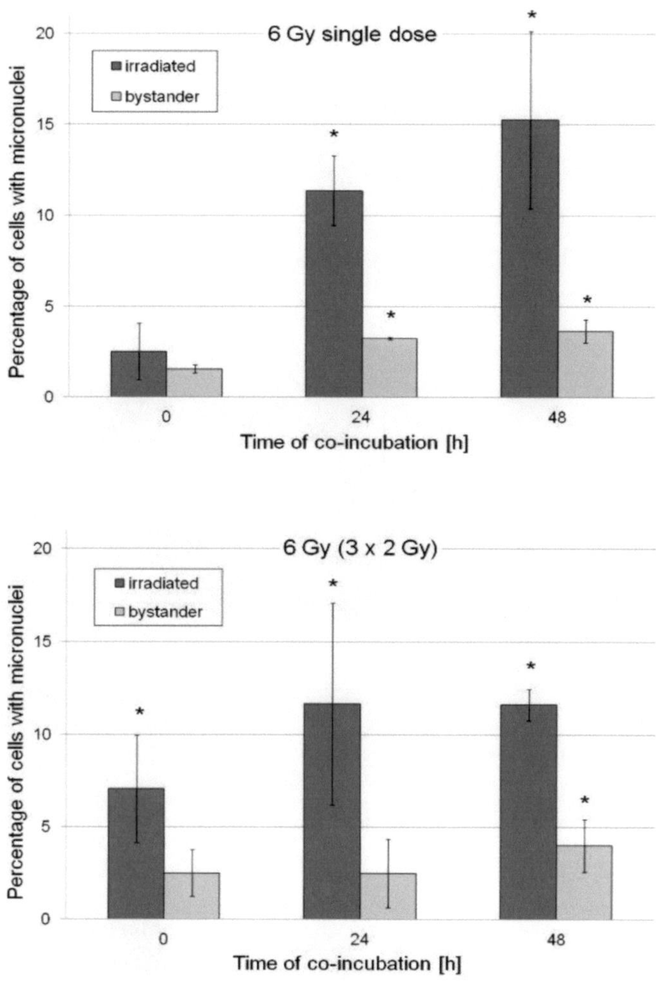

(* denotes statistical difference from corresponding control, $p<0.05$, Student's t-test).

Fig. 2. Yield of micronuclei induction in Me45 melanoma cells irradiated with single or fractionated doses, in comparison with bystander cells. Data show means ± standard deviation and were obtained from three independent experiments

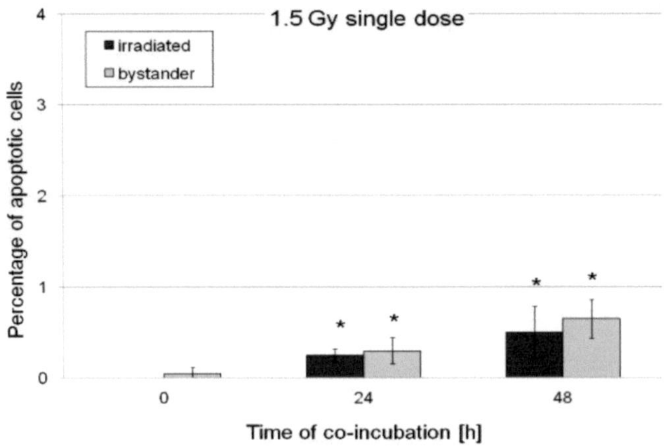
1.5 Gy single dose
irradiated
bystander
Percentage of apoptotic cells
4
3
2
1
0
*
*
*
*
0
24
48
Time of co-incubation [h]

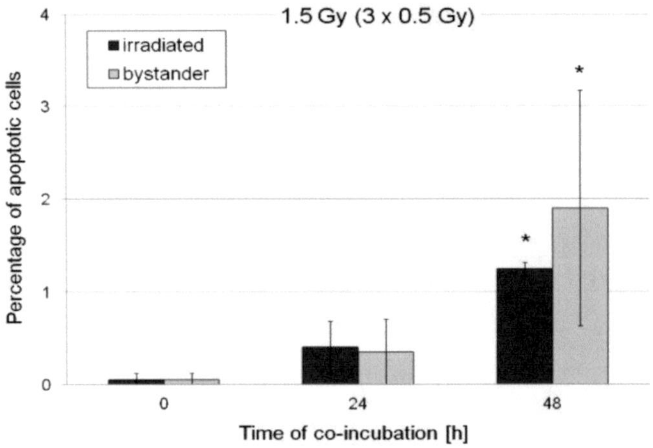
1.5 Gy (3 x 0.5 Gy)
irradiated
bystander
Percentage of apoptotic cells
4
3
2
1
0
*
*
0
24
48
Time of co-incubation [h]

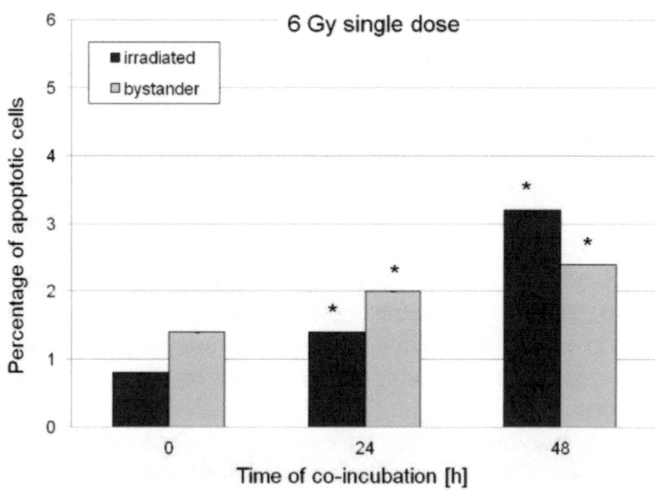

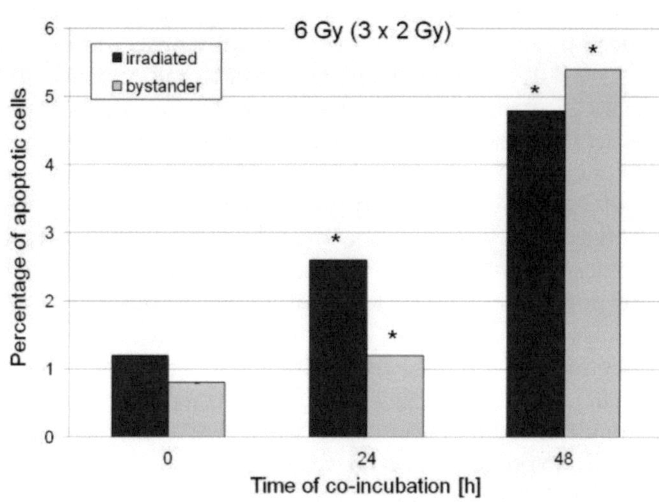

(* denotes statistical difference from corresponding control, $p<0.05$, Student's t-test).

Fig. 3. Yield of apoptosis induction in Me45 melanoma cells irradiated with single or fractionated doses, in comparison with bystander cells. Data show means ± standard deviation and were obtained from three independent experiments

Our data indicate that the bystander effect may play some role during fractionated radiotherapy and should be regarded as an important part of ionizing radiation effect on living cells. Although fractionated irradiation was also applied *in vivo* to study the bystander effect on the level of DNA epigenetic changes in the non-exposed spleen of cranial irradiated mice (Ilnytskyy et al., 2009), the fraction doses used were far below those clinically applied. However, the authors observed that acute irradiation induced more pronounced bystander effect than fractionated irradiation.

8. Radionuclide induced bystander effect

Induction of the bystander effect is prevalent at low radiation doses and low dose rates (Seymour and Mothersill, 2000), the characteristic features of targeted radionuclide treatment of cancer. Thus, one could expect that bystander effects induced by targeted radionuclides could have a strong impact on radiotherapeutic and diagnostic treatment (Prise and O'Sullivan, 2009). An increasing body of data indicates the involvement of bystander phenomenon after radionuclide application under experimental conditions. It can appear as damaging or protective effects in dependence on dose and dose rate. A very low dose of photon radiation (~ 30 keV) emitted by iodine-125 radioisotope (4mGy dose/day to 1,4 mGy/day) during a three month exposure of hybrid HeLa cells with human fibroblasts caused resistance of these cells to neoplastic transformation when they were challenged by subsequent irradiation with 3 Gy of ^{137}Cs gamma rays (Elmore et al., 2008). Lowering of dose rate below 1mGy/day abolished the adaptive answer, suggesting that low dose-rate above a certain threshold is responsible for this type of radio-adaptation. The damaging bystander effect induced by radionuclide is also frequently observed in *in vitro* experiments. Various type of cells may differ however in response to radionuclide induced bystander signals. Chen et al. (2008) using ^{125}I seeds irradiated two lung cancer cell lines that had different sensitivities to HDR gamma-ray irradiation and investigated the bystander effect of DNA DSBs as histone H2AX phosphorylation, and micronuclei formation. They found that the proportion of bystander cells with micronuclei and number of γH2AX foci was higher in radiosensitive NCI-H446 cell line than in more radioresistant A549 cell line. Interesting from clinical point of view was the observation that bystander effect compensated for the nonuniform distribution of radiation dosage in their experimental system. However, radionuclide induced bystander effect depends on the linear energy transfer (LET) of radionuclide emitters, being either damaging, or protective (Boyd et al., 2006, Mairs et al., 2007). Cells exposed to media collected from gamma-irradiated cells exhibited a dose-dependent reduction in survival fraction at low dosage and a plateau in cell-kill at >2 Gy. Cells exposed to media from metaiodobenzylguanidine-treated cells ([^{131}I]MIBG, a low LET β-emitter), demonstrated a dose-response relationship with respect to clonogenic cell death and no annihilation of this effect at high radiopharmaceutical dosage. Contrarily, cells exposed to media from cultures treated with meta-^{211}At-astatobenzylguanidine ([^{211}At]MABG, a high LET α-emitter) exhibited dose-dependent toxicity at low dose, but elimination of cytotoxicity with increasing radiation dose, i.e. U-shaped survival curves (Mairs et al., 2007). Biologically similar analogs of halobenzylguanidines radiolabeled with radionuclides emitting β-particles (^{131}I-MIBG), α-particles (^{211}At-MABG), or Auger electrons emitting ^{123}I-MIBG, were also tested in experiments performed by the same group (Boyd et al., 2006) on a human glioma cell line (UVW) and a cell line derived from human bladder transitional carcinoma (EJ138), both transfected with neurotransmiter (NAT) gene that enabled greater MIBG uptake. A similar U-shaped bystander phenomenon was observed for clonogenic cell-survival curve in case of high-LET alpha and Auger-electron emitters. No corresponding plateau in toxicity was observed after exposure of cells to the medium from β-irradiated cells. The reason for such behavior is not clear as yet. However, identification of the pathways involved in this process might pinpoint ways of manipulating the bystander effect for therapeutic purposes, i.e. to gain selective increase in tumor cell killing, accompanied by reduced side effects in normal tissue. Proliferative bystander responses have been also observed *in vitro* after irradiation with β-particles emitted by tritiated thymidine (^{3}HTdR). The rat liver epithelial cells (WB-F344 line) not treated with tritiated thymidine (unlabeled cells), in the presence of radiolabeled cells that received absorbed doses from 0.14 - 1.7 Gy, showed statistically significant increase of cell growth by 9-10% in comparison to control (Gerashchenko and Howell, 2004). The mean energy of β-particles is only 5.7 keV, (a range of ca. 1 μm in water). Thus, the probability that β-particles emitted from radiolabeled cells will target the nucleus of adjacent unlabeled cells in non-confluent co-culture used in the study is very low, because the majority of unlabeled cells were far beyond the range of β-particles emitted from radiolabeled cells. The authors compared ^{3}HTdR (the β-emitter) results with their earlier results obtained for γ-rays and found that a much lower dose of radionuclide (0.14 Gy) induced maximum response of bystander cells, whereas the maximum bystander response to γ-rays was not seen, even up to 1 Gy. According to the authors it is possible that the differences in the bystander dose response between γ-rays and ^{3}HTdR may be related to higher relative biological effectiveness (RBE) that has been observed for ^{3}HTdR, as compared to γ-rays.

The presence of bystander effect initiated by *in vivo* decay of radionuclide was demonstrated by Xue et al. (2002). When human colon LS174T adenocarcinoma cells prelabeled with lethal dose of Auger electron–emitting 5-[^{125}I]iodo-2-deoxyuridine (^{125}IUdR) were subcutaneously co-injected with LS174T unlabeled cells into nude mice, a considerable inhibition of tumor growth was observed. Since the ^{125}I present within the cells is DNA-bound, and 99% of the electrons emitted by the decaying ^{125}I atoms have a subcellular range (<0.5 μm), and since the overall radiation dose deposited by radiolabeled cells in the unlabeled cells within the growing tumor is less than 10 cGy, these authors concluded that the results obtained are a consequence of a bystander effect generated *in vivo* by factor(s) present within and/or released from the ^{125}IUdR-labeled cells. Radionuclides differ in their physical characteristics such as type of decay, the mean energy, the half-life and range of penetration. However, in spite of the identical decay, the Auger electrons for both, ^{123}I (half–life, 13.3 h) and ^{125}I (half-life, 60.5 d) they differ in mean energy which is 1.234 MeV and 179 keV for ^{123}I and ^{125}I respectively (Prise, 2008). The bystander effect induced *in vivo* by co-injection of radiolabeled and unlabeled LS174T cells was totally different (Kishikawa et al., 2006). ^{125}I labeled cells stimulated tumor growth, and inversely, ^{123}I labeled cells inhibited tumor growth after subcutaneous co-injection of cell mixture into nude mice. Similar pattern of response was observed in experiment *in vitro*. These contrasting effects were accompanied by different biochemical events; supernatants from cultures with ^{125}I-labeled cells were positive for tissue inhibitors of metalloproteinases (TIMP1 and TIMP2), and those from cultures with ^{123}I-labeled cells were positive for angiogenin (Kishikawa et al., 2006). These all studies demonstrate the potential of internalized radionuclides to generate bystander effects *in vivo* for therapeutic treatment, however many question remain in regard to bystander signaling evoked by application of different radionuclide, as pointed out in the review of Sgouros et al. (2007). E.g. are the protective or damaging effects specific for different types of radionuclides or type of cell? Is the in vivo bystander effect restricted to the damage to DNA by ionization secondary to Auger-elctron cascade or is it also possible when radionuclides deposit their energies within the cell cytoplasm or membrane? Additional studies are required to fully understand the bystander effects in radionuclide therapy.

9. *In vivo* bystander effect

Bystander effect in tissues distant from the radiation field, named "abscopal effect", was observed more than 50 years ago as haematological changes of bone marrow in children, who were given radiotherapy to the spleen in the treatment of leukemia (Parsons et al., 1954). Until recently, the abscopal effect was referred to the distant effects seen after local radiation therapy. Although the abscopal effect is potentially important for tumor control, it is still extremely controversial. However, it inspired *in vitro* and *in vivo* studies. It is believed mediated through cytokines and/or the immune system and results from loss of growth stimulatory or immunosuppressive factors from the tumor (Kaminski et al., 2005). The observation that irradiation of a murine tumor caused growth inhibition of another tumor outside of the radiation field was explained as the effect of immune system activation (Demaria et al., 2004). Interestingly, growth inhibition of tumors remote from the radiation field was tumor-specific. Camphausen et al. (2003) observed an abscopal effect as significant growth delay of distally implanted Lewis lung carcinoma and T241 fibrosarcoma cells in mice when they irradiated the non-tumorbearing legs. Furthermore, the authors compared this effect after fractionated irradiation with five 10-Gy fractions or twelve 2-Gy fraction and found dose dependent inhibition of tumor growth, being greater with higher fraction dose. Persuasive evidence of the bystander effect presence *in vivo* comes from experiments on rats in which the bases of the lungs were exposed to 10 Gy, while the remaining 70% of lungs were protected (Khan et al., 2003). A considerable increase in the DNA damage (micronuclei) was observed in the shielded lung. In addition, various parts of the lungs differed in the micronuclei frequency in response to direct irradiation, or only to bystander signals. The protective effects of two radical scavengers, superoxide dismutase (SOD) and nitro-L-arginine methyl ester (L-NAME), suggest that inflammatory cytokines induced by the irradiation may be involved in the initiation of a reaction generating reactive oxyradicals and nitric oxide that cause indirect DNA damage, both in and out of the radiation field (Khan et al, 2004). The mediators of bystander effect *in vivo* may be macrophages and inflammatory cytokines. Calveley et al. (2005) showed that activation of macrophages and expression of inflammatory cytokines fluctuated in a cyclic pattern in the directly irradiated and bystander regions of the same lung tissues. Cytokines including IL-1a, IL-1 IL-6, TNF-a and TGF-β were expressed to a similar degree in both, radiation targeted and non targeted lung tissues when measured on RNA levels. The results of animal studies involving irradiation of one side of the mouse body with 1 Gy X-ray showed DNA DSBs induction and increase in the levels of Rad51 (DSBs repairing protein) in non exposed skin (completely protected by lead shield). Furthermore, the levels of two methyl-binding proteins known to be involved in transcriptional silencing, MeCP2 and MBD2, were also increased in bystander tissue suggesting that radiation induced bystander effect may be epigenetically regulated. Global DNA hypomethylation is a typical feature of cancer cells. The methylation is one of the many types of histone modification processes which include, phosphorylation, acetylation, and ubiquitination, referred to epigenetic changes. Pogribny et al. (2004, 2005) investigated the effect of low-dose radiation exposure on the accumulation of DNA lesions and alterations of DNA methylation and histone H4-Lys20 trimethylation in the thymus tissue using an *in vivo* murine model. They found that fractionated whole-body application of 0.5 Gy X-ray leads to decrease in histone methylation and DNA damage accumulation in the thymus gland. The radiation-induced global genome DNA methylation changes were shown to be dose-dependent, sex- and tissue specific and long-persistant. Tissue specificity of bystander responses within the same organism has also been examined by Ilnytskyy et al. (2009). They analyzed changes in global DNA methylation in spleen of mice whole-body or cranial exposed to single 0.5 Gy of X-rays or to the same dose given in five 0.1 Gy fractions. After acute cranial exposure the major changes were observed in the animal spleen such as a significant loss of global DNA methylation 6 hr, 96 hr, and 14 days after irradiation, resembling those induced in whole body irradiated rats. These changes also include DNA binding protein methylation, expression of methylotransferases and the methyl group binding retrotransposomal element LINE-1, and overexpression of micro RNA, miR-194. Therefore, these transcriptionally regulated epigenetic changes seem undoubtedly to be related to the radiation induced bystander effect, although they may be specific to certain tissues, because similar changes were absent in the dermal tissue (Koturbash et al, 2007, Ilnytskyy et al., 2009). *TP53* overexpression, change of proliferation rate measured as Ki67 antigen expression, as well as the increase in the percentage of apoptosis and DNA double strand breaks, the marker of which was the histone H2AX phosphorylation were also observed in bystander spleen of mice exposed to 1 Gy X-rays to their heads. These changes persisted from 24 hours to seven months (Koturbash et al, 2008). All of those experiments indicate that cells and tissues irradiated *in vivo* send signals which are transmitted by paracrine and endocrine systems and are able to induce damage in DNA, apoptosis, clastogenic effects, and epigenetic changes that lead to genetic instability. The consequence of the long-persisting changes may be the late effects including mutation and induction of second primary cancer. In good agreement with data presented above are results of the elegant study on bystander effect in human tissue models, which preserve the three-dimensional structure and communication of cells present in tissues *in vivo* (Sedelnikova et al. 2007). The artificial skin which is able to survive 2-3 weeks in culture was irradiated with microbeam helium ions (7 MeV ^{4}He, range in tissue 31 μm). The beam size was restricted to a 1 to 2 nuclei width along the line of irradiation. Bystander effect was studied on histological slides prepared at various time post irradiation (up to seven days). The authors observed increases in bystander cells the double strand breaks formation, followed by increased levels of apoptosis and micronucleus frequency, hypomethylation of nuclear DNA, and by an increased fraction of senescent cells. These findings point out the DNA DSBs induced by bystander signals as precursors of different cellular consequences in human tissues.

10. The potential clinical consequences of radiation induced bystander effect

Although direct extrapolation of data from *in vitro* experiments to *in vivo* radiotherapy is not possible, (three-dimensional structure of tissues), one could assume that the bystander effect implies a risk of post-radiation complications in healthy tissues. It is suggested that genetic instability, which takes the form of delayed reproductive death (DRD), can participate in late side effects in patients treated with radiotherapy, because of damage, increased cell loss and longer recovery (Hendry, 2001). Increased level of chromosome aberrations and micronuclei was detected in the head and neck cancer patients undergoing radiotherapy within a year post treatment (Gamulin et al., 2008). DRD phenomenon associated with the presence of an increased percentage of stable and unstable chromosome aberrations in lymphocytes was detected in patients irradiated because of ankylosing spondylitis even several years after radiotherapy. Furthermore, increased mortality was related to single treatment course of X-rays because of this diseases (Smith & Doll, 1982). However, other studies performed in adults many years after radiotherapy in childhood haven't shown genetic instability (Tawn et al., 2005). Neither was it shown in persons having professional contact with radiation, who have suffered internal plutonium contamination at least 10 years previous to the study (Whitehouse and Tawn, 2001). Furthermore, creation of mutator phenotype as a result of genetic instability seems to increase the probability of induction of tumors. It has been shown on an animal model that ionizing radiation induces genetic instability emerging as delayed *TP53* mutations and more frequent transformation of mammary gland epithelial cells, leading to the development of cancer (Ulrich and Ponnaiya, 1998). Compared to healthy persons, irradiated cancer patients show increased incidence of second-wave primary cancers (Boice et al., 1985, Brenner et al., 2000), although the bystander effect does not need to be the only cause of such events. It is well-known that genetic predispositions and environmental factors may have significant influence on the formation of tumors (Mohandas, 2001).

Together with modern techniques of irradiation, such as three-dimensional conformal radiation therapy (3D-CRT) or intensity-modulated radiation therapy (IMRT), the purpose of which is to reduce radiation dose delivered to healthy tissues, there is an increased risk of adverse effects resulting from a possible bystander effect, especially because in these techniques larger volumes of normal tissues are exposed to a small dose (Hall, 2006). The risk of secondary cancers is increased especially in prostate cancer (Brenner et al., 2000) and cervical cancer (Boice et al., 1985, Kleinerman et al., 1995, Chaturvedi et al., 2008, Trott, 2009). Prostate surgery and radiotherapy are methods having comparable efficacy, therefore any late consequences in the form of secondary tumors should be taken into account, especially in younger people with a perspective of long-time survival. Brenner et al. (2000) compared the incidence of second-wave primary cancers in prostate cancer patients treated with surgery only (more than 50 000) to that in patients treated by radiotherapy (more than 70 000) and observed a statistically significant, although small, increase in the risk of secondary cancers in the latter group (6%, $p = 0.02$). This risk was associated with dose and latent time and grew with increasing survival time, amounting to 15% for patients surviving over 5 years and to 34% for those surviving over 10 years. The emerging cancers were solid tumors, such as bladder, bowel and lung carcinomas and sarcomas, the latter within the field of irradiation. The authors did not observe leukemia cases. The risk of secondary cancers after radiotherapy of cervical cancer is comparable to that of prostate cancer. Kleinerman et al. (1995) compared the risk of secondary cancers in radiation-treated, invasive cervical cancer patients (almost 50 000) with that in a group of non-irradiated patients surviving more than 30 years and showed a 12% increase in newly-diagnosed secondary cancers, where the increase was 15% after 10 years and 26% after 20 years post radiotherapy. Cancers of colon, bladder, rectum, vagina and ovary were within the fields covered by the high-dose radiation, but there were also few cases of leukemia. However, half of secondary neoplasms accounted for lung cancer. Occurrence of cancer of the lung, the organ relatively distant from the original tumor irradiation field, in which the radiation dose was estimated at ca. 0.6 Gy (Brenner et al., 2000), appears to have been associated with the bystander effect induced by signaling molecules in the neighborhood, and with potentially mutagenic carcinogens generated by irradiated cells, although environmental factors, genetic background and patients' lifestyle could also significantly contribute.

Calculations of the equivalent whole-body dose in the case of high-energy IMRT irradiation technique (Followill et al., 1997), indicate that, in comparison with conventional radiotherapy, the risk of secondary solid cancers has increased considerably. This increase is dependent on the X-ray energy and is 1% for 6 MV, 4.5% for 18 MV and 8.4% for 25 MV compared with 0.4, 1.6 and 3%, respectively, for those same radiation energy of X-rays given in a conventional way. Furthermore, as a conclusion from this study it appears that the risk of leukemia also increases after IMTR technique. The question of secondary tumors, as a succession of radiotherapy, was investigated in several recent studies [Suit et al., 2007, Trott K-R., 2009, Tubiana M., 2009, Xu et al., 2008]. Based on epidemiological and experimental radiobiological data, Suit et al. (2007) concluded that the relationship of tumor induction risk and dose is complex and differs not only between species of animals, between individuals of the species concerned, but it may also be different for various tissues and organs. Specifically, the risk increases with dose in the 1-45 Gy range for gastric and pancreatic cancer, but is stable in the 1-60 Gy dose range for bladder cancer, and even negative for colon cancer. These phenomena are difficult to explain. They could more likely be the result of genetic instability than the effect of bystander signals at lower doses, as well as result from inhibition of signals originating from cells lethally damaged by higher doses.

It seems that bystander effect can have beneficial consequences, particularly in radionuclide therapy as described above and probably in brachytherapy (Brans et al., 2006) in which tumor cells irradiated by intake or absorption of isotope energy are in the immediate vicinity of non-irradiated cells inducing in them the effect. The bystander effect can also increase damage to cancer cells during treatment with boron neutron capture therapy (BNCT) (Barth et al., 2005). As previously described, the "abscopal effect" is also an example of manifestation of the beneficial effects of irradiated cells, even at a distance from their location [Kaminski et al., 2005]. However, it is also possible that pro-survival signals, sent by lethally damaged cells, may increase the chances of survival of other, less damaged tumor cells within the field of irradiation and may pose a risk of local recurrence (Mackonis et al. (2007). The mutual communication between normal and cancer cells leading to radioprotective effect to radiation targeted cancer cells, as presented above, can also be taken into consideration. Furthermore, one can expect that the individuals exposed internally to radionuclides for routine diagnostic nuclear medical procedures might be at risk of bystander effect however, prediction whether it will be damaging or protective requires further studies.

11. Conclusion

Radiation induced bystander effect (RIBE) is unquestionable biological phenomenon which elicits in cells not directly irradiated but being in the neighborhood of targeted cells, or being exposed to molecular signals disclosed by irradiated ones. It has been found in variable *in vitro* and *in vivo* systems. RIBE predominate at externally applied low doses and low dose-rate, although many data confirm its presence at clinically used doses and radionuclide exposure. It may be either detrimental or potentially beneficial event depending on dose, dose-rate, means of irradiation, cell types and environmental conditions pointing out to its very composed nature. The bystander effect induced by radionuclide intake seems to be the most susceptible to modulation of bystander signaling for clinical purposes aimed to improvement of the therapeutic ratio. However, the potential and real clinical consequences of bystander effect are, as yet, not predictable. We are not able to predict whether and what form, damaging or radioprotective, will the bystander effect take in the patient without knowledge of patients/tumor response to low or high dose-rate irradiation, tumor vasculature and normal cells infiltration. The more, we are not yet able to modulate the response of the patient to the signals generated in the process. Therefore, additional studies are required to address these questions.

12. Acknowledgment

This paper was supported by the *grant No N N518 497 639 from the Polish Ministry of Science and Higher Education.*

13. References

Azzam E.I., Raaphorst G.P. & Mitchel R.E. (1994). Radiation-induced adaptive response for protection against micronucleus formation and neoplastic transformation in C3H 10T1/2 mouse embryo cells. *Radiation Research*, Vol.138, No.1 (Suppl.), (April 1994), pp. S28-31, ISSN:0033-7587

Azzam E.I., de Toledo, S.M., Raaphorst G.P. & Mitchel R.E. (1996). Low-dose ionizing radiation decreases the frequency of neoplastic transformation to a level below the spontaneous rate in C3H 10T1/2 cells. *Radiation Research,* Vol.146, No.4, (October 1996), pp. 369-373, ISSN:0033-7587

Azzam E.I., de Toledo S.M., Gooding T. & Little J.B. (1998). Intercellular communication is involved in the bystander regulation of gene expression in human cells exposed to very low fluency of alpha particles. *Radiation Research,* Vol.150, No.5, (November 1998), pp. 497-504, ISSN:0033-7587

Azzam E.I., de Toledo S.M., Spitz D.R. & Little J.B. (2002). Oxidative metabolism modulates signal transduction and micronucleus formation in bystander cells from alpha-particle-irradiated normal human fibroblast cultures. *Cancer Research,* Vol.62, No.19, (October 2002), pp. 5436-5442, ISSN:0008-5472

Azzam E.I., de Toledo S.M. & Little J.B. (2003). Oxidative metabolism, gap junctions and the ionizing radiation-induced bystander effect. *Oncogene,* Vol.22, No.45, (October 2003), pp. 7050-70577, ISSN:0950-9232

Barth R.F., Coderre J.A., Vicente M.G. & Blue T.E. 2005. Boron neutron capture therapy of cancer: current status and future prospects. *Clinical Cancer Research,* Vol.11, No.11, (June 2005), pp. 13987-4002, ISSN:1078-0432

Belyakov O.V., Folkard M., Mothersill C., Prise K.M. & Michael B.D. (2002). Bystander-induced apoptosis and premature differentiation in primary urothelial explants after charged particle microbeam irradiation. *Radiation Protection Dosimetry,* Vol.99, No.1-4, pp. 249–251, ISSN:0144-8420

Belyakov O.V., Mitchell S.A., Parikh D., Randers-Pehrsom G., Marino S.A., Amundson S.A., Geard C.R. & Brenner D.J. (2005). Biological effects in unirradiated human tissue induced by radiation damage up to 1 mm away. *Proceeding of the National Academy of Sciences of the United States of America,* Vol.102, No.40, (October 2005), pp. 14203–14208, ISSN:0027-8424

Belyakov O.V., Folkard M., Mothersill C., Prise K.M. & Michael B.D. (2006). Bystander-induced differentiation: A major response to targeted irradiation of a urothelial explant model. *Mutation Research,* Vol. 597 No.1-2, (May 2006), pp. 43–49, ISSN:0027-5107

Bishayee A., Rao D.V. & Howell R.W. (1999). Evidence for pronounced bystander effects caused by nonuniform distributions of radioactivity using a novel three-dimensional tissue culture model. *Radiation Research,* Vol.152, No.1, (July 1999), pp. 88-97, ISSN:0033-7587

Bishayee A., Hill H.Z., Stein D., Rao D.V. & Howell R.W. (2001). Free radical-initiated and gap junction-mediated bystander effect due to nonuniform distribution of incorporated radioactivity in a three-dimensional tissue culture model. *Radiation Research,* Vol.155, No.2, (February 2001), pp. 335–344, ISSN:0033-7587

Boice J.D., Day N.E., Anderson A. Brinton L.A., Brown R., Choi N.W., Clarke E.A., Coleman M.P., Curtis R.E., Flannery J.T., et al. (1985). Second cancers following radiation treatment for cervical cancer: an international collaboration among cancer registries. *Journal of the National Cancer Institute,* Vol.74, No.5, (May 1985), pp. 955-975, ISSN:0027-8874

Boyd M., Ross S.C., Dorrens J., Fullerton N.E., Tan K.W., Zalutsky M.R., & Mairs R.J. (2006). Radiation-induced biologic bystander effect elicited in vitro by targeted radiopharmaceuticals labeled with α-, β-, and Auger electron–emitting radionuclides. *Journal of Nuclear Medicine,* Vol.47, No.6, (June 2006), pp. 1007–1015, ISSN:0161-5505

Brans B., Linden O., Giammarile F., Tennvall J. & Punt C. (2006). Clinical applications of newer radionuclide therapies. *European Journal of Cancer,* Vol.42, No.8, (May 2006), pp. 994-1003, ISSN:0959-8049

Brenner D.J., Curtis R.E., Hall E.J. & Ron E. (2000). Second malignancies in prostate carcinoma patients after radiotherapy compared with surgery. *Cancer,* Vol.88, No.2, (January 2000), pp. 398-406, ISSN:0008-543X

Burdak-Rothkamm S., Short S.C., Folkard M., Rothkamm K. & Prise K.M. (2007). ATR-dependent radiation-induced γH2AX foci in bystander primary human astrocytes and glioma cells. *Oncogene,* Vol.26, No.7, (February 2007), pp. 993-1002, ISSN:0950-9232

Burdak-Rothkamm S., Rothkamm K. & Prise K.M. (2008). ATM acts downstream of ATR in the DNA damage response signaling of bystander cells. *Cancer Research,* Vol.68, No.17, (Septeber 2008), pp. 7059-7065, ISSN:0008-5472

Calveley V.L., Khan M.A., Yeung I.W., Vandyk J. & Hill R.P. (2005). Partial volume rat lung irradiation: temporal fluctuations of in-field and out-of-field DNA damage and inflammatory cytokines following irradiation. *International Journal of Radiation Biology,* Vol.81, No12, (December 2005), pp. 887-899, ISSN:0955-3002

Camphausen K., Moses M.A., Menard C., Sproull M., Beecken W-D., Folkman J. & O`Reilly M.S. (2003). Radiation abscopal antitumor effect is mediated through p53. *Cancer Research,* Vol.63, No.8, (April 2003), pp. 1990-1993, ISSN:0008-5472

Chang W.P. & Little J.B. (1992). Evidence that DNA double strand breaks initiate the phenotype of delayed reproductive death in Chinese hamster ovary cells. *Radiation Research*, Vol.131, No.1, (July 1992), pp. 53-59, ISSN:0033-7587

Chapman K.L., Kelly J.W., Lee R., Goodwin E.H. & Kadhim M.A. (2008). Tracking genomic instability within irradiated and bystander populations. *The Journal of Pharmacy and Pharmacology*, Vol.60, No.8, (August 2008), pp. 959-68, ISSN:0022-3573

Chaturvedi A.K., Engels E.A., Gilbert E.S., Chen B.E., Storm H., Lynch C.F., Hall P., Langmark F., Pukkala E., Kaijser M., Andersson M., Fosså S.D., Joensuu H., Boice J.D., Kleinerman R.A. & Travis L.B. (2007). Second cancers among 104,760 survivors of cervical cancer: evaluation of long-term risk. *Journal of the National Cancer Institute*, Vol.99, No.21, (November 2007), pp. 1634-1643, ISSN:0027-8874

Chaudhry M.A. (2006). Bystander effect: Biological endpoints and microarray analysis. *Mutation Research*, Vol.97, No.1-2, (May 2006), pp. 98-112, ISSN:0027-5107

Chen H.H., Jja R.F., Yu L., Zhao M.J., Shao C.L. & Cheng W.Y. (2008). Bystander effects induced by continuous low-dose-rate 125I seeds potentiate the killing action of irradiation on human lung cancer cells in vitro. International Journal of Radiation Oncology, Biology and Physics, Vol.72, No.5, (December 2008), pp. 1560–1566, ISSN:0360-3016

Chipuk J.E. & Green D.R. (2006). Dissecting p53-dependent apoptosis. *Cell Death and Differentiation*, Vol.13, No.6, (June 2006), pp. 994-1002, ISSN:1350-9047

Clutton S.M., Townsend K.M., Walker C., Ansell J.D. & Wright E.G. (1996). Radiation-induced genomic instability and persisting oxidative stress in primary bone marrow cultures. *Carcinogenesis*, Vol.17, No.8, (August 1996), pp. 1633-1639, ISSN:0143-3334

Demaria S., Ng B., Devitt M.L., Babb J.S., Kawashima N., Liebes L. & Formenti S.C. (2004). Ionizing radiation inhibition of distant untreated tumors (abscopal effect) is immune mediated. *International Journal of Radiation Oncology, Biology and Physics*, Vol.58, No.3, (March 2004), pp. 862-870, ISSN:0360-3016

Deshpande A., Goodwin E.H., Bailey S.M., Marrone B.L. & Lehnert B.E. (1996). Alpha-particle-induced sister chromatid exchange in normal human lung fibroblasts: evidence for an extranuclear target. *Radiation Research*, Vol.145, No.3, (March 1996), pp. 260-267, ISSN:0033-7587

Elmore E., Lao XY., Kapadia R., Giedzinski, E., Lizoli C. & Redpath J.L. (2008). Low doses of very low-dose-rate low-LET radiation suppress radiation-induced neoplastic transformation in vitro and induce an adaptive response. *Radiation Research*, Vol.169, No.5, (March 2008), pp. 311-318, ISSN:0033-7587

Followill D., Geis P. & Boyer A. (1997). Estimates of whole-body dose equiwalent produced by beam intensity modulated conformal therapy. *International Journal of Radiation Oncology, Biology and Physics*, Vol.38, No.3, (June 1997), pp. 667-672, ISSN:0360-3016

Gamulin M., Kopjar N., Grgić M., Ramić S., Bisof V. & Garaj-Vrhovac V. (2008). Genome damage in oropharyngeal cancer patients treated by radiotherapy. *Croatian Medical Journal*, Vol.49, No.4, (August 2008), pp. 515-527, ISSN:0353-9504

Gerashchenko B.I, & Howell R.W. (2004). Proliferative response of bystander cells adjacent to cells with incorporated radioactivity. *Cytometry A*, Vol.60, No.2, (August 2004), pp. 155–164, ISSN:1552-4922

Gerashchenko B.I. & Howell R.W. (2003). Flow cytometry as a strategy to study radiation-induced bystander effects in co-culture systems. *Cytometry A*; Vol.54, No.1, (July 2003), pp. 1–7, ISSN:1552-4922

Ghandhi S.A., Yagohoubian B. & Ammundson S.A. (2008). Global gene expression analyses of bystander and alpha particle irradiated normal human lung fibroblasts: Synchronous and differential responses. *BMC Medical Genomics*, Vol.1, pp. 63-76, ISSN:1755-8794

Gorgojo L. & Little J.B. (1989). Expression of lethal mutations in progeny of irradiated mammalian cells. *International Journal of Radiation Biology*, Vol.55, No.4, (April 1998), pp. 619-630, ISSN:0955-3002

Hall EJ. (2006). Intensity-modulated radiation therapy, protons, and the risk of second cancers. *International Journal of Radiation Oncology, Biology and Physics*, Vol.65, No.1, (May 2006), pp. 1-7, ISSN:0360-3016

Harada T., Kashino G., Suzuki K., Matsuda N., Kodama S. & Watanabe M. (2008). Different involvement of radical species in irradiated and bystander cells. *International Journal of Radiation Biology*, Vol.84, No.10, (October 2008), pp. 809-814, ISSN:0955-3002

Harper K., Lorimore S.A. & Wright E.G. (1997). Delayed appearance of radiation-induced mutations at the Hprt locus in murine hemopoietic cells. *Experimental Hematology*, Vol.25 No.3, (March 1997), pp. 263-269, ISSN:0301-472X

He M., Zhao M., Shen B., Prise K.M. & Shao C. (2010). Radiation-induced intercellular signaling mediated by cytochrome-c via a p53-dependent pathway in hepatoma cells. Oncogene. (2010 December 6) [Epub ahead of print], ISSN:0950-9232

Hendry J.H. (2001). Genomic instability: potential contributions to tumour and normal tissue response, and second tumours, after radiotherapy. *Radiotherapy and Oncology: Journal of the European Society for Therapeutic Radiology and Oncology*, Vol.59, No.2, (May 2001), pp. 117-126, ISSN:0167-8140

Hickman A.W., Jaramillo R.J., Lechner J.F. & Johnson N.F. (1994). Alpha particle-induced p53 protein expression in rat lung epithelial cell strain. *Cancer Research* Vol.54, No.22, (November 1994), pp. 5797-5800, ISSN:0008-5472

Hollowell J.G. & Litefield L.G. (1968). Chromosome damage induced by plasma of X-rayed patients: an indirect effect of X-ray. *Proceedings of the Society for Experimental Biology and Medicine*, Vol.129 No.1, (October 1968), pp. 240-244, ISSN:0037-9727

Holmberg K., Meijer A.E., Auer G. & Lambert B.O. (1995). Delayed chromosomal instability in human T-lymphocyte clones exposed to ionizing radiation. *International Journal of Radiation Biology*, Vol.68, No.3, (September 1995), pp. 245-255, ISSN:0955-3002

Hu B., Wu L., Han W., Zhang L., Chen S., Xu A., Hei T.K. & Yu Z. (2006). The time and spatial effects of bystander response in mammalian cells induced by low dose radiation. *Carcinogenesis*, Vol.27, No.2, (February 2006), pp. 245-51 ISSN:0143-3334

Huang L., Kim P.M., Nickoloff J.A. & Morgan W.F. (2007). Targeted and nontargeted effects of low-dose ionizing radiation on delayed genomic instability in human cells. *Cancer Research*, Vol.67, No.3, (February 2007), pp. 1099–1104, ISSN:0008-5472

Hussain S.P., Hofseth L.J. & Harris C.C. (2003). Radical causes of cancer. *Nature Reviews. Cancer*, Vol.3, No.4, (April 2003), pp. 276-285, ISSN:1474-175X

Ilnytskyy Y., Koturbash, I. & Kovalchuk O. (2009). Radiation-induced bystander effects in vivo are epigeneticcally regulated in a tissue specific manner. *Environmental and Molecular Mutagenesis*, Vol.50, No.2, (March 2009), pp. 105-113, ISSN:0893-6692

Iwakawa M., Hamada N., Imadome K., Funayama T., Sakashita T., Kobayashi Y. & Imai, T. (2008). Expression profiles are different in carbon ion-irradiated normal human fibroblasts and their bystander cells. *Mutation Research*, Vol.642, No.1-2, (July 2008), pp. 57-67, ISSN:0027-5107

Iyer R. & Lehnert B.E. (2000). Factors underlying the cell growth-related bystander responses to alpha particles. *Cancer Research,* Vol.60, No.5, (March 2000), pp. 1290–1298, ISSN:0008-5472

Iyer R., Lehnert B.E. (2002). Low dose, low-LET ionizing radiation-induced radioadaptation and associated early responses in unirradiated cells. *Mutation Research,* Vol.503, No.1-2, (June 2002), pp. 1-9, ISSN:0027-5107

Kadhim M.A., Lorimore S.A., Townsend K.M., Goodhead D.T., Buckle V.J. & Wright E.G. (1995). Radiation induced genomic instability: Delayed cytogenetic abberations and apoptosis in primary human bone marrow cells. *International Journal of Radiation Biology,* Vol.67, No.3, (March 1995), pp. 287-293, ISSN:0955-3002

Kaminski J.M., Shinohara E., Summers J.B., Niermann K.J., Morimoto A. & Brousal J. (2005). The controversial abscopal effect. *Cancer Treatment Reviews,* Vol.31, No.3, (May 2005), pp. 159-172, ISSN:0305-7372

Kashino G., Suzuki K., Matsuda N., Kodana S., Ono K., Watanabe M. & Prise K.M. (2007). Radiation induced bystander signals are independent of DNA damage and DNA repair capacity of the irradiated cells. *Mutation Research,* Vol.619, No.1-2, (June 2007), pp.134-138, ISSN:0027-5107

Khan M.A., Van Dyk J., Yeung I.W. & Hill R.P. (2003). Partial volume rat lung irradiation: assessment of early DNA damage in different lung regions and effect of radical scavengers. *Radiotherapy and Oncology: Journal of the European Society for Therapeutic Radiology and Oncology,* Vol.66, No.1, (January 2003), pp. 95-102, ISSN:0167-8140

Kishikawa H., Wang K., Adelstein S.J. & Kassis A.I. (2006). Inhibitory and stimulatory bystander effects are differentially induced by Iodine-125 and Iodine-123. *Radiation Research,* Vol.165, No.6,(June 2006), pp. 688-694, ISSN:0033-7587

Kleinerman R.A., Boice J.D. Jr, Storm H.H., Sparen P., Andersen A., Pukkala E., Lynch C.F., Hankey B.F. & Flannery J.T. (1995). Second primary cancer after treatment for cervical cancer. An international cancer registries study. *Cancer,* Vol.76, No.3, (August 1995), pp. 442-452, ISSN:0008-543X

Konopacka M. & Rzeszowska-Wolny J. (2006). The bystander effect-induced formation of micronucleated cells is inhibited by antioxidants, but the parallel induction of apoptosis and loss of viability are not affected. *Mutation Research,* Vol.593, No.1-2, (January 2006), pp. 32-38, ISSN:0027-5107

Koturbash I., Rugo R.E., Hendricks C.A., Loree J., Thibault B., Kutanzi K., Pogribny I., Yanch J.C., Engelward B.P. & Kovalchuk O. (2006). Irradiation induces DNA damage and modulates epigenetic effectors in distant bystander tissue in vivo. *Oncogene,* Vol.25, No.31, (July 2006), pp. 4267-4275, ISSN:0950-9232

Koturbash I., Boyko A., Rodriguez-Juarez R., McDonald R.J., Tryndyak P., Kovalchuk I., Pogribny I.P. & Kovalchuk O. (2007). Role of epigenetic effectors in maintenance of the long-term persistent bystander effect in spleen in vivo. *Carcinogenesis,* Vol.28, No.8, (August 2007), pp. 1831-1838, ISSN:0143-3334

Koturbash I., Zemp F.J., Kutanzi K., Luzhna L., Loree, J., Kolb B. & Kovalchuk O. (2008a). Sex-specific microRNAome deregulation in the shielded bystander spleen of cranially exposed mice. *Cell Cycle*, Vol.7. No.11, (June 2008), pp. 1658-1667, ISSN:1538-4101

Koturbash I., Loree J., Kutanzi K., Koganow C., Pogribny I. & Kovalchuk O. (2008b). In vivo bystander effect: cranial X-irradiation leads to elevated DNA damage, altered cellular proliferation and apoptosis, and increased p53 levels in shielded spleen. *International Journal of Radiation Oncology, Biology and Physics* , Vol.70, No.2, (February 2008), pp. 554-562, ISSN:0360-3016

Koyama S., Kodama S., Suzuki K., Matsumoto T., Miyazaki T. & Watanabe M. (1998). Radiation-induced long-lived radicals which cause mutation and transformation. *Mutation Research*, Vol.421, No.1, (October 1998), pp. 45-54, ISSN:0027-5107

Kramer-Marek G., Serpa C., Szurko A., Widel. M., Sochanik A, Snietura M, Kus P, Nunes RM, Arnaut LG, Ratuszna A. (2006). Spectroscopic properties and photodynamic effects of new lipophilic porphyrin derivatives: efficacy, localisation and cell death pathways. *Journal of Photochemistry and Photobiology B, Biology*, Vol.84, No.1, (July 2006), pp. 1-14, ISSN:1011-1344

Lala P.K. & Chakraborty C. (2001). Role of nitric oxide in carcinogenesis and tumour progression. The *Lancet Oncology*, Vol.2, No.3, (March 2001), pp. 149-156, ISSN:1470-2045

Little J.B., Gorgojo L. & Vetrovs H. (1990). Delayed appearance of lethal and specific gene mutations in irradiated mammalian cells. *International Journal of Radiation Oncology Biology and Physics*, Vol.19, No.6, (December 1990), pp. 1425-1429, ISSN:0360-3016

Little J.B., Nagasawa H., Pfenning T. & Vetrovs H. (1997). Radiation-induced genomic instability: delayed mutagenic and cytogenetic effects of X rays and alpha particles. *Radiation Research*, Vol.148, No.4, (October 1997), pp. 299-307, ISSN:0033-7587

Little J.B. (1999). Induction of genetic instability by ionizing radiation. *Comptes rendus de l'Académie des sciences. Série III, Sciences de la vie, cad. Sci.* III, Vol.322, No.2-3, (February-March 1999), pp. 127-134, ISSN:0764-4469

Little J.B. (2006a). Lauriston S. Taylor Lecture: Nontargeted Effects of Radiation: Implications for low-dose exposures. *Health Physics*, Vol.91, No.5, (November 2006), pp. 416-426, ISSN:0017-9078

Little JB. (2006b). Cellular radiation effects and the bystander response. Mutatation *Research*, Vol.597, No.1-2, (May 2006), pp. 113-118, ISSN:0027-5107

Lorimore S.A., Kadhim M.A., Pocock D.A., Papworth D., Stevens D.L., Goodhead D.T. & Wright E.G. (1998). Chromosomal instability in the descendans of unirradiated surviving cells after alpha-particle irradiation. *Proceedings of the National Academy of Sciences of the United States of America*, Vol.95, No.10, (May 1998), pp. 5730-5733, ISSN:0027-8424

Lorimore S.A., Coates P.J. & Wright E.G. (2003). Radiation-induced genomic instability and bystander effects: Inter-related nontargeted effects of exposure to ionizing radiation. *Oncogene*, Vol.22, No.45, (October 2003), pp. 7058–7069, ISSN:0950-9232

Luce A., Courtin A., Levalois C., Altmeyer-Morel S., Romeo P-H., Chevillard S. & Lebeau J. (2009). Death receptor pathways mediate targeted and non-targeted effects of ionizing radiations in breast cancer cells. *Carcinogenesis*, Vol.30, No.3, (March 2009), pp. 432-439, ISSN:0143-3334

Lyng F.M., Seymour C.B. & Mothersill C. (2000). Production of a signal by irradiated cells which leads to a response in unirradiated cells characteristic of initiation of apoptosis. *British Journal of Cancer*, Vol.83, No.9, (November 2000), 1223-1230, ISSN:0007-0920

Lyng F.M., Maguire P., McClean B., Seymour C. & Mothersill C. (2006). The involvement of calcium and MAP kinase signaling pathways in the production of radiation-induced bystander effects. *Radiation Research*, Vol.165, No.4, (April 2006), pp. 400-409, ISSN:0033-7587

Mackonis E.C., Suchowerska N., Zhang M., Ebert M., McKenzi D.R. & Jackson M. (2007). Cellular response to modulated radiation fields. *Physics in Medicine and Biology*, Vol.52, No.18, (September 21), 5469-5482, ISSN:0031-9155

Mairs R.J., Fullerton N.E., Zalutsky M.R. & Boyd M. (2007). Targeted Radiotherapy: Microgray Doses and the Bystander Effect. *Dose-Response*, Vol.5, No.3, (April 2007), pp. 204–213, 2007, ISSN:1559-3258

Marder B.A. & Morgan W.F. (1993). Delayed chromosomal instability induced by DNA damage. *Molecular and Cellular Biology*, Vol.13, No.11, (November 1993), pp. 6667-6677, ISSN:0270-7306

Marnett L.J. (2000). Oxyradicals and DNA damage. *Carcinogenesis*, Vol.21, No.3, (March 2000), pp. 361-370, ISSN:0143-3334

Marples B. & Skov K.A. (1996). Small doses of high-linear energy transfer radiation increase the radioresistance of Chinese hamster V79 cells to subsequent X-irradiation. *Radiation Research*, Vol.146, No.4, (October 1996), pp 382-387, ISSN:0033-7587

Massagué J. & Chen Y.G. (2000). Controlling TGF-beta signaling. *Genes & Development*, Vol.14, No.6, (March 2000), pp. 627-644, ISSN:0890-9369

Matsumoto H., Hayashi S., Hatashita M., Ohnishi K., Shioura H., Ohtsubo, T., Kitai, R., Ohnishi, T. & Kano, E. (2001). Induction of radioresistance by a nitric oxide-mediated bystander effect. *Radiation Research*, Vol.155, No.3, (March 2001), pp. 387–396, ISSN:0033-7587

Matsumoto H., Hamada N., Takahashi A., Kobayashi Y. & Ohnishi T. (2007). Vanguards of paradigm shift in radiation biology: radiation-induced adaptive and bystander responses. *Journal of Radiation Research* (Tokyo), Vol.48, No.2, (March 2007), pp. 97-106, ISSN:0449-3060

Mendonca M.S, Kurohara W., Antoniono R. & Redpath J.L. (1989). Plating efficiency as a function of time postirradiation: evidence for the delayed expression of lethal mutations. *Radiation Research*, Vol.119, No.2, (August 1989), pp. 387-393, ISSN:0033-7587

Mohandas K.M. (2001). Genetic predisposition to cancer. In: Current Science, Vol. 81, No. 5, (10 SEPT 2001), pp.482-489, http://www.iisc.ernet.in/~currsci/sep102001/482.pdf

Morgan WF. (2003a). Non-targeted and delayed effects of exposure to ionizing radiation: I. Radiation-induced genomic instability and bystander effects in vitro. *Radiation Research,* Vol.159, No.5, (May 2003), pp. 567–580, ISSN:0033-7587

Morgan WF. (2003b). Non-targeted and delayed effects of exposure to ionizing radiation: II. Radiation-induced genomic instability and bystander effects in vivo, clastogenic factors and transgenerational effects. *Radiation Research,* Vol.159, No.5, (May 2003), pp. 581–596, ISSN:0033-7587

Mothersill C. & Seymour C.B. (1997). Medium from irradiated human epithelial cells but not human fibroblasts reduces the clonogenic survival of unirradiated cells.

International Journal of Radiation Biology, Vol.71, No.4, (April 1997), pp. 421-427, ISSN:0955-3002

Mothersill C. & Seymour C.B. (1998). Cell-cell contact during gamma-irradiation is not required to induce a bystander effect in normal human keratinocytes: evidence for release during irradiation of signal controlling survival into the medium. *Radiation Research,* Vol.149, No.3, (March 1998), pp. 256-262, ISSN:0033-7587

Mothersill C. & Seymour C. (2001). Radiation-induced bystander effects: past history and future directions. *Radiation Research,* Vol.155, No.6, (June 2001), pp. 759-767, ISSN:0033-7587

Mothersill C. & Seymour C.B. (2002a). Bystander and delayed effects after fractionated radiation exposure. *Radiation Research,* Vol.158, No.5, (November 2002), pp. 626–633, ISSN:0033-7587

Mothersill C. & Seymour C. (2002b). Characterization of a bystander effect induced in human tissue explant cultures by low LET radiation. *Radiation Protection Dosimetry,* Vol.99, No.1-4, pp. 163–167, ISSN:0144-8420

Nagasawa M. & Little J.B. (1992). Induction of sister chromatid exchanges by extremely low doses of alpha-particles. *Cancer Research,* Vol.52, No.22, (November 1992), pp. 6394-6396, ISSN:0008-5472

Nagasawa H., Huo L. & Little J.B. (2003). Increased bystander mutagenic effect in DNA double-strand break repair-deficient mammalian cells. *International Journal of Radiation Biology,* Vol.79, No.1, (January 2003), pp. 35-41, ISSN:0955-3002

Narayanan P.K., LaRue K.E.A., Goodwin E.H. & Lehnert B.E. (1999). Alpha particles induce the production of interleukin-8 by human cells. *Radiation Research,* Vol.152, No.1, (Jul 1999), pp. 57-63, ISSN:0033-7587

Nikjoo H. & Khvostunov I.K. (2003). Biophysical model of the radiation-induced bystander effect. *International Journal of Radiation Biology* Vol.79, No.1, (January 2003), pp. 43-52, ISSN:0955-3002

Nikjoo H. & Khvostunov I.K. (2006). Modelling of radiation-induced bystander effect at low dose and low LET. *International Journal of Low Radiation,* Vol.3, No.2-3, pp. 143-158, ISSN: 1477-6545

Ojima M., Furutani A., Ban N. & Kai M. (2011). Persistence of DNA double-strand breaks in normal human cells induced by radiation-induced bystander effect. *Radiation Research,* Vol.175, No.1, (January 2011), pp. 90–96, ISSN:0033-7587

Parsons W.B., Watkins C.H., Pease G.L. & Childs D.S. Jr. (1954). Changes in sternal marrow following roentgen-ray therapy to the spleen in chronic granulocytic leukemia. *Cancer,* Vol.7, No.1, (January 1954), pp. 179-189, ISSN:0008-543X

Pogribny I., Raiche J., Slovack M. & Kovalchuk O. (2004). Dose-dependence, sex- and tissue-specificity, and persistence of radiation-induced genomic DNA methylation changes. *Biochemicl. Biophysical Research Communication,* Vol.320, No.4, (August 2004), pp. 1253-1261, ISSN:0006-291X

Pogribny I., Koturbash I., Tryndyak V., Hudson D., Stevenson S.M., Sedelnikova O., Bonner W. & Kovalchuk O. (2005). Fractionated low-dose radiation exposure leads to accumulation of DNA damage and profound alterations in DNA and histone methylation in the murine thymus. *Molecular Cancer Research : MCR,* Vol.3, No.10, (October 2005), pp. 553-561, ISSN:1541-7786

Poon R.C., Agnihotri N., Seymour C. & Mothersill C. (2007). Bystander effects of ionizing radiation can be modulated by signaling amines. *Environmental Research,* Vol.105, No.2, (October 2007), pp. 200-211, ISSN:0013-9351

Prise K.M., Belyakov O.V., Folkard M. & Michael B.D. (1998). Studies of bystander effects in human fibroblasts using a charged particle microbeam. *International Journal of Radiation Biology,* Vol.74, No.6, (December 1998), pp. 793-798, ISSN:0955-3002

Prise K.M., Folkard M. & Michael B.D. (2006). Radiation-induced bystander and adaptive responses in cell and tissue models. *Dose-Response,* Vol.4, No.4, (September 2006), pp. 263-276, ISSN:1559-3258

Prise K.M. (2008). Bystander effect and radionuclide therapy, In: Targeted Radionuclide therapy Biological Aspects. Stigbrand T., Carlson J. & Adams G.P. Editors, pp. 311-319, DOI:10.1007/978-1-4020-8696-0_17

Prise K.M. & O'Sullivan J.M. (2009). Radiation-induced bystander signaling in cancer therapy. *Nature Reviews. Cancer,* Vol.95, No.5, (May 2009), pp. 351-360, ISSN:1474-175X

Przybyszewski W.M., Widel M., Szurko A., Lubecka B., Matulewicz L., Maniakowski Z., Polaniak R., Birkner E. & Rzeszowska-Wolny J. (2004). Multiple bystander effect of irradiated megacolonies of melanoma cells on non-irradiated neighbours. *Cancer Letters,* Vol.214, No.1, (October 2004), pp. 91-102, ISSN:0304-3835

Ryan L.A., Seymour C.B., Joiner M.C., and Mothersill C.E. (2009). Radiation-induced adaptive response is not seen in cell lines showing a bystander effect but is seen in lines showing HRS/IRR response. *International Journal of Radiation Biology,* Vol. 5, No.1, (January 2009), pp. 87-95, ISSN:0955-3002

Rzeszowska-Wolny J., Przybyszewski W.M. & Widel M. (2009a). Ionizing radiation-induced bystander effects, potential targets for modulation of radiotherapy. *European Journal of Pharmacology,* Vol.625, No.1-3, (December 2009), pp.156-167, ISSN:0014-2999

Rzeszowska-Wolny J., Herok R., Widel M. & Hancock R. (2009b). X-irradiation and bystander effects induce similar changes of transcript profiles in most functional pathways in human melanoma cells. *DNA Repair (Amst),* Vol.8, No.6, (June 2009), pp. 732-738, ISSN:1568-7864

Sawant S.G., Randers-Pehrson G., Metting N.F. & Hall E.J. (2001). Adaptive response and the bystander effect induced by radiation in C3H 10T(1/2) cells in culture. *Radiation Research,* Vol.56, No.2, (August 2001), pp. 177-180, ISSN:0033-7587

Sedelnikova O.A., Nakamura A., Kovalchuk O., Koturbash I., Mitchell S.A., Marino S.A., Brenner D.J. & Bonner W.M. (2007). DNA double-strand breaks form in bystander cells after microbeam irradiation of three-dimensional human tissue models. *Cancer Research,* Vol.67, No.9, (May 2007), pp. 4295-4302, ISSN:0008-5472

Seymour C.B. & Mothersill C. (2000). Relative contribution and targeted cell killing to the low-dose region of the radiation dose-response curve. *Radiation Research,* Vol.153, No.5, (May 2000), pp. 508–511, ISSN:0033-7587

Sgouros G., Knox S.J., Joiner M.C., Morgan W.F., Kassis A.I. (2007). MIRD continuing education: Bystander and low dose-rate effects: are these relevant to radionuclide therapy? Journal of Nuclear Medicine, Vol.48, No.10, (October 2007), pp. 1683-1691, ISSN:0161-5505

Shao C., Folkard M., Michael B.D., Prise K.M. (2004). Targeted cytoplasmic irradiation induces bystander responses. *Proceedings of the National Academy of Sciences of the*

United States of America, Vol.101, No.37, (September 2004), pp.13495-13500, ISSN:0027-8424

Shao C., Lyng F.M., Folkard M. & Prise K.M. (2006). Calcium fluxes modulate the radiation-induced bystander responses in targeted glioma and fibroblast cells. *Radiation Research*, Vol.166: No.3, (September 2006), pp. 479-487, ISSN:0033-7587

Shao C., Folkard M. & Prise K.M. (2008a). Role of TGF-β1 and nitric oxide in the bystander respose of irradiated glioma cells. *Oncogene*, Vol.27, No.4, (January 2008), pp. 434-440, ISSN:0950-9232

Shao C., Prise K.M. & Folkard M. (2008b). Signaling factors for irradiated glioma cells induced bystander responses in fibroblasts. *Mutation Research*, Vol.638, No.1-2, (February 2008), pp. 139-145, ISSN:0027-5107

Smith P.G. & Doll R. (1982). Mortality among patients with ankylosing spondylitis after a single treatment course with X rays. *British Medical Journal* , Vol.284, No.6314, (February 1982), pp. 449–460. ISSN: 0267-0623

Sokolov M.V., Dickey J.S., Bonner W.M. & Sedelnikova O.A. (2007). γ-H2AX in bystander cells. *Cell Cycle*, Vol.6, No.18, (September 2007), pp. 2210-2212, ISSN:1538-4101

Snyder A.R. & Morgan W.F. (2004). Gene expression profiling after irradiation: clues to understanding acute and persistant responses? *Cancer Metastasis Reviews*, Vol.23, No.3-4, (August-December 2004), pp. 259-268, ISSN:0167-7659

Spitz D.R., Azzam E.I., Li J.J. & Gius D. (2004). Metabolic oxidation/reduction reactions and cellular responses to ionizing radiation : a unifying concept in stress response biology. *Cancer Metastasis Review*, Vol.23, No.3-4, pp.311-322, ISSN:0167-7659

Suit H., Goldberg S., Niemierko A., Ancukiewicz M., Hall E., Goitein M., Wong W. & Paganetti H. (2007). Secondary carcinogenesis in patients treated with radiation: a review of data on radiation-induced cancers in human, non-human primate, canine and rodent subjects. *Radiation Research*, Vol.167, No.1, (January 2007), pp. 12-42, ISSN:0033-7587

Tartier L., Gilchrist L., Burdak-Rothkamm S., Folkard M. & Prise K.M. (2007). Cytoplasmic irradiation induces mitochondrial-dependent 53BP1 protein relocalization in irradiated and bystander cells. *Cancer Research*, Vol.67, No.12, (June 2007), pp. 5872-5879, ISSN:0008-5472

Tawn E.J., Whitehouse C.A., Winther J.F., Curwen G.B., Rees G.S., Stovall M., Olsen J.H., Guldberg P., Rechnitzer C., Schrøder H. & Boice J.D. Jr. (2005). Chromosome analysis in childhood cancer survivors and their offspring - no evidence for radiotherapy-induced persistent genomic instability. *Mutation Research*, Vol.583, No.2, (June 2005), pp. 198-206, ISSN:0027-5107

Tlsty T.D. (2002). Functions of p53 suppress critical consequences of damage and repair in the initiation of cancer. *Cancer Cell*, Vol.2, No.1, (July 2002), pp. 2-4, ISSN:1535-6108

Trott K-R. (2009). Can we reduce the incidence of second primary malignancies occurring after radiotherapy? *Radiotherapy and Oncology: Journal of the European Society for Therapeutic Radiology and Oncology*, Vol.91, No.1, (April 2009), pp. 1-3, ISSN:0167-8140

Tubiana M. (2009). Can we reduce the incidence of second primary malignancies occurring after radiotherapy? A critical review. *Radiotherapy and Oncology: Journal of the European Society for Therapeutic Radiology and Oncology*, Vol.91, No.1, (April 2009), pp. 4-15, ISSN:0167-8140

Ulrich R.L. & Ponnaiya B. (1998). Radiation-induced instability and its relation to radiation carcinogenesis. *International Journal of Radiation Biology,* Vol.74 No.6, (December 1998), pp. 747-754, ISSN:0955-3002

Vines A.M., Lyng F.M., McClean B., Seymour C., & Mothersill C. (2009). Bystander effect induced changes in apoptosis related proteins and terminal differentiation in in vitro murine bladder cultures. *International Journal of Radiation Biology,* Vol.85, No.1, (January 2009), pp. 48-56, ISSN:0955-3002

Weissenborn U. and Streffer C. (1989). Analysis of structural and numerical chromosomal aberrations at the first and second mitosis after irradiation of one-cell mouse embryos with X-rays and neutrons. *Radiation Research,* Vol.117, No.2, (February 1989), pp. 214-220, ISSN:0033-7587

Whitehouse C.A. & Tawn E.J. (2001). No evidence for chromosomal instability in radiation workers with in vivo exposure to plutonium. *Radiation Research,* Vol.156, No.5, (November 2001), pp. 467-475, ISSN:0033-7587

Widel M., Szurko A., Przybyszewski W. & Lanuszewska J. (2008). Non-irradiated bystander fibroblasts attenuate damage to irradiated cancer cells. *Radioprotection,* Vol.43 No.5, p.158, ISSN:0033-8451

Widel M., Szurko A., Przybyszewski W., Brodziak L., Lalik A. & Rzeszowska-Wolny J. (2009). *P53* gene function is not required for bystander signals induction and transmission. *37th Annual Meeting of the European Radiation Research Society* Prague, Czech Republic, 6-29th August 2009

Wolff S. (1996). Aspects of the adaptive response to very low doses of radiation and other agents. *Mutation Research,* Vol.358, No.2, (November 1996), pp. 135-142, ISSN:0027-5107

Xu W., Liu L.Z., Loizidou M., Achmed M. & Charles I. (2002). The role of nitric oxide in cancer. *Cell Research,* Vol.12, No.5-6, (December 2002), pp. 311-320, ISSN:1001-0602

Xu X.G., Bednarz B. & Paganetti H. (2008). A review of dosimetry studies on external-beam radiation treatment with respect to second cancer induction. *Physics in Medicine and Biology,* Vol.53, No.13, (July 2008), pp. R193-241, ISSN:0031-9155

Xue L.Y., Butler N.J., Makrigiorgos G.M., Adelstein S.J. & Kassis A.I. (2002). Bystander effect produced by radiolabeled tumour cells in vivo. *Proceedings of the National Academy of Sciences of the United States of America,* Vol.99, No.21, (October 2002), pp. 13765–13770, ISSN:0027-8424

Zhong W., He Q., Chan L.L., Zhou F., ElNaghy M., Thompson E.B. & Ansan N.H. (2001). Involvement of caspases in 4-hydroxy-alkenal-induced apoptosis in human leukemic cells. *Free radical biology & medicine,* Vol.30, No.6, (March 2001), pp. 699-706, ISSN:0891-5849

Zhou H., Randers-Pherson G., Waldren C.A., Vannais D., Hall E.J., Hei T.K. (2000). Induction of a bystander mutagenic effect of alpha-particles in mammalian cells. *Proceedings of the National Academy of Sciences of the United States of America,* Vol.97, No.5, (February 2000), pp. 2099-2104, ISSN:0027-8424

18

The Potential Of I-129 as an Environmental Tracer

Andrej Osterc and Vekoslava Stibilj
Institute Jožef Stefan,
Slovenia

1. Introduction

Iodine has two natural isotopes - the only stable iodine isotope is ^{127}I, whilst ^{129}I is the only radioactive iodine isotope that is formed in nature ($T_{1/2} = 1.57 \cdot 10^7$ years). However, the main sources of ^{129}I in the environment are anthropogenic from nuclear fuel reprocessing plants (NFRP) and nuclear accidents. Current levels of ^{129}I do not represent any radiological hazard to humans, but the liquid discharges of ^{129}I from reprocessing plants into the ocean makes it a unique oceanographic tracer to study the movement of water masses, transfer of radionuclides and marine cycles of stable elements such as iodine. The gaseous releases of ^{129}I from reprocessing plants can be used as an atmospheric and geochemical tracer (Hou, 2004).
^{129}I and ^{127}I have the same chemical properties and therefore it is expected that they also behave similar in environment. Lack of ^{129}I and ^{127}I speciation data makes it difficult to confirm or disprove this assumption. The main problem is the mobility - species of newly introduced and old – natural ^{129}I. The old ^{129}I is in equilibrium with ^{127}I - natural $^{129}I/^{127}I$ ratio and this is disturbed with ^{129}I from NFRP which is released to the environment in volatile form. As such it is rapidly transferred among surface compartments. Liquid discharges to oceans influence areas in accordance with marine currents. Wet and, to a lesser extent, dry depositions of atmospheric ^{129}I are the main sources for ^{129}I in terrestrial environment, which is distant from ^{129}I sources such as NFRP.
The biggest reservoir of iodine is the ocean with an average concentration of approximately 50-60 μg L^{-1} seawater. From marine environment is iodine transferred to the atmosphere by volatilization mainly as iodomethane (CH_3I) and then washed out to terrestrial environment by wet and dry deposition. It is accumulated in soils where it is strongly bound-adsorb to organic matter, and iron and aluminium oxides in soil (Fuge, 2005). In the accumulation processes of iodine in soil besides various physico-chemical parameters including soil type, pH, Eh, salinity, and organic matter content, soil microorganism - especially bacteria were found to play an important role (Muramatsu & Yoshida, 1999, Amachi, 2008). In this way the biogeochemical cycling of ^{129}I is strongly connected to processes in ocean and soil systems - the atmosphere being the bridge between them.

2. Sources, inventory and levels of ^{129}I in marine and terrestrial environment

All ^{129}I formed in the primordial nucleosynthesis decayed to stable ^{129}Xe. Two natural processes responsible for natural background levels of ^{129}I are spallation of cosmic rays on

atmospheric Xe (cosmogenic) in the upper atmosphere and spontaneous fission of ^{238}U (fissiogenic).
Although ^{129}I is produced naturally the main part is a consequence of human nuclear activities (Table 1). In this way the sources can be divided in natural and man-made or in pre-nuclear and nuclear era. From 1945 anthropogenic sources of ^{129}I were nuclear weapons testing, nuclear accidents (Chernobyl) and at present marine and atmospheric discharges from NFRP. Operating plants in Europe are located in England (Sellafield), France (La Hague) and Russia (Mayak), and outside Europe in China, India, Pakistan and Japan (Tokaimura, Rakkasho). ^{129}I is produced during the operation of a nuclear power reactor by nuclear fission of $^{235}U(n, f)^{129}I$ and $^{239}Pu(n, f)^{129}I$. It was estimated that about 7.3 mg of ^{129}I is produced per megawatt day. ^{129}I is released during reprocessing of nuclear fuel - mainly by PUREX process. The fuel is first dissolved with nitric acid and at this step iodine is oxidized to volatile I_2 and despite all efforts to trap and collect released iodine some part may be discharged from the NFRP (Reithmeir et al., 2006).

Source	**Inventory/release (kg)****	**$^{129}I/^{127}I$ ratio in environment**
Nature	250	$\sim 1 \cdot 10^{-12}$
Nuclear weapons testing	57	$1 \cdot 10^{-11}$–$1 \cdot 10^{-9}$
Chernobyl accident	1.3–6	10^{-8}–10^{-6} (in contaminated area)
Marine discharges from European NFRP* by 2007	5200	10^{-8}–10^{-6} (North Sea and Nordic Sea water)
Atmospheric releases from European NFRP* by 2007	440	10^{-8}–10^{-6} (in rain, lake and river water in West Europe) 10^{-6}–10^{-3} (in soil, grass near NFRP)
Atmospheric releases from Hanford NFRP*	275	10^{-6}–10^{-3} (in air near NFRP)

*NFRP…nuclear fuel reprocessing plant; **Marine discharges are sum discharges from La Hague and Sellafield NFRP, Atmospheric releases are sum releases from La Hague, Sellafield, Marcoul and Karlsruhe-WAK (after Hou et al., 2009)

Table 1. Sources and $^{129}I/^{127}I$ ratio in environment

Until the beginning of the 1990s the total annual discharges from two European NFRP, La Hague and Sellafield, remained below 20 kg year^{-1}. The discharges increased later considerable - up to 300 kg year^{-1} and accounted until 2000 for more than 95 % of the total inventory in the global ocean (Fig. 1) (Alfimov et al., 2004; Lopez-Gutierrez et al., 2004).
The natural, pre-nuclear $^{129}I/^{127}I$ isotopic ratio was significantly influenced by releases of anthropogenic ^{129}I to the environment. The estimated pre-nuclear $^{129}I/^{127}I$ isotopic ratio in marine environment was assessed with analysis of marine sediments and agreed to be $1.5 \cdot 10^{-12}$ (Table 2) (Moran et al., 1998; Fehn et al., 2000a; Fehn et al., 2007). For the terrestrial environment - pedosphere and biosphere no agreed data on pre-nuclear ratio exist. Human nuclear activity increased the $^{129}I/^{127}I$ ratio in marine environment to 10^{-11} - 10^{-10} and to 10^{-8} - 10^{-5} (Table 3) in the Irish Sea, English Channel, North Sea and Nordic Seas which are influenced by liquid discharges from European NFRP (Frechou & Calmet, 2003; Alfimov et al., 2004; Hou et al., 2007). In the terrestrial environment the $^{129}I/^{127}I$ ratio increased to 10^{-9} –

10^{-7}, even 10^{-6}–10^{-4} in the vicinity of nuclear fuel reprocessing plants (Table 4) (Duffa & Frechou, 2003; Frechou & Calmet, 2003).

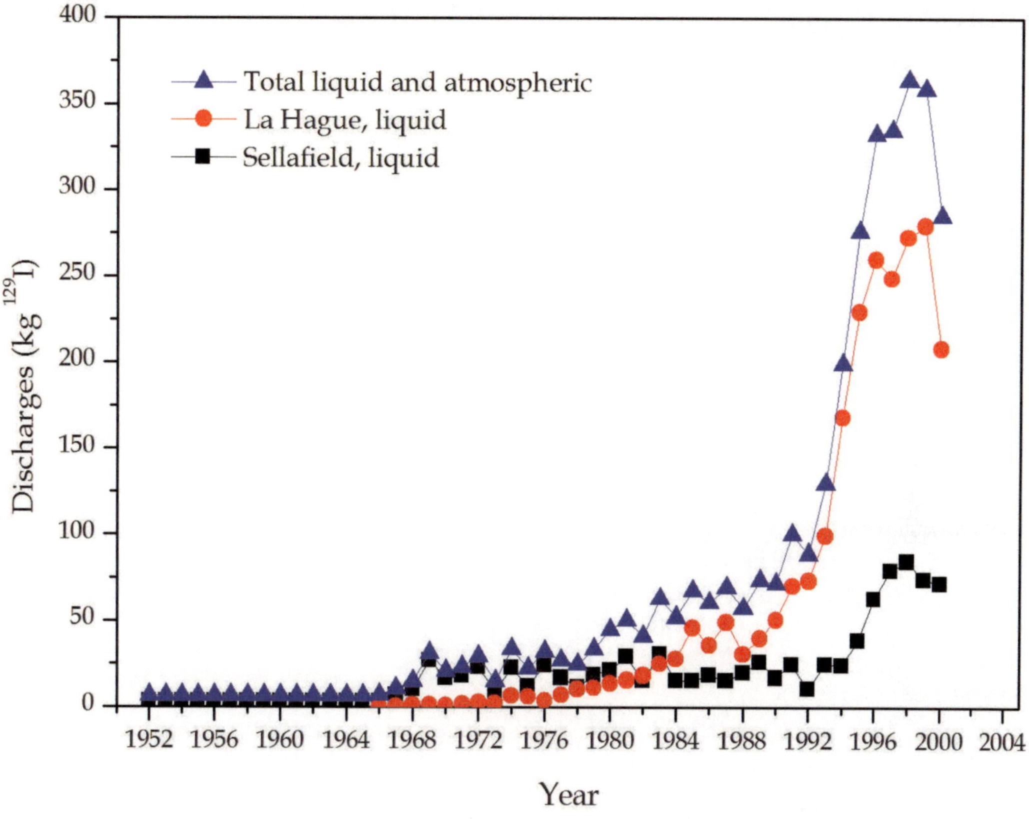

Fig. 1. Liquid and atmospheric releases of ^{129}I from NFRP in La Hague and Sellafield for period from 1952 to 2000 (compiled by Lopez-Gutierrez et al., 2004).

Atmospheric releases are not plotted, but they are considered in the total amount. Annual atmospheric releases ranged from 1.19 to 9.58 kg ^{129}I with a total amount of 235.5 kg in the period from 1952 to 2000.

Anthropogenic ^{129}I predominates in marine environment in biosphere and upper layers of the oceans and in terrestrial environment in soil, therefore it can be expected that the isotopic ratio ^{129}I/^{127}I is increasing in these compartments of the ecosystem. Precipitation and seawater are probably the main carriers for ^{129}I exchange among different compartments in marine and terrestrial environment. Data from literature clearly show that ^{129}I levels in marine sediment, marine algae and soil are several times higher than in seawater or precipitation. Meaning that ^{129}I is most probably chemically or biologically transformed to species which accumulate in those compartments (Tables 3 and 4).

To summarize, different values of ^{129}I/^{127}I isotopic ratios in environment are today envisaged as 10^{-12} for pre-nuclear era, 10^{-9} in slightly contaminated regions and 10^{-9}–10^{-6} in regions affected by the releases from NFRP. The highest ratios were found in the close vicinity of NFRP with values from 10^{-6} to 10^{-4} (Hou, 2009).

Sample	$^{129}I/^{127}I$ (10^{-12})	Reference
TERRESTRIAL ENVIRONMENT		
Soil		
Russia, Moscow, 1910	168	Szidat et al., 2000a
Russia, Bogoroditsk, 1909	25	
Russia, Lutovinovo, 1939	5.7	
Thyroid powder		
Species not given (USA), 1947	4.6	Szidat et al., 2000a
USA, Pig, 1947	58	
USA, Horse, 1947	1230	
MARINE ENVIRONMENT		
Sediment		
Peru, depth: 155-199 cm	1.50	Moran et al., 1998
Mexico, Baja peninsula, depth: 415-420 cm	1.48	
Ecuador, depth: 315-320 cm	1.05	
Algae		
Japan		Fehn et al., 2007
Laminaria Japonica		
Hokkaido, 1883	1.40	
Hokkaido, 1883	0.55	
Miyagi, 1883	0.52	
Miyagi, 1883	0.67	
Pelvita		
Miyagi, 1904	1.87	
Russia		Cooper et al., 1998
Laminaria digitata		
Novaya Zemlya, 1930	1.00	
Novaya Zemlya, 1931	3.69	
White Sea, 1938	1.35	
White Sea, 1930	1.37	
White Sea, 1938	1.92	

Table 2. $^{129}I/^{127}I$ isotopic ratios in pre-nuclear age environmental and biological samples

3. Factors affecting biogeochemical cycling of iodine

Iodine is a trace element present in the hydrosphere, lithosphere, atmosphere and biosphere at different concentrations and as different iodine species (Table 5). Speciation analysis of iodine was mainly done on stable ^{127}I (Hou et al., 1997; dela Veija et al., 1997; Sanchez & Szpunar, 1999; Hou et al., 2000c; Leiterer et al., 2001; Schwehr & Santschi, 2003; Shah et al., 2005; Gilfedder et al., 2008), with some studies on ^{129}I (Hou et al., 2001; Hou et al., 2003b; Schwehr et al., 2005; Englund et al., 2010b). Majority of researches performed on ^{127}I and ^{129}I are limited to fractionations of iodine - water soluble, exchangeable, bound to oxides, organic-inorganic fraction, etc. In general just the most abundant chemical forms of iodine - iodide (I^-) and iodate (IO_3^-) are determined and the rest of total iodine content is associated with organic iodine. It is well known that organic iodine fraction mainly consist of iodine

Sample	^{129}**I/**^{127}**I (**10^{-8}**)**	**Reference**
Sea water		
Germany, North Sea, 1999	153	Szidat et al., 2000a
Greenland, 1999 (n = 5)	0.07–0.24	Hou, 2004
England, Irish Sea, near Sellafield 2004-05 (n = 4)	89–820	Atarashi-Andoh et al., 2007
Scotland, Scottish Sea, influence of Sellafield, 2003-2005 (n = 14)	7.2–336	Schnabel et al., 2007
Israel, Sea of Galilee, June 1998	0.31	Fehn & Snyder, 2000b
Israel, Engedi, Dead Sea, June 1998	0.003	
Japan Sea, Toyama Bay, October 2006	0.0086	Suzuki et al., 2008
Japan Sea, Off Sekine, 2006-2007 (n = 2)	0.0063–0.0068	
Sediment		
Sweden, Baltic Sea, influence of La Hague and Sellafield, core sample -from 0 to 21 cm, 1997	0.34–1.06	Aldahan et al. 2007
Seaweed		
Greenland, 1997 (*Fucus distichus*, n = 7)	0.07–0.15	Hou et al., 2000a
Norway (Utsira), 1980-1995 (*Fucus vesiculosus*, n = 16)	1.88–18.5	
Denmark (influence of liquid discharges from NFRP) Roskilde Fjord and Bornholm, 1995-1998 (*Fucus vesiculosus*, n = 8)	2.50–9.12	
Klint, 1986-1999 (*Fucus vesiculosus*, n = 39)	3.54–37.5	
France (vicinity of La Hague)		Frechou et al., 2003
Goury, 1998-1999 (*Fucus vesiculosus*, n = 3)	1010–1940	
Goury, 1998-1999 (*Fucus serratus*, n = 3)	930–1210	
Goury, 1998-1999 (*Laminaria digita*, n = 2)	540–1270	
Goury and Dielette, 2003 (*Fucus serratus*, n = 12)	496–1960	Barker et al., 2005
Goury and Dielette, 2003 (*Laminaria digita*, n = 8)	349–960	
Ireland		Keogh et al., 2007
West and South coastline, *Fucus vesiculosus*		
1985, n = 7	0.08–0.73	
1994, n = 7	0.47–6.5	
2003, n = 9	0.21–5.0	
East coastline (influence of liquid discharges from NFRP), *Fucus vesiculosus*		
1985, n = 8	4.8–85	
1994, n = 7	0.83–30	
2003, n = 8	24–85	
Russia, *Laminaria digitata*		Cooper et al., 1998
Murmansk region, 1966	0.016	
Murmansk region, 1967	0.034	
White Sea, 1971	0.027	
Novaya Zemlya, 1989	0.48	
Novaya Zemlya, 1993	0.72	
Slovenia, Adriatic Sea, *Fucus virsoides*, September 2005, five locations	0.086–0.11	Osterc & Stbilj, 2008
Italy, Adriatic Sea, *Fucus virsoides*, June 2006, five locations	0.068–0.15	
Croatia, Adriatic Sea, *Fucus virsoides*, October 2006, three locations	0.15–0.31	

Table 3. $^{129}I/^{127}I$ isotopic ratios in nuclear age environmental and biological samples from marine compartments

Sample	$^{129}I/^{127}I$ (10^{-8})	Reference
Aerosol		
Spain, Seville, 2001 (n = 12)	0.29–2.72	Santos et al., 2005
Spain, Seville, 2001-2002	0.18–5.35	Santos et al., 2006
Sweden, Kiruna and Ljungbyhed, 1983-2008	0.5–147	Englund et al., 2010a
Gas		
Spain, Seville, 1993-1994 and 1998	0.01–0.80	Santos et al., 2006
Precipitation		
Germany, Hanover, 1986	16.6	Szidat et al., 2000a
Germany, Lower Saxony, 1997	83.4	
Germany, Upper Bavaria, 2003	14.6–38.6	Reithmeier et al., 2005
Spain, Seville, 1996-1997	0.23–52	Santos et al., 2006
Antartica, McMurdo Station, snowmelt 1999	0.004	Snyder et al., 2004
Antartica, Mt Erebus, snow, 2000	0.009	
Shallow ground water		
Germany, Lower Saxony, 1997	0.8	Szidat et al., 2000a
Lake water		
Denmark, 2000 (n = 7)	2.5–27.3	Hou, 2004
Lithuania, 1999 (n = 2)	6.6–7.3	Hou et al., 2002
England, lakes near Sellafield, 2004-2005 (n = 7)	24.8–638	Atarashi-Andoh et al., 2007
Germany, Munich, Kleinhesseloher See, July 1997	2.4	Fehn & Snyder, 2000b
Germany, Malchow, Malchow See, July 1997	8.6	
Germany, Harz, Okersee, June 1999	1.0	Snyder et al., 2004
USA, Oregon, Crater Lake, September 1996	0.9	
USA, Colorado, Navajo Lake, June 2000	0.25	Snyder et al., 2003a
Central America, Nicaragua, Lake Managua, 1998	0.029	Fehn & Snyder, 2000b
South America, Chile, Lago Verde, Februar 1999	0.24	Snyder et al., 2004
Australia, New South Wales, Lake George, 1997	0.53	Fehn & Snyder, 2000b
New Zealand, Lake Taupo, 1999	0.005	Snyder et al., 2004
Japan, Odanoike lake, May 2000	0.79	
Indonesia, Bali, Lake Beratan	0.032	
River water		
England, London, river Thames, March 1999	1.9	Snyder et al., 2004
England, Cambridge, river Granta, March 1999	1.0	
England, rivers near Sellafield 2004-2005 (n = 4)	158-825	Atarashi-Andoh et al., 2007
USA, Colorado, Pine River, June 2000	0.13	Snyder et al., 2003a
USA, Colorado, Animas River, June 2000	0.08	
India, Tista River, 1999	0.18	Snyder et al., 2004
India, Ganges River, 1999	0.03	
Central America, El Salvador, Rio Lempa, 1999	0.058	Snyder et al., 2003b
Africa, Botswana, Thamakkane river, May 2000	0.10	Snyder et al., 2004
Japan, Kugino river, May 2000	0.04	
Mongolia, Tuyu Gol River, January 2000	0.068	
Thyroid		
France, vicinity of La Hague (1–30 km), 1980-1999 (Bovine, n = 19)	100–25068*	Frechou et al., 2003
China (Tianjin), 1994-1995 (Human, male; n = 4)	0.04–0.09	Hou et al., 2000b
China (Tianjin),1995 (Human, female; n = 2)	0.16–0.20	

*The highest isotopic ratio ($2.5\ 10^{-4}$) was obtained for an animal coming from Digulleville, a village 3 km to the north-east of the NFRP

Table 4. $^{129}I/^{127}I$ isotopic ratios in nuclear age environmental and biological samples from terrestrial compartments

Compartment	Main iodine species	Reference
	Concentration range	
Atmosphere	particle associated (aerosol); inorganic gaseous: I_2, HI, HIO; organic gaseous: CH_3I, CH_2I_2, $CH_3CH_2CH_2I$, etc.	Hou et al., 2009
	1–100 ng m^{-3}	Wershofen & Aumann, 1989; Yoshida & Muramatsu, 1995
Hydrosphere		
oceans	inorganic: I^-, IO_3^-; organic: CH_3I	Hou et al., 2001
	45–60 ng mL^{-1}	Hou et al., 2009
fresh water	1–3 ng mL^{-1}	Hou et al., 2009
precipitation	1–6 ng mL^{-1}	Yoshida & Muramatsu, 1995; Hou, 2004
Lithosphere		
soil	inorganic: I^-, IO_3^-, bound to metal oxides, carbonates and minerals; organic: bound to humic and fulvic acids	Schmitz & Aumann, 1995
	0.5–40 µg g^{-1}	Muramatsu & Yoshida, 1999
surface sea sediment	1–2000 µg g^{-1}	Muramatsu & Wedepohl, 1998
metamorphic and magmatic rocks	<0.1 µg g^{-1}	
Biosphere		
seaweed	inorganic: I^-, IO_3^-; organic: iodo-amino acids (*Laminaria japonica*); bound to proteins, pigments, polyphenols*	Hou et al., 1997 Hou et al., 2000c Shah et al., 2005
	10–6000 µg g^{-1}	Hou & Yan, 1998; Osterc & Stibilj, 2008
plants (terrestrial)	<1 µg g^{-1}	Hou et al., 2009
thyroid gland	inorganic: I^- organic: iodo-amino acids → iodo-thyronine and iodo-tyrosine	dela Vieja et al., 1997
	500–5000 µg g^{-1}	Hou et al., 2003a
milk (bovine)	inorganic: I^- organic: bound to proteins*	Leiterer et al., 2001
	0.017–0.49 µg mL^{-1}	

*species not identified

Table 5. Concentrations of stable iodine in environmental compartments

bound to proteins - but these are still not identified for most environmental and biological samples, not for ^{127}I and certainly not for ^{129}I. The main problem is lack of appropriate standards for speciation analysis and very small amounts of ^{129}I in environmental and biological samples.

Iodine is released from marine environment to the atmosphere partly as aerosols formed from the sea spray - inorganic iodide and iodate - and mainly as volatile organic iodine compounds (VOIC) such as iodomethane (Baker et al., 2000; Leblanc et al., 2006, Chance et al., 2009). Bacteria, phytoplankton and brown algae present in marine environment are capable to reduce the most thermodynamically stable form of iodine, the iodate to iodide. On the other hand microalgae and macroalgae-seaweed accumulate iodide and transform it into VOIC - the most important are CH_3I, CH_2I_2, CH_2BrI and CH_2ClI (Leblanc et al., 2006). The emitted organic iodine is decomposed by sunlight into inorganic iodine compounds. The photolytic lifetimes of VOIC differ; CH_2I_2 has a lifetime of 5 minutes, followed by CH_2BrI with a lifetime of 45 minutes and CH_2ClI with a lifetime of 10 h (Stutz, 2000). The longest photolytic lifetime of 14–18 days has CH_3I (Stutz, 2000). During this process of photolization reactive iodine oxides such as HOI, I_2O_2 and IO_2 form, which either form condensable vapours as nuclei for aerosols or react with ozone. From the atmosphere iodine enters the marine and terrestrial environment by processes of wet and dry deposition. In the iodine terrestrial cycle interactions between water and soil are most important (Santschi & Schwehr, 2004). Beside physical and chemical factors, biological processes especially promoted by microorganism influence the cycling of iodine. Microorganisms are involved in environmental processes as primary producers and also as consumers and decomposers. They have bioremedial and biotransformable potential and in this way affect the mobility of elements. Oxidation and reduction mechanisms contribute to transformations between soluble and insoluble forms. Experiments with ^{125}I tracer showed the importance of microbial participation in iodine accumulation - sorption and desorption processes - in soil. Muramatsu et al. (1996) observed desorption of iodine from flooded soil during cultivation of rice plants. Microorganisms created reducing conditions in the flooded soil and iodine once adsorbed on the soils was desorbed (Muramatsu et al., 1996). Amachi et al. (2001) reported a wide variety of terrestrial and marine bacteria that are capable to produce CH_3I under oligotrophic conditions. Aerobic bacteria showed significant production of CH_3I, whereas anaerobic did not produce it. The methylation of iodide was catalysed enzymatically with S-adenosyl-$_L$-methionine as the methyl donor.

The biding of iodine by organic matter and/or iron and aluminium oxides has the potential to modify the transport, bioavailability and transfer of iodine isotopes to man (Santschi & Schwehr, 2004). Because of the same chemical properties ^{129}I and ^{127}I should behave similar in environmental processes. Major pathways are the volatilization of organic iodine compounds into the atmosphere, accumulation of iodine in living organisms, oxidation and reduction of inorganic iodine species, and sorption of iodine by soils and sediments. These processes are influenced or even controlled by microbial activities (Amachi, 2008).

^{129}I is gradually released in trace quantities into the atmosphere and aquatic environment from reprocessing plants. It is then physically transported in the air or water media under the influence of chemical and biological processes. Newly introduced ^{129}I from NFRP is in volatile form and as such more mobile compared to ^{127}I. By taking this aspect into account one cannot be sure that biogeochemical behaviour of ^{129}I and ^{127}I is the same. Even more, Santschi & Schwehr (2004) discussed that biogeochemical behaviour of iodine and its isotopes appears to be different in North American and European waters.

4. Measurement of ^{129}I

^{129}I decays by emitting beta particles ($E_{\beta max}$ = 154.4 keV), gamma rays (E_{γ} = 39.6 keV) and X-rays (29–30 keV) to stable ^{129}Xe (Tendow, 1996). Therefore it can be measured by gamma and X-ray spectrometry and by beta counting using liquid scintillation counters (LSC).
Another method for determination of ^{129}I is neutron activation analysis (NAA) that is based on neutron activation of $^{129}I(n, \gamma)^{130}I$, which is measured by gamma spectrometry (E_{γ} = 536 keV (99 %). In recent year's mass spectrometry - such as accelerator mass spectrometry (AMS) and inductively coupled plasma mass spectrometry (ICP-MS) are also used.
For determination of ^{129}I levels in environmental samples only two analytical methods are available, radiochemical neutron activation analysis (RNAA) and AMS. The main advantage of the AMS is the detection limit that is close to 10^{-14} expressed as $^{129}I/^{127}I$ ratio. RNAA can only measure ^{129}I at elevated levels - nuclear era. AMS enables measurement of ^{129}I in all environmental samples, also the natural, pre-nuclear levels, and the needed amount of sample is 10-100 times smaller than in the case of RNAA. Detection limits for ^{129}I using different analytical methods are compared in Table 6.

Analytical method/Sample	**Detection limit** $g\ g^{-1}$ (10^{-12})	$^{129}I/^{127}I$ (10^{-12})	**Reference**
γ-X spectrometry			
seaweed (400 g)	300	not given	Lefevre et al., 2003
LSC			
radioactive waste (coolant, 1 L)	23	not given	Gudelis et al., 2006
ICP-MS			
Aqueous solution	100	1000000	Muramatsu et al., 2008
Aqueous solution	0.8	not given	Izmer et al., 2003
Aqueous solution (groundwater)	5	not given	Brown et al., 2007
Sediment	30	not given	Izmer et al., 2003
Sediment	0.4	not given	Izmer et al., 2004
RNAA			
soil (100 g)	0.05	5000	Osterc et al., 2007
soil (100 g)	0.015	10000	Muramatsu & Yoshida, 1995
soil (80 g)	0.27	not given	Michel et al., 2005
soil	0.13	410	Szidat et al., 2000b
AMS			
commercial AgI	not given	0.44	Suzuki et al., 2006
blank sample	not given	0.50	Gomez-Guzman et al., 2011
blank sample	not given	0.17	Muramatsu et al., 2008
soil (1 g)	0.0015	40	
soil (80 g)	0.00015	5	Michel et al., 2005
Woodward Iodine*	not given	0.023	Reithmeier et al., 2005
Woodward Iodine	not given	0.04	Buraglio et al., 2001
oil and gas hydrates	not given	0.20	Alfimov & Synal, 2010
soil	0.000023	0.75	Szidat et al., 2000b

*Woodward Iodine is elemental iodine mined by Woodward Iodine Corp. in Oklahoma for which the lowest ratio is reported.

Table 6. Limits of detection for ^{129}I in various samples using different analytical methods

4.1 Direct gamma and X-ray spectrometry

Direct gamma-X spectrometry ($E_\gamma = 39.6$ keV; X-rays, 29–30 keV) is a non-destructive technique that is rapid and can be applied to different matrices. It is used for monitoring of environmental samples collected in vicinity of NFRP such as thyroid, urine, seaweed, and for nuclear waste by using high purity Ge or plenary Si detector (Suarez et al., 1996; Bouisset et al., 1999; Frechou et al., 2001; Lefevre et al., 2003; Frechou & Calmet, 2003; Barker et al., 2005). To lower the detection limits normally big samples (50–500 g) are used, which induces considerable attenuation at low energies. The attenuation depends on the matrix composition of the sample and geometric parameters of the container. Therefor the mass energy-attenuation coefficient (self-absorption correction) at a given energy must be measured for all sample matrices with respect to that of the standard source. Experimentally obtained self-absorption correction factors are used to obtain accurate results (Bouisset et al., 1999; Lefevre et al., 2003, Barker et al., 2005). To quantify self-absorption correction factors ^{210}Pb (46.5 keV) and ^{241}Am (59.6 keV), with gamma lines close to ^{129}I are used. Detection limits as low as 2 Bq kg^{-1} dry mass can be reached for *Fucus sp.* samples (Bouisset et al., 1999).
Chemical separation of ^{129}I from the sample matrix and interfering radionuclides - destructive method - improves the detection limit when using direct gamma-X spectrometry (Suarez et al., 1996).
By using direct gamma-X spectrometry ^{129}I was determined in seaweed sample FC-98 Seaweed, which was prepared by Frechou et al. (2001), by using direct gamma -X spectrometry (Osterc & Stibilj, 2008).

4.2 Liquid Scintillation Counting (LSC)

Liquid scintillation counting is based on emissions of beta particles from radionuclides - beta decay ($E_{\beta max} = 154.4$ keV). ^{129}I has to be separated from the sample matrix and other radionuclides and dissolved or suspended in a scintillation cocktail containing an organic solvent and a scintillator. Beta particles emitted from the sample transfer energy to the solvent molecules, which in turn transfer their energy to the scintillator which relaxes by emitting light - photons. In a liquid scintillation counter each beta emission (ideally) results in a pulse of light, which is amplified in a photomultiplier and detected.
Recently extraction chromatographic resins for the separation and determination of ^{36}Cl and ^{129}I have been developed. First results show a promising potential to use the resins within the context of the monitoring of nuclear installations - during operation and especially during decommissioning (Zulauf et al., 2010).

4.3 Inductively Coupled Plasma Mass Spectrometry (ICP-MS)

ICP-MS has been used to determine ^{129}I in contaminated environmental samples with high level ^{129}I content such as sediments, groundwater samples, soil and seaweed (Izmer et al., 2003; Izmer et al., 2004; Becker, 2005; Brown et al., 2007; Li et al., 2009). The lowest detection limit of the method reported as ^{129}I/^{127}I isotopic ratio is 10^{-7}.
The method is based on iodine separation and injection to the machine as solution or gaseous iodine, I_2. Iodine is decomposed into iodine atom and ionized to positive iodine ion at a temperature ~6000–8000 K. It is then extracted from the plasma into a high vacuum of the mass spectrometer via an interface. The extracted ions are separated by mass filters of

either quadropole type time-of-flight or combination of magnetic and electrostatic sector and measured by an ion decetor (Hou et al., 2009).
Difficulties encountered when determining ^{129}I with ICP-MS are low ^{129}I quantities present with high ^{127}I concentrations, isobaric and molecular ions interferences ($^{129}Xe^+$, $^{127}IH_2^+$), memory effects and tailing of ^{127}I. To improve $^{129}I/^{127}I$ determination it was found that introduction of helium gas into collision cell reduces peak tail of a high-abundant isotope, ^{127}I by up to three orders of magnitude. Detection limits have been improved by applying oxygen as collision gas for selective reduction of ^{129}Xe (Izmer et al., 2003, Hou et al., 2009).

4.4 Neutron Activation Analysis (NAA)

NAA enables determination of ^{129}I in environmental samples at 10^{-10} $^{129}I/^{127}I$ isotopic ratios. The concentration levels of ^{129}I in environmental samples are very low and chemical separation/pre-concentration procedures have to be developed which can be used for a wide variety of matrices.
Neutron activation analysis is based on induction of ^{129}I with thermal neutrons - irradiation in a nuclear reactor via following nuclear reaction:

$$^{129}\mathrm{I}(\mathrm{n},\ \gamma)^{130}\mathrm{I}\ \left(T_{½} = 12.36\ \text{hours}, E_{\gamma} = 536.1\ \text{keV}\right) \tag{1}$$

^{129}I is determined by measuring of ^{130}I activity on a high purity Ge detector. Interfering nuclear reactions induced during irradiation of sample from other nuclides resulting in ^{130}I production can influence the correct determination of ^{129}I. These undesired nuclides are ^{235}U, ^{128}Te and ^{133}Cs and nuclear reactions: $^{235}U(n, f)^{129}I(n,\gamma)^{130}I$, $^{235}U(n,f)^{130}I$, $^{128}Te(n,\gamma)^{129m}Te(\beta^-)^{129}I(n,\gamma)^{130}I$ and $^{133}Cs(n,\alpha)^{130}I$ (Hou et al., 1999). They have to be removed from the sample before irradiation to avoid nuclear interferences.
During irradiation radioactivity in sample is produced mainly due to the radioisotopes $^{23}Na(n,\gamma)^{24}Na$ ($T_{½}$ = 14.96 hours), $^{41}K(n,\gamma)^{42}K$ ($T_{½}$ = 12.36 hours) and $^{81}Br(n,\gamma)^{82}Br$ ($T_{½}$ = 35.30 hours) present in sample, which renders the direct measurement of ^{130}I after irradiation and radiochemical separation of induced ^{130}I after irradiation is necessary. Solvent extraction with CCl_4 or $CHCl_3$ are normally used to extract iodine (Osterc & Stibilj, 2005; Osterc et al., 2007).
In first step pre-concentration of iodine from large amounts of sample is performed. Solid samples, such as soil, sediment, vegetation, biological samples can be decomposed by alkaline fusion (Hou et al., 1999, Osterc et al., 2007). The sample is mixed with potassium hydroxide/alkali solution and then gradually heated to 600 °C. Iodine is leached from the decomposed sample with hot water, isolated with solvent extraction and precipitated as PdI_2 or MgI_2 or trapped on activated charcoal (Fig. 2) (Hou et al., 1999, Osterc et al., 2007). Another method to separate iodine from solid samples is combustion at high temperature, ~1100 °C (Muramatsu & Yoshida, 1995). Released iodine is trapped in an alkaline solution or adsorbed on activated charcoal.
The pre-concentrated iodine is than irradiated for up to 12 hours simultaneously with a $^{129}I/^{127}I$ standard. After radiochemical separation the ^{130}I induced from ^{129}I (see nuclear reaction 1) is counted on a high purity Ge detector and compared to standard of known activity and corrected for chemical yield (Osterc et al., 2007).

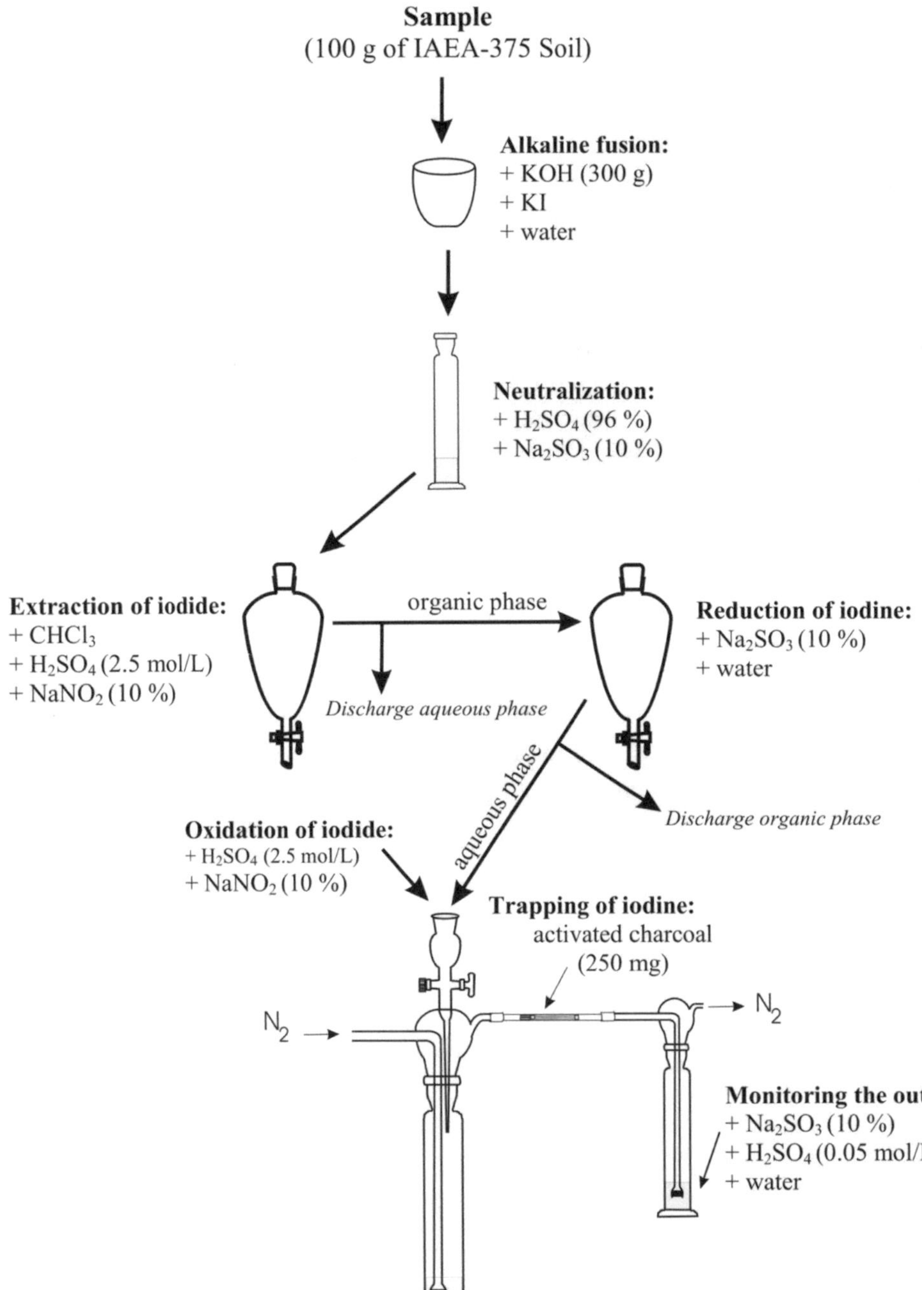

Fig. 2. The scheme for pre-concentration of iodine from solid samples (Osterc et al., 2007)

For liquid samples, such as milk, urine and water samples anion exchange method using anion exchange resins can be applied. Adsorbed iodide is eluted and isolated from the eluate with solvent extraction and precipitated as PdI_2 or MgI_2 (Parry et al., 1995; Hou et al., 2001; Hou et al., 2003a).

4.5 Accelerator Mass Spectrometry (AMS)

An AMS facility is set up off injector and analyser linked with a tandem accelerator. The detector is either a combination of time-of-flight and silicon charged particle detector or gas ionization energy detector. Iodine has to be separated from the sample with same techniques as used for NAA, such as pyrohydrolysis at 1000 °C, and prepared as AgI targets (Muramatsu et al., 2008). Negative iodine ions are produced from AgI targets by Cs sputter ion source and injected into the tandem accelerator. The formed $^{129}I^-$ and $^{127}I^-$ ions are accelerated to positive high-voltage terminal converting negative ions to I^{3+}, I^{5+} or I^{7+}. The positively charged ions pass through a magnetic analyser where ions of ^{129}I and ^{127}I based on charge state and energy are selected and directed to a detector. AMS measures the $^{129}I/^{127}I$ isotopic ratio and the ^{129}I absolute concentration is calculated by the ^{127}I content determined in the sample and the chemical yield for separation of iodine from sample - preparation of AgI targets (Hou et al., 2009).

AMS is the only technique that enables measurement of pre-nuclear age samples and samples with low ^{129}I content, below 10^{-10} $^{129}I/^{127}I$ isotopic ratio (Moran et al., 1998; Fehn et al., 2000a; Buraglio et al., 2001; Alfimov et al., 2004; Santschi & Schwehr, 2004; Snyder & Fehn, 2004; Michel et al., 2005; Fehn et al., 2007; Hou et al., 2007; Keogh, et al., 2007; Muramatsu et al., 2008; Gomez-Guzman et al., 2011). Instrumental background of 10^{-14} $^{129}I/^{127}I$ has been obtained (Buraglio et al., 2000). But the detection limit depends on the chemical separation before measurement and especially on addition of iodine carrier. When carrier and chemical processing are included the typical reported blank $^{129}I/^{127}I$ isotopic ratio is $1 \cdot 10^{-13}$ (Buraglio et al., 2000). For environmental samples with a very low $^{129}I/^{127}I$ isotopic ratio Hou et al. (2010) reported a method for preparation of carrier free AgI targets based on co-precipitation of AgI with AgCl to exclude the influence of interferences from ^{129}I and ^{127}I in the carrier. They calculated a detection limit of 10^5 atoms, which corresponds to $2 \cdot 10^{-16}$ g of ^{129}I.

4.6 Quality assurance of ^{129}I analyses

To be able to determine ^{129}I by RNAA in environmental samples from nuclear era pre-concentration of iodine from large amounts of sample (up to 150 g) is needed. In this pre-concentration step contamination of sample with ^{129}I is possible. It is important to make a blank control when establishing a new method and verify the method by reference materials to evaluate possible contamination during the entire analytical process; including pre-concentration, irradiation, radiochemical separation and gamma activity measurement.

Also analysis of ^{129}I by AMS requires intensive and continuous control - control charts of the analytical blank and verification of accuracy by analysis of reference materials, which has to be continued periodically also during routine operation (Szidat et al., 2000a). Influence of sample mass - AgI targets on accuracy of ^{129}I determination was studied by Lu et al. (2007). They found that samples with masses above 0.3 mg did not show an influence on accuracy - ion current of the sample was constant, but it fell strongly for samples with masses below 0.3 mg. Samples wit masses below 0.1 mg did not produced sustainable currents for ^{129}I determination. Presence of 5000 ^{129}I atoms or 50 µg in the target is sufficient for a successful ^{129}I determination.

To validate and or evaluate an analytical method, to run a laboratory inter-comparison, to check accuracy of analytical method, and ensure globally comparable and traceable results to stated references, as the SI units, certified reference materials are needed. Environmental samples represent a huge variety of different combinations of substances to be analysed and the matrices in which they are embedded. This countless combinations of substances -

elements, radionuclides, contaminants - and matrices means that certified reference materials always lack.

The only reference material with a recommended value for ^{129}I available on the market was the reference material IAEA-375 Soil - Radionuclides and Trace Elements in Soil. Top soil to a depth of 20 cm was obtained from the "Staryi Viskov" collective farm in Novozybkov, Brjansk, Russia in July 1990. Unfortunately this reference material is now out of stock.

Only informative and not certified values for ^{129}I, determined in one laboratory, are reported for NIST SRM 4357 - Ocean Sediment Environmental Radioactivity Standard, which is a blend of ocean sediments collected off the coast of Sellafield, UK, and in the Chesapeake Bay, USA, and NIST SRM 4359 - Seaweed Radionuclide Standard, which is a blend of seaweed collected off the coast of Ireland and the White Sea.

Recently a new reference material, with a certified value for ^{129}I, IAEA-418: I-129 in Mediterranean Sea Water was characterised in an interlaboratory comparison exercise. The used method was AMS (accelerator mass spectrometry).

Another new reference material for radionuclides in the mussel *Mytilus galloprovincialis* from Mediterranean Sea, IAEA-437 was characterised. They reported for the mussel sample collected in 2003 at Anse de Carteau, Port Saint Louis du Rhône, France an informative average massic activity of 0.8 ± 0.1 mBq kg^{-1} dry mass (Pham et al., 2010).

5. Applications of I-129 as an environmental tracer

Use of ^{129}I as an intrinsic tracer for natural iodine kinetics was discussed as early as 1962 (Edwards, 1962). Already at that time two reprocessing plants, one for military purposes in Marcoule, France (from 1958) and one for nuclear fuel in Thurso, United Kingdom (from 1958) existed.

To be able to use ^{129}I as an environmental tracer certain conditions have to be met. These are: (1) ^{129}I must trace a single environmental process with a defined time scale; (2) ^{129}I must be equilibrated with ^{127}I; (3) The predominant chemical species of ^{129}I and their geochemical properties must be known (Santschi & Schwehr, 2004); (4) Conservative behaviour, meaning relatively constant concentration in a reservoir over time, is desirable. The natural $^{129}I/^{127}I$ ratio has been strongly shifted by continuous additions from anthropogenic sources, which still persists. To trace existing and future global changes in inventories of anthropogenic ^{129}I continuous monitoring and revised budget calculation are indispensable (Aldahan et al. 2007a). Recently also a prediction model system to better understand the dispersion of ^{129}I from point sources (Sellafield and La Hague) to the northern North Atlantic Ocean has been developed (Orre et al. 2010).

United Nations Scientific Committee on the Effects of Atomic Radiation (UNSCEAR, 2000) identifies as globally dispersed radionuclides ^{3}H, ^{14}C and ^{129}I. Because of its very long half live is ^{129}I one of the most important radionuclides in long-term radiological assessment of its discharges from nuclear fuel reprocessing plants. ^{129}I is present in the environment in low quantities (in traces) and its increase in a particular compartment of the ecosystem can be instantly recognized.

5.1 ^{129}I as an oceanographic tracer

Transport, circulation and exchange of water masses in the Northeast Atlantic and Arctic Oceans has long been studied by using radionuclides such as ^{137}Cs, ^{134}Cs, ^{90}Sr, ^{125}Sb and ^{99g}Tc originating from reprocessing of spent nuclear fuel. In recent years ^{129}I became interesting as an oceanographic tracer, because the discharges from NFRP in La Hague and Sellafield increased since 1990 and highly sensitive analytical method, AMS, developed for analysis (Hou, 2004).

Concentrations and species of ^{129}I and ^{129}I/^{127}I isotopic ratio were determined in many environmental and biological samples from marine environment, especially in areas influenced by NFRP. Results for Northeast Atlantic, Arctic and Baltic Seas indicate a strong influence of liquid discharges from NFRP in La Hague and Sellafield. Hou et al. (2000a) determined ^{129}I concentrations in archived time series seaweed *Fucus vesiculosus* samples from Danish, Norwegian and Northwest Greenland coast collected in a period from 1980 to 1997 (Table 3). They used the ^{129}I/^{99}Tc ratio to estimate the origin of and transit times of ^{129}I. Transit times were estimated to be 1–2 years from La Hague, 3–4 from Sellafield, to Denmark (Klint) and Norway (Utsira), and 9–14 years from La Hague, 11–16 from Sellafield, to NW Greenland.

Iodine exists in seawater mainly as dissolved iodate and iodide, and a small amount of organic iodine (Wong, 1991). Chemical speciation of ^{129}I can be used to investigate the transport, dispersion, and circulation of the water masses - especially at the boundary of two or more sources. (Hou et al., 2001).

5.2 ^{129}I as a geochemical tracer

^{129}I was used in geochemical studies as a tracer for determining ages and migration of brines (Muramatsu et al., 2001, Snyder et al., 2003a, Fehn et al., 2007). Isolated system contain lower or close to estimated pre-nuclear ^{129}I/^{127}I ratio, $1.5 \cdot 10^{-12}$. For correct interpretation of results - age calculation based on ^{129}I one must consider the effect of possible fissiogenic production and initial concentration on isotopic ratios. The estimated pre-nuclear ratio can be disturbed along continental margins with lower isotopic ratios likely caused by releases of methane-rich fluids with high stable iodine concentrations derived from old organic sources, where ^{129}I already partly decayed. The isotopic ratio of the open ocean is not disturbed, justifying the use of estimated pre-nuclear ratio (Fehn et al., 2007).

5.3 ^{129}I in precipitation

Atmospheric releases of ^{129}I from European and Hanford NFRP were much higher than from nuclear weapons tests and Chernobyl accident together (Table 1). Measurement of ^{129}I in atmosphere and precipitation can be used to investigate the transport pathways of ^{129}I from point sources, such as NFRP. But it is important to be aware that ^{129}I levels in atmosphere and precipitation can originate either directly from atmospheric releases from NFRP, and from volatilization from seawater and terrestrial environment. To study transport pathways of ^{129}I all of this aspects have to be considered and obtained results for atmospheric and precipitation samples compared to reported releases from NFRP in particular timescale. Many precipitation and atmospheric samples have to be measured continuously to establish a pattern or trend.

5.4 ^{129}I for reconstruction of ^{131}I dose

The same chemical and physical properties of isotopes of particular element enable to use ^{129}I as a tool for the reconstruction of ^{131}I doses after a nuclear accident. This was done after the nuclear accident in Chernobyl. Levels of ^{129}I were determined in soils and from the measured ^{129}I/^{131}I ratio, 12–19 (Kutschera et al., 1988; Mironov et al., 2002), the long-lived

^{129}I can be used to reconstruct ^{131}I dose to thyroids. This method is limited only to areas that were relatively strong contaminated by fallout from Chernobyl like areas in Ukraine and Belarus (Michel et al., 2005; Straume et al., 2006).

6. Radiological hazard of ^{129}I for man

Transport pathways of iodine to human are ingestion and inhalation. Iodine present in food is adsorbed into blood in small intestine - inhaled iodine from the air is also transferred into blood. More than 80 % of iodine absorbed into the blood is concentrated in the thyroid gland, which is therefore the target organ of iodine - also radioactive ^{129}I. Due to low beta and gamma energy of ^{129}I and long half-life the radiation toxicity of ^{129}I is mainly related to long term and low dose internal exposure of the thyroid to the beta radiation of ^{129}I. An average iodine content in human thyroid is 10–15 mg. ^{129}I and ^{127}I are taken up by thyroid indiscriminately. The highest reported ^{129}I/^{127}I ratio was 10^{-4} in close vicinity of NFRP, which corresponds to 10^{-6} g or 6.64 Bq at 10 mg stable iodine content in thyroid. The corresponding annual radiation dose to thyroid would be 0.1 mSv year^{-1}, which is 2.5 times higher than the dose regulation limit of 0.04 mSv year^{-1} set by the U.S. NRC for combined beta and photon emitting radionuclide to the whole body or any organ (Hou et. al., 2009). An annual thyroid equivalent dose of 1 mSv, which is comparable to the level of natural back-ground radiation, would only be reached by ratios exceeding $1.5 \cdot 10^{-3}$ (Michel, 1999).
Current concentrations of ^{129}I in the environment do not represent any radiological hazard for man, even in the vicinity of nuclear fuel reprocessing plants. But to assess environmental impact and potential risk and consequences during long-term exposition information on the distribution and radionuclide species, speciation analysis, influencing the mobility, biological uptake and accumulation of radionuclides is needed (Salbu, 2007). Speciation analysis provides crucial information for evaluation of radionuclide transport mechanism in the environment and to the human body and accurate risk assessments (Hou et al., 2009).

7. Conclusion

Anthropogenic ^{129}I considerable enriched pre-nuclear environmental levels. Presently the main sources of ^{129}I in the environment are nuclear fuel reprocessing plants (NFRP). Global distribution of ^{129}I is not uniform - concentrations are elevated near NFRP - but anthropogenic ^{129}I was detected in remote areas such as Antarctic.
Before the onset of nuclear age ^{129}I and ^{127}I were in equilibrium. Analysis of pre-nuclear material and deep layer of marine sediment gave the best estimated value for natural ^{129}I/^{127}I ratio in surface reservoirs to be $(1.5 \pm 0.15) \cdot 10^{-12}$.
In transport and exchange of ^{129}I among different compartments marine and soil ecosystems influenced by present biota - microorganisms play major role. Biogeochemical cycling of iodine is influenced by its strong association with organic material - ocean is the main reservoir of mobile iodine, where it is rapidly exchanged between biota, hydrosphere and atmosphere.

8. References

Aldahan, A., Alfimov, V., Possnert, G. (2007a). ^{129}I anthropogenic budget: Major sources and sinks. *Applied Geochemistry,* Vol. 22, No. 3, pp. (606-618), ISSN 0883-2927

Aldahan, A., Englund, E., Possnert, G., Cato, I., Hou X.L. (2007). Iodine-129 enrichment in sediment of the Baltic Sea. *Applied Geochemistry,* Vol. 22, No. 3, pp. (637-647), ISSN 0883-2927

Alfimov, V., Aldahan, A., Possnert, G., Winsor, P. (2004). Anthropogenic iodine-129 in seawater along a transect from the Norwegian coastal current to the North Pole. *Marine Pollution Bulletin,* Vol. 49, No. 11-12, pp. (1097-1104), ISSN 0025-326X

Alfimov, V., Synal, H.A. (2010). ^{129}I AMS at 0.5 MV tandem accelerator. *Nuclear Instruments and Methods in Physics Research B,* Vol. 268, No. 7-8, pp. (769-772), ISSN 0168-583X

Amachi, S., Kamagata, Y., Kanagawa, T., Muramatsu, Y. (2001). Bacteria mediate methylation od iodine in marine and terrestrial environments. *Applied and Environmental Microbiology,* Vol. 67, No. 6, pp. (2718-2722), ISSN 0099-2240

Amachi, S. (2008). Microbial Contribution to Global Iodine Cycling: Volatilization, Accumulation, Reduction, Oxidation, and Sorption of Iodine. *Microbes and Environments,* Vol. 23, No. 4, pp. (269-276), ISSN 1342-6311

Atarashi-Andoh, M., Schnabel, C., Cook, G., MacKenzie, A.B., Dougans, A., Ellam, R.M., Freeman, S., Maden, C., Olive, V. Synal, H.-A., Xu, S. (2007). ^{129}I/^{127}I ratios in surface waters of the English Lake District. *Applied Geochemistry,* Vol. 22, No. 3, pp. (628-636), ISSN 0883-2927

Baker, A.R., Thompson, D., Campos, M.L.A.M., Parry, S.J., Jickells, T.D. (2000). Iodine concentration and availability in atmospheric aerosol. *Atmospheric Environment,* Vol. 34, No. 25, pp. (4331-4336), ISSN 1352-2310

Barker, E., Masson, M., Bouisset, P., Cariou, N., Germain, P., Siclet, F. (2005). ^{129}I determination by direct gamma-X spectrometry and its application to concentration variations in two seaweed species. *Radioprotection,* Vol. 40, No. 1, pp. (581-587)

Becker, J.S. (2005). Inductively coupled plasma mass spectrometry (ICP-MS) and laser ablation ICP-MS for isotope analysis of long-lived radionuclides. *International Journal of Mass Spectrometry,* Vol. 242, No. 2-3, pp. (183-195), ISSN 1387-3806

Bouisset, P., Lefevre, O., Cagnat, X., Kerlau, G., Ugron, A., Calmet, D. (1999). Direct gamma-X spectrometry measurement of ^{129}I in environmental samples using experimental self-absorption corrections. *Nuclear Instruments and Methods in Physics Research A,* Vol. 437, No. 1, pp. (114-127), ISSN 0168-9002

Brown, C.F., Geiszler K.N, Lindberg, M.J. (2007). Analysis of ^{129}I in groundwater samples: Direct and quantitative results below the drinking water standard. *Applied Geochemistry,* Vol. 22, No. 3, pp. (648-655), ISSN 0883-2927

Buraglio, N., Aldahan, A., Possnert, G. (2000). ^{129}I measurements at the Uppsala tandem accelerator. *Nuclear Instruments and Methods in Physics Research B,* Vol. 161, pp. (240-244), ISSN 0168-583X

Buraglio, N., Aldahan, A., Possnert, G., Vintersved, I. (2001). ^{129}I from the nuclear reprocessing facilities traced in precipitation and runoff in Northern Europe. *Environmental Science and Technology,* Vol. 35, No. 8, pp. (1579-1586), ISSN 0013-936X

Chance, R., Baker, A.R., Küpper, F.C., Hughes, C., Kloareg, B., Malin, G. (2009). Release and transformations of inorganic iodine by marine macroalgae. *Estuarine, Coastal and Shelf Science,* Vol. 82, No. 3, pp. (406-414), ISSN 0272-7714

Cooper, L.W., Beasley, T.M., Zhao, X.L., Soto, C., Vinogradova, K.L., Dunton, K.H. (1998). Iodine-129 and plutonium isotopes in Arctic kelp as historical indicators of transport of nuclear fuel-reprocessing wastes from mid-to-high latitudes in the Atlantic Ocean. *Marine Biology,* Vol. 131, No. 3, pp. (391-399), ISSN 0025-3162

Duffa, C., Frechou, C. (2003). Evidence of long-lived I and Pu isotopes enrichment in vegetation samples around the Marcoule nuclear reprocessing plant (France). *Applied Geochemistry,* Vol. 18, No. 12, pp. (1867-1873), ISSN 0883-2927

Edwards, R.R. (1962). Iodine-129: Its Occurrenice in Nature and Its Utility as a Tracer. *Science,* Vol. 137, No. 3533, pp. (851-853)

Englund, E., Aldahan, A., Hou X.L., Possnert, G., Söderström C. (2010a). Iodine (^{129}I and ^{127}I) in aerosols from northern Europe. *Nuclear Instruments and Methods in Physics Research B,* Vol. 268, No. 7-8, pp. (1139-1141), ISSN 0168-583X

Englund, E., Aldahan, A., Hou, X.L., Petersen, R., Possnert, G. (2010b). Speciation of iodine (^{127}I and ^{129}I) in lake sediments. *Nuclear Instruments and Methods in Physics Research B,* Vol. 268, No. 7-8, ISSN 0168-583X

Fehn, U., Snyder, G., Egeberg, P.K. (2000a). Dating of Pore Waters with ^{129}I: Relevance for the Origin of Marine Gas Hydrates. *Science,* Vol. 289, No. 5488,pp. (2332-2335), ISSN 0036-8075

Fehn, U., Snyder, G. (2000b). ^{129}I in the Southern Hemisphere: Global redistribution of an anthropogenic isotope. *Nuclear Instruments and Methods in Physics Research B,* Vol. 172, No. 1-4, pp. (366-371), ISSN 0168-583X

Fehn, U., Moran, J.E., Snyder, G.T., Muramatsu, Y. (2007). The initial ^{129}I/I ratio and the presence of 'old' iodine in continental margins, *Nuclear Instruments and Methods in Physics Research B,* Vol. 259, No. 1, pp. (496-5029), ISSN 0168-583X

Frechou, C., Calmet, D., Bouisset, P., Piccot, D., Gaudry, A., Yiou, F., Raisbeck, G. (2001). ^{129}I and ^{129}I/^{127}I ratio determination in environmental biological samples by RNAA, AMS and direct γ-X spectrometry measurements. *Journal of Radioanalytical and Nuclear Chemistry,* Vol. 249, No. 1, pp. (133-138)

Frechou, C., Calmet, D. (2003). ^{129}I in the environment of the La Hague nuclear fuel reprocessing plant-from Sea to land. *Journal of Radioanalytical and Nuclear Chemistry,* Vol. 70, No. 1-2, pp. (43-59)

Fuge, R. (2005). Soils and iodine deficiency, In: *Essentials of Medical Geology,* O. Selinus (Ed.), 417-433, Elsevier, ISBN 0-12-636341-2, Amsterdam, The Netherlands.

Gilfedder, B.S., Lai, S.C., Petri, M., Biester, H., Hoffmann, T. (2008). Iodine speciation in rain, snow and aerosols. *Atmospheric Chemistry and Physics,* Vol. 8, No. 20, pp. (6069-6084), ISSN 1680-7316

Gomez-Guzman, J.M., Lopez-Gutierrez, J.M., Holm, E., Pinto-Gomez, A.R. (2011). Level and origin of ^{129}I and ^{137}Cs in lichen samples (*Cladonia alpestris*) in central Sweden. *Journal of Environmental Radioactivity,* Vol. 102, No. 2, pp. (200-205), ISSN 0265-931X

Gudelis, A., Lukšiene, B., Druteikiene, R., Gvozdaite, R., Kubarevičiene, V. (2006). Applications of LSC for the determination of some radionuclides in waste matrices from the Ignalina NPP. *Proceedings of the 2005 International Liquid Scintillation Conference,* Arizona Board on behalf of the University of Arizona, pp. (343-353), Katowice, Poland, October 17-21, 2005

Hou X., Chai C., Qian Q., Yan X., Fan X. (1997). Determination of chemical species in some seaweeds (I). *Science of Total Environment,* Vol. 204, No. 3, pp. (215-221), ISSN 0048-9697

Hou, X.L., Yan, X.J. (1998). Study on the concentration and seasonal variation of inorganic elements in 35 species of marine algae. *Science of the Total Environment,* Vol. 222, No. 3, pp. (141-156), ISSN 0048-9697

Hou, X., Dahlgaar., H., Rietz, B., Jacobsen, U., Nielsen, S.P., Aarkrog, A. (1999). Determination of ^{129}I in seawater and some environmental materials by neutron activation analysis. *Analyst,* Vol. 124, No. 7, pp. (1109-1114), ISSN 0003-2654

Hou, X.L., Dahlgaard, H, Nielsen, S.P. (2000a). Iodine-129 time series in Danish, Norwegian and northwest Greenland coast and the Baltic Sea by seaweed. *Estuarine Coastal and Shelf Science,* Vol. 51, No. 5, pp. (571-584), 0272-7714

Hou, X., Dahlgaard, H., Nielsen, S.P., Ding, W. (2000b). Iodine-129 in human thyroids and seaweed in China. *Science of the Total Environment,* Vol. 246, No. 2-3, pp. (285-291), ISSN 0048-9697

Hou X., Yan X., Chai C. (2000c). Chemical species of iodine in some seaweeds II. Iodine-bound biological macromolecules. *Journal of Radioanalytical and Nuclear Chemistry,* Vol. 245, No. 3, Vol. (461-467), ISSN 0236-5731

Hou, X., Dahlgaard, H., Nielsen S.P. (2001). Chemical speciation analysis of ^{129}I in seawater and a preliminary investigation to use it as a tracer for geochemical cycle study of stable iodine. *Marine Chemistry,* Vol. 74, No. 2-3, pp. (145-155), ISSN 0304-4203

Hou, X.L., Dahlgaard, H., Nielsen, S.P., Kucera, J. (2002). Level and origin of Iodine-129 in the Baltic Sea. *Journal of Environmental Radioactivity,* Vol. 61, No. 3, pp (331-343), ISSN 0265-931X

Hou X., Malencheko A.F., Kucera J., Dahlgaard H., Nielsen S.P. (2003a). Iodine-129 in thyroid and urine in Ukraine and Denmark. *Science of the Total Environment,* Vol. 302, No. 1-3, pp. (63-73), ISSN 0048-9697

Hou, X.L., Fogh, C.L., Kucera, J., Andersson, K.G., Dahlgaard, H., Nielsen, S.P. (2003b). Iodine-129 and Caesium-137 in Chernobyl contaminated soil and their chemical fractionation. *Science of the Total Environment,* Vol. 308, No., 1-3, pp. (97-109), ISSN 0048-9697

Hou, X. (2004). Application of ^{129}I as an environmental tracer. *Journal of Radioanalytical and Nuclear Chemistry,* Vol. 262, No. 1, pp. (67-75), ISSN 0236-5731

Hou, X.L., Aldahan, A., Nielsen, S.P., Possnert, G., Nies, H., Hedfors, J. (2007). Speciation of I-129 and I-127 in seawater and implications for sources and transport pathways in the North Sea. *Enironmental Science and Technology,* Vol. 41, No. 17, pp. (5993-5999), ISSN 0013-936X

Hou, X., Hansen, V., Aldahan, A., Possnert, G., Lind, O.C., Lujaniene, G. (2009). A review on speciation of iodine-129 in the environmental and biological samples. *Analytica Chimica Acta,* Vol. 632, No. 2, pp. (181-196), ISSN 0003-2670

Hou, X., Zhou, W., Chen, N., Zhang L., Liu, Q., Lou, M., Fan, Y., Liang, W., Fu, Y. (2010). Determination of Ultralow Level I-129/I-127 in Natural Samples by Separation of Microgram Carrier Free Iodine and Accelerator Mass Spectrometry Detection. *Analytical Chemistry,* Vol. 82, No. 18, pp. (7713-7721), ISSN 0003-2700

Izmer, A.V., Boulyga, S.F., Becker, J.S. (2003). Determination of $^{129}I/^{127}I$ isotope ratios in liquid solutions and environmental soil samples by ICP-MS with hexapole collision cell. *Journal of Analytical Atomic Spectrometry,* Vol. 18, No. 11, pp. (1339-1345), ISSN 0267-9477

Izmer, A.V., Boulyga, S.F., Zoriy, M.V., Becker, J.S. (2004). Improvement of the detection limit for determination of ^{129}I in sediments by quadrupole inductively coupled plasma mass spectrometer with collision cell. *Journal of Analytical Atomic Spectrometry,* Vol. 19, No. 9, pp. (1278-1280), ISSN 0267-9477

Keogh, S.M., Aldahan, A., Possnert, G., Finegan, P., Vintro, L. L., Mitchell P.I. (2007). Trends in the spatial and temporal distribution of ^{129}I and ^{99}Tc in coastal waters sorrouding Ireland using *Fucus vesiculosus* as bio-indicator. *Journal of Environmental Radioactivity,* Vol 95, No. 1, pp. (23-38), ISSN 0265-931X

Kutschera, W., Fink, D., Paul, M., Hollos, G., Kaufman, A. (1988). Measurement of the I-129/I-131 ratio in Chernobyl fallout. *Physica Scripta,* Vol. 37, No. 2, pp. (310-313), ISSN 0281-1847

Leblanc, C., Colin, C., Cosse, A., Delage, L., La Barre, S., Morin, P., Fiévet, B., Voiseux, C., Ambroise, Y., Verhaeghe, E., Amouroux, D., Donard, O., Tessier, E., Potin, P. (2006). Iodine transfers in the coastal marine environment: the key role of brown algae and of their vanadium-dependent haloperoxidases. *Biochimie,* Vol. 88, No. 11, pp. (1773-1785), ISSN 0300-9084

Lefevre, O., Bouisset, P., Germain, P., Barker, E., Kerlau, G., Cagnat, X. (2003). Self-absorption correction factor applied to ^{129}I measurement by direct gamma-X spectrometry for *Fucus serratus* samples. *Nuclear Instruments and Methods in Physics Research A,* Vol. 506, No. 1-2, pp. (173-185), ISSN 0168-9002

Leiterer, M., Truckenbrodt, D., Franke, K. (2001). Determination of iodine species in milk using ion chromatographic separation and ICP-MS detection. *European Food Research and Technology,* Vol. 213, No. 2, pp. (150-153), ISSN 1438-2377

Li, K., Vogel, E., Krähenbühl, U. (2009). Measurement of I-129 in environmental samples by ICP-CRI-QMS: possibilities and limitations. *Radiochimica Acta,* Vol. 97, No. 8, pp. (453-458), ISSN 0033-8230

Lopez-Gutierrez, J.M. Garcia-Leon, M., Schnabel, Ch., Suter, M., Synal, H.A., Szidat, S., Garcia-Tenorio, R. (2004). Relative influence of ^{129}I sources in a sediment core from the Kattegat area. *Science of The Total Environment,* Vol. 323, No. 1-3, pp. (195-210), ISSN 0048-9697

Lu, Z., Fehn, U., Tomaru, H., Elmore, D., Ma, X. (2007). Reliability of $^{129}I/I$ ratios produced from small sample masses. *Nuclear Instruments and Methods in Physics Research B,* ol. 259, No. 1, pp. (359-364), ISSN 0168-583X

Michel, R. (1999). Long-lived radionuclides as tracers in terrestrial and extraterrestrial matter. *Radiochimica Acta,* Vol. 87, No. 1-2, pp. (47-73), ISSN 0033-8230

Michel, R., Handl, J., Ernst, T., Botsch, W., Szidat, S., Schmidt, A., Jakob, D., Beltz, D., Romantschuk, L.D., Synal, H.A., Schnabel, C., López-Gutiérrez, J.M. (2005). Iodine-129 in soils from Northern Ukraine and theretrospective dosimetry of the iodine-131 exposure after the Chernobyl accident. *Science of the Total Environment,* Vol. 340, No. 1-3, pp. (35-55), ISSN 0048-9697

Mironov, V., Kudrjashov, V., Yiou, F., Raisbeck G.M. (2002). Use of I-129 and Cs-137 in soils for the estimation of I-131 deposition in Belarus as a result of the Chernobyl accident. *Journal of Environmental Radioactivity,* Vol. 59, No. 3, pp. (293-307), ISSN 0265-931X

Moran, J.E., Fehn, U., Teng, R.T.D. (1998). Variations in $^{129}I/^{127}I$ ratios in recent marine sediments: evidence for a fossil organic component. *Chemical Geology,* Vol. 152, No. 1-2, pp. (193-203), ISSN 0009-2541

Muramatsu, Y, Yoshida, S. (1995). Determination of ^{129}I and ^{127}I in environmental samples by neutron activation analysis (NAA) and inductively coupled plasma mass spectrometry (ICP-MS). *Journal of Radioanalitycal and Nuclear Chemistry,* Vol. 197, No. 1, pp. (149-159), ISSN 0236-5731

Muramatsu, Y., Yoshida, S., Uchida, S., Hasebe, A. (1996). Iodine desorption from rice paddy soil. *Water, Air and Soil Pollution,* Vol. 86, No. 1-4, pp. (359-371), ISSN 0049-6979

Muramatsu, Y., Wedepohl, K.H. (1998). The distribution of iodine in the earth's crust. *Chemical Geology,* Vol. 147, No. 3-4, pp. (201-216), ISSN 0009-2541

Muramatsu, Y., Yoshida, S. (1999). Effects of microorganisms on the fate of iodine in the soil environment. *Geomicrobiological Journal,* Vol. 16, No. 1, pp. (85-93), ISSN 0149-0451

Muramatsu, Y., Fehn, U., Yoshida, S. (2001). Recycling of iodine in fore-arc areas: evidence from the iodine brines in Chiba, Japan. *Earth and Planetary Science Letters,* Vol. 192, No. 4, pp. (583-593), ISSN Recycling of iodine in fore-arc areas: evidence from the iodine brines in Chiba, Japan

Muramatsu, Y., Takada, Y., Matsuzaki, H., Yoshida, S. (2008). AMS analysis of ^{129}I in Japanese soil samples collected from background areas far from nuclear facilities. *Quaternary Geochronology,* Vol. 3, No. 3, pp. (291-297), ISSN 1871-1014

Orre, S., Smith, J.N., Alfimov, V., Bentsen, M. (2010). Simulating transport of ^{129}I and idealized tracers in the northern North Atlantic Ocean. *Environmental Fluid Mechanisms,* Vol. 10, No. 1-2, pp. (213-233), ISSN 1567-7419

Osterc, A., Stibilj, V. (2005). Measurement uncertainty of iodine determination in radiochemical neutron activation analysis. *Accreditation and Quality Assurance,* Vol. 10, No. 5, pp. (235-240), ISSN 0949-1775

Osterc, A., Jaćimović, R., Stibilj, V. (2007). Development of a method for ^{129}I determination using radiochemical neutron activation analysis. *Acta Chimica Slovenica,* Vol. 54, No. 2, pp. (273-283), ISSN 1318-0207

Osterc, A., Stibilj, V. (2008). ^{127}I and $^{129}I/^{127}I$ isotopic ratio in marine alga *Fucus virsoides* from the North Adriatic Sea. *Journal of Environmental Radioactivity,* Vol. 99, No. 4, pp. (757-765), ISSN 0265-931X

Parry, S.J., Bennett, B.A., Benzig, R., Lally, A.E., Birch, C.P., Fulker, M.J. (1995). The determination of 129I in milk and vegetation using neutron activation analysis. *Science of the Total Environment,* Vol. 173-174, No. 1, pp. (351-360), ISSN 0236-5731

Pham, M.K., Betti, M., Povinec, P.P., Benmansour, M., Bojanowski, R., Bouisset, P., Calvo, E.C., Ham, G.J., Holm, E., Hult, M., Ilchmann, C., Kloster, M., Kanisch, G., Köhler, M., La Rosa, J., Legarda, F., Llauradó, M., Nourredine, A., Oh, J.-S., Pellicciari, M., Rieth, U., Rodriguez y Baena, A.M., Sanchez-Cabeza, J.A., Satake, H., Schikowski, J., Takeishi, M., Thebault, H., Varga, Z. (2010). A new reference material for radionuclides in the mussel sample from the Mediterranean Sea (IAEA-437). *Journal of Radioanalyticaland and Nuclear Chemistry,* Vol. 283, No. 3, pp. (851-859), ISSN 0236-5731

Reithmeier, H., Lazarev V., Kubo, F., Rühm, W., Nolte, E. (2005). ^{129}I in precipitation using a new TOF system for AMS measurements. *Nuclear Instruments and Methods in Physics Research B,* Vol. 239, No. 3, pp. (273-280), ISSN 0168-583X

Reithmeier, H., Lazarev, V., Rühm, W., Schwikowski, M., Gäggeler, H., Nolte, E. (2006). Estimate of European ^{129}I Releases Supported by ^{129}I Analysis in an Alpine Ice Core. *Environmental Science and Technology,* Vol. 40, No. 19, pp. (5891-5896), ISSN 0013-936X

Salbu, B. (2007). Speciation of radionuclides - analytical challenges within environmental impact and risk assessments. *Journal of Environmental Radioactivity,* Vol. 96, No. 1-3, pp. (47-53), ISSN 0265-931X

Sanchez, L.F., Szpunar, J. (1999). Speciation analysis for iodine in milk by size-exclusion chromatography with inductively coupled plasma mass spectrometric detection (SEC-ICP MS). *Journal of Analytical Atomic Spectrometry,* Vol. 14, No. 11, pp. (1679-1702), ISSN 0267-9477

Santos, F.J., Lopez-Gutierrez, J.M., Garcia-Leon, M., Suter, M., Synal H.A. (2005). Determination of ^{129}I/^{127}I in aerosol samples in Seville (Spain). *Journal of Environmental Radioactivity,* Vol. 84, No. 1, pp. (103-109), ISSN 0265-931X

Santos, F.J., Lopez-Gutierrez, J.M., Chamizo, E., Garcia-Leon, M., Synal H.A. (2006). Advances on the determination of atmospheric ^{129}I by accelerator mass spectrometry (AMS). *Nuclear Instruments and Methods in Physics Research B,* Vol. 249, No. 1-2, pp. (772-775), ISSN 0168-583X

Schmitz, K., Aumann, D.C. (1995). A study on the association of two iodine isotopes, of natural ^{127}I and of the fission product ^{129}I, with soil components using a sequential extraction procedure. *Journal of Radioanalytical and Nuclear Chemistry,* Vol. 198, No. 1 pp. (229-236)

Santschi, P.H., Schwer, K.A. (2004). ^{129}I/^{127}I as a new environmental tracer or geochronometer for biogeochemical or hydrodynamic processes in the hydrosphere and geosphere: the central role of organo-iodine. *Science of the Total Environment,* Vol. 321, No. 1-3, pp. (257-271), ISSN 0048-9697

Schnabel, C., Olive, V. Atarashi-Andoh, M., Dougans, A., Ellam, R.M., Freeman, S., Maden, C., Stocker, M., Synal, H.A., Wacker, L., Xu, S. (2007). ^{129}I/^{127}I ratios in Scottish coastal surface sea water: Geographical and temporal responses to changing emissions. *Applied Geochemistry,* Vol. 22, No. 3, pp. (619-627), ISSN 0883-2927

Schwehr, K.A., Santschi, P.H. (2003). Sensitive determination of iodine species, including organo-iodine, for freshwater and seawater samples using high performance liquid chromatography and spectrophotometric detection. *Analytica Chimica Acta,* Vol. 482, No. 1, pp. (59-71), ISSN 0003-2670

Schwehr, K.A., Santschi, P.H., Elmore, D. (2005). The dissolved organic iodine species of the isotopic ratio of ^{129}I/^{127}I: A novel tool for tracing terrestrial organic carbon in the estuarine surface waters of Galveston Bay, Texas. *Limnology and Oceanography: Methods,* Vol. 3, pp. (326-337)

Shah, M., Wuilloud, R.G., Kannamkumaratha, S.S., Caruso, J.A. (2005). Iodine speciation studies in commercially available seaweed by coupling different chromatographic techniques with UV and ICP-MS detection. *Journal of Analytical Atomic Spectrometry,* Vol. 20, No. 3, pp. (176-182), ISSN 0267-9477

Straume, T., Anspaugh, L.R., Marchetti, A.A., Voigt, G., Minenko, V., Gu, F., Men, P., Trofimik, S., Tretyakevich, S., Drozdovitch, V., Shagalova, E., Zhukova, O., Germenchuk, M., Berlovich, S. (2006). Measurement of I-129 and Cs-137 in soils from Belarus and construction of I-131 deposition from the Chernobyl accident. *Health Physics,* vol. 91, No. 1, pp. (7-19), ISSN 0017-9078

Stutz, J., Hebestreit K., Alicke, B., Platt, U. (2000). Chemistry of Halogen Oxides in the Troposphere: Comparison of Model Calculations with Recent Field Data. *Journal of Atmospheric Chemistry,* Vol. 34, No. 1, pp. (65-85)

Suarez, J. A., Espartero, A. G., Rodriguez, M. (1996). Radiochemical analysis of ^{129}I in radioactive waste streams. *Nuclear Instruments and Methods in Physics Research A,* Vol. 369, No. 2-3, pp. (407-410), ISSN 0168-9002

Suzuki, T., Kitamura, T., Kabuto, S., Togawa, O., Amano, H. (2006). High sensitivity measurement of iodine-129/iodine-127 ratio by accelerator mass spectrometry. *Journal of Nuclear Science and Technology,* Vol. 43, No. 44, pp. (1431-1435), ISSN 0022-3131

Suzuki, T., Kabuto, S., Amano, H., Togawa, O. (2008). Measurement of iodine-129 in seawater samples collected from the Japan Sea area using accelerator mass spectrometry: Contribution of nuclear fuel reprocessing plants. *Quaternary Geochronology,* Vol. 3, No. 3, pp. (268-275), ISSN 1871-1014

Snyder, G.T., Riese, W.C., Franks, S., Fehn, U., Pelzmann, W.L., Gorody, A.W., Moran, J.E. (2003a). Origin and history of waters associated with coalbed methane: ^{129}I, ^{36}Cl, and stable isotope results from the Fruitland Formation, CO and NM. *Geochimica et Cosmochimica Acta,* Vol. 67, No. 23, pp. (4529-4544), ISSN 0016-7037

Snyder, G., Poreda, R., Fehn, U., Hunt, A. (2003b). Sources of nitrogen and methane in Central American geothermal settings: Noble gas and I-129 evidence for crustal and magmatic volatile components. *Geochemistry Geophysics Geosystems,* Vol. 4, Article No. 9001, ISSN 1525-2027

Snyder, G., Fehn, U. (2004). Global distribution of I-129 in rivers and lakes: implications for iodine cycling in surface reservoirs. *Nuclear Instruments and Methods in Physics Research Section B,* Vol. 223-224, pp. (579-586), ISSN 0168-583X

Szidat, S., Schmidt, A., Handl, J., Jakob, D., Botsch, W., Michel, R., Synal, H.A., Schnabel, C., Suter, M., López-Gutiérrez, J.M., Städe, W. (2000a). Iodine-129: Sample preparation, quality control and analyses of pre-nuclear materials and of natural waters from Lower Saxony, Germany. *Nuclear Instruments and Methods in Physics Research B,* Vol. 172, No. 1-4, pp. (699-710), ISSN 0168-583X

Szidat, S., Schmidt, A., Handl, J., Jakob, D., Michel, R., Synal, H.A., Suter, M. (2000b). Analysis of iodine-129 in environmental materials: Quality assurance and applications. *Journal of Radioanalyticaland and Nuclear Chemistry,* Vol. 244, No. 1, pp. (45-50), ISSN 0236-5731

Tendow, Y. (1996). Nuclear Data Sheets for A = 129. *Nuclear Data Sheets,* Vol. 77, No. 4, pp. (631-770)

UNSCEAR Report (2000). Sources and effects of ionizing radiation, Vol. I: Sources, Annex A: Dose Assessment Technologies, United Nations Scientific Committee on the Effects of Atomic Radiation, pp. 63

dela Vieja, A., Calero, M., Santisteban, P., Lamas, L. (1997). Identification and quantitation of iodotyrosines and iodothyronines in proteins using high-performance liquid chromatography by photodiode-array ultraviolet-visible detection. *Journal of Chromatography B,* Vol. 688, No. 1, pp. (143-149), ISSN 0378-4347

Wershofen, H., Aumann, D.C. (1989). Iodine-129 in the environment of a nuclear fuel reprocessing plant: VII. Concentrations and chemical forms of ^{129}I and ^{127}I in the atmosphere. *Journal of Environmental Radioactivity,* Vol. 10, No. 2, pp. (141-156)

Wong, G.T.F. (1991). The marine geochemistry of iodine. *Reviews in Aquatic Sciences,* Vol. 4, pp. (45-73)

Yoshida, S., Muramatsu, Y. (1995). Determination of organic, inorganic and particulate iodine in the coastal atmosphere of Japan. *Journal of Radioanalytical and Nuclear Chemistry,* Vol. 196, No. 2, pp. (295-302), ISSN 0236-5731

Zulauf, A., Happel, S., Mokili M.B., Bombard, A., Jungclas, H. (2010). Charactarization of an extraction chromatographic resin for the separation and determination of ^{36}Cl and ^{129}I. *Journal of Radioanalyticaland Nuclear Chemistry,* Vol. 286, No. 2, pp. (539-546), ISSN 0236-5731

19

Hydrodynamic Characterization of Industrial Flotation Machines Using Radioisotopes

Juan Yianatos[1] and Francisco Díaz[2]
[1]*Department of Chemical Engineering, Santa Maria University,*
[2]*Nuclear Applications Dept., Chilean Commission of Nuclear Energy, Chile*

1. Introduction

1.1 Objective and organization of chapter

In order to study the hydrodynamic behaviour of large flotation machines, the radioactive tracer technique has been used to measure a number of internal characteristics such as:

- Residence time distribution (RTD) of liquid, solid and gas, in industrial cells and columns. Actual mean residence time evaluation.
- Mixing regime in single cells, banks of cells and pneumatic columns.
- Froth mean residence time of liquid, floatable and non-floatable solids.
- Mixing time and internal pulp circulation in large industrial self-aerated cells.
- Gas holdup and gas residence time distribution in flotation machines.
- Direct measurement of gangue entrainment.
- Industrial flotation cell scanning with gamma ray.
- Pulp flowrate distribution in parallel flotation banks.
- Flotation rate distribution.

1.2 Relevance to industrial flotation machines

Industrial flotation cells need to accomplish several functions such as: air bubble dispersion, solid suspension as well as to provide the best conditions for bubble-particle collision, aggregate formation and froth transport. For this reason, cells are typically provided with mechanical agitation systems which generate well mixed conditions for the pulp and air bubbles. In an industrial mechanical cell, however, the mixing condition prevents that particles have the same opportunity to be collected because a significant fraction of them actually spent a very short time in the cell (in a well-mixed condition almost 40% of particles stay in the cell for less than a half of the mean residence time). Because of the large short circuit in single continuous cells, the industrial flotation operation considers the arrangement of cells in banks. Thus, banks of 5-10 cells in series are commonly used in plant practice. The largest flotation cells presently used in industrial flotation operation are 130, 160, 250 and 300 m^3. Figure 1 show the main characteristics of a self-aerated mechanical flotation cell, where the feed pulp circulates upwards through a draft tube by the rotor. Also, the air is self-aspirated from the upper part of the cell by the rotor.

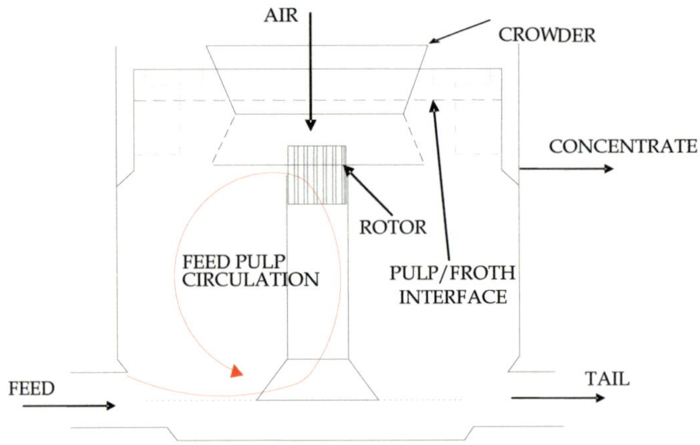

Fig. 1. Large mechanical flotation cell.

In pneumatic columns the mixing is primarily due to convective pulp recirculation and dispersion due to the turbulence caused by bubble motion, as well as the pulp feed and air entrances near the top and bottom, respectively. Mixing characteristics are also a key aspect related to scale-up from laboratory batch flotation cells, or pilot columns, to large size continuous flotation operations. In the last decade a dramatic increase in cell and column sizes has been observed, while more than 2 billion tons per year of ore are presently treated by flotation processes in the world. Thus far, however, the design and scale-up of flotation devices are still mainly based on empirical rules (Yianatos, 2010a).

1.3 Impulse response method

In a multiphase system with segregation, estimation of the mean residence time of each phase is related to the effective volume occupied by each phase. In plant practice, the volume occupied by each phase (liquid, solid or gas) is unknown and varies with operational conditions. Alternatively, in order to evaluate the effective residence time of the liquid and solid in flotation machines, the impulse response method has been used. This method is a dynamic identification procedure which consists of introducing a small amount of tracer (close to an impulse) into the system operating at steady state, and to register the transient response (tracer concentration). The response corresponds to the transfer function of the system, and it is useful for hydrodynamic characterization, dynamic identification (order, noise) in control systems studies, as well as for dynamic modelling (short-circuiting, recirculation) (Yianatos and Bergh, 1992, Yianatos et al., 2002).

1.4 Process modelling

1.4.1 RTD modelling of mechanical flotation cells

In order to model a flotation bank of cells (Mavros, 1992; Yianatos et al., 2001, 2005a), the following equation has been used,

$$E(t) = \frac{(t)^{N-1} e^{-t/(\tau/N)}}{(\tau/N)^N \Gamma(N)} \tag{1}$$

which describes the continuous operation of "N" perfect mixed tank in series. For description of the RTD of a single large flotation cell, the LSTS model (large and small tank in series) has been used. This model gives a better fit to the actual flotation process of a single cell, which consists of one large perfect mixer (residence time τ_L) and one small perfect mixer in series (residence time τ_S) represented by the following equation,

$$E(t) = \frac{e^{-t/\tau_S} - e^{-t/\tau_L}}{\tau_S - \tau_L} \tag{2}$$

where the overall mean residence time τ is given by,

$$\tau = \tau_S + \tau_L \tag{3}$$

Both models can be normalized in terms of the dimensionless time (θ=t/τ), as it was described by Yianatos et al. (2008c).

1.4.2 RTD modelling of pneumatic flotation columns

A typical approach to describe the mixing condition in column flotation operations is the use of the axial dispersion model (Dobby and Finch, 1985). Also, the liquid RTD of industrial flotation columns can be described using a model of less than two perfect mixers in series. However, a better fit has been obtained using a model of one large perfect mixer (residence time τ_L) and two small perfect mixers in series (residence time τ_S) represented by the following equation (Yianatos et al., 2005b),

$$E(t) = \frac{(-t/\tau_S - \alpha)e^{-t/\tau_S} + \alpha e^{-t/\tau_L}}{\tau_L - \tau_S} \tag{4}$$

where

$$\alpha = \tau_L / (\tau_L - \tau_S) \tag{5}$$

1.5 Experimental method

The hydrodynamic characteristics of industrial flotation machines have been evaluated from residence time distribution RTD measurements using radioactive tracers. This technique allows for non-invasive tracer detection, and is also adaptable to different kinds of equipment (Goodall and O'Connor, 1991, Niemi, 1995). However, the application of the radioactive tracer technique for industrial flotation characterization is rather scarce (Yianatos and Bergh, 1992, Lelinski et al., 2002). The procedure consists of selecting a liquid or solid tracer that allows on-line RTD data acquisition. The way the tracer is injected into the feed is critical in order to generate a pulse signal (closer to impulse). For this purposes, a pneumatic system of high reliability has been developed in order to introduce a small amount of radioactive tracer (around 100 mL of liquid, pulp with solids, or gas) at the feed pulp entrance (Díaz and Yianatos, 2010). Then, the time response of the radioactive tracer was measured on-line using non-invasive sensors located directly in different points of the cell or cell discharge. Activity (cps) was measured by scintillating crystal sensors of NaI(Tl) of 1"x1.5", Saphymo Srat, thus allowing the simultaneous data acquisition of up to 12 control points, with a minimum period of 50 milliseconds. Br-82 in solution was used as liquid tracer, while mineral gangue was used as non-floatable irradiated solid tracer. The solid tracer was also tested at three size classes (coarse: +150, intermediate: -150+45 and, fine -45 microns) in order to evaluate solids transport and segregation in mechanical cells and pneumatic columns. Floatable irradiated solid tracer was used to evaluate the RTD of floatable minerals recovered into the concentrate. Also, Kripton-85 and Freon 13B1 have been used as gaseous tracers for industrial flotation columns and mechanical cells testing. Liquid, solid and gaseous tracers were irradiated at the nuclear reactor of the Chilean Commission of Nuclear Energy in Santiago, Chile. An advantage of using the radioactive tracer technique is the direct testing of the actual solid particles (similar physical and chemical properties, size distribution, shape, etc.). Tracer injection is almost instantaneous, because only a small amount of radioactive tracer is required. Another advantage is its capability for on-line measurements at various points inside the system without disturbances related to process sampling.

2. Results and applications

Residence time distribution measurements have been developed for flotation cells of 45, 100, 130, 160, 250 and 300 m^3; this has allowed the evaluation of the mean residence time for liquid and solid per size classes. Also, internal properties such as mixing regime, mixing time, froth mean residence time and mineral entrainment in large industrial flotation cells have been characterized by using radioactive tracers. The measurement of RTD in parallel industrial flotation circuits allowed the identification of uneven pulp flowrate distribution. Applications and results of using radioisotopes to characterize the hydrodynamic behaviour of industrial flotation machines are presented as follows.

2.1 RTD of liquid, solid and gas, in industrial cells and columns: Actual mean residence time evaluation

2.1.1 Mechanical cells

Measurements of RTD using radioactive tracer have been conducted in self-aerated and forced air mechanical cells, as well as in banks with different number of cells in rougher, cleaner and scavenger flotation circuits.

2.1.1.1 Liquid and solid tracer tests

Figure 2 shows an example of the liquid residence time distribution after 1, 3, 5 and 7 cells in a bank of self-aerated flotation cells of 130 m^3, at El Teniente, Codelco-Chile. Here, it can be clearly observed the significant decrease in pulp short-circuiting by increasing the number of cells in series in the flotation bank arrangement (Diaz and Yianatos, 2010). Also, the continuous lines show the good agreement between the data points and the RTD model, described by Eq.(1), for 3, 5 and 7 cells in series. While, for the first cell the best fit was found using the LSTS model, Eq.(2), which confirmed that the pulp zone in a single large flotation cell was not perfectly mixed. The mean residence time calculated from the RTD data was in good agreement with the estimation of the effective mean time according to the pulp flowrate, froth depth and gas holdup measurements.

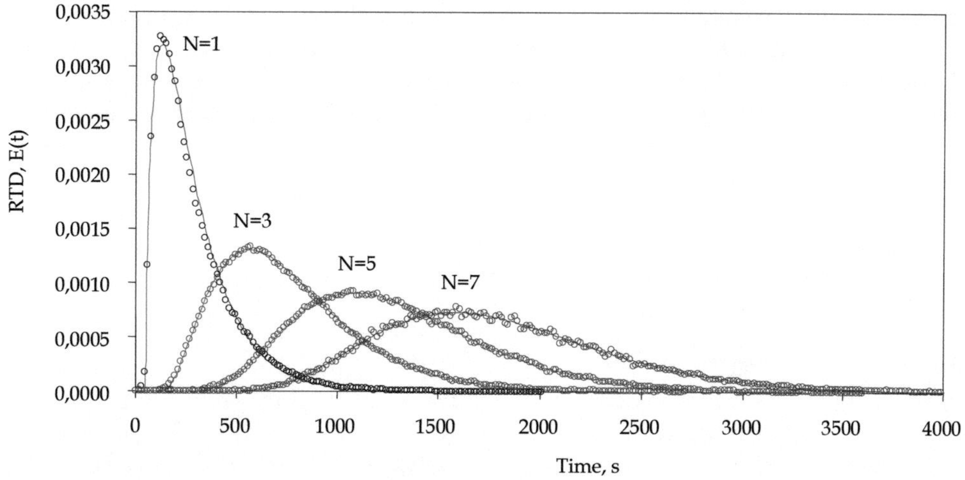

Fig. 2. Liquid RTD in a flotation bank after 1, 3, 5 and 7 cells of 130 m^3 (Diaz and Yianatos, 2010).

Figure 3(a) shows the experimental results of the solid residence time distribution in a rougher flotation bank at El Salvador, Codelco-Chile, consisting of nine 42.5 m^3 self-aerated mechanical cells in series (Yianatos et al., 2002). Activated mineral gangue (final tailing) was used as solid non-floatable tracer. For this operation the tank-in-series model, Eq. (1) in dimensionless form, considering N=9, showed a very good agreement with experimental data.

2.1.1.2 Mineral segregation: Effect of particle size on RTD

Figure 3(b) shows the effect of particle size on the RTD in a rougher flotation circuit, consisting of nine cells in series, 42.5 m^3 each, at El Salvador, Codelco-Chile (Yianatos et al., 2003). It can be appreciated that mixing characteristics are similar for the different particle sizes. However, it was found that the mean residence time of solid was approximately 5% lower than liquid, thus showing a minor segregation mainly related with coarser particles (+100μm).

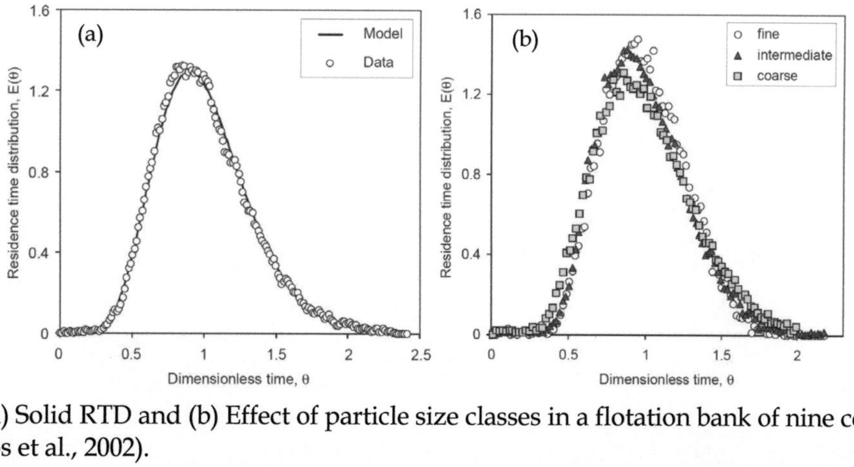

Fig. 3. (a) Solid RTD and (b) Effect of particle size classes in a flotation bank of nine cells (Yianatos et al., 2002).

Industrial testing developed in large size cells (160 and 300 m³) at Chuquicamata, Codelco-Chile, have shown that the global solid residence time was about 90% of the liquid residence time (Morales et al., 2009), as shown in Fig. 4. This effect, however, is less significant than the one observed in pneumatic flotation columns, where the solid residence time of coarse particles (100 μm) was only a half (50%) of the liquid residence time (Yianatos and Bergh., 1992), also shown in Fig. 4 for a 0.91m diameter column.

2.1.2 Pneumatic flotation columns

The RTD of different industrial flotation columns has been measured under normal plant operating conditions, using radioactive tracer tests.

2.1.2.1 Liquid and solid tracer tests

Figure 5(a) shows the liquid RTD data for an industrial column, 2x6x13 m, operating as a single cleaner stage circuit at El Salvador, Codelco-Chile, as well as the good agreement between the data and the LSTS model, Eq. (4) (Yianatos et al., 2005a). Figure 5(b) shows the solid RTD data for the same industrial column, operating as a single cleaner and the good agreement between the data and the LSTS model, Eq. (4).

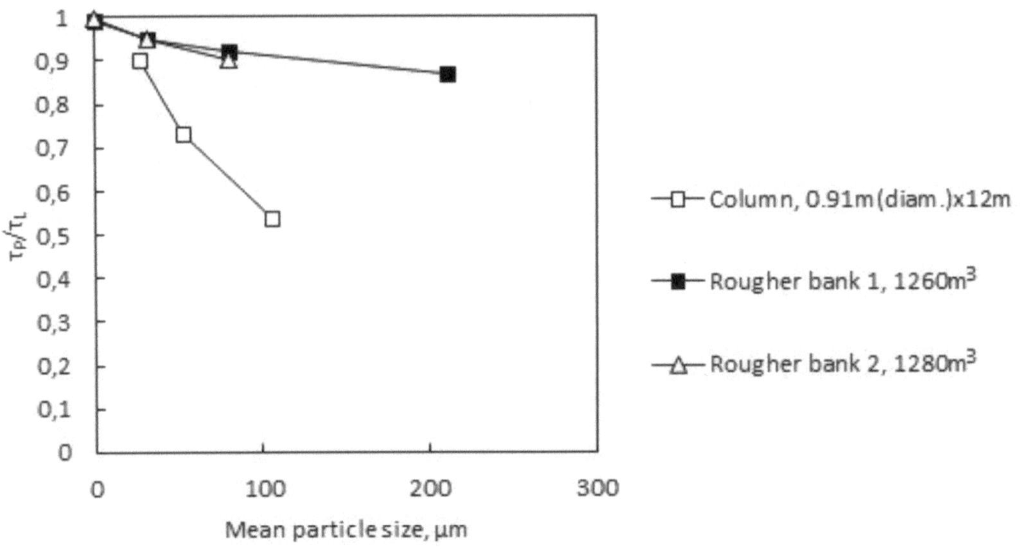

Fig. 4. Particle size effect on solid/liquid relative residence time.

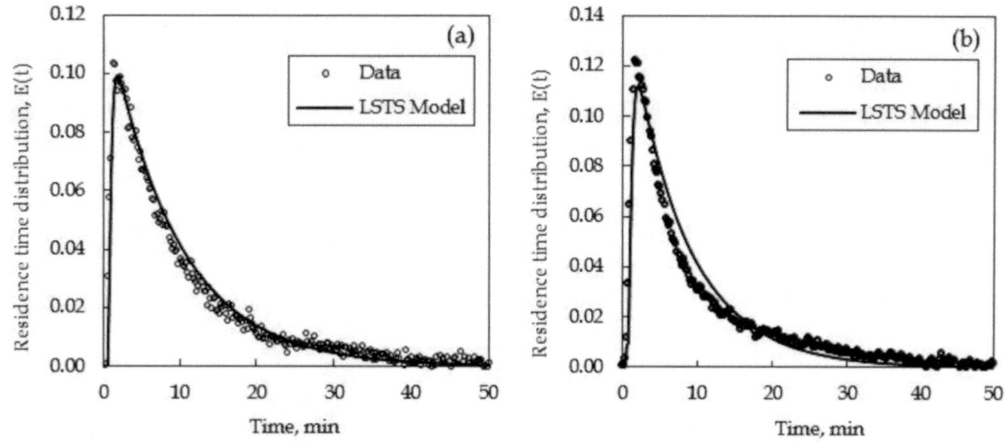

Fig. 5. (a) Liquid and (b) Solid RTD in industrial flotation column 2x6x13m (Yianatos et al., 2005a).

2.1.2.2 Mineral segregation: Effect of particle size on RTD

Figure 6 shows the solid residence time distribution at three size classes: fine (-39 μm), medium (-75+38 μm) and coarse (-150+75 μm), in a 0.91 m diameter, 12 m height column, located at San Francisco concentrator, Compañía Minera Disputada de Las Condes, Chile (Yianatos and Bergh, 1992). It was found that liquids and solids are reasonably well mixed in industrial flotation columns. Solids segregation by gravity in column was more significant than in mechanical cells, with an overall solids residence time 11% smaller than the liquid, despite the average particle size was finer, 25% + 24 μm. This result was in good agreement with previous work reporting the effect of particle size on solid residence time in industrial columns (Dobby and Finch, 1985). It was observed that in flotation columns, a mineral with an average size of 100 microns has a residence time equal to a half of the liquid residence time, as it was shown in Fig. 4. The significant effect of particle size on particle residence time is due to the gravitational transport of solids along the column, plus the effect of the bubbles moving upwards, which generates additional recirculation and classification of solids (Yianatos and Bergh, 1992).

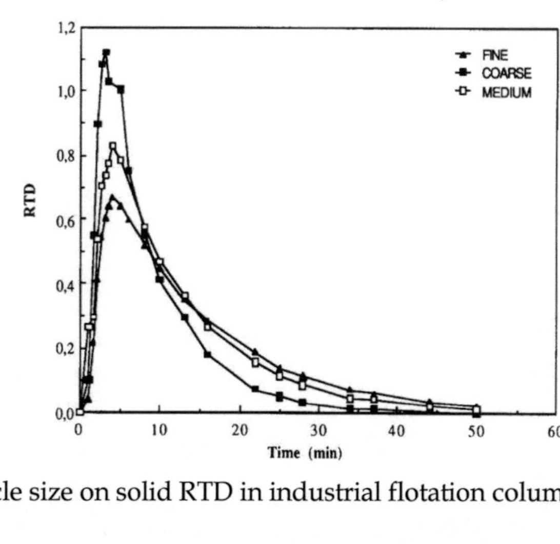

Fig. 6. Effect of particle size on solid RTD in industrial flotation column, (Yianatos and Bergh, 1992).

2.2 Mixing regime in single cells, banks of cells and pneumatic columns

Mixing characteristic of industrial flotation equipment can be evaluated from residence time distribution measurement using radioactive tracers. Figure 7 shows the experimental RTD data for the fine solid (-45 μm) in the first cell of the rougher bank at El Teniente, Codelco-Chile, consisting of seven 130 m^3 self-aerated mechanical cells in series. Also, the LSTS model fit showed the best agreement in describing the data trend along the response time, Eq. (2) in dimensionless form. Similar results were found for the medium and coarse particles (Yianatos et al., 2008c). This result shows that single self-aerated industrial flotation cells do not operate like a perfect mixer. For this reason, flotation cells are commonly arranged in banks of 5–10 cells in series in order to compensate the pulp short-circuiting.

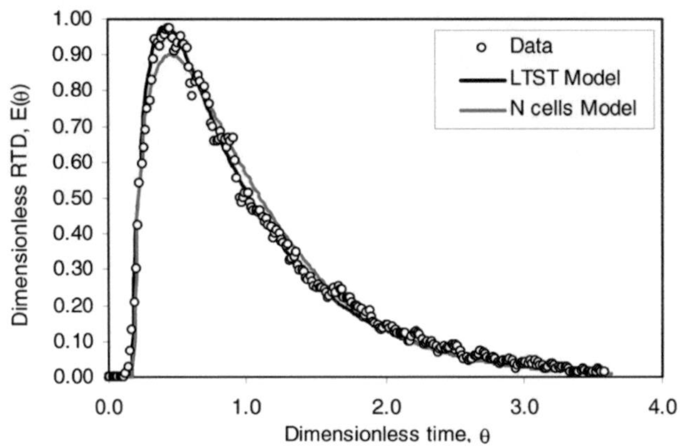

Fig. 7. Fine (-45 μm) non-floatable solid RTD in first cell of the rougher bank (Yianatos et al., 2008c).

Figure 8 shows the RTD of single 130 m^3 and 250m^3 self-aerated mechanical cells, both experimental data and LTST model, Eq. (2). A comparison of the RTD shows that, despite the difference in cells size was nearly twice, the mixing conditions are similar, thus allowing for an effective scale-up in terms of the hydrodynamic behavior. The LTST model showed an excellent fit to describe the hydrodynamic behavior of both cells.

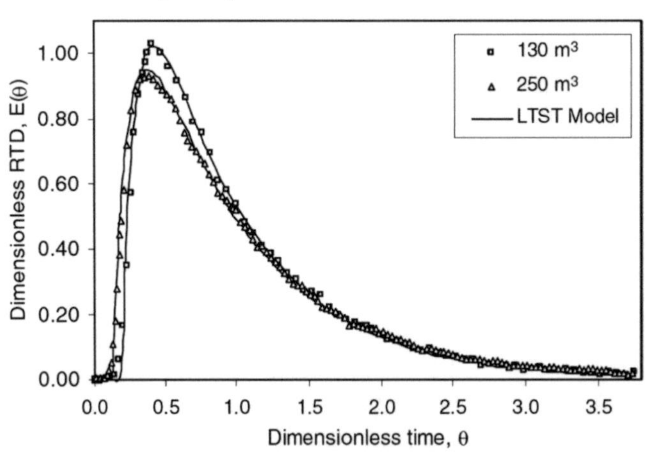

Fig. 8. Comparison of liquid RTD in 130 m^3 and 250 m^3 self-aerated mechanical cells Yianatos et al, 2008c).

Figure 9 shows the liquid RTD of single 160 m³ and 300 m³ forced air mechanical cells, at dimensionless time scale, for mixing regime comparison (Morales et al. 2009). It can be seen that despite the difference in size between both cells, almost twice, the residence time distribution was similar.

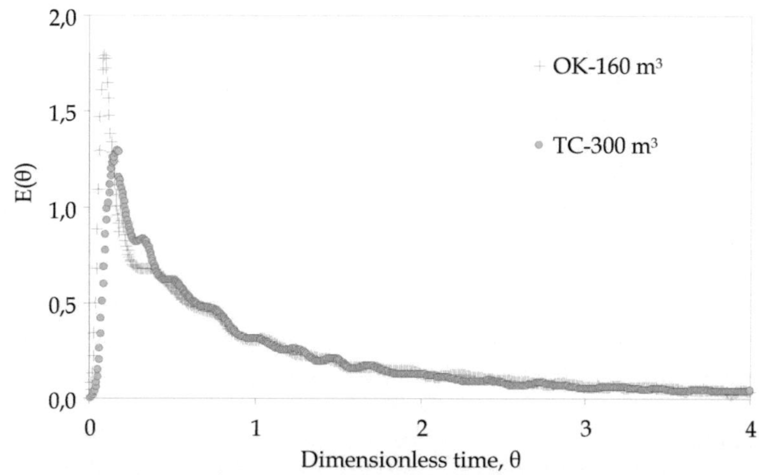

Fig. 9. Comparison of liquid RTD in 160 m³ and 300 m³ forced air mechanical cells (Morales et al, 2009).

2.3 Froth mean residence time of liquid, floatable and non-floatable solids

Froth plays an important role in flotation processes preventing the pulp transport to the concentrate (short-circuit). Thus, it contributes to increasing the concentrate grade by gravity drainage of entrained particles, back into the pulp. Key parameters affecting the froth performance are the mean residence times of solids, liquid and gas in the froth. Large mechanical flotation cells are provided with a froth crowder, a concentric inverted cone located near the top, which accelerates the froth discharge to the concentrate overflow. Also, large flotation cells are provided with internal radial launders which decrease the distance of horizontal transport in the froth, as shown in Fig. 10.

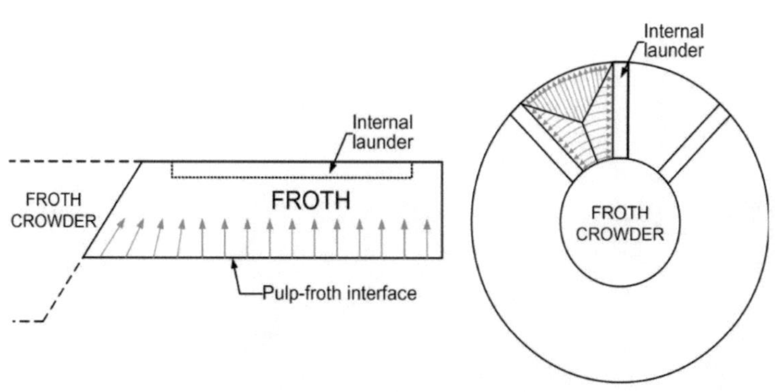

Fig. 10. Cut and top view of the transport paths in the froth of large flotation cells (Yianatos et al., 2008b)

The froth mean residence times were evaluated from direct measurements of liquid and solid time responses in the froth of self-aerated copper flotation cells of 130 m^3 (Yianatos et al., 2008b). For this purpose the radioactive tracer technique was applied, using ^{82}Br as liquid tracer, and non-floatable mineral particles in three size classes (+150; -150 +45; -45 μm) as solid tracers. All tracers were injected at the cell feed entrance, Fig. 11, which allowed the tracer to circulate first through the rotor, and become well distributed over the whole cross-sectional area before entering the froth.

Each tracer time response was measured on-line 10 cm below the pulp/froth interface (sensor S2: input signal) and at the concentrate overflow discharge (sensor S1: output signal). The froth mean residence time was then obtained by difference between the average times of the froth input and output tracer signals. Sensors S3 and S4 were installed at 65 and 120 cm below the pulp–froth interface, respectively, to verify the axial transport of tracer along the quiescent zone below the froth. A reasonable well mixed condition was normally observed below the pulp froth interface. Thus, sensor S2 was selected to represent the froth input composition. Figure 12 illustrate the input (sensor S2) and output (sensor S1) signals, for the non-floatable solid entering the froth at the pulp froth interface level and traveling up to the froth overflow lip level. Similar measurements were performed for the liquid and floatable mineral as well as for the non-floatable mineral at three size classes. For the copper rougher flotation, the froth mean residence time of non-floatable solids was 9–12 s, while, the froth mean residence times of liquid and floatable solid were significantly larger, 21 and 24 s, respectively.

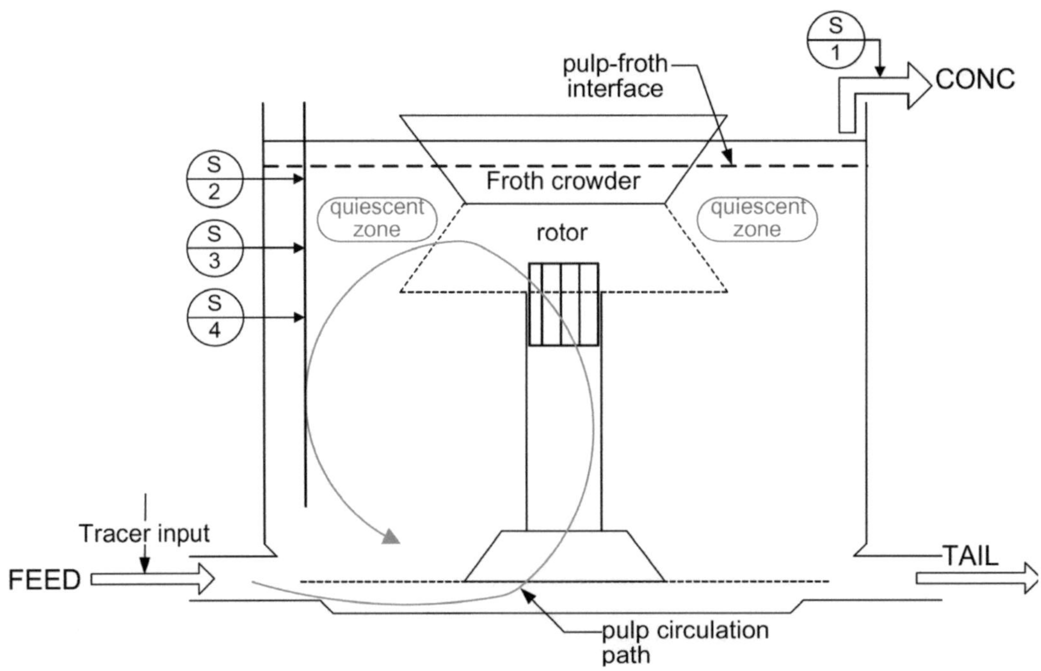

Fig. 11. Location of sensors in a 130 m^3 flotation cell (Yianatos et al., 2008b).

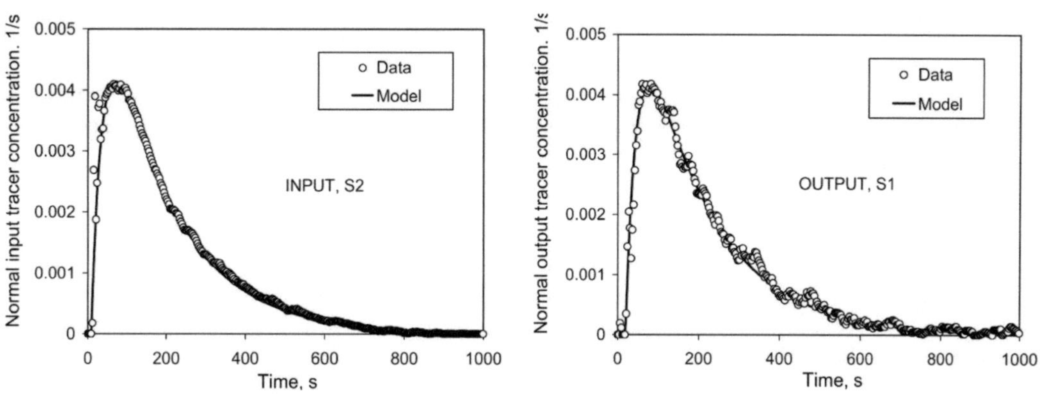

Fig. 12. Froth input and output signals for global non-floatable solid (Yianatos et al., 2008b).

The experimental results showed that mineral particles entering the froth, either attached to the bubbles or entrained, had a minimum residence time similar to the gas mean transport time in the froth, approximately 10-12 s. In this study, it was found that the radioactive tracer technique is a powerful tool for direct measurements of the liquid and solids (floatable and non-floatable) froth residence time.

2.4 Mixing time and internal pulp circulation in large industrial self-aerated cells

Short time mixing is relevant to the flotation operation because the efficiency of the process depends upon the probability of collision between particles and bubbles in order to create particle-bubble aggregates. Figure 13 shows a cell design, self-aspirating provided with a riser tube to promote the pulp circulation through the impeller located near the pulp-froth interface. The cell is also provided with a froth crowder (inverted cone) to improve the froth transport into the launders. Self-aspirated or forced air enters the cell from the top, through the annular section located around the rotational axis. Bubbles are generated at the impeller zone, also called the active flotation zone. It has been established that the main opportunity for an efficient particle-bubble contact occurs when pulp circulates through the impeller zone (Arbiter, 2000). Thus, two relevant parameters to describe the mixing condition in a big flotation cell are the number of pulp circulations through the impeller, before complete mixing takes place, and the number of pulp circulations before the pulp leaves the cell.

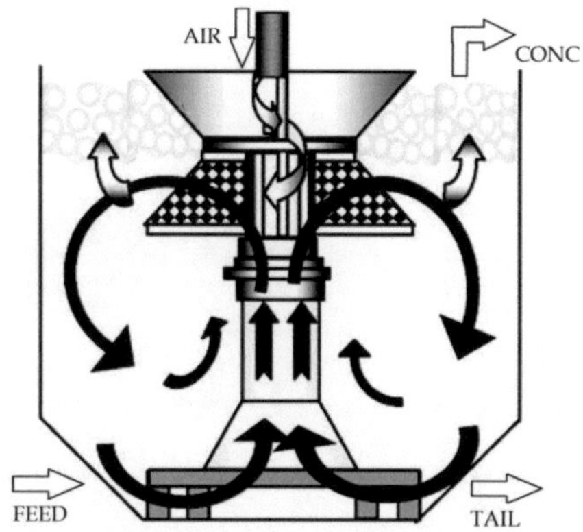

Fig. 13. Mechanical flotation cell with self-induced air.

A new approach to characterizing the mixing evolution and mass transport patterns in big flotation cells was developed (Yianatos et al., 2008a). The procedure consists of using a non-invasive radioisotope tracer technique which allows for the continuous measurement of the local concentration of liquid and solid phases at different points in the cell. Short-term mixing was experimentally characterized by using ^{82}Br in solution as liquid tracer and ^{24}Na was used to trace the solid, considering three particle size classes.

2.4.1 Mixing time

The mixing time in the 130 m³ flotation cell was estimated as the time where the four tracer detectors, S1, S2, S3 and S4, located on the cell wall, as shown in Fig. 14, reached a similar (equal) tracer concentration level within a minimum periodic oscillation.

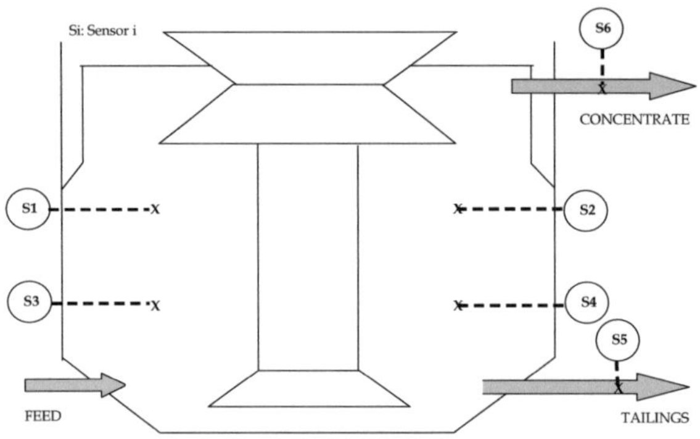

Fig. 14. Location of sensors in a 130 m³ flotation cell (Yianatos et al., 2008a).

Figure 15 shows the tracer concentration at the four symmetric locations inside the cell, after a feed impulse injection consisting of fine non-floatable particles of less than 45 µm. Here, it was observed that after a period of 100 s, the feed became almost fully mixed. A similar result was observed for the solid mineral, of different particle sizes, and liquid tracers.

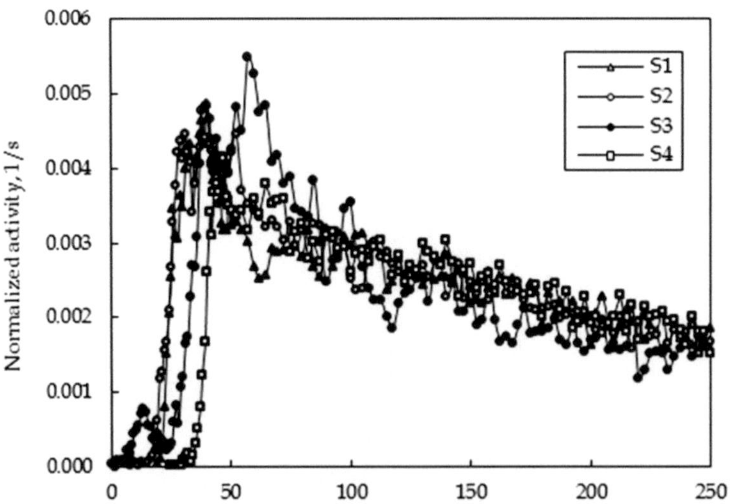

Fig. 15. Short-term mixing of fine solid (−45 µm) in a big cell, 130 m³ (Yianatos et al., 2008a).

2.4.2 Internal circulating ratio

The mixing condition in the big self-aerated cells is dictated by the pulp suction capacity (pumping capacity) of the impeller moving the pulp upwards through the riser pipe into the impeller zone. Thus, in order to characterize the pulp circulation in the big cell, an internal circulating ratio R (%) was defined,

$$R(\%) = 100\left(\frac{Q_{imp}}{F}\right) \tag{6}$$

where, Q_{imp} (m^3/h) is the volumetric flowrate through the impeller of the cell, and F is the volumetric feed flowrate entering the cell. The internal pulp circulation was calculated from liquid and solid tracer measurements and adjusted mass balances. Experimental results showed a mixing time of around 100 s, for liquid and solids, while the pulp mean residence time was around 350 s. It was found that the feed pulp circulates 1.4 times through the impeller zone, in a 130 m^3 self-aerated flotation cell, before reaching a well-mixed condition. Also the feed pulp, on average, circulates 5.0 times through the impeller zone, before leaving the cell into the tailings flowrate. These results are relevant to identify the short term pulp circulation patterns, to better understand how the mixing occurs, and to evaluate the probability of particle-bubble contact near the impeller zone in a big flotation cell.

2.5 Gas holdup and gas RTD measurements in flotation machines

In flotation processes, the gas flowrate (typically air) is a key variable which provides the gas surface required for selective mineral particles capture and transport. The gas residence time distribution (RTD) measurement is a powerful tool because it allows the evaluation of the mean gas residence time as well as the effective gas holdup in the cell. Also, the presence of gas recirculation through the rotor and gas entrainment into tailings can be identified.

2.5.1 Mechanical cells

A suitable technique to measure the actual gas RTD, as well as to estimate the gas holdup, gas circulation and entrainment, in large size industrial flotation cells was developed and tested in a 130m^3 self-aerated mechanical flotation cell (Yianatos et al., 2010b). Bromine Tri-Fluor-Methane (CF_3Br), also called Freon 13B1, was selected as the gaseous tracer because it is an inert gas which only contains Bromine (Br) an activating element with a half-life of 36 hours, which is compatible with times required for preparation, activation, manipulation, transportation and gas application in the industrial plant (International Atomic Energy Agency, 1990). The gas was stored in a stainless steel tank, and then activated by direct irradiation in a 5MW Nuclear Reactor, RECH-1, at the Chilean Commission of Nuclear Energy. After neutron irradiation in the nuclear reactor, the radioactive gaseous tracer was put into a specially designed stainless steel cylinder for the radioactive tracer transport. The injection system, shown in Fig. 16, consists of a cylinder where the gas contained in the transport container was transferred by means of a valve system which allows the regulation of the proper charge of radioactive gas tracer for each experiment, using mechanical vacuum and cooling. For example, 10 mCi (0.37 GBq) of Br-82 was required in Freon 13B1.

Fig. 16. Gas injection system (Yianatos et al., 2010b).

The radioactive tracer technique consists of the injection of a gas impulse signal through the gas (air) inlet, consisting of a 25.4 cm (12 in.) pipe located at the top of the cell, Fig. 17, which allowed the tracer to circulate through the rotor, thus being well distributed over the whole cross-sectional area. Figure 18 shows the sensors (S1, S2, S3) location inside the cell, as well as sensor S4 (entrainment) and sensor S5 (RTD) located outside the cell.

2.5.1.1 Gas holdup

The tracer concentration inside the cell, as well as the presence of tracer leaving the cell at the concentrate (on top of froth) and tailings streams, was recorded on-line by non-invasive sensors. Also, the actual gas holdup was directly measured at the level of sensor S1 (as reference), and the gas holdup at the level of sensors S2 and S3 was scaled from sensor S1.

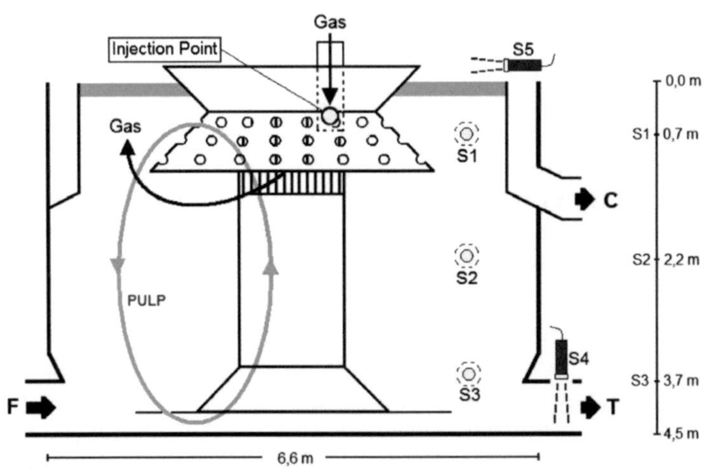

Fig. 17. Side view of the industrial flotation cell (F: Feed, C: Concentrate, T: Tailings) (Yianatos et al., 2010b).

Figure 18 shows the gas holdup profile estimated along the pulp zone in a $130m^3$ flotation cell, relative to the local gas holdup measurement (reference) near sensor S1. The total gas radiation intensity measured by sensor 4, located on the tailing discharge pipe (see Fig. 17), was almost negligible. This result confirms that the gas entrainment into tailings in the $130m^3$ cell was nil, which is different from previous experiments of significant gas entrainment into tailings in industrial flotation columns (Yianatos et al., 1994).

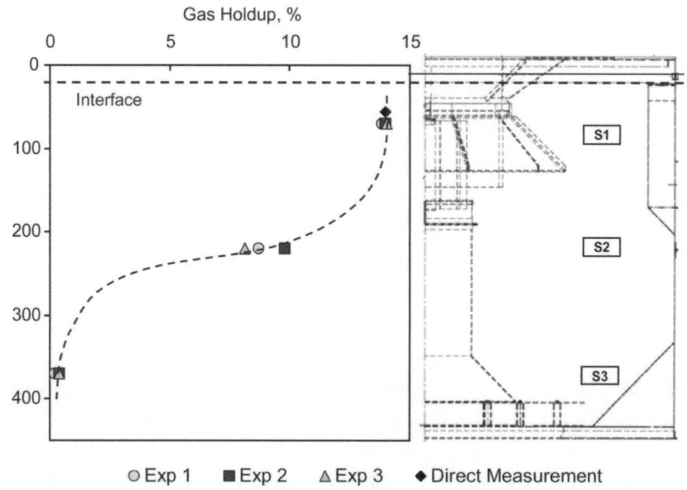

Fig. 18. Estimated gas holdup profile in industrial flotation cell (Yianatos et al., 2010b).

2.5.1.2 Gas residence time distribution

Figure 19 shows the normalized data, registered by sensor S5, located on top of the froth, during the gas residence time distribution measurements. Also the good fit of the LSTS model, Eq. (4) was observed for the gas RTD in a mechanical cell.

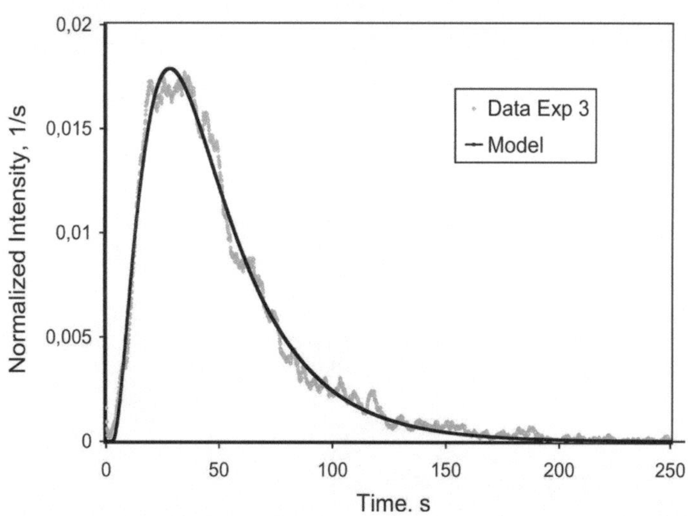

Fig. 19. Gas residence time distribution data and model fit (sensor S5) (Yianatos et al., 2010b).

Once obtained the mean gas residence time, the effective gas volume in the industrial cell can be directly calculated from the gas flowrate measurement.

2.5.2 Flotation columns

In a flotation column the pulp feed enters the collection zone below the interface level and moves downward by gravity, thus contacting a bubble swarm generated from the bottom through a gas sparger. The gas phase residence time distribution of an industrial column of 0.91 m diameter and 15 m height, operating in a molybdenite cleaning circuit, was investigated experimentally by the impulse response method using a radioactive gaseous tracer Kripton-85 (Yianatos et al., 1994). The radioactive gaseous tracer Krypton-85 was selected because is a beta emitter and also has a low gamma radiation emission (514 KeV, 0.41%). Another important property of the Kr-85 is the large half-life (10.7 year), which allows for a large storage time. On the other hand, the disadvantage is that after the tracer discharges to the atmosphere it has a slow decay. Fortunately, this is not critical because the amount of tracer used for process testing is very small and there is a large dilution into the air. This aspect was also quantified. It was assumed that the response signal should be 10 times the background noise. Results showed an activity requirement in the order of 300 mCi per injection for the industrial column. The experimental methodology consisted of introducing an impulse of radioactive gas inside the air sparger using a specially designed device, and on-line measurement of the transient response at various levels in the column. Figure 20 shows the location of the gamma radiation sensors in the industrial column. Sensor 1 was located just above the gas sparger in front of the tracer input. Sensor 2 was located in the froth 65 cm below the lip level, while sensor 3 was located 15 cm above the top of the froth in the industrial column. Sensor 4 was located in front of the tailings line, to register the gas entrainment. According to this arrangement activities were calculated in order to measure the tracer presence from outside the column at different sensor locations. In order to insure the proper removal and dilution of the gaseous tracer from the top of the column, an extraction unit was installed above each column.

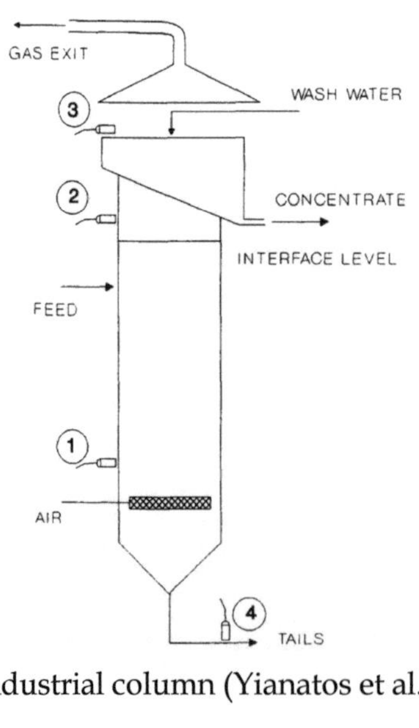

Fig. 20. Sensors Location in industrial column (Yianatos et al., 1994).

The system was arranged like an inverted funnel and was made of polyethylene and provided with a gas extractor to discharge the gaseous tracer outside the building. All the system was on-line monitored with portable radiation sensors during tests.

2.5.2.1 Gas injection system

The gas sparger of the industrial column consists of 8 parallel rubber tubes. The Kr-85 gas injection system, Fig. 21, was connected into the air line entering one central rubber tube, from the air manifold. The tracer was first transferred from the storage tank to the injection cylinder under vacuum. The gaseous tracer was then diluted with air until it reached the pressure of the air line in the cylinder. Finally, the Kr- 85 was instantaneously injected into the gas sparger by means of a nitrogen overpressure. The gas residence time in the sparger was about 3-6(s).

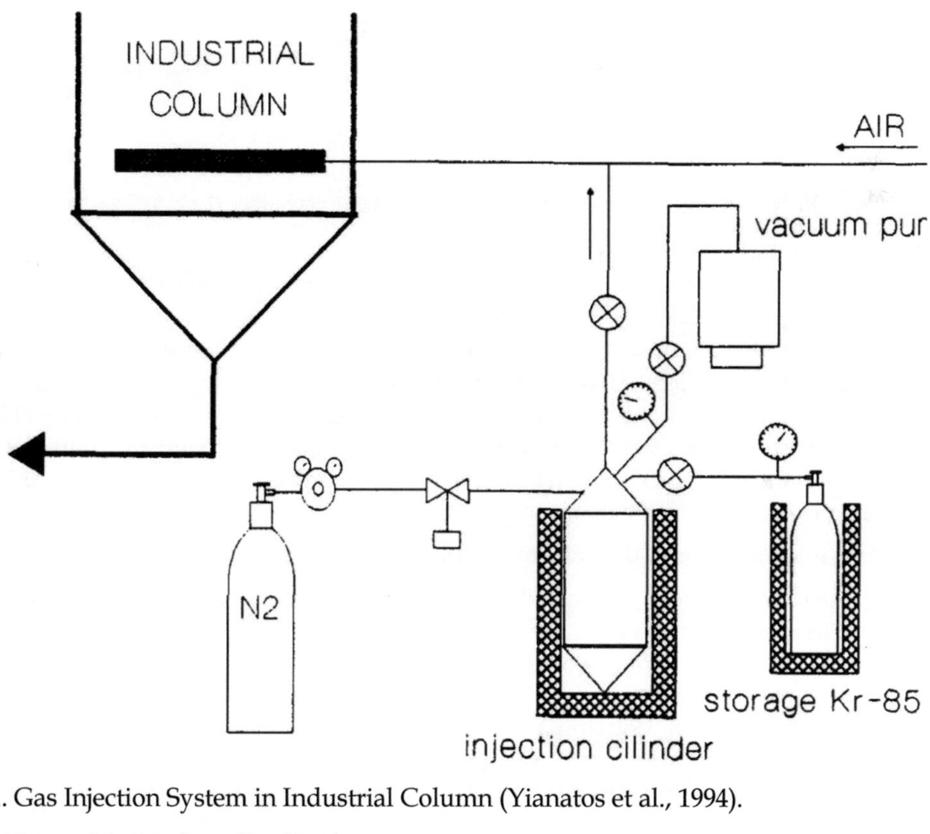

Fig. 21. Gas Injection System in Industrial Column (Yianatos et al., 1994).

2.5.2.2 Gas residence time distribution

The column study showed that tracer injection (sensor 1) was closer to an impulse, before the signal became contaminated by internal gas circulation in the column. Figure 22 shows the time response curves, observed after the impulse injection at time zero. Sensors 2, in the froth, and sensor 3, at the gas exit, show the dispersion of tracer in the froth zone and leaving the froth zone are similar with a time delay. Sensor 3 typically showed a pulsating response, about 2 min period. The observation from sensor 3 corresponds to the overall gas residence time distribution RTD.

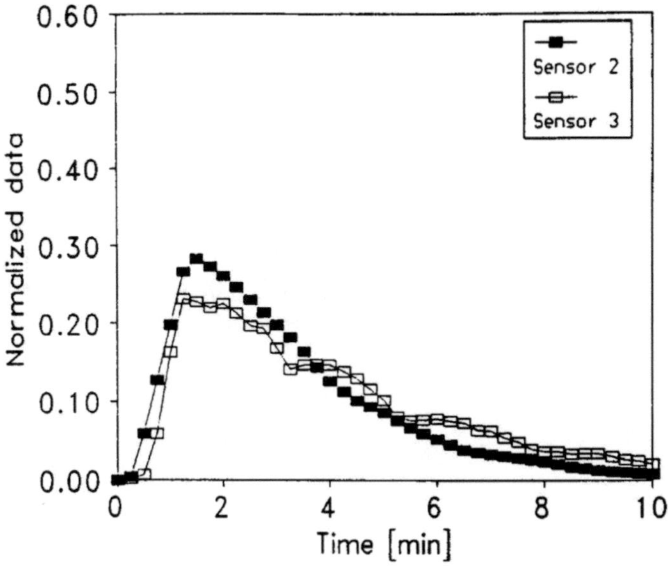

Fig. 22. Time Response Curves from Industrial Column (Yianatos et al., 1994).

Furthermore, an independent estimate of the gas residence time was developed from direct measurements of gas flowrate and gas holdup. These results showed a reasonable agreement which validates the RTD data. Average gas residence time from the industrial flotation column was about 4-5 (min). The gas phase in the froth zone behaved closer to a plug flow while operating at superficial gas rates lower than 1.5 cm/s and superficial wash water rates of 0.2-0.4 cm/s. Also, it was found that transport of floatable minerals along the froth was very similar to that of the gas, showing a similar dispersion and time delay. On the other hand, the residence time distribution of the floatable minerals reporting to the tails showed a behavior similar to that of the gangue.

2.5.2.3 Gas entrainment into the column tailings

Data obtained from sensor 4, located in front of the tailing flow (Fig. 20), showed a significant presence of gaseous tracer in the tailings stream. Thus, the industrial column design favors the entrainment of finer bubbles from the gas sparger to the bottom exit pipe. In summary, it was found that the radioactive tracer technique provides an effective way to evaluate the gas RTD in flotation machines where other techniques, such as thermal conductivity, gas spectrometry, and FID-gas chromatography, typically used at laboratory scale, are less suitable for industrial scale measurements in large size equipment.

2.6 Direct measurement of gangue entrainment

In a flotation machine the effective separation occurs at the pulp/ froth interface and during the froth transport into the concentrate launder. Particles enter the froth zone by two mechanisms; forming particle-bubble aggregates (true flotation) or by entrainment. Fig. 23 shows a two-stage model consisting of the pulp zone, related to the collection process, and the froth separation zone. The mass flowrate (tph) in the mineral transport streams is denoted as, F: feed, C: concentrate, T: tailings, B: bubble-particle aggregate, E: entrainment and D: drop-back [2].

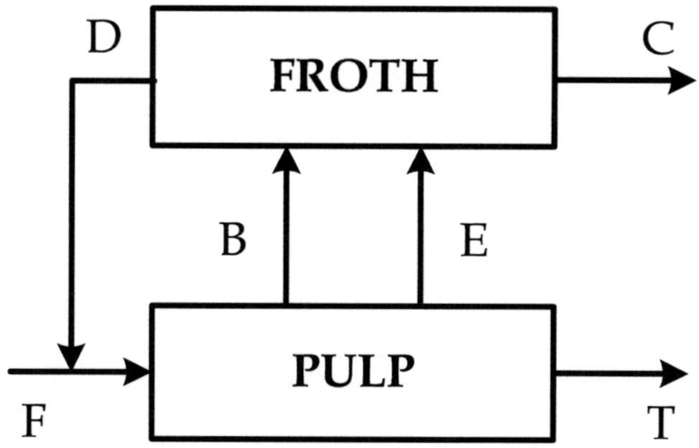

Fig. 23. Mineral transport streams in a flotation cell.

From experimental testing, it has been observed that water recovery is the main responsible for the non-selective fine particles transport by entrainment, from the pulp-froth interface up to the concentrate launder. Also, the gangue entrainment is generally not significant in coarser particle size classes (i.e. in the case of silica equivalent, for particles larger than 50 μm). The mineral feed characteristics and conditioning determine the grade of the particles attached directly to the surfaces of the bubbles by true flotation, while the operating conditions, such as gas rate, bubble size, froth depth and others, determine the amount of gangue recovered by entrainment, which finally decreases the concentrate grade. In this aim, the recovery of liquid and solids by entrainment was evaluated by direct measurement of the fraction of liquid and solids reported to the concentrate in a 130 m^3 mechanical flotation cell (Yianatos et al., 2009). The liquid and solids entrainment, per size classes (+150; −150+45; −45 μm), was measured by the radioactive tracer technique. The procedure consisted of introducing a tracer impulse at the cell feed entrance. The tracer time response was monitored on-line at the concentrate overflow and at the tailings discharge. Also, in order to obtain the quantitative distribution of the feed, samples were taken periodically from the concentrate and tailings streams, for a period of 4 residence times, during the tracer tests. This allowed the quantification of the mass of tracer reporting to both streams.

2.6.1 Tracer sampling and on-line detection

The radioactive tracer technique consists of the injection of an impulse signal (water or gangue) at the feed pulp entrance of the cell. Also, discrete samples were taken periodically from the concentrate and tailings streams. Figure 24 shows the sensor location and the sampling points. On-line radioactive tracer detection at the concentrate and tailings streams was used to obtain a smooth and almost continuous (minimum period of 50 ms) signal to estimate the residence time distribution of the gangue leaving the cell in each stream. However, even though both signals are proportional to the corresponding mass flowrates, they are not directly comparable. Consequently, a second measurement was required to provide a quantitative estimate of the tracer concentration during the impulse tracer test. For this purpose, discrete sampling of the concentrate and tailing streams was performed, in order to measure (off-line) the tracer concentration during the impulse time response.

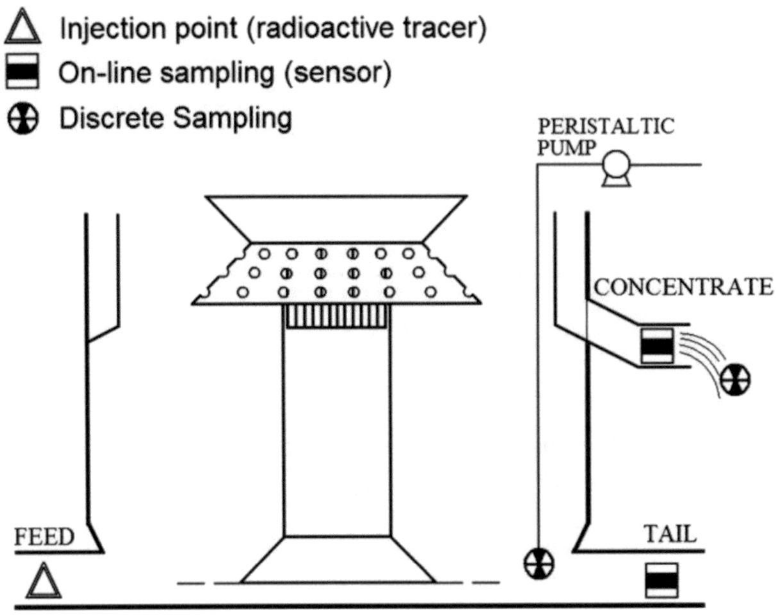

Fig. 24. Sensors location and sampling points (Yianatos et al., 2009).

Samples from the concentrate stream were taken directly from the overall concentrate discharge using a standard manual cutter, while samples of tailing stream were obtained by pumping the sample from the cell bottom discharge, see Fig. 24. Thus, the sampled signals allowed obtaining a quantitative description of the tracer signals, but they have fewer data points and are noisier.

2.6.2 Discrete sampling (off-line) for tracer RTD measurement

Discrete samples were obtained periodically and simultaneously from the concentrate and tailings streams. In order to measure the mineral tracer content, each sample of 250 [mL] was introduced in a lead vessel (to avoid external radiation) which contains a 3 in.×3 in. NaI(Tl) radiation detector. The signal from the radiation sensor was associated to a multi-channel-analyzer system, model Nomad from ORTEC, and connected to a notebook provided with software for spectrum analysis. Figure 25 shows the comparison between the fine gangue (−45 μm) tracer data in white circles, which was sampled while leaving the cell at the concentrate output. Also, this figure shows the best fit of the model derived from on-line measurements, which was scaled to describe the residence time distribution observed from the sampled data. Thus, the area under the curve corresponds to the time integral and represents the total amount of tracer in the respective stream. Similarly, considering the tailings and concentrates areas for the water tracer, obtained after RTD modelling of on-line data and model scaling to fit the sampled data under the same pattern curve, the quantitative comparison for mass split estimation of the liquid tracer was obtained.
In summary, the recovery of liquid and solids by entrainment was evaluated by direct measurement, of the fraction of liquid and solids reported to the concentrate in a 130 m^3 flotation cell. Thus, the strong dependence of entrainment on fine particle sizes of less than 45 μm, was confirmed. Also, recovery of coarse particles (larger than 150 μm) by entrainment was 0.05 %.

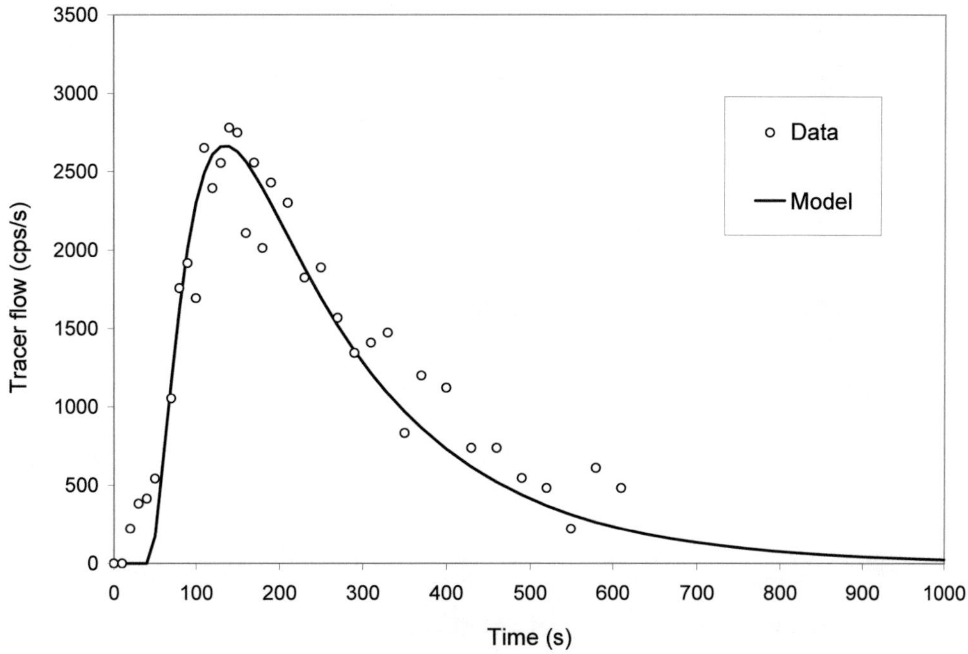

Fig. 25. Sampled data fit by on-line RTD derived model for concentrate (Yianatos et al., 2009).

2.7 Industrial flotation cell scanning with gamma ray

A new observation of the internal flotation machine characteristics has been developed and evaluated in an industrial 130 m^3 rougher flotation cell (Yianatos et al., 2008a). The measurement consists of gamma ray scanning, using a neutron backscatter technique across a vertical plane to measure relative density as a function of height. The plane was selected as the vertical projection of a chord, to avoid the impeller, internal baffles and other disturbances. This technique easily reveals pulp-froth interface level, sanding-up due to particle settling and any other disturbing condition in the cell. The experimental device consisted of a specially designed machine which provides an automatic collimated displacement (up and down) of the emission source and a detector, at a constant velocity.

Figure 26 shows the cell scanner output, where no significant disturbances can be appreciated along the pulp zone, from the bottom up to the pulp/froth interface. In the first 0.2 m from the bottom, the activity (270–300 cps) corresponds to the cell shell with no pulp inside (open bottom). Above 0.2 m and up to 1.3 m, the activity was attenuated (150-130 cps) by the presence of the corner (inclined) baffle and the increasing cross-section of pulp. Closer to 3 m from the bottom, the presence of the internal circular launder decreases the activity (100 cps), and then from 3-4 m the activity increases because the circular launder was not full of pulp and froth. Finally, above 4 m, the presence of the internal radial launders not fully loaded with pulp and the pulp-froth interface, above which the air holdup increases significantly (15%-90%), which increased the gamma ray activity to a level similar to the clear bottom (270–300 cps).

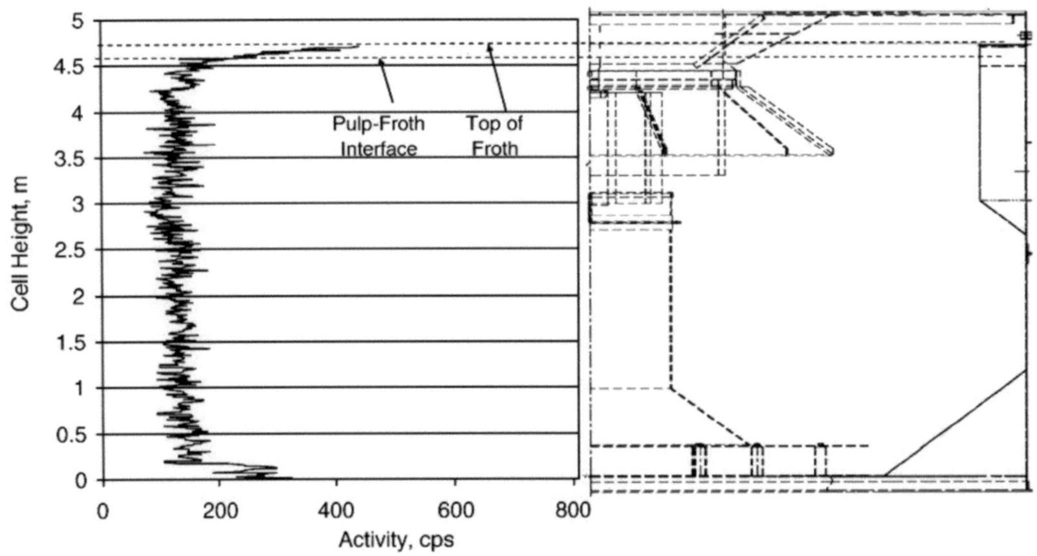

Fig. 26. Gamma ray scanner of a big flotation cell, 130 m³ (Yianatos et al., 2008a).

2.8 Pulp flowrate distribution in parallel flotation banks

Radioactive tracer measurements allowed the calculation of the effective mean residence time of liquid and solids, as well as the mass flowrate distribution in parallel lines of flotation machines. Table 1 shows a summary of the degree of segregation observed in Cu/Mo collective and selective rougher flotation circuits (Morales et al., 2010), where the third line in both circuits shows a significantly higher feed flowrate percentage. Also, in the selective first cleaner it was found that line 1 processed 52% while line 2 only 48% of the first cleaner feed. Thus, unequal feed distributions were observed in the three evaluated circuits. The mean residence time can also be affected by solids partial embankment, which is also an abnormal operating condition.

Collective Rougher		Line 101	Line 102	Line 103
	Feed flow, %	31.6 ±0.7	31.8 ±0.5	36.6 ±0.6
Selective Rougher		Line 1	Line 2	Line 3
	Feed flow, %	31.2 ±2.3	29.2 ±0.7%	39.6 ±1.7
Selective First Cleaner		Line 1	Line 2	
	Feed flow, %	52.0	48.0	

Table 1. Solid and liquid mean distribution in rougher flotation (Morales et al., 2010).

Residence time distribution measurements were carried out at Chuquicamata concentrator, Codelco-Chile, consisting of three parallel rougher lines (Morales et al, 2009). This kind of measurements allowed the calculation of the effective mean residence time of liquid and solids, as well as the mass flowrate distribution in parallel lines. Table 2 shows the effective mean feed pulp distribution in lines 1 and 3, for liquid and solids (global and particle size classes), calculated by de-convolution between input and output signals. Tracer was injected into the feed distributor, or directly into the feed box of the first cell in each line. Here, results showed a high consistency in pulp distribution measurements for lines 1 and 3.

Rougher		Line 1	Line 2	Line 3
	Feed flow, %	33.6 ±0.4	36.0	30.4 ±1.7

Table 2. Feed pulp distribution in rougher flotation (Morales et al., 2009).

2.8.1 Effective volume of flotation cells

Considering simultaneously the mineral treatment (tph), measured in the conventional grinding plant in Colón, and the pulp mean residence time measured by radioactive tracer tests in the rougher Metso cells (42.5 m^3), an effective pulp volume of 82% of the total cell volume was determined. This result was similar to that observed for similar cells (78.8-80.4%), at El Salvador concentrator Codelco-Chile (Yianatos et al., 2001).

2.9 Flotation rate distribution

2.9.1 Introduction

The batch flotation process has been commonly characterized assuming a flotation rate distribution function F(k), e.g.: Dirac delta, Rectangular, Gamma or Weibull functions. The identification of F(k) for the collection zone of single continuous industrial cells, larger sizes, is more complex and a novel procedure to estimate the flotation rate distribution from the collection zone of industrial flotation cells, using the radioactive tracer technique, was recently described (Yianatos et al., 2010c). The approach consists of measuring the impulse response of the floatable mineral tracer concentration, and non-floatable gangue tracer concentration (Residence Time Distribution, RTD), in the cell tailings. Then, the floatable tracer concentration can be compared with the model prediction, using the Gamma function and the RTD of the non-floatable tracer. Thus, the F(k) distribution parameters were obtained by means of the least-square estimation. The new approach was successfully tested in two industrial rougher flotation cells of large size using radioactive mineral tracers.

2.9.2 Experimental procedure

Plant tests were performed in large size cells in two copper concentrators, and consisted of sampling the copper rougher circuits using the short-cut method (Yianatos and Henríquez, 2006), which is sampling of the first cell and the overall rougher bank, for mass balance adjustment and flotation rate estimation. The hydrodynamic characterization of the rougher flotation cells was carried out by the radioactive tracer technique (Yianatos et al., 2008a; 2009) which consists of introducing a tracer sample, like an impulse, in the cell feed. Liquid, solid and solid per size classes were used as tracers in order to measure the RTD as well as to estimate the effective liquid and solid residence time. The liquid tracer was a Br-82 solution and the actual non-floatable solid from final tails was used as the solid tracer for the estimation of the gangue RTD, while final concentrate (d_{80} = 45 microns) was used for tracing the floatable mineral RTD in cell tailings. Two case studies were carried out in mechanical cells of 130 m^3 and 300 m^3. For example, measurements of floatable and non-floatable tracer concentration, normalized, are shown in Fig. 27 for the 300 m^3 cell. A good flotation rate model should respect the parsimony principle (low number of parameters), showing a good fitting with experimental data.

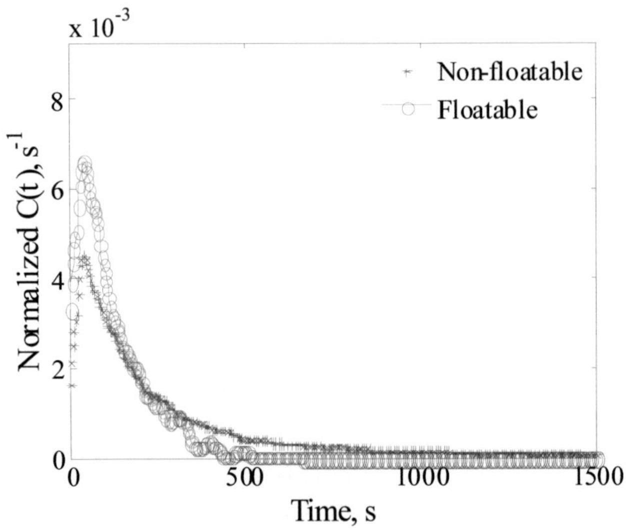

Fig. 27. Normalized mineral tracer concentration (Yianatos et al., 2010c).

Also, the function must be bounded by physical limits (e.g. the fraction of floatable mineral at zero rate constant must be zero). From these results the gamma model structure was selected for the plant flotation rate. Using the methodology to estimate the Gamma function parameters of the floatability distribution, the model fit shown in Fig. 28 was obtained.

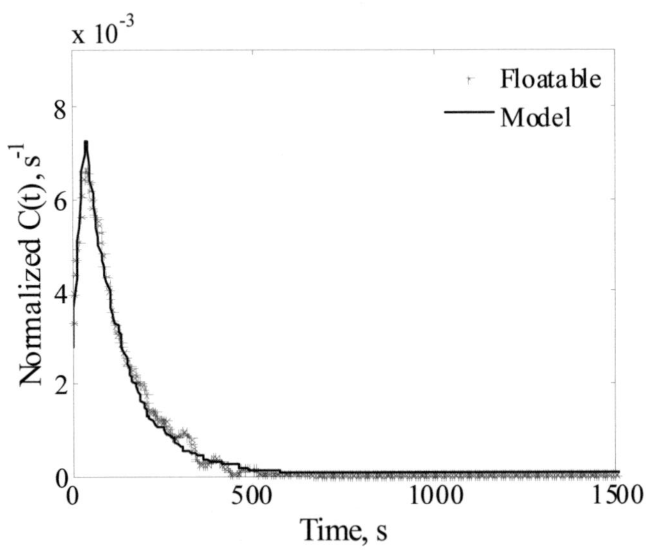

Fig. 28. Model fit of floatable tracer concentration (Yianatos et al., 2010c).

Plant results from two industrial cell operations showed that the flotation rate, predicted by the Gamma model, follow a rather normal distribution. The same result was observed in laboratory batch tests. The methodology has the advantage that allows the estimation of the flotation rate distribution of a single industrial cell under continuous operation. This was performed by generating a transient response, by means of floatable and non-floatable radioactive tracers, which allows the identification of the single cell parameters. The method requires additional measurements for mass balance around the cell as well as the froth recovery estimation.

3. Conclusions

The use of radioisotopes has proven to be a powerful tool to study the hydrodynamic behaviour of large flotation machines. Mean residence time in pulp and froth zones were evaluated from RTD measurements for liquid, solids and gas radioactive tracers. Relevant parameters such as mixing regime, mixing time, flow distribution in parallel flotation banks, gangue and gas entrainment, particles segregation, have been evaluated using non-invasive sensors for radiation detection without disturbing the flotation operation. This information is fundamental for improving flotation machines operation, control and optimization.

4. Acknowledgements

Funding for process modeling and control research is provided by CONICYT, Project Fondecyt 1100854, NEIM, Project P07-087-F, ICM-Mideplan, Santa Maria University, Project 271068, and the Chilean Commission of Nuclear Energy, CCHEN.

5. References

Arbiter, N. (2000). Development and scale-up of large flotation cells. *Mining Engineering* 52 (3), 28–33.

Díaz, F and Yianatos J. (2010). Residence time distribution in large industrial flotation cells. *Atoms for Peace - an International Journal*, Vol.3, N°1, pp.2-10.

Dobby, G.S. and J.A. Finch (1985). Mixing characteristics of industrial flotation columns. *Chemical Engineering Science*, 40, N°7, 1061-1068.

Goodall, C.M. and O'Connor, C.T. (1991). Residence time distribution studies in a flotation column. Part 1: the modelling of residence time distributions in a laboratory column flotation cell. *International Journal of Mineral Processing*, 31(1-2), 97-113.

International Atomic Energy Agency. Guidebook on Radioisotopes Tracers in Industry. Reports Series N° 16, Vienna (1990).

Lelinski, D. Allen, J., Redden, L., & Weber, A. (2002). Analysis of the residence time distribution in large flotation machines. *Minerals Engineering*, 15 (7), 499-505.

Mavros, P. (1992). Mixing and hydrodynamics in flotation cells. In: Mavros, P., Matis, K.A. (Eds.), Innovation in Flotation Technology, NATO Series, vol. E208. Kluwer Academic Publishers, The Netherlands, pp. 211–234.

Morales, P., Elgueta, H., Torres, C., Yianatos, J., Vinnett, L. and Diaz, F. (2010). Hydrodynamic and metalurgical characterization of flotation cells in a molybdenum plant. *VII International Mineral Processing Seminar, Procemin 2010*, Santiago, Chile.

Morales, P., Coddou, F., Yianatos, J., Contreras, F., Catalán, M., Díaz, F. (2009). Hydrodynamic Performance of the Division Codelco Norte Concentrator's Large Flotation Cells. in: P. Amenlunxen, W. Kracht, R. Kuyvenhoven (Eds.), *Proc. VI International Mineral Processing Seminar*, 2-4 Dec., Santiago, Chile, 2009, pp. 385-393.

Niemi, A.J. (1995). Role of kinetics in modelling and control of flotation plants. *Powder Technology*, 82, 69-77

Yianatos, J., Contreras, F., Morales, P., Coddou, F., Elgueta, H., Ortíz, J. (2010a). A novel scale-up method for mechanical flotation cells. *Minerals Engineering*, Vol.23, pp.877-884.

Yianatos J., Contreras F. and Díaz, F. (2010b). Gas holdup and RTD measurement in an industrial flotation cell. *Minerals Engineering*, Vol.23, pp.125-130.

Yianatos, J., Bergh, L., Vinnett, L., Contreras, F., Díaz, F. (2010c). Flotation rate distribution in the collection zone of industrial cells. *Minerals Engineering*, Vol.23, pp. 1030-1035.

Yianatos J.B., Contreras F., Díaz, F. and Villanueva, A. (2009). Direct measurement of entrainment in large flotation cells. *Powder Technology*, Vol.189, pp. 42-47.

Yianatos, J.B., Larenas, J., Moys, M., Díaz, F. (2008a). Short time response in a big flotation cell, *International Journal of Mineral Processing*, Vol. 89, pp. 1-8.

Yianatos J.B., Bergh, L.G., Tello, K., Díaz. F., Villanueva, A. (2008b). Froth mean residence time measurement in industrial flotation cells. *Minerals Engineering*, Vol.21, pp.982-988.

Yianatos J.B., Bergh, L.G., Tello, K., Díaz. F., Villanueva, A. (2008c). Residence time distribution in single big industrial flotation cells. *Minerals & Metallurgical Processing Journal*, Vol.25, N°1, pp. 46-52.

Yianatos J.B. and Henríquez, F.H. (2006). Short-cut method for flotation rates modelling of industrial flotation banks. *Minerals Engineering*, Vol.19, pp. 1336-1340.

Yianatos, J.B. , Bergh, L.G., Díaz, F. and J. Rodríguez (2005a). Mixing characteristics of industrial flotation equipments. *Chemical Engineering Science*. Vol. 60, N° 8/9, pp. 2273-2282.

Yianatos, J.B., Bucarey, R., Larenas, J., Henríquez, F. and Torres, L. (2005b), Collection zone kinetic model for industrial flotation columns, *Minerals Engineering*, Vol. 18, pp. 1373–1377.

Yianatos, J.B., Díaz, F., & Rodríguez, J. (2003). Mixing and effective pulp volume in flotation equipments, in : Gomez, C.O. & Barahona, C.A., (Eds.), *Proceedings of the 5th International Conference Copper 2003*, Vol. 3, Mineral Processing, Santiago, Chile, pp.179-194.

Yianatos, J.B. Díaz, F. and J. Rodríguez (2002). Industrial flotation process modelling: RTD measurement by radioactive tracer technique. *XV IFAC World Congress*, 21-26 July, Barcelona, Spain.

Yianatos, J.B., Bergh, L.G., Condori P. & Aguilera, J. (2001). Hydrodynamic and metallurgical characterization of industrial flotation banks for control purposes. *Minerals Engineering*, Vol.14(9), pp. 1033-1046.

Yianatos, J.B., L. G. Bergh, O.U. Durán, F.J. Díaz and N.M. Heresi (1994). Measurement of residence time distribution of the gas phase in flotation columns. *Minerals Engineering*, Vol.7, Nos 2/3, 333-344

Yianatos, J.B. and L.G. Bergh (1992). RTD studies in an industrial flotation column: Use of radioactive tracer technique. *International Journal of Mineral Processing*, 36, 81-91

Part 3

Radioisotopes in Power System Applications

20

Radioisotope Power: A Key Technology for Deep Space Exploration

George R. Schmidt[1], Thomas J. Sutliff[1] and Leonard A. Dudzinski[2]
[1]*NASA Glenn Research Center,*
[2]*NASA Headquarters*
USA

1. Introduction

Radioisotope Power Systems (RPS) generate electrical power by converting heat released from the nuclear decay of radioactive isotopes into electricity. Because all the units that have flown in space have employed thermoelectrics, a static process for heat-to-electrical energy conversion that employs no moving parts, the term, Radioisotope Thermoelectric Generator (RTG), has been more popularly associated with these devices. However, the advent of new generators based on dynamic energy conversion and alternative static conversion processes favors use of "RPS" as a more accurate term for this power technology.
RPS were first used in space by the U.S. in 1961. Since that time, the U.S. has flown 41 RTGs, as a power source for 26 space systems on 25 missions. These applications have included Earth-orbital weather and communication satellites, scientific stations on the Moon, robotic explorer spacecraft on Mars, and highly sophisticated deep space interplanetary missions to Jupiter, Saturn and beyond. The New Horizons mission to Pluto, which was launched in January 2006, represents the most recent use of an RTG. The former U.S.S.R. also employed RTGs on several of its early space missions. In addition to electrical power generation, the U.S. and former U.S.S.R. have used radioisotopes extensively for heating components and instrumentation.
RPS have consistently demonstrated unique capabilities over other types of space power systems. A comparison between RPS and other forms of space power is shown in Fig. 1, which maps the most suitable power technologies for different ranges of power level and mission duration. In general, RPS are best suited for applications involving long-duration use beyond several months and power levels up to one to 10 kilowatts.
It is important to recognize that solar power competes very well within this power level range, and offers much higher specific powers (power per unit system mass) for applications up to several Astronomical Units (AU) from the Sun. However, RPS offer the unique advantage of being able to operate continuously, regardless of distance and orientation with respect to the Sun. The flight history of RTGs has demonstrated that these systems are long-lived, rugged, compact, highly reliable, and relatively insensitive to radiation and other environmental effects. Thus, RTGs and the more capable RPS options of the future are ideally suited for missions at distances and extreme conditions where solar-based power generation becomes impractical. These include travel beyond the asteroid belt, operation within the radiation-intensive environments around Jupiter and close to the Sun,

extended operation within permanently shadowed and occulted areas on planetary surfaces, and general applications requiring robust, unattended operations.

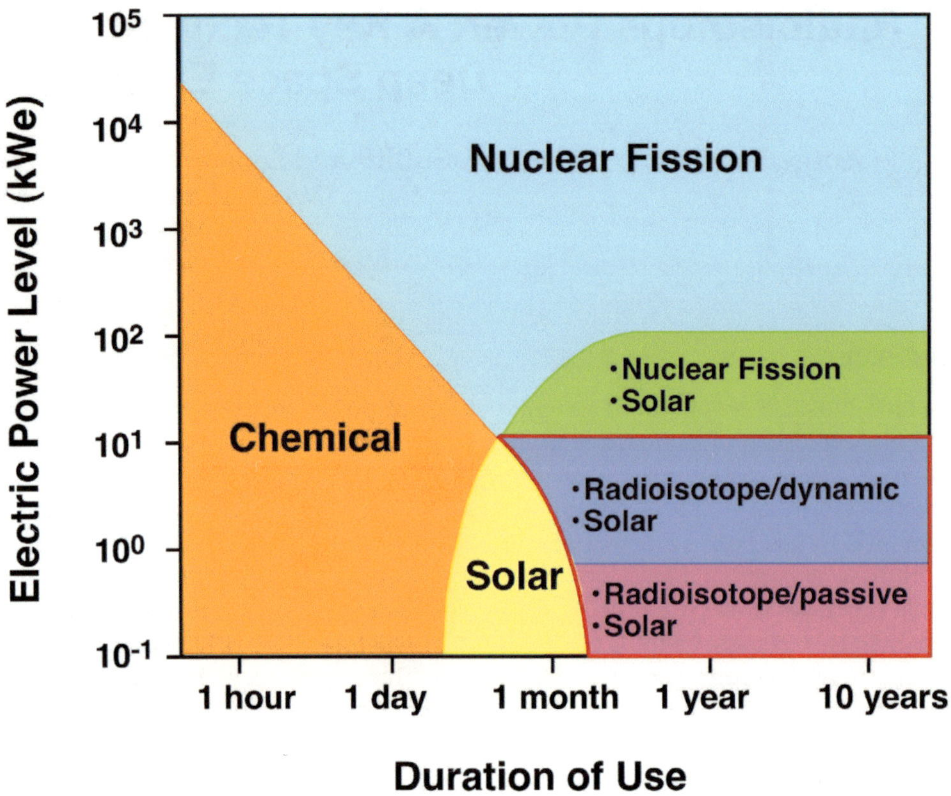

Fig. 1. Suitability of space power system technologies.

Table 1 presents a chronological summary of the U.S. missions that have utilized radioisotopes for electrical power generation. Although three missions were aborted by launch vehicle or spacecraft failures, all of the RTGs that flew met or exceeded design expectations, and demonstrated the principles of safe and reliable operation, long life, high reliability, and versatility of operating in hostile environments. All of the RTGs flown by the U.S. comprise seven basic designs: SNAP-3/3B, SNAP-9A, SNAP-19/19B, SNAP-27, TRANSIT-RTG, MHW-RTG and GPHS-RTG. The first four types were developed by the Atomic Energy Commission (AEC) under the auspices of its Systems for Nuclear Auxilliary Power (SNAP) program. Although the original objective was to provide systems for space, the SNAP program also developed generators for non-space, terrestrial applications.

The GPHS-RTG is the most recently developed unit, and has been the workhorse on all RPS missions since 1989. A cutaway view of the unit is shown in Fig. 2. NASA and the Department of Energy (DOE) are looking beyond this capability, and are currently developing two new units: the Multi-Mission RTG (MMRTG), which draws on the design heritage of the SNAP-19, and the new Advanced Stirling Radioisotope Generator (ASRG) with its much more efficient dynamic conversion cycle.

	Spacecraft/ System	Principal Energy Source (#)	Destination/ Application	Launch Date	Status
1	Transit 4A	SNAP-3B7 RTG (1)	Earth Orbit/ Navigation Sat	29 June 1961	RTG operated for 15 yrs. Satellite now shutdown.
2	Transit 4B	SNAP-3B8 RTG (1)	Earth Orbit/ Navigation Sat	15 Nov 1961	RTG operated for 9 yrs. Operation intermittent after 1962 high alt test. Last signal in 1971.
3	Transit 5BN-1	SNAP-9A RTG (1)	Earth Orbit/ Navigation Sat	28 Sep 1963	RTG operated as planned. Non-RTG electrical problems on satellite caused failure after 9 months.
4	Transit 5BN-2	SNAP-9A RTG (1)	Earth Orbit/ Navigation Sat	5 Dec 1963	RTG operated for over 6 yrs. Satellite lost navigational capability after 1.5 yrs.
5	Transit 5BN-3	SNAP-9A RTG (1)	Earth Orbit/ Navigation Sat	21 Apr 1964	Mission aborted because of launch vehicle failure. RTG burned up on reentry as designed.
6	Nimbus B-1	SNAP-19B2 RTG (2)	Earth Orbit/ Meteorology Sat	18 May 1968	Mission aborted because of range safety destruct. RTG fuel recovered and reused.
7	Nimbus III	SNAP-19B3 RTG (2)	Earth Orbit/ Meteorology Sat	14 Apr 1969	RTGs operated for over 2.5 yrs. No data taken after that.
8	Apollo 12	SNAP-27 RTG (1)	Lunar Surface/ Science Station	14 Nov 1969	RTG operated for about 8 years until station was shutdown.

	Spacecraft/ System	Principal Energy Source (#)	Destination/ Application	Launch Date	Status
9	Apollo 13	SNAP-27 RTG (1)	Lunar Surface/ Science Station	11 Apr 1970	Mission aborted. RTG reentered intact with no release of Pu-238. Currently located at bottom of Tonga Trench in South Pacific Ocean.
10	Apollo 14	SNAP-27 RTG (1)	Lunar Surface/ Science Station	31 Jan 1971	RTG operated for over 6.5 years until station was shutdown.
11	Apollo 15	SNAP-27 RTG (1)	Lunar Surface/ Science Station	26 July 1971	RTG operated for over 6 years until station was shutdown.
12	Pioneer 10	SNAP-19 RTG (4)	Planetary/Payload & Spacecraft	2 Mar 1972	Last signal in 2003. Spacecraft now well beyond orbit of Pluto.
13	Apollo 16	SNAP-27 RTG (1)	Lunar Surface/ Science Station	16 Apr 1972	RTG operated for about 5.5 years until station was shutdown.
14	Triad-01-1X	Transit-RTG (1)	Earth Orbit/ Navigation Sat	2 Sep 1972	RTG still operating as of mid-1990s.
15	Apollo 17	SNAP-27 RTG (1)	Lunar Surface/ Science Station	7 Dec 1972	RTG operated for almost 5 years until station was shutdown.
16	Pioneer 11	SNAP-19 RTG (4)	Planetary/Payload & Spacecraft	5 Apr 1973	Last signal in 1995. Spacecraft now well beyond orbit of Pluto.
17	Viking 1	SNAP-19 RTG (2)	Mars Surf/Payload & Spacecraft	20 Aug 1975	RTGs operated for over 6 years until lander was shutdown.
18	Viking 2	SNAP-19 RTG (2)	Mars Surf/Payload & Spacecraft	9 Sep 1975	RTGs operated for over 4 years until relay link was lost.

	Spacecraft/ System	Principal Energy Source (#)	Destination/ Application	Launch Date	Status
19	LES 8, LES 9	MHW-RTG (4)	Earth Orbit/ Com Sats	14 Mar 1976	Single launch with double payload. LES 8 shutdown in 2004. LES 9 RTG still operating.
20	Voyager 2	MHW-RTG (3)	Planetary/ Payload & Spacecraft	20 Aug 1977	RTGs still operating. Spacecraft successfully operated to Jupiter, Saturn, Uranus, Neptune, and beyond.
21	Voyager 1	MHW-RTG (3)	Planetary/ Payload & Spacecraft	5 Sep 1977	RTGs still operating. Spacecraft successfully operated to Jupiter, Saturn, and beyond.
22	Galileo	GPHS-RTG (2)	Planetary/Payload & Spacecraft	18 Oct 1989	RTGs continued to operate until 2003, when spacecraft was intentionally deorbited into Jupiter atmosphere.
23	Ulysses	GPHS-RTG (1)	Planetary/Payload & Spacecraft	6 Oct 1990	RTG continued to operate until 2008, when spacecraft was deactivated.
24	Cassini	GPHS-RTG (3)	Planetary/Payload & Spacecraft	15 Oct 1997	RTGs continue to operate successfully. Scientific mission and operations still continue.
25	New Horizons	GPHS-RTG (1)	Planetary/Payload & Spacecraft	Jan 19 2006	RTG continues to operate successfully. Spacecraft in transit to Pluto.

Table 1. U.S. Missions Using Radioisotope Power Systems (RPS)

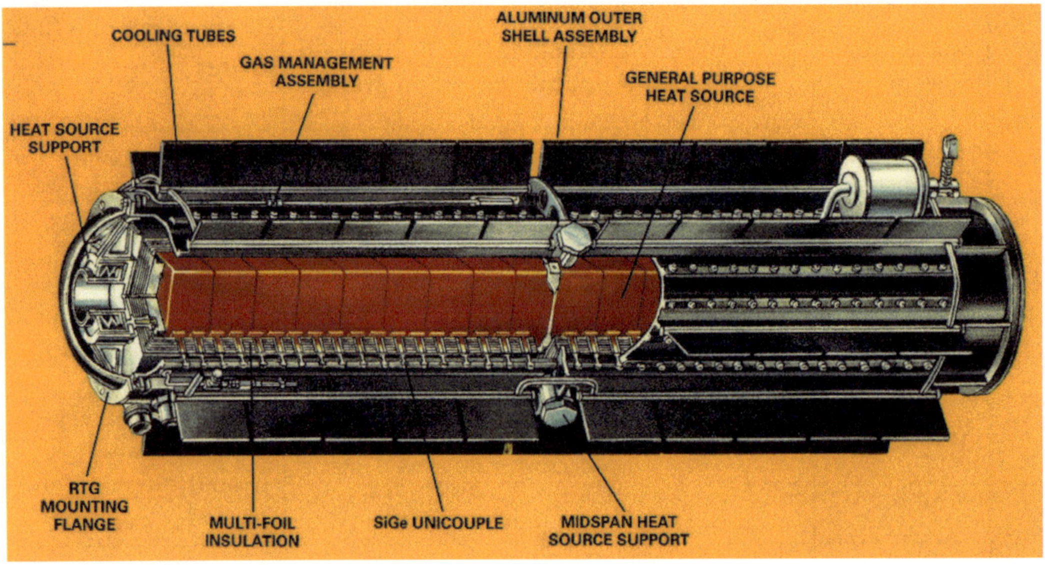

Fig. 2. GPHS-RTG.

2. RPS design

A typical RPS generator consists of two subsystems: a thermal source and an energy conversion system. The thermal source provides heat, which is produced by the decay process within the radioisotope fuel. This heat is partially transformed into electricity in the energy conversion system. Most of the remaining amount is rejected to space via radiators, although a small portion can be used to heat spacecraft components.

2.1 Thermal source

The performance characteristics of an attractive fuel include: a long half-life (i.e., the time it takes for one-half of the original amount of fuel to decay) compared to the operational mission lifetime; low radiation emissions; high specific power and energy; and a stable fuel form with a high enough melting point. The fuel must be producible in useful quantities and at a reasonable cost (compared to its benefits). It must be capable of being produced and used safely, including in the event of potential launch accidents.

Thermal source designs have been driven by aerospace nuclear safety standards, which have evolved considerably over time. For example, the fuel form for the early SNAP-3B and 9A systems was designed to burn up in the event of an atmospheric reentry, and disperse at high altitudes. Later systems, such as SNAP-19, were designed for fuel containment in the event of reentry. A key design feature now is to immobilize the Pu-238 fuel during all nominal and potentially abnormal phases of the mission, including launch abort, reentry into Earth's atmosphere, and post-reentry impact.

Establishing a fuel production and a fuel form fabrication capability is a very costly and time-consuming endeavor. Flight qualification of a new fuel form requires considerable effort in terms of costs and schedule. In addition, there are only a limited number of radioisotope fuels that meet the requirements for half-life, radiation, power density, fuel form, and availability for use in space power applications.

A variety of radioisotopes have been evaluated for space and terrestrial applications. The isotope initially selected for development was Cerium-144 (Ce-144), because it was one of the most plentiful fission products available from reprocessing defense reactor fuel at AEC's Hanford Site. Its short half-life (290 days) made Ce-144 compatible with the 6-month military reconnaissance satellite mission envisioned as the RPS application at that time. The cerium oxide fuel form and its heavy fuel capsule met all safety tests for intact containment of the fuel during potential launch abort fires, explosions, and terminal impacts. However, the high radiation field associated with the beta/gamma emission of Ce-144 complicated handling and caused problems with payload interaction, as well as safety issues upon reentry from orbit. Although Ce-144 was used to fuel SNAP-1, the first RTG, it was never used in space.

By the late 1950s, large amounts of Polonium-210 (Po-210) became available, also as a by-product of the nuclear weapons program. Po-210 is an alpha emitter with a very high power density (~1,320 W/cm3) and low radiation emissions. It is made by neutron irradiation of Bismuth-209 targets in a nuclear reactor. It was used in polonium-beryllium neutron sources. Po-210 metal was used to fuel the small (5 We) SNAP-3 RTG in order to demonstrate RTG technology. It was first displayed at the White House in January 1959. Several SNAP-3 RTGs were fueled with Po-210 and used in various exhibits. However, the short 138-day half-life of Po-210 makes it suitable for only limited duration space power applications.

In order to provide a longer-lived radioisotope fuel, Strontium-90 (Sr-90), an abundant fission product with a 28.6-year half-life, was recovered from defense wastes at Hanford. A very stable and insoluble fuel form, strontium-titanate, was developed and widely used in terrestrial power systems. Because Sr-90 and its daughter Yttrium-90 emit high-energy beta particles, they give off significant bremsstrahlung radiation and require heavy shielding. However, shield mass is not as critical for most terrestrial power systems as it is for space power applications.

By 1960, Plutonium-238 (Pu-238) had been identified as an attractive radioisotope fuel. It could be made by irradiating Neptunium-237 (Np-237) targets in defense production reactors. The availability of Pu-238 was extremely limited due to a shortage of Np-237 target material, which must be recovered from processing (and recycling) high burn-up, enriched uranium fuel. However, Pu-238 has all the desirable characteristics for a space power system fuel: long half-life (87.74 years), low radiation α-particle emissions, high power density and useful fuel forms (as the metal or the oxide form). Therefore, after flight qualification of its heat source, a Pu-238 fueled SNAP-3A RTG was launched on the Transit 4A Navy navigation satellite in June 1961 – the first use of nuclear power in space.

The first Pu-238 heat sources used in space were relatively small and employed Pu-238 metal or plutonium-zirconium alloy fuel forms contained in tantalum-lined superalloy (Haynes-25) fuel capsules. These heat sources withstood all postulated launch pad accident and downrange impact environments, but they were designed to burn-up and disperse throughout the upper atmosphere in the event of reentry from space. This type of accident happened during the fifth launch of an RTG (SNAP-9A aboard Transit 5BN-3) when the spacecraft failed to achieve orbit and the RTG burned up over the Indian Ocean in April 1964.

Subsequent Pu-238 fueled space power systems were designed to use progressively higher temperature fuel forms and containment materials with a progressively higher degree of containment of the fuel under all postulated accident conditions (including reentry). As the intact reentry heat source technology was developed, the fuel inventories (power levels) per launch also increased. A number of RTGs were launched on NASA and Navy missions with Pu-238 dioxide microsphere and plutonia-molybdenum-cermet (PMC) fuel forms in the late-1960s and early-1970s. Since the mid-1970s, pressed Pu-238 oxide fuel forms have been exclusively used in all RPS launched into space.

The amount of Pu-238 that could be produced has always been a limiting factor in its use in space missions. Therefore, several other radioisotopes have been thoroughly evaluated for space use over the years. Sr-90 and Po-210 fuels were considered for use in higher powered military satellite constellations for which there were insufficient quantities of Pu-238 available. These programs were cancelled before they were completed, so these fuels were never used in space by the U.S.

Curium-242 (Cm-242) was selected to fuel an isotope power system for the 90-day Surveyor mission to the Moon. Both the SNAP-11 RTG and SNAP-13 thermionic generators were developed for the Surveyor mission. Cm-242 is produced by reactor irradiation of Americium-241 (Am-241) targets. Cm-242 has a short half-life of 162 days, which is acceptable for a 90 day mission, and has a very high power density, which is necessary for a thermionic heat source. It also has a high melting point oxide fuel form capable of the high operating temperature necessary for thermionic energy conversion. A Cm-242 demonstration heat source was produced for the SNAP-13 engineering unit. However, it was decided that the Surveyor program would not use isotope power units, and Cm-242 fueled power systems have never been used in space. Due to its short half-life, Cm-242 is not suitable for the longer durations required by most space missions.

At one time, Curium-244 (Cm-244) was investigated as a potential alternative to Pu-238, because it was expected to become available in significant quantities from the U.S. program to develop breeder reactor fuel cycles. Cm-244 was considered an attractive space fuel because it has a relatively long half-life (18.2 years), a power density five times greater than that of Pu-238 and has a very stable, high temperature oxide fuel form. However, higher neutron and gamma emissions due to the higher rate of spontaneous fission of Cm-244 would increase shielding requirements for handling and for protection of spacecraft instrumentation. The increase weight of shielding and power flattening equipment required with Cm-244 makes it less desirable than Pu-238, especially for long duration missions. Cm-244 is also more difficult to produce, requiring successive neutron captures starting with Pu-239. Many years ago, several kilograms of Cm-244 were made as a target material for the Californium-252 program, but there is currently no practical production or processing capability for large quantities of Cm-244.

In the final analysis, Pu-238 is clearly superior to other radioisotope fuels for use in long duration space missions. The technology for producing and processing Pu-238 fuel forms has been refined over the past 50 years. Pu-238 fueled heat sources have been through rigorous flight qualification testing and have performed reliably in all of the RPS employed in the U.S. space program to date.

The most significant issue with Pu-238 is its limited availability. For the past 50 years the production and processing of Pu-238 fuel has been accomplished as a by-product of the production of materials for nuclear weapons. The discontinuation of this production in the 1990s eliminated the traditional means for producing Pu-238. During the 2000s, the U.S. began to purchase Pu-238 from Russia. However, this supply is also limited, so in the long-term, resumption of production is necessary.

2.2 Fuel encapsulation and containment

Encapsulation is an important aspect of the design of the thermal source and consists of several elements, each of which serves one or more important functions in the safe handling and use of the fuel. The state-of-the-art in fuel encapsulation and containment is the General Purpose Heat Source (GPHS) module shown in Fig. 3.

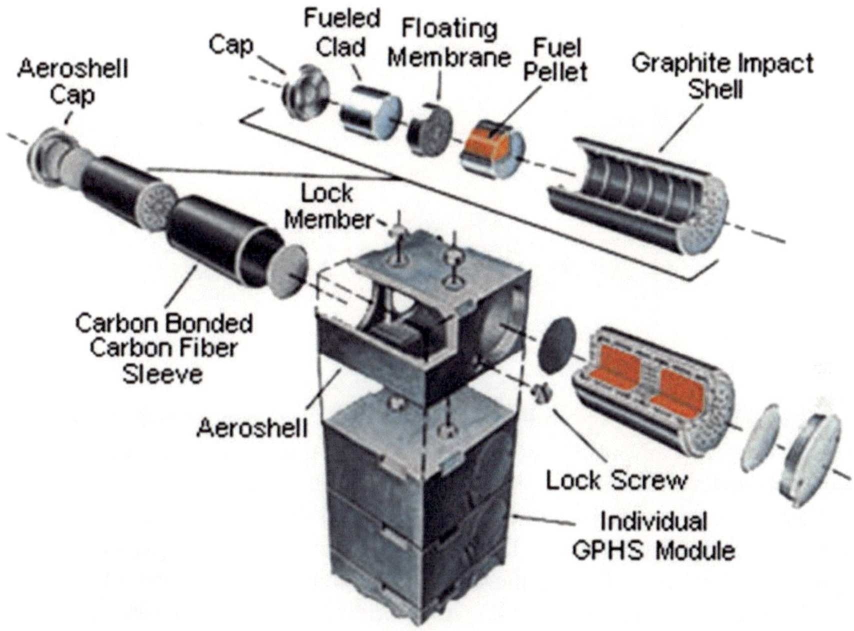

Fig. 3. GPHS module assembly.

The GPHS is modular in design, thus allowing it to be stacked into variable thermal source configurations. Eighteen of these modules are stacked together to serve as the thermal source for the GPHS-RTG, shown in Fig. 2. It is being used in an 8-module stack for the recently developed MMRTG, and two individual GPHS will be used as the heat source for the new ASRG, currently under development.

Safety is the principal design driver for the GPHS. The main objective is to keep the fuel contained or immobilized to prevent inhalation or ingestion by humans. Each module is composed of five main elements: the fuel; the fuel cladding; the graphite impact shell (GIS); the carbon-bonded carbon fiber (CBCF) insulation; and the Fine Weave Pierced Fabric (FWPFTM) aeroshell. Each GPHS module contains four fuel pellets made of a high-temperature PuO2 ceramic with a thermal inventory of approximately 62.5 Wt (Watts-thermal) per pellet and 250.0 Wt per module. Each module has a total mass of about 1.43 kg.

During its development program in the late 1970's, the GPHS went through a number of exacting engineering tests to assess its performance under operating conditions, including vibration and operating temperature. An extensive safety testing and analyses program was conducted to assess the GPHS performance under a range of postulated accident conditions such as launch pad explosions, projectile impacts, propellant fires, impacts, and atmospheric reentry.

The fuel pellets, one of which is shown in Fig. 4, are individually encapsulated in a welded iridium alloy clad.

Fig. 4. GPHS PuO2 Fuel Pellet.

The alloy is capable of resisting oxidation in a hypothetical post-impact environment while also being chemically compatible with the fuel and graphitic components during high-temperature operation and postulated accident environments.

Two fueled clads are encased in a cylindrical graphite impact shell (GIS) made of FWPFTM, a carbon-carbon composite material. The GIS is designed to provide protection to the fueled clads for postulated impact. Two of these GIS assemblies, each containing two fueled clads, are located in each FWPFTM aeroshell. A carbon-bonded carbon fiber (CBCF) insulator surrounds each GIS within the aeroshell to limit the peak temperature of the fueled clad during inadvertent reentry and to maintain a sufficiently high temperature to ensure its ductility upon the subsequently postulated impact.

The aeroshell serves as the primary structural member of the GPHS module as it is stacked inside the RPS unit. The aeroshell is designed to contain the two graphite impact shell assemblies under a wide range of postulated reentry conditions and to provide additional protection against postulated impacts on hard surfaces at terminal velocity. FWPFTM was selected because its composite structure gave it a high margin of safety against the thermal stresses associated with postulated atmospheric reentries. The aeroshell also provides protection for the fueled clads from postulated launch vehicle explosion overpressures and fragment impacts and it can provide protection in the event of a propellant fire.

2.3 Power conversion systems

A portion of the heat generated from the thermal source is converted to useful electrical energy in the power conversion system. There are two general classes of energy conversion systems: static and dynamic. Static systems include thermoelectric, thermionic, and thermophotovoltaic conversion devices which can convert heat to electricity directly with no moving parts. Dynamic systems involve heat engines with working fluids that transform heat to mechanical energy which in turn is used to generate electricity. Dynamic systems include Stirling, Brayton and Rankine cycle engines that operate with various types of working fluids.

After passing through the energy conversion system, the unconverted waste heat must be rejected to the environment at lower temperatures. For space power systems some of the waste heat can be utilized to control the temperature of the spacecraft equipment, but ultimately the waste heat must be radiated to the space vacuum environment.

Thus, the operating temperatures for an RPS are set on the hot side by the heat source and conversion system material limitations (Thot) and on the cold side by the size, weight, and heat sink conditions of the radiator (Tcold). The overall efficiency of the energy conversion system is limited to something less than the Carnot efficiency of (Thot - Tcold)/Thot. Higher efficiencies can significantly reduce fuel usage, which has many implications for cost, availability, size, weight, and safety.

Conversion system reliability is another important consideration. Since mission success depends on having sufficient electrical power over the life of the mission, conversion system selection must be consistent with mission power levels and lifetimes. For instance, it makes little sense to combine an unreliable or short-lived energy conversion unit with a 100% reliable, long-lived isotope heat source. Graceful power degradation over the life of a mission is acceptable as long as it is within predictable limits.

Other important considerations in selecting a system include mass, size, ruggedness to withstand shock and vibration loads, survivability in hostile particle and radiation environments, scalability in power levels, flexibility in integration with various types of spacecraft (and launch vehicles), and versatility to operate in the vacuum of deep space or on planetary surfaces with or without solar energy.

2.4 Thermoelectric energy conversion

All of the RPS units flown in space have utilized thermoelectric energy conversion. Thermoelectric converters are useful over a very wide range of power levels (from milliwatts to kilowatts) and their operating temperatures are ideally suited for radioisotope heat sources. Thermoelectric converters are reliable over operational lifetimes of several decades, compact, rugged, radiation resistant, easily adapted to a wide range of applications, and produce no noise, vibration or torque during operation. Thermoelectric converters require no start-up devices to operate, and begin producing electrical power (direct current and voltage) as soon as the heat source is installed. Power output is easily regulated at design level by maintaining a matched resistive load on the converter. The only disadvantage of thermoelectrics is their relatively low conversion efficiencies , which is typically less than 10%.

Thermoelectric materials, when operating over a temperature gradient, produce a voltage due to the Seebeck effect. When connected in series with a load, the internally generated voltage causes a current to flow through the load producing useful power. The Seebeck effect was discovered in 1825, but had little practical use, except in measuring temperatures with dissimilar metal thermocouples. With the advent of semiconductor materials in the 1950s, application of thermoelectrics has expanded dramatically.

Power is produced in a thermoelectric element by placing it between a heat source and a heat sink. Good thermoelectric semiconductor materials have large Seebeck voltages in combination with a relatively high electrical conductivity and low thermal conductivity (in contrast to most metals). By proper doping, n and p type elements can be formed so that current will flow in the same or opposite directions as the heat. By electrically joining the n and p elements through a hot shoe, a thermocouple is formed which can be connected to other thermocouples at the cold shoe to form a converter with the desired output voltage and current. Thermocouples can be connected in a series-parallel arrangement to enhance reliability by minimizing the effect on total power due to an open circuit or short circuit failure in a single thermocouple. Typically, thermoelectric couples are low voltage, high current devices so a number of them must be connected in series to produce normal load voltages.

The most widely used thermoelectric materials in order of increasing temperature capability, are: Bismuth Telluride (BiTe); Lead Telluride (PbTe); Tellurides of Antimony, Germanium and Silver (TAGS); Lead Tin Telluride (PbSnTe); and Silicon Germanium (SiGe). All except BiTe have been used in space RTG applications. Many more materials have been, and are still being, investigated in hopes of finding that ideal thermoelectric material from which to produce higher efficiency, lower mass, and more stable performance over longer operating lifetimes.

The telluride materials are limited to a maximum hot junction temperature of 550 C. Due to the deleterious effects of oxygen on these materials and their high vapor pressure, the tellurides must be operated in a sealed generator with an inert cover gas to retard sublimation and vapor phase transport within the converter. Bulk-type, fibrous thermal insulation must be used due to the presence of the cover gas. Buildup of helium gas from α-particle emission must be controlled by using a separate container around the heat source or permeable seals in the generator design. Gas management considerations in the generator housing design and the use of bulk insulation materials increase the size and weight of the generator. However, this type of RTG is equally useful for space vacuum or for planetary atmospheric applications.

SiGe materials can be operated at hot junction temperatures up to 1,000 C. Their sublimation rates and oxidation effects, even at these higher temperatures, can be controlled by use of sublimation barriers around the elements and an inert cover gas within the generator during ground operation. A pressure release device, designed to open upon reaching orbitral altitude, opens the generator to space vacuum for operation on deep space missions. This allows use of multifoil thermal insulation and also vents the helium to space as it is generated. A SiGe RTG is usually smaller and lighter than is a telluride RTG of similar power level.

The overall efficiency of the two types of thermoelectric generators are comparable. Although the tellurides have a higher material efficiency than SiGe, the SiGe operates over a larger temperature gradient. Cold junction temperatures are determined more by radiator weight than by efficiency considerations for space RTGs and are normally in the range of 200-300 C. Although various convectively cooled radiator systems have been developed (e.g., heat pipes), conductively coupled finned radiators attached to the generator housing are normally more weight efficient for low-powered RTGs of up to 300 We.

2.5 Stirling energy conversion

For higher power levels of 100 We and above, the more efficient dynamic power conversion technologies enable better use of the limited radioisotope fuel, offer systems with a higher power-to-weight ratio and make it easier to integrate the radioisotope power system with the spacecraft compared to the number of RTGs required to produce kilowatts of power. Dynamic heat-to-electricity conversion efficiencies of 25% and more are achievable, which reduce the radioisotope inventory by at least one-quarter of that for RTGs. This reduces mass, cost, and potential safety risks for higher-powered radioisotope systems.

Stirling cycle engines use a light working gas that expands by absorption of heat on the hot side and contracts by rejection of heat on the cold side causing rapidly changing pressure cycles across a piston forcing it to move in a reciprocating fashion. The movement of the piston can drive a linear alternator to produce electricity.

Traditional Stirling engines use a rhombic drive mechanism to convert the reciprocating motion into a rotary motion that drives an ordinary rotating alternator. This requires lubrication of a gear-box and seals to separate the working gas from the lubricating oil. The engine housing cannot be hermetically sealed because of the penetration of the rotary power shaft. Such Stirling engines have been widely used throughout the world.

A more recent development is the Free Piston Stirling engine which requires no lubricating fluids and produces electricity by means of a linear alternator within the hermetically sealed engine housing. The piston moves back and forth at a resonant frequency on a cushion of working gas between it and the surrounding cylinder wall. Piston displacement is controlled by gas pressure across the piston. A permanent magnet is attached to the power piston and produces electrical currents in surrounding alternator coils as it vibrates back and forth. Since the reciprocating motion of the piston would cause unbalanced vibration loads, these Stirling engines usually are designed in pairs with dynamically opposed pistons so that no net load is transmitted to the engine mounts.

Heat is also exchanged between the hot and cold gas flowing from one side of the piston to the other to enhance the conversion efficiency. Due to the limited volume of working gas within the Stirling engine, heat transfer between the heat source and the heater head of the engine, between the hot and cold gas, and between the cold gas and a radiator system are the most challenging requirements for an optimum engine design. The Stirling cycle provides the highest conversion efficiencies of any dynamic cycles at the same cycle temperatures. Therefore, efficiencies of 30% or more are possible at operating temperatures achievable with isotope heat sources and oxidation-resistant superalloy structural materials. The Stirling engine also promises to retain its high performance characteristics at lower power levels compared to the other dynamic systems, which is also attractive for radioisotope power systems.

2.6 Other energy conversion technologies

Research and development of other energy conversion technologies has been an important aspect of RPS programs in the past. Although thermoelectrics and Stirling have received the most attention, there are several other technologies that could achieve higher heat-to-electric conversion efficiencies and considerably lower masses than the systems in use today.

One of these is Thermophotovoltaics (TPV), which is another static form of electrical power conversion. A thermophotovoltaic (TPV) converter transforms the energy from infrared photons emitted by a hot surface into electricity using photovoltaic (PV) cells. TPV converters use advanced PV cells, spectrally-tuned to optimize conversion of the emitted photon energy. Controlling the frequency of photon energy impinging on the PV cells by means of selective emitters, PV cell materials and filter properties are key to achieving high performance. Studies in the past have suggested the possibility of achieving efficiencies of up to 20% with TPV.

On the dynamic side, the Brayton is another thermodynamic cycle consisting of a turbine/alternator, compressor and heat exchangers. An additional recuperator heat exchanger is often used to transfer heat within the cycle and improve cycle efficiency. An inert gas working fluid, typically a mixture of helium and xenon, is sequentially heated in one heat exchanger, expanded through the turbine, passed through a gas cooler and pressurized by the compressor thus completing the cycle. A rotary alternator attached to the turbine shaft produces alternating current (AC) electrical power.

3. The early years

The history of RPS began in the early years of the Cold War, when surveillance satellites were a major impetus for the early space race. The Manhattan project and the years leading up to it had yielded a wealth of knowledge on nuclear physics, particularly the radio-decay properties of actinides and other alpha particle-producing materials. The energy released from the radioactive decay of different elements had become well characterized, and it was recognized early on that radioisotopes could provide power for military satellites and other remote applications. An early study by the North American Aviation Corporation had considered radioisotopes for space power. Then a RAND Corporation report in 1949 evaluated options for space power, and concluded that a radioactive cell-mercury vapor system could feasibly supply 500 We (watts-electric) for up to one year. In 1952, RAND issued a report with an extensive discussion on radioisotope power for space applications, which spurred interest in applying the technology on satellites.

Recognizing the viability of nuclear power for reconnaissance satellites, the Department of Defense (DOD) requested in August 1955 that the Atomic Energy Commission (AEC) perform studies and limited experimental work toward developing a nuclear reactor auxiliary power unit for an Air Force satellite system concept. AEC agreed, but wanted to broaden its examination to both radioisotope and reactor heat sources. This marked the beginning of the SNAP program, which was structured into parallel power plant efforts with two corporations. Odd-numbered SNAP projects focused on RPS and were spearheaded by the Martin Company, while even-numbered SNAP projects using reactors were performed by the Atomics International Division of North American Aviation, Inc.

In these early days, efforts focused on dynamic energy conversion. The work of the Martin Company progressed through an early SNAP-1 effort that used the decay heat of Cerium-144 to boil Mercury and drive a small turbine in a Rankine cycle. In early 1954, a new simpler static energy conversion method was conceived by Kenneth Jordan and John Birden of the AEC's Mound Laboratory in Miamisburg, Ohio. Having been frustrated in their efforts to use radioisotope heat sources to generate electricity via steam turbines, these two researchers considered using two metals with markedly different electrical conductivities to generate electricity directly from an applied heat load. This thermoelectric method was patented by Jordan and Birden, and has remained the basis for all RTGs to the present day.

In 1958, work began on two thermoelectric demonstration devices at Westinghouse Electric and 3M, while AEC contracts with other companies explored the development of demonstration thermionic units.

The project to develop a generator based on thermoelectric energy conversion was given the designation, SNAP-3. The 3M Company delivered a workable converter to the Martin Company in December 1958. Shortly thereafter, a complete radioisotosope-powered generator was delivered to the AEC as a proof-of-principle device, producing 2.5 We with a half charge of Polonium-210 (Po-210) fuel.

That SNAP-3 actually never flew in space, but it became an invaluable showpiece for RPS and the SNAP program. President Eisenhower, who had been keenly interested in developing nuclear power for U.S. surveillance satellites, was shown this breakthrough device in January 1959, when the SNAP-3 was displayed on his desk in the Oval Office (Fig. 5). Eisenhower used the opportunity to emphasize his view of "peaceful uses" of nuclear technology, and it afforded him an opportunity to issue a challenge to NASA to develop missions that could exploit the device's potential. The SNAP-3 continued its marketing role, and was shown at several foreign capitals as part of the U.S.'s "Atoms for Peace" exhibits.

Fig. 5. SNAP-3 presentation to President Eisenhower.

4. Flight systems

4.1 SNAP-3B

The first successful use of RTGs in space took place with the U.S. Navy's Transit satellite program. Also known as the NNS (Navy Navigation Satellite), the Transit system was used by the Navy to provide accurate location information to its ships. It was also used for general navigation by the Navy, as well as hydrographic and geodetic surveying, and was the first such system to be used operationally. The Johns Hopkins Applied Physics Laboratory (APL) developed the system, starting in 1957. Many of the technologies developed under the Transit program are now in use on the Global Positioning System (GPS).

Several of the Transit developers had been considering the use of RPS since the beginning of the program. Although solar cells and batteries had powered the first six Transit satellites, there was concern that the battery hermetic seals would not meet the five-year mission requirement. Thus, APL accepted an offer from the AEC to include an auxiliary nuclear power source on the satellite. At that time, however, the radioisotope fuel of choice, Plutonium-238 (Pu-238), was unavailable due to AEC restrictions, and APL refused to use beta-decaying Strontium-90 because of the excessive weight associated with its necessary shielding. The AEC eventually acquiesced and agreed to provide the Pu-238 fuel. The SNAP-3 was converted from use of Po-210 to Pu-238, and acquired the new designation, SNAP-3B. The SNAP-3B RTGs on board these spacecraft supplemented solar cell arrays and demonstrated operation of nuclear systems for space power applications.

A schematic of the SNAP-3B generator is shown in Fig. 6. Each unit had a mass of 2.1 kg and an initial power output of 2.7 We, and was designed to last five years. Although this power level was quite low, the RTG performed the critical function of powering the crystal oscillator that was the heart of the electronic system used for Doppler-shift tracking. It also powered the buffer-divider-multiplier, phase modulators and power amplifiers. The heat source produced approximately 52.5 Wt from 92.7 grams of encapsulated plutonium metal, which had an isotopic mass composition of 80% Pu-238, 16% Pu-239, 3% Pu-240, and 1% Pu-241. The power conversion assembly consisted of 27 spring-loaded, series-connected pairs of Lead-Telluride (Pb-Te) thermoelectric elements operating at a hot-juncture temperature of about 783 K and a cold-juncture temperature of about 366 K. The power system had a power-conversion efficiency of 5 to 6 percent and a specific power of 1.3 We/kg.

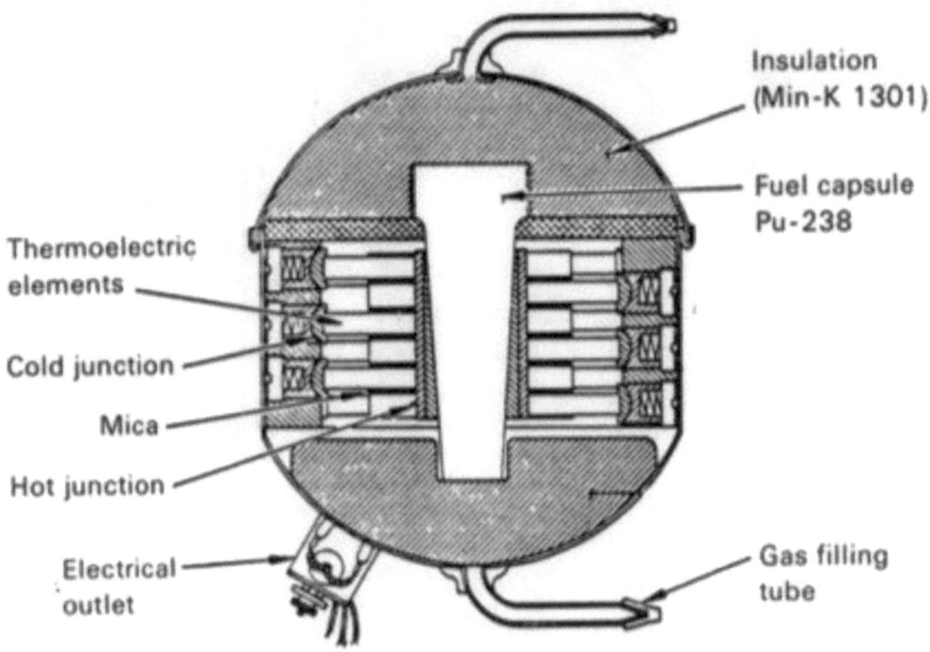

Fig. 6. SNAP-3B Schematic.

Transit 4A was launched, along with two other satellites (Fig. 7), on June 29, 1961 aboard a Thor-Able rocket. Transit 4B was launched soon afterward on November 15, 1961. Even for this first use of nuclear power in space, there was controversy stemming from concerns over launch safety. The State Department, in particular, expressed concern with its trajectory over Cuba and South America. As part of the aerospace nuclear safety philosophy at that time, the generators were designed for burnup and high altitude fuel dispersal to concentrations below the background radiation attributed to atmospheric nuclear weapons testing. In addition, the spacecraft were placed into 1,100-km orbits, which provided orbital lifetimes (>1,000 years) sufficient for the fuel to decay to these background levels. The Transit 4A generator operated for 15 years, and was shutdown in 1976. The last reported signal from Transit 4B was in April 1971.

Fig. 7. Integrated Transit payload. Transit satellite is positioned at bottom of stack.

4.2 SNAP-9A

After the success of SNAP-3B, the team consisting of the AEC, Martin, 3M, Mound Laboratory and APL proceeded to develop the SNAP-9A for the next series of Transit satellites. There was also a growing demand for isotope power for terrestrial applications. For instance, the SNAP-7 series of devices was under development for the Navy, Coast Guard, and Weather Bureau for navigation lights and weather stations on Earth.

DOD decided to continue using RTGs for its navigational satellites because of their resistance to radiation. A high-altitude nuclear explosive test in 1962 had adversely impacted the solar cells of earlier Transit satellites, and DOD was concerned with their susceptibility to radiation and other space effects in the future. The SNAP-9A was essentially an expanded version of the SNAP-3B, and was the first RTG employed as the primary spacecraft power source. Its power capability of 26.8 We at beginning of mission (BOM) was nearly an order of magnitude greater than the SNAP-3B.

Each 12.3-kg SNAP-9A was designed to provide continuous power for five years in space after one year of storage on Earth. The thermal inventory of 525 Wt (watts-thermal) was supplied by Pu-238 metal encapsulated in a heat source of six fuel capsules maintained in a segmented graphite heat-accumulator block. As shown in Fig. 8, the main body was a sealed cylindrical magnesium-thorium shell containing six heat-dissipating magnesium fins. The unit was 26.7 cm tall and had a fin-to-fin diameter (fin span) of 50.8 cm. The 70 pairs of series-connected Pb-Te thermoelectric couples were assembled in 35 modules of two couples each. Hot junction temperature was calculated at about 790 K at beginning of life. Some waste heat from the RTG was used to maintain electronic instruments in the satellite at a temperature near 293 K.

The SNAP-9A missions in 1963 also marked the beginning of a formal launch safety review process. Although the launches were for DOD systems, NASA was invited to participate in the reviews, which were made a responsibility of the joint AEC/NASA Space Nuclear Power Office. It was during these early launches that efficient and comprehensive review and approval procedures were developed. As early as January 1963, a model charter had been developed for an ad-hoc interagency review committee. Eventually this became known as the INSRP (Interagency Nuclear Safety Review Panel).

Fig. 8. SNAP-9A RTG.

After a period of program delays, Transit 5BN-1 (Fig. 9) was launched successfully on September 28, 1963, followed by Transit 5BN-2 on December 5, 1963. The third and last launch of the Transit 5BN-3 on April 21, 1964 was not as successful. A mission abort occurred after the payload had reached an altitude of 1,000 miles over the South Pole. Preliminary data indicated that the payload reentered the atmosphere over the Mozambique Channel at a steep angle. The Pu-238 fuel was designed to burn up into particles of about one millionth of an inch in diameter and disperse widely so as not to constitute a health hazard. Balloon samples taken over the next few years confirmed that the generator's fuel had indeed burned up as expected after the spacecraft failed to achieve orbit.

Fig. 9. Transit 5BN-1.

Although there was a commitment to fly higher power NASA missions, the loss of Transit 5BN-3 led to concerns that the dispersion approach would be unsafe with larger inventories of fuel. Thus, the basic safety concept changed from designing for burn-up and dispersion to designing for intact reentry. By the time that new approach was integrated into an RTG-powered space mission, however, the mechanisms for interagency review and meticulous safety analysis were well established. Another change was the mobilization and decentralization of technical and administrative support so as to directly involve more of the laboratories and facilities of both AEC and NASA.

4.3 SNAP-19 – Nimbus

Noting the success of the SNAP-3A, NASA requested the AEC to evaluate the feasibility of a 50-We RTG for an upcoming Nimbus weather satellite. Nimbus was the first U.S. weather satellite system to make day and night global temperature measurements at varying levels in the atmosphere, and all earlier satellites had been powered exclusively by solar cells. The request led to design and integration studies by the AEC and establishment of the SNAP-19 technology improvement program. With Nimbus, the SNAP program received its first opportunity to test and demonstrate an RTG on a NASA spacecraft.

The unit that eventually flew on Nimbus, SNAP-19B, was used as an auxiliary system. As shown in Fig. 10, each Nimbus satellite carried two SNAP-19B RTGs, which provided about 20% of the total power delivered to the spacecraft bus. This extra continuous power enabled full-time operation of a number of extremely important atmospheric-sounder experiments. Without the RTGs, the total delivered power would have fallen below the load line about two weeks into the mission.

Fig. 10. Nimbus III. First NASA application of Radioisotope power.

SNAP-19B was very similar to the SNAP-9A in terms of configuration and performance. It had a height of 26.7 cm and a fin span of 53.8 cm. It's mass of 13.4-kg and BOM power level of 23.5 We yielded a specific power of 2.1 We/kg, the same as SNAP-9A.

The SNAP-19B was unique in its use of a new 645 Wt heat source, called the Intact Impact Heat Source (IIHS), in conjunction with an array of 90 Pb-Te thermocouples. The IIHS was designed to contain the fuel under normal operating conditions and to limit probability of contaminating the environment in the event of a launch abort or accident. In contrast to the SNAP-9A fuel design, the fuel form for SNAP-19B was changed from Plutonium metal to small Plutonium oxide (PuO2) microspheres carried in capsules. Even in a worst-case scenario involving release and dispersal of the microspheres, the particles would be too big for inhalation. Additional safety design requirements included survival upon reentry and containment/immobilization of the fuel upon impact.

Launch of the Nimbus-B-1 took place on May 18, 1968. Unfortunately an error in setting a guidance gyro caused Nimbus-B-1 to veer off course. The Range Safety Officer sent the destruct signal 120 seconds into flight, thus blowing up the Agena stage at an altitude of 100,000 feet. The upper portion of the stage, including the satellite, fell into water depths of 300 to 600 feet about two to four miles to the north of San Miguel Island in the Santa Barbara Channel. The unit was found in September 1968, and was sent back to the Mound Laboratory for reuse. A second Nimbus satellite (Nimbus III or Nimbus-B-2) was launched and successfully placed into orbit on April 14, 1969. The SNAP-19B RTGs used here had slightly more fuel than their predecessors due to the use of less efficient but more stable thermoelectrics. The units operated fine for approximately 20,000 hours (2.5 years) until they experienced a sharp degradation in performance. This decline was attributed to the sublimation of thermoelectric material and loss of the hot junction bond due to internal cover gas depletion.

Nimbus was the first and last time RTGs were used in Earth orbit by NASA. At that time, solar photovoltaics were still relatively new. With advancement in this area, NASA did not feel that RTGs were warranted for applications where solar cells could work. In addition with the more structured launch safety review process, it was much more cost effective to use solar cells whenever possible.

4.4 SNAP-19 – Pioneer and Viking

The successful demonstration of Nimbus III encouraged NASA to commit to use of SNAP-19 on the Pioneer and Viking missions, arguably NASA's most exciting science missions of the 1970's. The SNAP-19 design for these applications (Fig. 11), however, had to be modified. For Pioneer, this was driven by the need for a mission life of up to six years. Other modifications were required to deliver a higher power, and to withstand the unique environments of Mars and deep space. For Pioneer, the most significant modification was incorporation of TAGS/Sn-Te thermoelectric elements (thermocouple legs consisting of Tellerium, Antimony, Germanium, Silver and Tin), which increased efficiency, lifetime and power performance. The generator height was also increased to 28.2 cm, and the fin span was reduced to 50.8 cm. This yielded a power output of 40.3 We. The resultant specific power of 3.0 We/kg was nearly 50% higher than the Nimbus design.

Pioneers 10 and 11 were launched on 2 March 1972 and 6 April 1973, respectively. Pioneer 10 was the first spacecraft to travel through the asteroid belt and to make direct observations of Jupiter, which it encountered on 3 December 1973. According to some definitions, Pioneer 10 became the first artificial object to leave the solar system, on 13 June 1983. Pioneer 11 also encountered Jupiter, and in addition to conducting measurements, the spacecraft used a Jupiter gravity assist maneuver to alter its trajectory toward Saturn. After nearly five years, Pioneer 11 encountered Saturn in September 1979, and provided the first local measurements of this planet and its rings before it followed an escape trajectory out of the solar system.

The most noteworthy aspect of the SNAP-19s used for these missions (Fig. 12) was the extremely long time the units continued to operate past their primary tasks and baseline mission lifetimes. Both of these spacecraft continued to transmit data far beyond the orbit of Pluto, and more than fulfilled the original expectations for their operation.

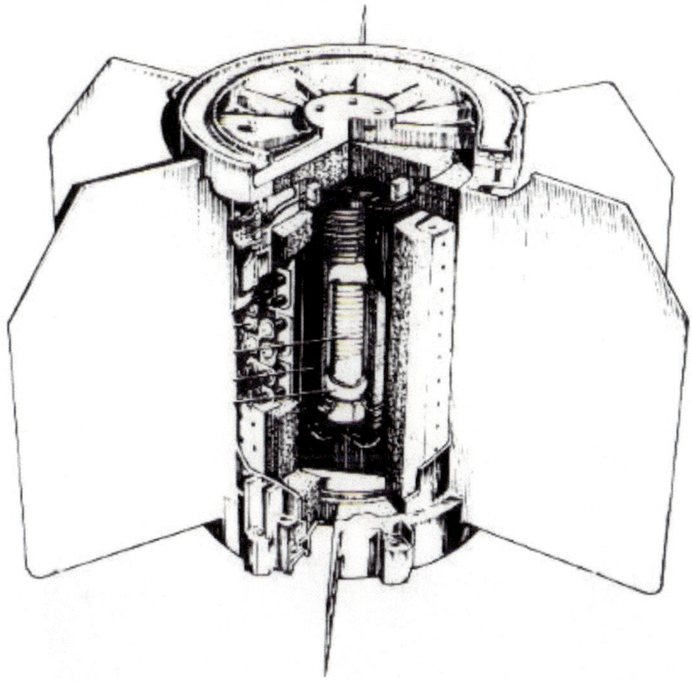

Fig. 11. Pioneer SNAP-19.

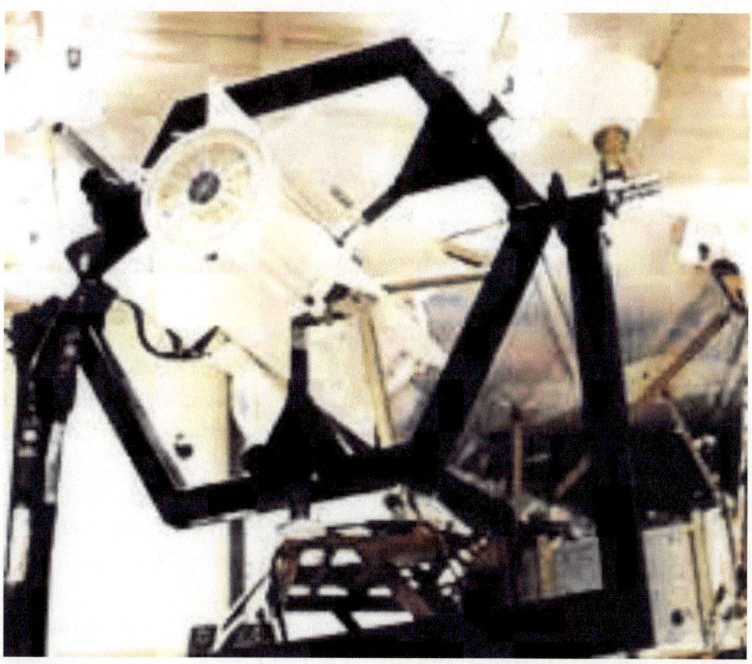

Fig. 12. SNAP-19s installed on Pioneer.

The modifications for Viking went further to ensure the RTG, which is shown in Fig. 13, could withstand high temperature sterilization procedures in support of the planetary quarantine protocol, storage during the flight to Mars, and the severe temperature extremes of the Martian surface.

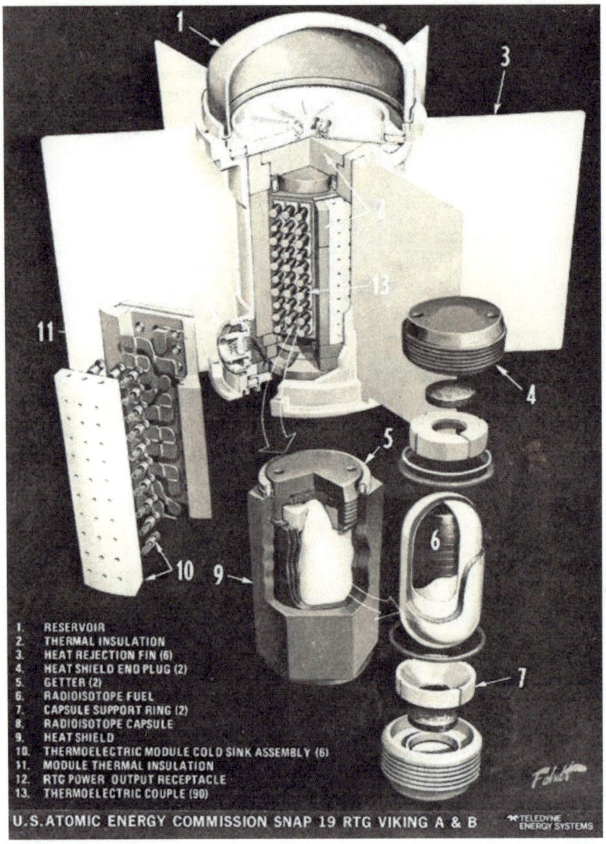

Fig. 13. Viking SNAP-19.

The landers were sterilized before launch to prevent contamination of Mars by terrestrial microorganisms. Among the modifications to the Pioneer SNAP-19 design was the addition of a dome reservoir to allow a controlled interchange of gases. This minimized heat source operating temperatures prior to launch, while maximizing electrical power output at the end of mission. This resulted in the Viking SNAP-19 being slightly larger and more massive than the version used on Pioneer (40.4 cm tall, 58.7 cm fin span, 15.2 kg mass, and 2.8 We/kg specific power).

Vikings 1 and 2 were identical spacecraft (Fig. 14), each of which consisted of a Lander, with a robot laboratory to study the nature of the surface, and an Orbiter, designed to serve as a communications relay to Earth. Each Lander carried two SNAP-19s. Viking 1 was launched on 20 August 1975 from Cape Canaveral. It reached Mars orbit on 19 June 1976, and reached the surface on 20 July 1976 on the western slope of Chryse Planitia. Viking 2 was launched on 9 September 1975, and it touched down on the surface on 3 September 1976 at Utopia Planitia.

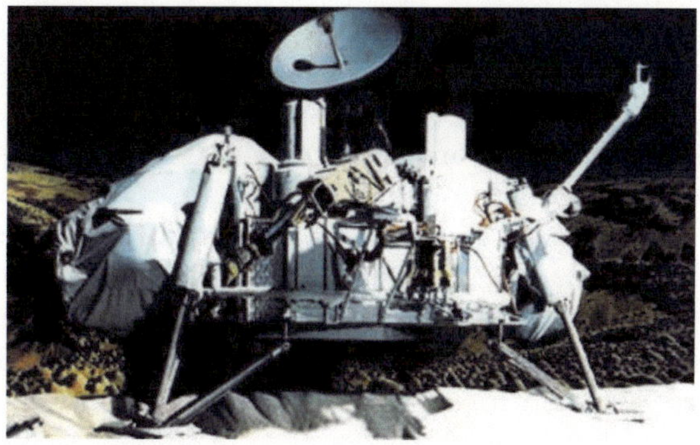

Fig. 14. Viking Lander.

The Viking missions were a complete success. In addition to characterization of the Mars environment, the Landers provided over 4,500 high quality images of the Martian landscape. All four SNAP-19 RTGs easily met their original 90-day requirement, thus allowing the Viking Landers to operate for years until other system failures led to a loss of data. When the last data were received from Viking 1 in November 1982, it had been estimated that the RTGs were capable of providing sufficient power for operation until 1994, 18 years beyond the original mission requirement.

4.5 Transit-RTG (TRIAD)

Interest in RTGs for Navy navigation satellites continued after the earlier Transit missions. The next DOD application of RTGs took place with TRIAD, the first in a series of three experimental spacecraft designed to test and demonstrate improvements to the NNS. These were all developed under the Transit Improvement Program (TIP), which was established in 1969 to provide a radiation-hardened satellite that could maintain its correct position for over five days without an update from the ground.

The Transit-RTG was designed to serve as the primary power source for the satellite, with auxilliary power provided by four solar-cell panels and a 6 Amp-hr Nickel Cadmium battery. The 13.6-kg Transit RTG was modular in design, and was 36.3 cm tall and approximately 61 cm across its lower attachment (Fig. 15). The RTG delivered 35.6 We at BOM, and used a SNAP-19 heat source. The Transit RTG was the first to employ radiative heat coupling between its heat source and thermocouples, although this was accomplished at some loss in efficiency.

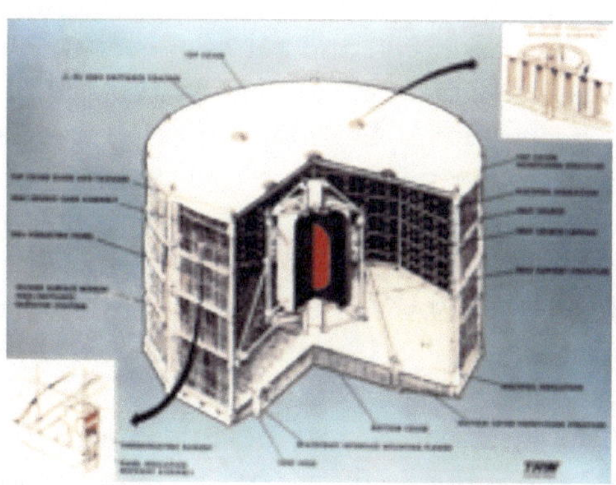

Fig. 15. Cutaway of TRANSIT RTG.

The 12-sided converter used Pb-Te thermoelectric "Isotec" panels operated at a low hot-side temperature of 673 K in a vacuum, thus eliminating the need for hermetic sealing and a cover gas to inhibit thermoelectric material sublimation. Each of the 12 Isotec panels contained 36 Pb-Te thermocouples arranged in a series-parallel matrix with four couples in a row in webbed, magnesium-thorium corner posts with Teflon insulators.

The TRIAD satellite (Fig. 16) was launched on September 2, 1972 from Vandenburg Air Force Base into a 700 to 800 km orbit. The short-term objectives of the TRIAD satellite were successfully demonstrated, including a checkout of RTG performance. However, a telemetry-converter failure onboard the spacecraft caused a loss of telemetry data about a month into the mission. This, in turn, precluded measuring the Transit-RTG power level versus time. However, the TRIAD satellite continued to operate normally for some time and provided magnetometer data using power from the RTG.

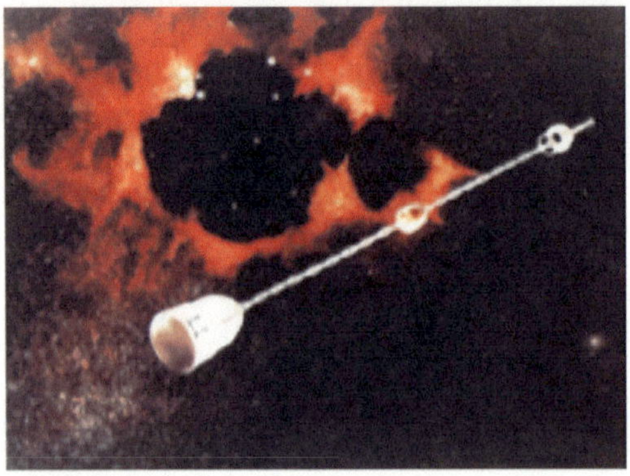

Fig. 16. Transit TRIAD Satellite.

4.6 SNAP-27

During the 1960's, scientists involved with the Apollo program envisioned placing scientific stations on the lunar surface that could transmit data long after the astronauts returned to Earth. They were interested in many measurements, including fluctuations in solar and terrestrial magnetic fields, changes in the low concentrations of gas in the lunar atmosphere, and internal structure and composition of the Moon. These ideas culminated in the Apollo Lunar Surface Experiment Package (ALSEP), led by Bendix Aerospace Systems Division. The requirement for multi-year operation and survival over many 14-day lunar day/night cycles favored use of RPS as the primary power source for ALSEP. Although NASA looked at using the new SNAP-19 for this application, ALSEP power requirements would have necessitated multiple SNAP-19s per mission and considerable effort in deployment by the Apollo crew. Instead, the AEC was requested to develop a new RTG, called the SNAP-27 (Fig. 17).

Special features were added to the SNAP-27 to ensure safety and facilitate its deployment by the astronauts on the lunar surface (Fig. 18). Chief of these was the separate storage of the heat source in a graphite lunar module fuel cask (GLFC) carried on the Lunar Excursion Module (LEM). The GLFC enclosed the fuel module during the trip to the Moon, and provided thermal and blast protection in the event of a launch pad explosion, launch abort, or reentry into the Earth's atmosphere and ground impact.

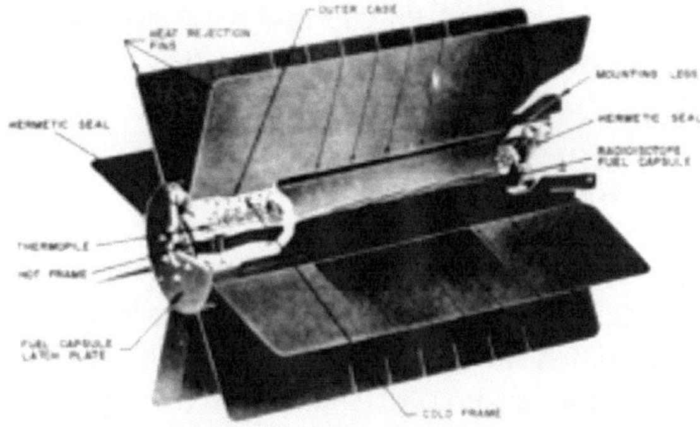

Fig. 17. SNAP-27.

Fig. 18. Use of SNAP-27 on the Moon. Alan Bean deploying SNAP-27 on Apollo 12.

Thermal energy from the fuel capsule was transferred to the generator hot frame by radiative coupling. When deployed on the lunar surface, the fuel capsule operated at 1005 K, while the Inconel 102 alloy hot frame was 880 K. The hot junction temperature ranged between 855 K and 865 K, reflecting an overall temperature drop of 15 to 25 K. On the Moon's surface, where temperatures can vary from 350 K during the lunar day to a frigid 100 K during the lunar night, the generator's cold side temperature operated at 545 K. Pb-Te served as the TE material and the couples were assembled in a series-parallel electrical arrangement to prevent string loss. The power capability for the 19.6 kg RTG was at least 63.5 We at 16 Vdc for one year after lunar emplacement. The converter was 46 cm tall and 40 cm wide across the fins. The specific power was greater than 3.2 We/kg, which represented a 10% increase over the Pioneer SNAP-19.

The five units deployed on the lunar surface from 1969 to 1972 operated flawlessly. Telemetry data from their operation stopped in 1977 when the ALSEPs were intentionally shutdown. Until then, their degradation in performance matched all predictions.

The only potential problem with SNAP-27 occurred with the Apollo-13 mission, when there was concern over the SNAP-27 onboard the LEM reentering the Earth's atmosphere. Normal reentry trajectory and velocity were achieved as had been assumed in the pre-launch review accounting for this type of event. The detached LEM broke up on reentry, as anticipated, while the graphite-encased Pu-238 fuel cask survived the breakup and went down intact in the 20,000 foot deep Tonga Trench, as had been projected for an aborted mission in a lifeboat mode situation.

4.7 Multihundred Watt (MHW) RTG

In anticipation that NASA would require higher power RTGs for increasingly ambitious robotic science missions in the future, the AEC contracted with GE to conduct a technology readiness effort for an RTG with a power capability in the range of several hundred We. Development of this unit, which later became known as the MHW-RTG, was initiated in anticipation that NASA would conduct a Grand Tour mission of the planets. This was realized with the Voyager missions launched in 1977. At the same time, the DOD also had a requirement for a hundred watt-class RTG, and requested the AEC to develop such a unit for two communication satellite technology demonstrators built by MIT's Lincoln Laboratory. These Lincoln Experimental Satellites (LES) 8 and 9 were launched together in 1976.

The MHW-RTG represented a dramatic advancement in RTG technology with its use of Silicon-Germanium (Si-Ge) thermoelectric materials and a much higher temperature heat source. The higher hot-side temperature translated to greater power conversion efficiency, and, most importantly, enabled radiation of waste heat at higher temperatures. This allowed a substantial reduction in radiator size and a significant increase in specific power over its Pb-Te/TAGS predecessors. Thermocouples made of Si-Ge can operate over a broad temperature range, up to 1,000 C, much higher than telluride-based thermocouples. Plus with a Silicon Nitride coating, Si-Ge does not sublimate significantly, and allows operation without a cover gas in the vacuum of space.

The MHW-RTG had a length of 58.3 cm and fin span of 39.7 cm (Fig. 19). The converter housing consisted of a beryllium outer shell and pressure domes, with unicouples attached directly to the outer shell. Like SNAP-19, the heat source was designed to immobilize and contain the fuel in the event of a launch abort. It was shaped as a right circular cylinder, and contained twenty-four 3.7-cm diameter fuel containers of PuO_2 (Fig. 20). Each fuel container produced 100 Wt, and had a metallic iridium shell containing the PuO_2 fuel and a graphite impact shell, which provided the primary resistance to mechanical impact loads.

Fig. 19. MHW-RTG. Cutaway view on left. Installation in test fixture on right.

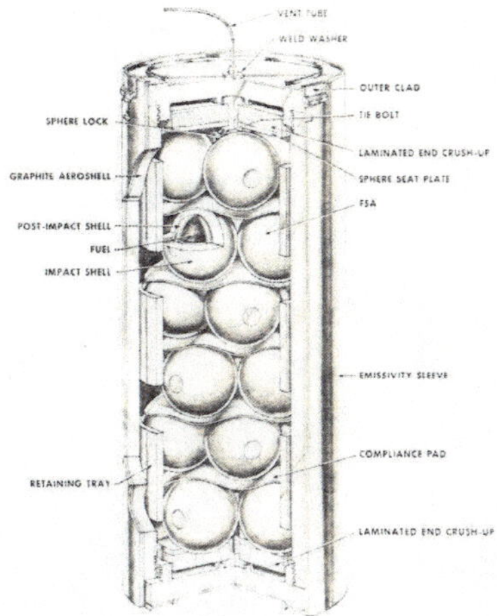

Fig. 20. MWH-RTG heat source.

The power converter contained 312 Si-Ge unicouples arranged in 24 circumferential rows with each row containing 13 couples. The MHW-RTGs flown on LES 8 and 9 had an average mass of 39.7 kg, BOM power of 154 We, and specific power of 3.9 We/kg. The six RTGs for Voyager were modified to yield a higher specific power of 4.2 We/kg, based on an average mass of 37.7 kg and BOM power of 158 We.

LES 8 and 9 were launched together aboard a Titan IIIC launch vehicle on 15 March 1976, and were deployed to a geosynchronous orbit altitude of approximately 36,000 km (Fig. 21). Each LES used two MHW generators (Fig. 19), which provided primary power for all spacecraft systems. The MHW-RTGs more than met the mission goals for lifetime. They also enabled the demonstration of improved methods for maintaining voice or digital data circuits among widely separated mobile communications terminals. Although its RTGs were still providing usable electric power, LES-8 was turned off on 2 June 2004 due to control difficulties. LES-9, however, continues to operate over 30 years after launch.

Fig. 21. LES-8 and 9 in orbit.

The Voyager 2 spacecraft launched on 20 August 1977 aboard a Titan-Centaur launch vehicle (Fig. 22). Each Voyager probe carried three MHW generators. Voyager 1 followed on 5 September 5, also aboard a Titan-Centaur rocket.

Fig. 22. Voyager spacecraft.

The Voyager spacecraft explored the most territory of any mission in history, including all the giant planets of the outer solar system, 48 of their moons, and the unique system of rings and magnetic fields those planets possess. The final planetary encounter was conducted by Voyager 2, which had its closest approach with Neptune on 25 August 1989. Although Pioneers 10 and 11 were the first spacecraft to fly beyond all the planets, Voyager 1 passed Pioneer 10 to become the most distant human-made object in space. As of 11 August 2007, the power generated by the spacecraft had dropped to about 60% of the power at launch. This is better than the pre-launch predictions based on a conservative thermocouple degradation model. As the electrical power decreases, spacecraft loads must be turned off, eliminating some spacecraft capabilities.

4.8 General Purpose Heat Source (GPHS) RTG

Following the successful launches of the Voyager spacecraft, DOE turned its focus on developing a new selenide-based RTG for NASA's planned International Solar Polar Mission (ISPM) and the Jupiter Orbiter Probe, which later became the Ulysses and Galileo missions, respectively. Nuclear power was required for these missions, since they would both operate in the vicinity of Jupiter with its low solar energy flux, cold temperatures and intense radiation environment. Both missions were to be launched in the mid-1980s aboard the then under development U.S. Space Shuttle.

Upon determining that selenide thermoelectrics would not be suitable for long-duration missions, DOE went back to Si-Ge technology and considered modifying flight spares of the MHW-RTG for use on Galileo. However, the joint NASA-ESA ISPM team requested a new, larger, more powerful RTG for their spacecraft. When the Galileo project saw the benefits of the planned ISPM RTG they requested two for the Galileo spacecraft. As a result the ISPM RTG was renamed the GPHS-RTG.

The GPHS-RTG used the same Si-Ge alloy unicouples used in the MHW-RTG. Because production of the unicouples had been stopped after the Voyager program there was a need to restart production. However, the rest of the design was very different. For one, the converter housing was made of a less expensive and more manufacturable Aluminum 2219-T6 alloy, instead of the beryllium used in the MHW-RTG. Another big difference was the heat source, which employed an assembly of newly developed General Purpose Heat Source (GPHS) modules. This modular approach to heat source design opened the door for developing RTGs of different sizes and powers in the future, but it required an extensive development and qualification program to replace the fuel sphere assemblies used in the MHW-RTG. Finally, DOE had decided to move the RTG assembly and testing work from its RTG contractors to DOE's Mound Laboratory, which necessitated a rapid buildup of the infrastructure at a new location.

The GPHS-RTG, shown in Fig. 2, was composed of two main elements: a linear stack of 18 GPHS modules and the converter. The converter surrounds the heat source stack, and consists of 572 radiatively-coupled Si-Ge unicouples, which operate at a hot side temperature of 1,275 K and a cold side/heat rejection temperature of 575 K. The outer case of the RTG provides the main support for the converter and heat source assembly, which is axially preloaded to withstand the mechanical stress environments of launch and to avoid separation of GPHS modules. The converter also provides axial and mid-span heat source supports, a multifoil insulation packet and a gas management system. The latter provides an inert gas environment for partial power operation on the launch pad, and also protects the multifoil and refractory materials during storage and ground operations.

The complete GPHS-RTG has an overall length of 114 cm and a fin span of 42.2 cm. Its mass of 55.9 kg and BOM power level of up to 300 We provides a specific power of 5.1 to 5.3 We/kg, far greater than any of its predecessors.

The Galileo spacecraft (Fig. 23) was launched on 18 October 1989 on the Space Shuttle, after a 3.5-year delay caused by the Challenger accident. Forced to take a long, circuitous trajectory involving Earth and Venus gravity assists, Galileo arrived at Jupiter in December 1995, The Orbiter spacecraft investigated the Jupiter and its Galilean satellites from space, while the Galileo Probe, which was battery-powered but kept warm via a number of small radioisotope heater units, entered Jupiter's atmosphere on 7 December 1995. Both GPHS-RTGs met their end of mission (EOM) power requirements, thus allowing NASA to extend the Galileo mission three times. However on 21 September 2003, after eight years of service in orbit about Jupiter, the mission was terminated by intentionally forcing the orbiter to burn up in Jupiter's atmosphere. This was done to avoid any chance of contaminating local moons, especially Europa, with micro-organisms from Earth.

The Ulysses (Fig. 24) was launched nearly a year later by the Space Shuttle on 6 October 1990. The mission included a Jupiter gravity assist performed on 8 February 1992 in order to place the spacecraft in a trajectory over the polar regions of the Sun. The single GPHS-RTG performed flawlessly and exceeded its design requirement. As a result, the Ulysses mission was extended beyond its original planned lifetime goal, thus allowing it to take measurements over the Sun's poles for the third time in 2007 and 2008. However after it became clear that the power output from the RTG would be insufficient to operate science instruments and keep onboard hydrazine propellant from freezing, the decision was made to end the mission on 1 July 2008.

Fig. 23. Galileo spacecraft. Pre-launch assembly on left. Artist concept of spacecraft in orbit around Jupiter on right.

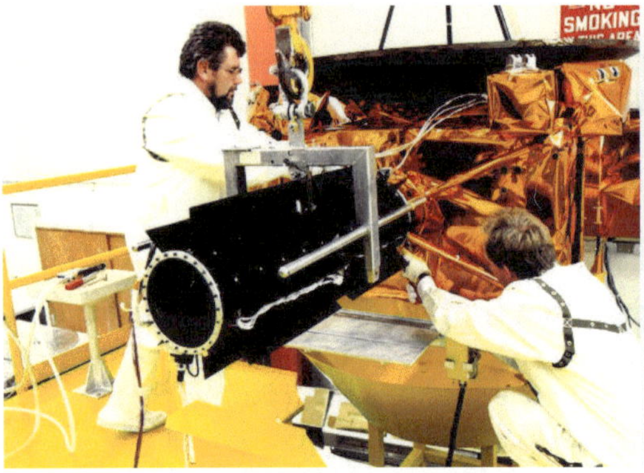

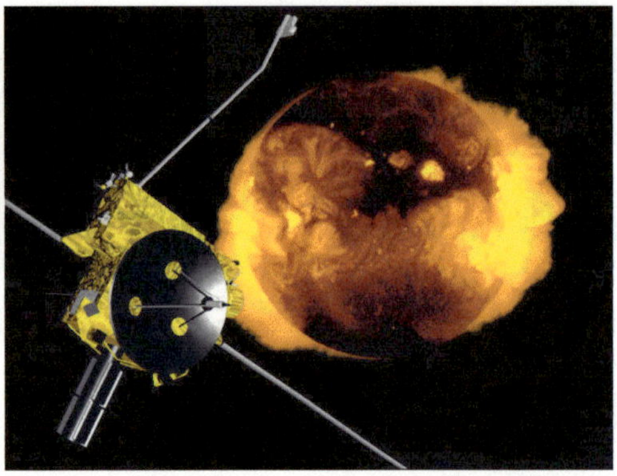

Fig. 24. Ulysses spacecraft. Installation and checkout of RTG on left. Artist concept of vehicle on right.

The third mission to use the GPHS-RTG was Cassini (Fig. 25), which was launched, along with the ESA-built Huygens Titan Probe, on 15 October 1997 aboard a Titan IV/Centaur launch vehicle. Cassini achieved Saturn orbit insertion on 1 July 2004 after a 6.7-year transit involving gravity assists about Venus and Earth. The Huygens probe, which carried the same radioisotope heater units as Galileo, successfully landed on Titan and provided the first close-up views of that enigmatic world. Because of mission complexity, Cassini needed more power than used on previous flagship-class missions. The three GPHS-RTGs that were used have so far operated flawlessly and have exceeded their expected power output. The mission has now been approved for an extension to 2017.

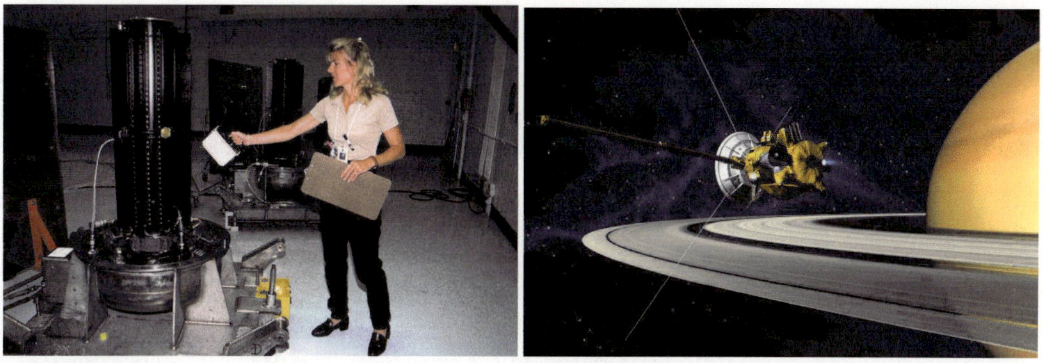

Fig. 25. Cassini spacecraft. Pre-launch checkout of RTG on left. Artist concept of vehicle on right.

The most recent mission to use a GPHS-RTG is the New Horizons mission to Pluto (Fig. 26), which was launched on 19 January 2006 aboard an Atlas V 551. The spacecraft is currently on a 9.5-year transit to Pluto and Charon. At encounter, which is expected in July 2015, New Horizons will characterize and map the surfaces of Pluto and Charon and their atmospheres. From 2016 to 2020, the spacecraft will continue to conduct encounters with one or two Kuiper Belt Objects. So far, it is anticipated that the RTG will exceed its power and lifetime requirements.

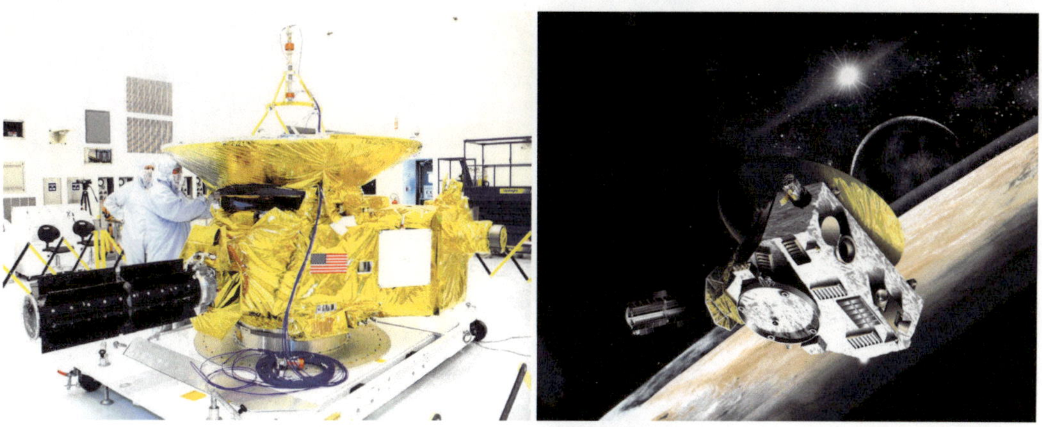

Fig. 26. New Horizons spacecraft. Pre-launch integration with spacecraft on left. Artist concept of New Horizons flyby of Pluto and Charon on right.

4.9 Multi-mission RTG (MMRTG)

Although the GPHS-RTG served well on Ulysses and Galileo and continues to meet requirements for Cassini and New Horisons, it is not suitable for future missions on Mars and other planetary bodies with atmospheres. The GPHS-RTG was only designed to function effectively in a vacuum environment. Furthermore, its relatively large size and power level limit its modularity and ease of integration on future small to mid-size spacecraft.

DOE and NASA are currently developing a new generation of RPS generators that could be used for a variety of space missions. One is the Multi-Mission RTG (MMRTG), which has been designed to operate on planetary bodies with atmospheres, such as Mars, as well as in the vacuum of space. The MMRTG's smaller size of about 110 We is more modular in design and flexible in meeting the needs of a broader range of different missions as it generates electrical power in smaller increments. The design goals for the MMRTG include ensuring a high degree of safety and reliability, optimizing power levels over a minimum lifetime of 14 years, and minimizing mass.

The MMRTG (Fig. 27) is designed to use a heat source consisting of eight Step 2 GPHS modules. These Step 2 modules have additional material in the GPHS aeroshell that improves structural integrity and performance. Although the Pb-Te/TAGS thermoelectric materials are the same as those used on SNAP-19, and represent a thoroughly flight proven technology, the physical dimensions and material changes to improve performance have resulted in different degradation compared to the SNAP-19. The MMRTG generator has a fin span of 64 cm, a length of 66 cm, and a mass of about 45 kg. Its BOM power level of approximately 110 We yields a specific power that is less than the SNAP-19. However, the purpose in pursuing this unit is not to advance state-of-the-art in specific power, but to minimize development risk, while providing an RPS capable of operating in different mission environments.

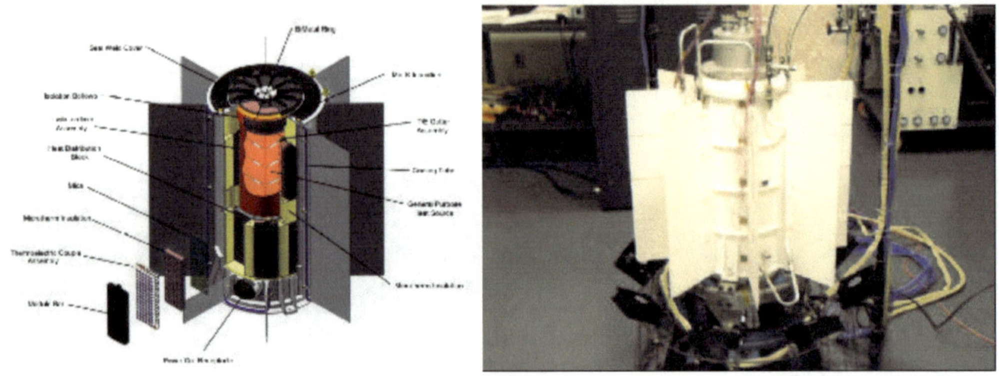

Fig. 27. Multi-Mission RTG (MMRTG). Cutaway schematic of power unit on left. MMRTG Qualification Unit undergoing tests on right.

The MMRTG is being developed to serve as the primary power source on the Mars Science Laboratory (MSL), a concept of which is shown in Fig. 28. This mission is currently planned for launch in 2011, and is anticipated to land on Mars in 2012.

Fig. 28. Mars Science Laboratory.

MSL is considerably larger than the Mars Exploration Rovers that landed on the planet in 2004. It will carry more advanced scientific instruments than any other Mars mission to date, including analysis of samples scooped up from the soil and drilled powders from rocks. It will also investigate the past and present ability of Mars to support life. The MSL rover will use power from an MMRTG to supply heat and electricity for its components and science instruments. A coolant loop and heat exchanger coupled with the MMRTG radiators will transport waste heat to the electronics, thus extending operation of the rover into the Martian night and winter season. The goal is to operate for at least one Martian year (i.e., two Earth years) over a wide range of possible landing sites.

The MMRTG could be used on a number of other potential missions in the future. One exciting prospect is to use the MMRTG as the principal electrical power and heat source for a Titan aerobot/balloon mission (Fig. 29). In this scenario, the considerable waste heat produced by the MMRTG would be used to heat a gas and generate buoyancy for a balloon carrying a long-lived payload, in addition to providing electrical power to onboard instruments.

Fig. 29. Titan Aerobot.

4.10 Advanced Stirling Radioisotope Generator (ASRG)

When the potential of radioisotope power became apparent in the 1950s, the original focus was on development of dynamic power conversion systems. Most of these activities concentrated on applying the high efficiencies achievable with Brayton and Rankine cycles, in expectation that systems would evolve to larger power levels in the future.

Although thermoelectric technology supplanted this approach and became the dominant power conversion option for every RPS flown in space, work on Dynamic Isotope Power Systems (DIPS) continued at various times throughout the intervening decades. The principal focus of these efforts was on eventual development of power systems capable of producing up to tens of kilowatts of power. These higher power technologies would be used in conjunction with the ambitious crewed missions anticipated in the future. The studies of DIPS pointed to its excellent suitability for lunar and planetary surface exploration, particularly surface rovers, remote science stations and backup power supplies to central base power.

Interest in DIPS was particularly high during the Space Exploration Intitiative (SEI) of the early-1990s. However with the demise of that effort in 1992, the focus shifted to determine how dynamic power conversion could benefit radioiosotope power systems in the multi-hundred watt range. During the 1990s, several advanced dynamic and static conversion technologies were researched and evaluated. Several technologies that had appeared promising initially proved to be ill-suited for the unique demands of deep space missions. In the end, it became apparent that the free-piston Stirling engine offered the best hope of advancing the efficiency of future generators, while offering lifetimes up to a decade or two. Unlike previous DIPS designs, which featured turbomachinery-based conversion technologies (e.g. Brayton), small Stirling DIPS could be advantageously scaled down to multihundred-watt unit size while preserving size and mass competitiveness with RTGs.

In 2002, NASA and DOE began a Stirling Radioisotope Generator (SRG) project focused on evaluating and demonstrating a unit for flight development. The work was initiated to provide a back-up RPS for the MSL mission. The unit used Stirling convertors built and tested under a technology development effort funded by DOE. Although the SRG could achieve a four-fold reduction in fuel requirements for the same power, the final system specific power of the unit was only slightly better than the MMRTG.

In less than two years, it became apparent that the MMRTG would be selected by NASA's Mars program, so that the rover could make use of the significant waste heat produced by that unit. Finally, a small business technology project initiated in the early 2000s with Sunpower Technologies in Athens, Ohio, indicated that convertors with much better mass performance could be developed and substituted into an SRG-based design. Such a unit could potentially achieve specific powers of about 7 We/kg. With the advancement in Stirling generator heater head materials and with improved temperature margin and higher temperature operation, units with specific powers greater than 8 We/kg may be possible..

In 2005, the decision was made to redirect efforts toward development of an Advanced SRG (ASRG) technology demonstration Engineering Unit (EU). The effort drew upon the work that had gone on previously with the controller, housing and insulation systems for the SRG, but incorporated use of the higher specific power Sunpower generators. In addition to high specific power, the ASRG would likely achieve an efficiency over 30%. This is four to five times higher than that from a GPHS-RTG, and is particularly important for conserving the very limited worldwide supply of Pu-238 fuel.

The ASRG, which is shown in Fig. 30, is being developed under the joint sponsorship of the U.S. Department of Energy (DOE) And NASA. The eventual flight units are expected to produce over 130 We in a space environment and to have a mass of 32 kg or less. The prime contractor is Lockheed-Martin Corporation of Valley Forge, PA, with Sunpower, Inc. of Athens, Ohio as the main subcontractor. NASA Glenn Research Center (GRC) is supporting the technology development, along with evaluation and testing of the Stirling convertors used in the device. In addition to improving fuel utilization efficiency over previous RPS, the ASRG is being designed for multi-mission use in deep space, and within the atmosphere of Mars and possibly Titan.

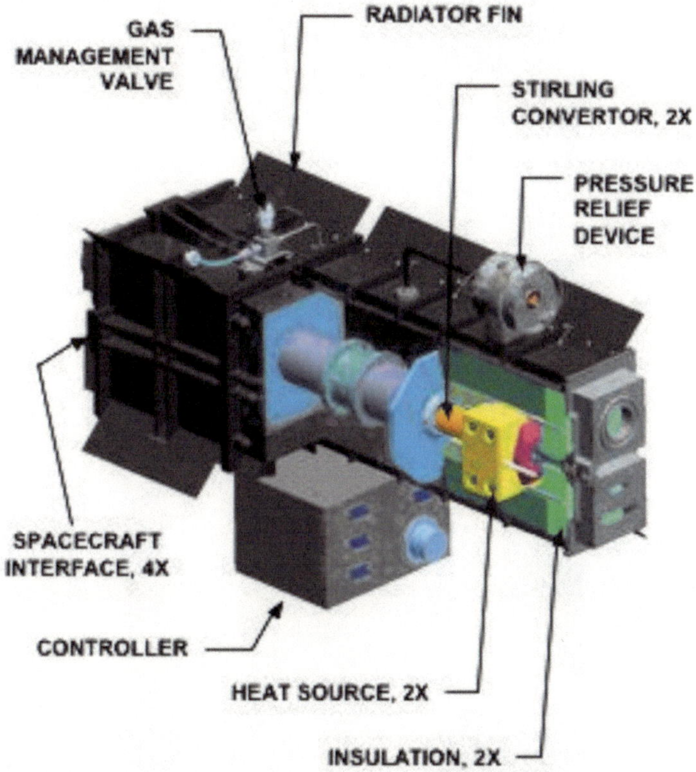

Fig. 30. Advanced Stirling Radioisotope Generator (ASRG).

Activities are focused on developing and testing the ASRG-EU in thermal and vibrational environments that closely approximate qualification-level tests (Fig. 31).

The ASRG-EU uses two axially-opposed Advanced Stirling Convertors (ASCs), operating at a hot-end temperature of 650 deg C, producing about 140 We. Sunpower is developing the ASC under a 2002 NASA Research Announcement (NRA) with GRC. The low mass of the ASC is key to the ASRG's high overall system specific power.

The ASRG has achieved a TRL 6 (system demonstration in a relevant environment) with operation at qualification level thermal and dynamic environments. Tests on the ASRG-EU were completed in June 2008 at the Lockheed-Martin Space System Company in King of Prussia, PA. These evaluations included thermal balance, thermal performance, mechanical disturbance, sine transient, random vibration, simulated pyrotechnic shock and electromagnetic interference and magnetic field emission tests. Over 1,000 hours of successful EU operating time with numerous startup and shutdown cycles were accumulated during the testing at Lockheed-Martin. The ASRG-EU is now undergoing extended/multi-year duration testing at NASA GRC. It has achieved over 11,000 hours of successful operation as of April 2011, and is expected to exceed 14,000 hours of operation by the end of 2011.

Ongoing ASRG-EU tests use electrical resistance heaters that simulate the heating characteristics of the actual GPHS module. Avoiding use of nuclear materials during early phases of development greatly facilitates testing and evaluation of the ASRG subsystems.

Fig. 31. ASRG Engineering Unit.

5. Other potential applications

MMRTG and ASRG should satisfy most RPS mission requirements well beyond 2010, particularly for those applications involving several hundred watts of power. However, there will likely be a demand for additional types of units in the future. One potential need identified by the space science community is for small RPS units ranging in power from ~10 milliwatts (mW_e) to ~20 W_e. These so-called 'milliwatt' and 'multiwatt-class' power supplies could extend the capability of small, low cost missions supported through NASA's small to mid-size programs, and augment human missions involving deployment of monitoring stations and autonomous devices. They would likely utilize the GPHS or other existing heat sources. Although flight-qualified systems in this size range do not presently exist, the promise of RPS has led NASA and DOE to evaluate the possible development of a small RPS unit in the future.

Nuclear Electric Propulsion (NEP) has been studied since the early 1960's because of its potential for future high-energy space missions. Almost all NEP assessments to date have assumed fission as the nuclear energy source. Unlike solar-powered electric propulsion (SEP) systems, NEP operation is generally independent of distance and orientation with respect to the Sun. Over the last decade, several studies have pointed to Radioisotope Power Systems (RPS), instead of reactor power sources, as the best way of implementing NEP. Radioisotope-based NEP, also known as Radioisotope Electric Propulsion (REP), has been evaluated before, but has not been seriously considered for flight due to the low specific power range of traditional RPS (e.g., 3 to 5 We/kg). However, the prospects for REP have improved substantially with the advent of the ASRG and its likely improvement in specific power.

In this capacity, REP would principally be used as an interplanetary stage for long-duration deceleration and acceleration in deep space. At remote destinations, REP would perform deceleration, orbit insertion and maneuvers around outer planets and other planetary bodies. REP-based spacecraft could also provide ample power at destination for sophisticated science instruments and communications, but it would fit better within the relatively modest kilowatt-scale power requirements of the space science community.

6. Conclusion

Radioisotope power systems will continue to play an important role in NASA's exploration efforts. These systems also have the potential for use in a variety of new applications, which would benefit from the technology's versatility in a broad range of space and planetary environments. In the near-term, the MMRTG will expand the capability for conducting science on the surface of Mars. The ASRG will enable even higher performance missions. These units will also enable more ambitious exploration of other planetary surfaces and provide a reliable means of powering spacecraft in deep space. Current activities would also allow the potential development of new systems that could expand application of RPS to smaller science missions. The key to successful implementation of RPS is to maintain close ties with potential users and the science community at large. With these advancements, radioisotope power systems and technology will offer tremendous benefits for future exploration endeavors.

7. References

Angelo, J.A. and Buden, D., *Space Nuclear Power*, Orbit Book Co., Malabar, FL, 1985, pp. 133-157.

Bennett, G.L., "Space Nuclear Power: Opening the Final Frontier," AIAA-2006-4191, 4th International Energy Conversion Engineering Conference, June 2006.

Bennett, G.L., and Skrabek, E.A., "Power Performance of U.S. Space Radioisotope Thermoelectric Generators," in proceedings of *15th International Conference on Thermoelectrics*, June 1996, pp. 357-372.

Chan, J., Hill, D., Hoye, T., and Leland, D., "Development of Advanced Stirling Radioisotope Generator for Planetary Surface and Deep Space Missions," AIAA-2009-5768, 6th International Energy Conversion Engineering Conference, July 28-30, 2009.

Engler, R.E., "Atomic Power in Space - A History," U.S. Department of Energy Report DE-AC01-NE32117, March 1987.

Furlong, R.R. and Wahlquist, E.J., "U.S. Space Missions Using Radioisotope Power Systems," *Nuclear News*, April 1999, pp. 26-34.

Hammel, T.E., and Osmeyer, W.E., "The Selendie Isotope Generators," AIAA-1997-498, Conference on the Future of Aerospace Power Systems, March 1977.

Hammel, T.E., Bennett, R., Otting, W., and Fanale, S., "Multi-Mission Radioisotope Thermoelectric Generator (MMRTG) and Performance Prediction Model," AIAA-2009-4576, 7th International Energy Conversion Engineering Conference, August 2-5, 2009.

Lange, R.G. and Mastal, E.F., "A Tutorial Review of Radioisotope Power Systems," in *A Critical Review of Space Nuclear Power and Propulsion*, edited by M.S. El-Genk, American Institute of Physics, Melville, New York, 1994, pp. 1-20.

National Research Council, *New Frontiers in the Solar System - An Integrated Exploration Strategy*, National Academies Press, Washington, DC, 2003a.

National Research Council, *The Sun to the Earth - and Beyond: Panel Reports*, National Academies Press, Washington, DC, 2003b.

Schmidt, G., Wiley, R., Richardson, R. and Furlong, R., "NASA's Program for Radioisotope Power System Research and Development," in proceedings of *Space Technology and Applications International Forum (STAIF-2005)*, edited by M.S. El-Genk, American Institute of Physics, Melville, New York, 2005.

Schmidt, G., Abelson, R., and Wiley, R., "Benefit of Small Radioisotope Power Systems for NASA Exploration Missions," in proceedings of *Space Technology and Applications International Forum (STAIF-2005)*, edited by M.S. El-Genk, American Institute of Physics, Melville, New York, 2005.

Surampudi, R., Carpenter, R., El-Genk, M., Herrera, L., Mason, L., Mondt, J., Nesmith, B., Rapp, D. and Wiley, R., *Advanced Radioisotope Power Systems Report*, Jet Propulsion Laboratory, Pasadena, CA, March 2001.

Wiley, R. and Carpenter, R., "Small Radioisotope Power Source Concepts," in proceedings of *Space Technology and Applications International Forum (STAIF-2004)*, edited by M.S. El-Genk, American Institute of Physics, Melville, New York, 2004.

U.S. Department of Energy, *Atomic Power in Space*, Prepared by Planning & Human Systems, Inc. under Contract DE-AC01-NE32117, 2nd ed., Springer-Verlag, New York, 1983, Chaps. 7, 14.

21

Radioisotope Power Systems for Space Applications

Antonio Sanchez-Torres
Universidad Politécnica de Madrid, Escuela Técnica Superior de Ingenieros Aeronáutcos, Departamento de Física Aplicada, Spain

1. Introduction

At the beginning of the Space Age, both propulsion and power generation in the spacecraft has been the main issue for consideration. Considerable research has been carried out on technologies by several Space Agencies to reach outer planets and generate electric power for the systems and subsystems in the spacecraft (SC). Various types of power source such as solar photovoltaic, Radioisotope power systems (RPS) have been used by Space Agencies. New technology such as reactor based, electric solar sail and electrodynamic bare tethers might be used in the future for both propulsion and power generation. Mainly, both NASA and Russian Agency worked separately using nuclear technology to obtain more efficiency in their systems for deep space exploration.

Radioisotope Power Systems (RPS), is a nuclear-powered system to generate electric power to feed communication and scientific systems on a spacecraft. Radioisotope Thermoelectric Generators (RTGs), a type of Radioisotope Power System, were used in the past as electric power supplies for some navigational and meteorological missions, and most outer-planet missions. Radioisotope power systems use the natural decay of radionuclides produced by a nuclear reactor. The expensive, man-made Plutonium-238 (^{238}Pu) is the appropriate source of energy used in RPS fueling; its long half-life (~87 years) guarantees long time missions. The limited avability of Plutonium-238 is inadequate to support scheduled NASA mission beyond 2018. After the Cold War, throughout the Non-Proliferation of Nuclear Weapons Treaty, the production and processing of these resources have been severally reduced. There is a high-priority recommendation to reestablish production to solve the severe ^{238}Pu demand problem (National Reseach Council, 2009).

The isotope initially selected for terrestrial and space power applications was Cerium-144 because it is one of the most useful fission products available from nuclear reactor (Furlog, 1999; Lange, 2008). Its short half-life (about 290 days) made Cerium-144 compatible with a possible short-time mission. However, the high radiation associated with a powerful beta/gamma emission produces several problems with the payload interaction and safety in the case of reentry orbit. The development of RTGs was assigned to The Atomic Energy Commission in 1955. The first system developed for space situation was the System for Nuclear Auxiliary Power (SNAP). The Cerium-144 fueled SNAP-1 power system was never used in space. The first flight with a RTG was SNAP-3 in 1961 delivering 11.6 kW over a 280 days period, using as fueling Polonium-210 (Po-210) isotope. Po-210 is an alpha emitter with a very high power density and low radiation emissions. Since Po-210 has short half-life (138 days), space missions are highly limited. The early RTGs developed a specific power slightly larger than 1 W/kg. SNAP-9A system reached 20 W/kg whereas later systems such as Galileo developed 5.4 W/kg (Brown, 2001; Griffin, 2004). Several past missions have used RPS as it shown in Table 1. Table 2 shows several future missions that will use RPS as main power system.

Power source (number)	Spacecraft	Mission type	Launch
SNAP-3 RTG (1)	Transit 4A	Navigational	1961
SNAP-3 RTG (1)	Transit 4B	Navigational	1961
SNAP-9A RTG (1)	Transit 5BN-1	Navigational	1963
SNAP-9A RTG (1)	Transit 5BN-2	Navigational	1963
SNAP-9A RTG (1)	Transit 5BN-3	Navigational	1964
SNAP-10A Reactor	Snapshot	Experimental	1965
SNAP-19B RTG (2)	Nimbus B-1	Meteorological	1968
SNAP-19B RTG (2)	Nimbus III	Meteorological	1969
ALRH Heater	Apollo 11	Lunar	1969
SNAP-27 RTG (1)	Apollo 12	Lunar	1969
SNAP-27 RTG (1)	Apollo 13	Lunar	1970
SNAP-27 RTG (1)	Apollo 14	Lunar	1971
SNAP-27 RTG (1)	Apollo 15	Lunar	1971
SNAP-19 RTG (4)	Pioneer 10	Planetary	1972
SNAP-27 RTG (1)	Apollo 16	Lunar	1972
Transit-RTG (1)	Triad-01-1X	Navigational	1972
SNAP-27 RTG (1)	Apollo 17	Lunar	1972
SNAP-19 RTG (4)	Pioneer 11	Planetary	1973
SNAP-19 RTG (2)	Viking 1	Planetary	1975
SNAP-19 RTG (2)	Viking 2	Planetary	1975
MHW-RTG (4)	LES 8, LES 9	Communication	1976
MHW-RTG (3)	Voyager 2	Planetary	1977
MHW-RTG (3)	Voyager 1	Planetary	1977
GPHS-RTG (2) RHU Heater	Galileo	Planetary	1989
GPHS-RTG (1)	Ulysses	Planetary	1990
RHU Heater (3)	Mars Pathfinder	Planetary	1996
GPHS-RTG (2) RHU Heater	Cassini	Planetary	1997
RHU Heater (8)	Mars MER Spirit	Mars rover	2003
RHU Heater (8)	Mars MER Opportunity	Mars rover	2003
GPHS-RTG (1)	New Horizons	Planetary	2006

Table 1. US spacecraft with RPS

The RTG fuel must be produced in adequate quantities with appropriate nuclear safety requirements for space missions. There are only a limited number of radioisotopes available for space power system applications. Using isotopes with pure low-energy beta emission would eliminate the requirements to shield against gamma radiation. Low energy particles would also generate low energy bremstrhlung x rays that is easy to shield against. This suggests isotopes such as T^3, Ni^{63}, Sr^{90}, Tc^{99}, Pw^{147}, Curium-242 and Curium-244 are other possibilities.

When solar panels cannot be used efficiently for planetary missions, RPS becomes the best available alternative. Typical RTG structure consists basically on a couple of metallic conductor, with hot and cold end-connectors. The system operates under thermoelectric generation principle, the so-called Seebeck effect. Heating one end from the natural decay of a radioactive isotope and the other end keeping cold, the gradient of tempreature between two ends will produce a voltage drop. Connecting the terminals through a resistive load causes an amount of current flowing in the electric external circuit, and then generating electric power.

Considerable research has been carried out to develop new technologies to improve RTG efficiency using more efficient thermoelectric materials with low thermal conductivity. The dynamic conversion systems, which convert partially the thermal energy in the fluid into mechanical work to drive an alternator to produce electricity, would provide higher electric power per unit mass, reducing the amount of Plutonium-238 required.

Power source (number)	Spacecraft	Mission type	Launch
RHU Heater (1-4)	Europa Impactor Micro-Lander	Planetary	2015
RHU Heater (1-4)	Titan Micro-Rover	rover	2015
GPHS (1)	Europa Lander	Planetary	2015
GPHS (1)	Titan Moon Lander	Planetary	2015
GPHS (1)	Ganymede Lander	Planetary	2015
GPHS (1)	Callisto Lander	Planetary	2015
GPHS (1)	Titan Rough Lander	Planetary	2015
GPHS (1)	Europa Rough Lander	Planetary	2015
GPHS (1)	Callisto Orbiter Subsatellite	Planetary	2015
GPHS (1)	Ganymede Orbiter Subsatellite	Planetary	2015
GPHS (1)	Europa Orbiter Subsatellite	Planetary	2015
GPHS (1)	Outer Planets Magnetosphere Subsatellite	Planetary	2015
GPHS (2-4)	Titan Rover	Rover	2015
GPHS (1)	Titan Amphibius Rover	Amphibius Rover	2015
GPHS (1-3)	Lander Amorphor. Rover Array Mini-Lander	Planetary	2020
MMRTG-ASRG	Jupiter Europa Orbiter	Planetary	2020
RHU Heater (7-9)	Prospecting Asteroid Mission Micro-Sat	Planetary	2020-2030
RHU Heater (7-9)	Saturn Autonomous Ring Array Micro-Sat	Planetary	2020-2030

Table 2. Several US future missions with RPS

In section 2 we review the fuel requirements for an optimal RPS. The main part of the RPS, the well-known General Purpose Heat Source, is described in section 3. In section 4 we study the static conversion energy (without movable parts), analyzing thermoelectric effects in the conductors. Additionally, we describe both RTG and Multi-Mission-RTG structures and principle characteristics. The dynamic conversion energy is reviewed in detail in section 5, focusing on Stirling and Brayton power systems. Due to Planetary Protections Requirements, some tentative outer-planets missions like Jovian moons exploration in Europa Jupiter System or re-entry missions which use RPS have to be safety enough. In section 6, we review the safety models for possible RPS accidents. In section 7, RTG will be compared with solar arrays. Conclusions are written in section 8.

2. Radioisotopes for power generation

At least 1300 radioisotopes, both natural and man-made, are available for terrestrial and space applications. Many are generated in both nuclear reactors and particle accelerators. The initial activity of the isotope is

$$A_0 = \lambda N_0 \ [\mathrm{Bq}] \tag{1}$$

where N_0 is the initial isotope amount and $\lambda = Ln(2/t_{1/2})$ is the decay constant of the isotope for a $t_{1/2}$ half-life. Table 3 shows several characteristics of useful radioisotopes for RPS. The specific electrical power generated by the heat of the source is given by

$$P_0 = 1.6 \cdot 10^{-13} \eta \times \frac{E[\mathrm{MeV}]\ \lambda\left[\mathrm{s}^{-1}\right] N_A[\mathrm{nuclei/mol}]}{M[\mathrm{amu}]} \tag{2}$$

where η is the conversion efficiency from thermal energy to electricity, E is the energy release per decay, N_A is the Avogadro's number and M the atomic mass.

Isotope	Radiation emission	$t_{1/2}$	Specific Power (W/g)
Tritium-3	β^- ,no ϒ	12.3 years	0.26
Cobalt-60	β^- , ϒ	83.8 days	17.70
Nickel-63	β^- ,no ϒ	100.1 years	0.002
Krypton-85	β^- , ϒ	10.7 years	0.62
Stronium-90	β^- ,no ϒ	29.0 years	0.93
Ruthenium-108	β^- ,no ϒ	1.0 years	33.10
Cesium-137	β^- , ϒ	30.1 years	0.42
Cerium-144	β^- , ϒ	284.4 days	25.60
Promethium-147	β^- , ϒ	2.6 years	0.33
Polonium-210	α, ϒ	136.4 days	141.00
Plutonium-238	α, ϒ	87.7 years	0.56
Americium-241	α, ϒ	432 years	0.11
Curium-242	α, ϒ	162.8 days	120.00
Curium-244	α, ϒ	18.1 years	2.84

Table 3. Characteristics of isotopes useful for RPS. Notice both high $t_{1/2}$ and specific power of the Plutonium-238

The radioisotope fuel must be not be very expensive. Additionally, the radionuclide proposed has to be easily shielded against deep penetration radiation, as gamma radiation, avoiding the destruction of the electronic components on the spacecraft onboard. The fuel capsule must withstand impact against the ground at high velocity in case of a rocket launch failure, and an Earth-reentry situation. These accidents will be described in section 6.
High P_0 and $t_{1/2}$ half-life values are required for space applications, reducing the valuable radioisotopes. Isotopes without powerful radiation such as gamma or beta is also required. The negative beta emitters can be recovered abundantly from fission fuel reprocessing plants. The alpha emitters with weak gammas are easier to shield. However, they are more expensive than the beta emitters.
The radionuclide most used in RPS, Plutonium-238, is produced by the isotope Np-237. Using U^{238} in the nuclear reactor, the isotope Np^{238} is produced by decay reaction

$$n^1 + U_{92}^{238} \rightarrow 2n^1 + U_{92}^{237}$$

$$U_{92}^{237} \quad \rightarrow e + Np_{93}^{237} .$$

Separating Np^{237} from reactor fuel and further irradiated in a neutron flux, the plutonium required is generated by

$$n^1 + Np_{93}^{237} \rightarrow \gamma + Np_{93}^{238}$$

$$Np_{93}^{238} \quad \rightarrow e + Pu_{94}^{238} .$$

The Plutonium-238 is selected for both high $t_{1/2}$ and specific power, producing heat by emitting alpha particles. The fuel is prepared in the form of pure plutonium oxide (PuO_2) with 0.7 ppm Plutonium-238 and less than 0.5 percent Thorium-238 and Uranium-232.

3. General purpose heat source

The appropriate isotope combined with other components create a heat source that efficiently transfer the isotope heat to electrical power. The most used system for space missions is the general purpose heat source (GPHS). Fig. 1 shows the GPHS structure used in missions such as Galileo, Ulysses, and Cassini. Each module is designed to produce about 250 W at the beginning of mission. Its weight is about 1.43 kg, and its size and shape are selected to survive orbital reentry and post-impact into the ground at high terminal velocity. Typical dimensions are 9.72 cm × 9.32 cm × 5.31 cm.
Each GPHS module contains four pressed PuO_2 fuel pellets. Both diameter and length of the cylindrical fuel pellet is about 2.75 cm. An iridium alloy containment shell and clad made of 0.05 cm aluminum thickness encapsulate the fuel pellet. The iridium alloy is made to resist oxidation in a post-impact environment scenario. The fueled clad is the combination of fuel pellet and cladding.
Two of these clads are confined in a Graphite Impact Shell (GIS) made of carbon material. The GIS structure is designed to decrease the damage to the iridium clads during a possible free-fall accident. Two GISs are inserted into an aeroshell that is composed in graphite material. A thermal insulation layer of carbon-fiber cover each GIS decreasing the high temperature supported to the clads during atmospheric reentry heating. The aeroshell

provides protection against surfaces. Step 1 GPHS module, which is used on the New Horizons eexploration mission to Pluto, improves the initial GPHS device including an aeroshell between the two GISs. A second aeroshell improvement, known as Step 2 GPHS module, gives additional protection in the clads for hipervelocity reentry into the atmosphere (Benett, 2006; Brown, 2001; Griffin, 2004; Hastings, 2004).

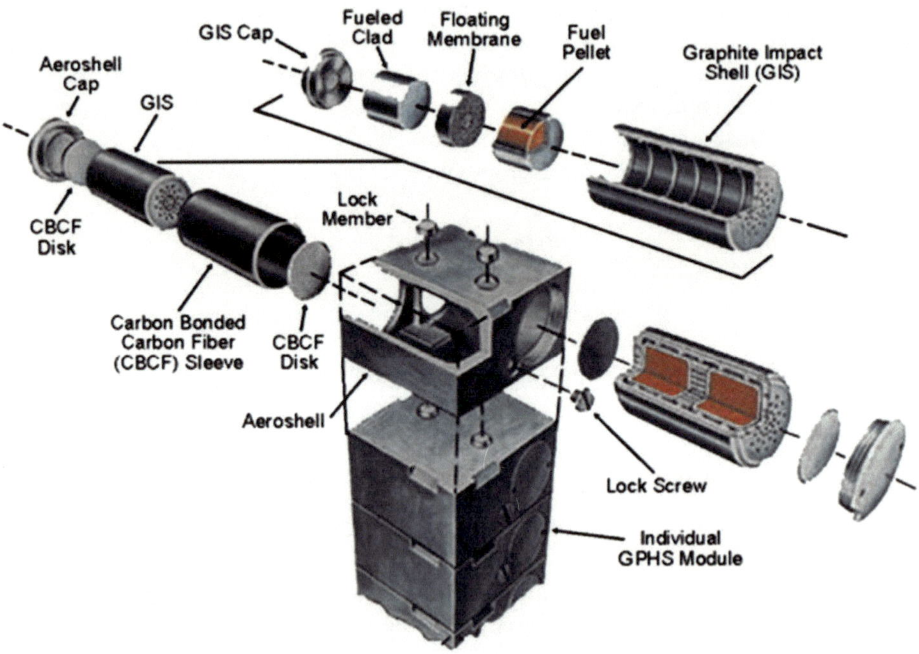

Fig. 1. General purpose heat source (GPHS) structure. (Source NASA/DOE/JPL)

4. Static conversion energy

The static conversion energy use the well-known thermoelectric or Seebeck effect. The thermoelectric effects in metals depend on the electronic structure of the materials. A temperature difference between two points in a conductor or semiconductor results in a voltage difference between these two regions. The Seebeck coefficient gives the magnitude of this effect. The thermoelectric voltage generated per unit temperature difference in a conductor is called the Seebeck coefficient.

Consider a metallic rod that is heated at one end and cooled at the other end as represented in Fig. 2. Since the electrons in the hot region are more energetic with greater velocities than those in the cold region, the electrons from the hot end diffuses toward the cold part. This situation prevails until the electric field developed between the positive ions in the hot region and the excess electrons in the cold region prevents further electron motion from the hot to cold end. A voltage is therefore gathered between the hot and cold ends with hot end at positive potential. The Seebeck coefficient S is given by the potential-to-temperature difference ratio

$$S = \frac{\Delta V}{\Delta T}, \tag{3}$$

where ΔV is the potential difference across a piece of metal due to a temperature difference ΔT. The sign of the Seebeck coefficient represents the potential of the cold side with respect to the hot side. For electrons diffusing from hot to cold end, the cold side is negative with respect to the hot side, making $S < 0$. Since the Seebeck coefficient depends on temperature, the voltage between two hot/cold regions is

$$\Delta V = \int_{T_i}^{T} S\,dT\,. \tag{4}$$

Using the Fermi-Dirac distribution, the average energy E_{av} per electron in a metal is given by

$$E_{av} = \frac{3}{5}E_{F0}\left[1+\frac{5\pi^2}{12}\left(\frac{kT}{E_{F0}}\right)^2\right], \tag{5}$$

where E_{F0} is the Fermi energy at 0 K. The average energy in the hot end is greater, and energetic electrons in the hot end diffuse toward the cold region until the potential prevents further diffusion. Notice that the average energy in Eq. (5) also depends on the material through E_{F0}.

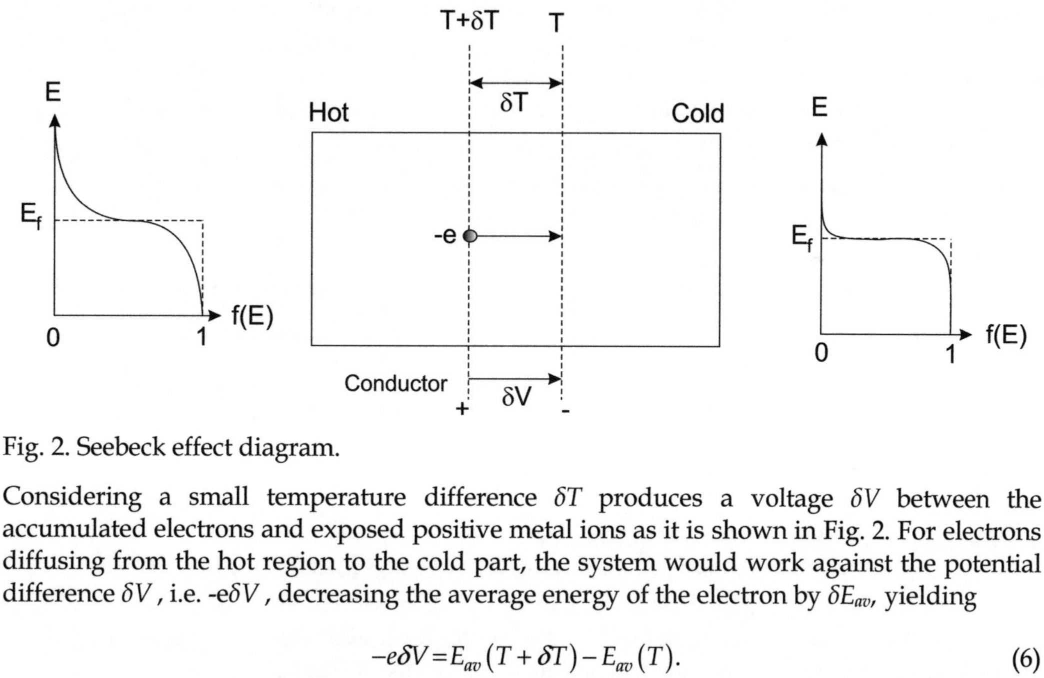

Fig. 2. Seebeck effect diagram.

Considering a small temperature difference δT produces a voltage δV between the accumulated electrons and exposed positive metal ions as it is shown in Fig. 2. For electrons diffusing from the hot region to the cold part, the system would work against the potential difference δV, i.e. $-e\delta V$, decreasing the average energy of the electron by δE_{av}, yielding

$$-e\delta V = E_{av}(T+\delta T) - E_{av}(T). \tag{6}$$

Using Eq. (5) in (6), and expanding $T+\delta T$, neglecting δT^2 term we obtain,

$$-e\delta V \approx \frac{\pi^2 k^2 T}{2E_{F0}}. \tag{7}$$

the Seebeck coefficient reads

$$S \approx -\frac{\pi^2 k^2 T}{2eE_{F0}}. \tag{8}$$

Table 1 shows typical experimental values for the Seebeck coefficient for several metals. Notice that some metals have positive S such as copper. The sign means that the electrons moves from cold to hot end of a copper rod.

Considering an aluminum rod heated at one end and cooled at the other end, the voltage difference reads

$$V_{AB} = \int_{T_0}^{T} (S_A - S_B)\, dT, \tag{9}$$

where S_A-S_B is the thermoelectric power for the thermoelectric couple given by both rods joined in a closed circuit. The voltage produced by the thermocouple pair depends on the metal used. Some conductor doped by the addition of impurities can produce deficiencies or an excess of electrons providing greater efficiency. The power extracted of the thermoelectric material is a function of its operating temperature. Elements with high enough thermal conductivity produce energy looses. Heat entering into the hot end would escape without much conversion to electricity. For a thermoelectric generator the thermoelectric rating, $Z = S^2/RK$, depends on the characteristic of the material, i.e. the voltage produced for the difference of temperature. Both R and K are electrical resistivity and thermal conductivity of the material, respectively. The thermoelectric generator will be more efficient with high Z values, i.e. high S, $1/R$ and $1/K$. Ordinary metals like cooper are very good heat conductors.

Metal	S at 0° C (μV K^{-1})	S at 27° C (μV K^{-1})	E_{F0} (eV)
Al	-1.60	-1.80	11.6
Cu	1.70	1.84	7.0
Ag	1.38	1.51	5.5
Au	1.79	1.94	5.5

Table 4. Seebeck coefficients for several metals.

4.1 Radioisotope thermoelectric generator

The typical static conversion system used in all outer planet mission is the well-known RTG (see Fig. 3), which is composed by a stack of 18 GPHS modules. The joined module GPHS-RTG, operates at normal voltage output of 28 V-dc. Both diameter and length of the RTG are 0.42 and 1.14 meters, respectively, and its weigh is about 55.9 kg.

The heat source assembly is surrounded by 572 silicon germanium (SiGe) thermocouples, known as unicouples. The unicouples are connected in two series-parallel electric wiring circuits providing the full output voltage. The induced magnetic field by the wires in the RTG is minimized, rearranging the electrical wiring (Abelson, 2004; Lange, 2008). The most recent use of a GPHS-RTG module was built for the New Horizons mission, launched in January 2006 to reach Pluto in 2015.

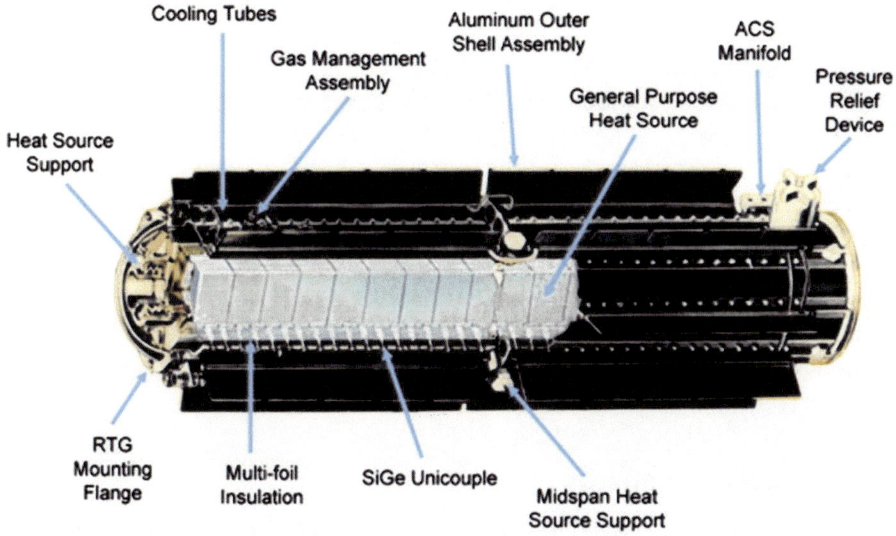

Fig. 3. Radioisotope Thermoelectric generator (RTG). (Source NASA/DOE/JPL)

4.2 Multi-mission radioisotope thermoelectric generator

The multi-mission radioisotope power generation (MMRTG) is the next generation of space RTGs (see Fig. 4). MMRTG is being developed by The Department of Energy (DOE) for planetary missions.

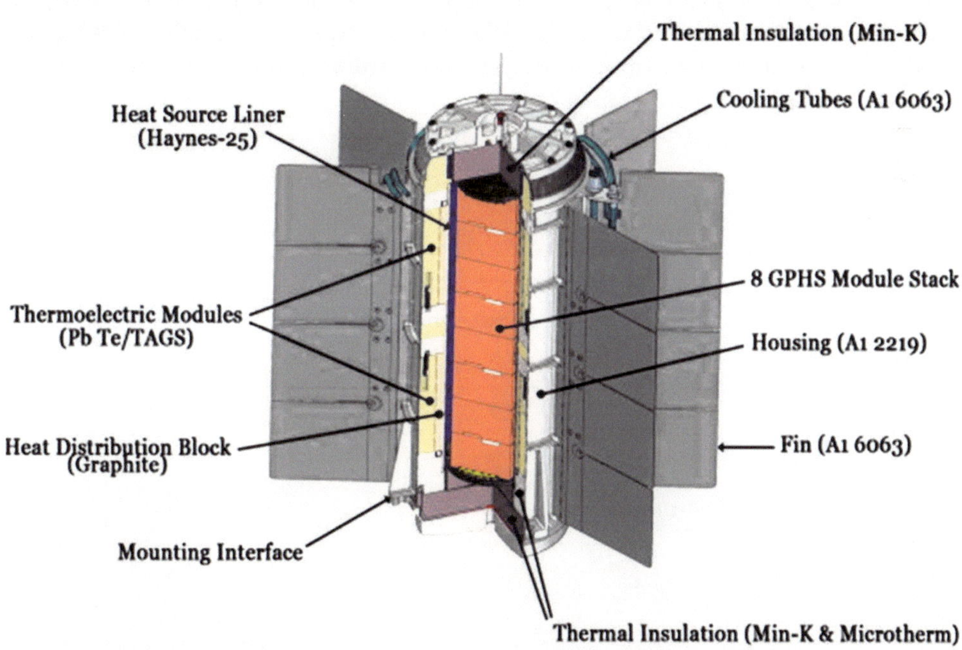

Fig. 4. Multi-mission Radioisotope Thermoelectric Generator (MMRTG). (Source NASA/DOE/JPL)

The MMRTG will generate 120 W of power at launch from a Pu-238 heat source assembly containing a stack of 8 Step 2 GPHS modules, which are described in section 3. The MMRTG operates at a normal output voltage of 28 V-dc. Both diameter and length of MMRTG are 64 cm diameter and 66 cm, respectively. The central heat source cavity is separated from the thermoelectric converter by a helium isolation liner. The helium generated by the Pu-238 is dumped to the environment by diffusion through an elastomeric gasket seal. The thermoelectric converter cavity can operate in both atmospheric environment or space vacuum (Ritz, 2004; Lange, 2008).
The thermocouples are connected in a series/parallel electrical circuit to improve the efficiency up 6.8%. Waste heat is radiated from the eight radial fins. These fins are made of aluminum alloys coated with a high-emissivity to disintegrate and release the GPHS modules in the case of reentry into the Earth's atmosphere. The MMRTG is both lighter and smaller than RTG system.

5. Dinamic conversion energy

For dynamic systems the conversion mechanism consists on that the thermal energy is partially transformed into mechanical work, moving an alternator to produce electric power. Rankine, Brayton and Stirling systems use this conversion mechanism. Typical cycle diagram is shown in Fig. 5. The isotope heats an inert gas working fluid which is expanded through a turbine. The high-efficiency Brayton cycle is capable to recuperate part of the energy. The turbine discharge gas is cooled, first in the recuperator, then in the radiator. The resulting low pressure gas is passed though the compressor, compressed to the highest cycle pressure and heated at essentially constant pressure in the recuperator before being returned to the heat source. The recuperator recovers a significant amount of heat, which would otherwise be dissipated through the radiator resulting in a higher cycle efficiency (Abelson, 2004; Benett, 2006; Lange, 2008).

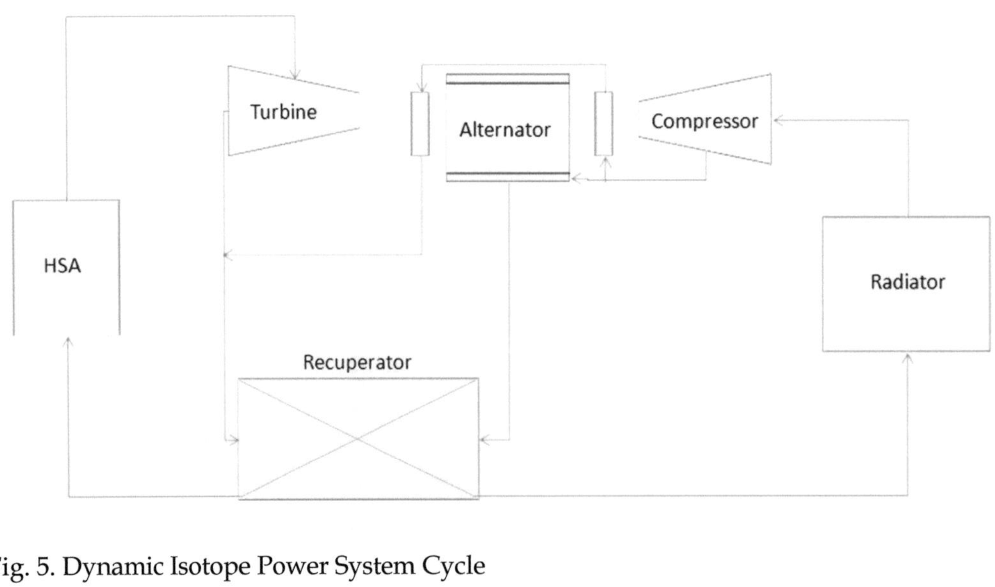

Fig. 5. Dynamic Isotope Power System Cycle

High efficiency in the RPS would both reduce system mass and fuel requirement, decreasing the total cost. Normally, RPS are tested at the ground in a vacuum chambers for a long time (>1000 hours). This would demonstrate that the design work also in the space with high power conversion efficiency.
Since the Brayton cycle is useless for power generation under 0.5 kW, missions with lower power requirement need an auxiliary electrical power generator.
The Stirling power system is based on a kinematics engine driving a three phase alternator. The initial design only worked during six months. It would not be appropriate for long-term missions. using a free piston combined with a linear alternator would be a promising technology for space power applications. The Stirling system, at difference of Brayton type, might provide low power. Further, Stirling engines operating in reverse have already used in space to provide cryogenic cooling for imaging sensors. The device has reciprocating pistons and displacers as it shown in Fig. 6. The motion of the components depends on physical springs or gas and on the cycle pressure swing of the engine. Typically, the engine contains only two moving parts.

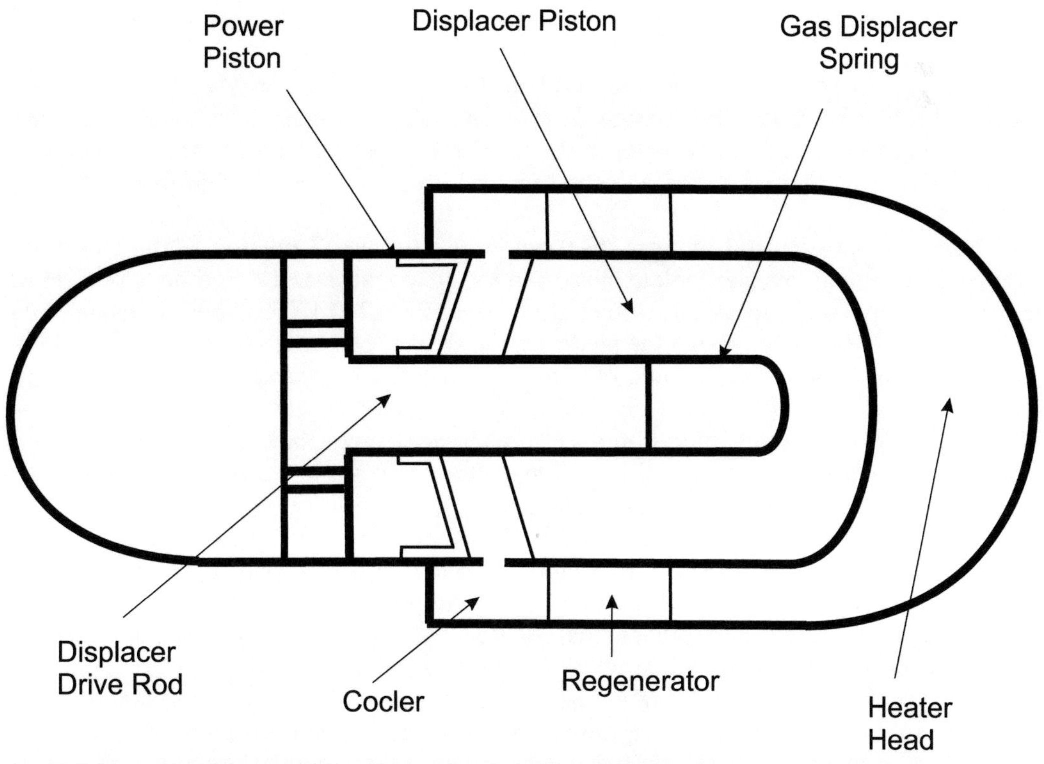

Fig. 6. Schematic diagram of the Stirling Isotope Power System

Two types of free pistons, linear alternator, Stirling machines are under development for mission in the range between 10 and 100 watts. The first type, under development by NASA. The power conversion efficiency for the Stirling producing 100 W is about 30%. The second type uses flexural spring to support the moving component to prevent friction and to provide enough axial springing for the free piston movement. The engine relies on the gap between the cylinder and the displacer to serve as regenerator for the system. As with

the Brayton cycle, heat regeneration is essential to achieve high efficiencies in Stirling engines. The design of a 10-W system has been tested, using fossile fuel combustion with a 20 % efficiency.
One of the problem of these systems is the attitude dynamic effects over the spacecraft. Both Brayton and Stirling systems have accumulated many time of testing. However, more tests would be required for outer planet missions that are expected to take more than 5 years.

6. RPS safety and accident evaluation

The Department of Energy (DOE) has worked to improve the safety of the RPS under all accident conditions, including accidents occurring near the launch pad and for orbital reentry accidents. The Pu-238 fuel for was changed from a metal to a more stable pressed oxide (PuO_2).
On April 21, 1964 the Transit-5-BN-3 mission was aborted because of a launch vehicle failure resulting in burn-up of the RTG during reentry, in keeping with the RTG design at the time. Some amount of the plutonium fuel was dropped in the upper atmosphere. The RTG design was changed to provide for survival of the fuel modules during orbital reentry.
A second accident occurred when the Nimbus B-1 was launched on May 18, 1968. It was aborted shortly after launch by a range destruction safety. The heat sources were recovered intact in about 90 meters under water in the California coast without release of plutonium. The fuel capsules were reworked and the fuel was used in a later mission (Abelson, 2004; Furlog, 1999).
The third incident occurred in April 1970, when the Apollo 13 mission to the moon was aborted following an oxygen tank explosion in the spacecraft service module. Upon return to Earth, the Apollo 13 lunar excursion module with a SNAP-27 RTG on board reentered the atmosphere and broke up above the south Pacific Ocean. The heat source module fell into the ocean. Atmospheric and oceanic monitoring showed no evidence of release of nuclear fuel.
The ceramic form covering plutonium-238 dioxide is heat-resistant and limits the rate of vaporization in fire or reentry conditions. The material also has low solubility in water. This material does not disperse though the environment.
More than 35 years have been researched in the engineering concepts and testing of RPS systems. Multiple layers of protective materials, including iridium capsules (or platinium-shodium capsules for RHUs) and high strength, heat-resistant graphite blocks are used to protect the radionuclide and prevent its release. Iridium is a strong, corrosion-resistant metal that is chemically compatible with plutonium dioxide. In addition, graphite is used because it is lightweight and highly heat-resistant. Several test for potential accident scenarios to know how RTG responses has been developed. Results of the failure mechanisms provide the basis for the determination of the source terms which are the characterization of plutonium releases including their quantity, location and particle size distribution. Recent large fragment tests in the GPHS safety test program have demonstrated in Solid Rocket Boosters (SRB) accident case, fragments impacting the full RTG system will not breach the fueled clads at velocities up to 0.12 km/s.
The multi-layer containment concept employed for the systems is designed to contain the radioisotope but even if the containment is breached, the ceramic pellet has been designed to limit dispersal of the material into the environment.

Several accidents can occur in a space missions. Typical phases for deep space exploration missions (interplanetary mission) consists on: phase 1, called as ascent, begins with litoff of the Space Shuttle vehicle from launch pad, and then continues until the Solid Rocket Boosters are jettisoned some time after; phase 2, Second stage. This phase includes the first burn of the Orbital Maneuvering System (OMS) engines. The Shuttle main engine cutoff is included in this phase; phase 3, on Orbit, starting with the first burn of the OMS (OMS-1) and ends when the payload are deployed form the Orbiter. The phase include the first and second burns of the OMS (OMS-1 and OMS-2) for following the correct orbit and circularization; phase 4, Payload deploy, when reach the Earth escape velocity; phase 5, Maneuvers. To make possible some outer missions, is needed Gravitational Assist Maneuver, to obtain an impulse on the Spacecraft using the rotation energy of the planet. Critical issue is an Earth Gravity Assist, because the SC come back to the Earth; and a possible reentry (phase 6), exclusively for missions which ends with an spacecraft on an Earth reentry.

Various consequences could result from the accident environments that have been defined for the safety evaluation in the Final Safety Analysis Report (FSAR). In phase 1, the possible accidents resulting from Solid Rocket Booster (SRB) failures, either self induced or resulting from Range Safety destruct, can in certain instances lead to damaged GPHS modules with subsequent release of fuel due to: impact by SBR case fragments and subsequent impact agains ground surfaces or launch pad structures. In phase 2, vehicle breakup resulting from orbiter failures can result in reentry of the RTG and breakup of the GPHS modules on hard ground surfaces. In both phases 3 and 4, Shuttle failures can result in reentry of the SC (and RTGs) with subsequent breakup and release of the GPHS modules to impact on ground surfaces. In the case of the spacecraft should fail to reach escape velocity it would reenter into the Earth atmosphere. The heat of reentry would release the heat source from the generator and allow it to impact to the ground. The capsule would be exposed to reentry heating, Earth impact, and oxidation. If the heat shield were to fail, the unprotected capsule could fail in reentry and expose the bare fuel disks to the reentry and impact conditions (JPL, 1994; Richins, 2007). Additionally, in an Earth gravitational maneuver scenario, SC might reenter at very high velocity due to a spacecraft failure or a mission failure, such as puncture of the SC propellant tank by a micrometeoroid (space debris).

7. RTGs versus solar arrays

In regions on the space near Sun, NASA has historically used a few solar electric power systems such as solar panels. Several mission such as Mars Observer, the Viking Orbiters and Mariners missions were solar powered missions. For improving the systems efficiency, the Mars Global Surveyor used solar power with gallium-arsenide cells (JPL, 1994).

For outer planet missions, NASA has used radioisotope thermoelectric generators for the Cassini spacecraft. High electrical power for mission science requirements in powering the instruments and communication systems makes the RTG systems better option than solar arrays. The low efficiency of the solar cells for distances beyond Jupiter is an important drawback. Further, the spacecraft must be as lighter as possible. The size of the theoretical arrays of solar panels to obtain the power required for all sciences systems would be very large, increasing the spacecraft mass.

As regards on the solar cell technology, the actual production efficiencies of advanced solar cells have historically lower than research findings. The high-efficiency ESA solar cell devices are relative thick and heavy compared to the usual solar cells. Further, these advanced cells would be radiation sensitive. Solar-powered Juno mission will be launched in August on 2011, to study Jupiter. The spacecraft avoid the intense radiation belts using a innovative polar orbit, obtaining a great visibility for both the solar light arriving from the sun and communications.

Large solar arrays would severely impact the design, mass and operation of the spacecraft. This structure would have to be deployable, i.e. it could fit inside the rocket payload, and then unfold once the SC reached the outer planet. The mechanical components to fold and unfold the arrays would increase notably the size and mass on the SC. The long solar arrays would also severely complicate the stability on the trajectory and the attitude for scientific observations and data transmission to the Earth. Large spacecraft size, indeed, would make the maneuvers slower, which is critical for scientific data collection.

The electrical power requirements of the spacecraft for science instruments and telecommunications, lunch mass, and mission lifetime are all of critical concern in choosing the electrical power source.

8. Conclusion

In space application, Radioisotope Power Systems takes some advantages over solar panels. In several space operations there are long periods of darkness, and RPS will be the best actual technology. For outer planet missions, RTGs are more useful than solar panels to generate electric power for feeding communication systems and scientific instruments on the spacecraft. Additionally, there are new space technologies that use natural resources with/without radioisotope power systems. Future mission such as Europa Jupiter System Mission (EJSM), which is a joined NASA/ESA mission, will intend to study Jovian system, focusing two particular Jovian moons. NASA-led will use one type of RPS on Jupiter Europa Orbiter (JEO) to reach Europa, whereas ESA will consider solar arrays for Ganymede exploration. NASA's Juno mission will use solar panels for Jovian system exploration, in spite of the low solar light reaching Jupiter. The JEO spacecraft is designed to meet the planetary protection requirements. The flight system will use five multi-mission RTGs (MMRTG) to generate ~ 540 W of electrical power at the end of the mission. The high radiation environment (>50 the dose supported of Juno mission) makes the RPS more useful than solar array, because of the low solar wind reaching Jupiter. Waste heat from the MMRTGs would be used for thermal control in order to reduce electrical power.

Safety analysis of RPS requires a combination of deterministic and probabilistic steps to accurately predict the probability of system failure. The system failure is defined as rupture of one or more of the internal containment capsules surrounding the radioisotope fuel. To reduce the accident probability, we would have to identify among credible accidents, and analyze typical accident scenarios and consequences for overall flight phases of the spacecraft. The Launch Accident Scenario Evaluation Program (LASEP) computer program analyze the overall response of the GPHS-RTG in the various on-pad and near-pad launch accidents.

Actual high-magnitude earthquakes events occurred in Japan in 2011, has severally damaged the Fukushima reactor. This marks the difficult to change the public opinion about nuclear energy. Besides, the low disposal of Plutonium-238 is a serious drawback. The reestablishment of this man-made radioisotope production will be more difficult with these events. For using less plutonium than required, RPS efficiency must improve. Using low-conductivity materials and high thermoelectric rating, Z, RPS efficiency would improve. A high-efficiency Stirling-type system would give an apparent mass/power benefit, as well as using less plutonium for a similar power output. If we want to continue using RPS with Plutonium-238 as fuelling, we have to develop more high-efficiency systems, avoiding vibrations on the attitude on the spacecraft, as itself occurs with dynamic-conversion system. The current RPS power conversion efficiency is not too high. It is also required lower cost power systems.

Tethers might be used as alternative to solve the severe power generation problem. An electrodynamic tether, which is a very long wire capable to generate the suggested power, might radiate waves to satisfy communication requirements itself (Sanchez-Torres et al., 2010). The large electromotive force produced by the tether moving in some plasma ambient near the planet generate induced current and then electric power (Sanmartin et al., 1993). Tethers might be very useful for generating electric power both in Low Earth Orbit (high plasma density and moderate magnetic field) and in Jovian conditions (low plasma density and high magnetic field).

9. Acknowledgment

This work was supported by the Ministry of Science and Innovation of Spain (BES-2009-013319 FPI Grant).

10. References

Abelson, R. et al. (2004). Enabling Exploration with Small Radioisotope Power Systems, *JPL Pub 04-10,* Avalible from http://hdl.handle.net/2014/40856

Bennett, G. et al. (June 2006). Mission of Daring: The General-Purpose Heat Source Radioisotope Thermoelectric Generator, *4th International Energy Conversion Engineering Conference and Exhibit,* San Diego, California, Avalible from http://www.fas.org/nuke/space/gphs.pdf

Brown, C. (2001). Elements of Spacecraft Design, *AIAA Education Series,* Reston, Virginia, ISBN 1-56347-524-3

Furlog, R. & Wahlquist, E. (1999). U.S. Space Missions Using Radioisotope Power Systems, *Nuclear News,* pp. 26-34, Avalible from http://www.ans.org/pubs/magazines/nn/pdfs/1999-4-2.pdf

Griffin, M. & French, J. (2004). Space Vehicle Design, Second Edition, *AIAA Education Series,* Reston, Virginia, ISBN 1-56347-539-1

Hastings, D. & Garrett, H. (2004). Spacecraft-Environment Interactions, *Cambridge University Press,* Cambridge, UK, ISBN 0 521 60756 6

JPL (July 1994). Cassini Program Environmental Impact Statement Supporting Study. Volume 2: Alternate Mission and Power Study, *JPL Publication No. D-11777. Cassini Document No. 699-070-2,* Avalible from http://saturn.jpl.nasa.gov/spacecraft/safety/eisss2.pdf

Lange, R. & Carrol, W. (2008). Review of Recent Advances of Radioisotope Power Systems, *Energy Conversion and Management,* 49, pp. 393-401, ISSN: 0196-8904

National Reseach Council. Aeronautics and Space Engineering Board, Space Studies Board, Engineering and Physical Sciences (2009), Radioisotope Power Systems: An Imperative for Maintaining US Leadership in Space Exploration, *Tech. Rep.*, The *National Academic Press*, Washington, D.C. , Avalible from http://www.nap.edu/catalog/12653.html

Richins, W. & Lcay, J. (June 2007). Safety Analysis for a Radioisotope Stirling Generator, *Proceedings of Space Nuclear Conference, Paper 2024,* Avalible from http://georgenet.net/misc/rtg/Safety%20Analysis%20for%20RTG.pdf

Ritz, F. & Peterson, G. (2004). Multi-Mission Radioisotope Thermoelectric Generator (MMRTG) Program Overview, pp. 2950-2957, *IEEE Aerospace Conference Proceedings,* ISSN: 1095-323X

Sanchez-Torres, A.; Sanmartin, J., Donoso, J., Charro, M. (2010). The radiation impedance of electrodynamic tethers in a polar Jovian orbit, *Advances in Space Reseach,* Vol. 45, pp. 1050-1057. ISSN: 0273-1177

Sanmartin, J.; Martinez-Sanchez, M., Ahedo, E. (19930). Bare Wire Anode for Electrodynamic Tethers, *J. of Propulsion and Power,* Vol 9, pp. 353-360, ISSN: 0748-4658

Sturgis, B. et al. (September 2006). Methodology Assessment and Recommandations for the Mars Science Laboratory Launch Safety Analysis, *Sandia Report Sand* 2006-4563. DOI: 10.2172/893553.

22

U.S. Space Radioisotope Power Systems and Applications: Past, Present and Future

Robert L. Cataldo[1] and Gary L. Bennett[2]
[1]*NASA Glenn Research Center*
[2]*Metaspace Enterprises*
USA

1. Introduction

Radioisotope power systems (RPS) have been essential to the U.S. exploration of outer space. RPS have two primary uses: electrical power and thermal power. To provide electrical power, the RPS uses the heat produced by the natural decay of a radioisotope (e.g., plutonium-238 in U.S. RPS) to drive a converter (e.g., thermoelectric elements or Stirling linear alternator). As a thermal power source the heat is conducted to whatever component on the spacecraft needs to be kept warm; this heat can be produced by a radioisotope heater unit (RHU) or by using the excess heat of a radioisotope thermoelectric generator (RTG).

As of 2010, the U.S. has launched 45 RTGs on 26 space systems. These space systems have ranged from navigational satellites to challenging outer planet missions such as Pioneers 10/11, Voyagers 1/2, Galileo, Ulysses, Cassini and the New Horizons mission to Pluto. In the fall of 2011, NASA plans to launch the Mars Science Laboratory (MSL) that will employ the new Multi-Mission Radioisotope Thermoelectric Generator (MMRTG) as the principal power source.

Hundreds of radioisotope heater units (RHUs) have been launched, providing warmth to critical components on such missions as the Apollo 11 experiments package and on the outer planet probes Pioneers 10/11, Voyagers 1/2, Galileo and Cassini.

A radioisotope (electrical) power source or system (RPS) consists of three basic elements: (1) the radioisotope heat source that provides the thermal power, (2) the converter that transforms the thermal power into electrical power and (3) the heat rejection radiator. Figure 1 illustrates the basic features of an RPS.

The idea of a radioisotope power source follows closely after the early investigations of radioactivity by researchers such as Henri Becquerel (1852-1908), Marie Curie (1867-1935), Pierre Curie (1859-1906) and R. J. Strutt (1875-1947), the fourth Lord Rayleigh. Almost 100 years ago, in 1913, English physicist H. G. J. Moseley (1887-1915) constructed the first nuclear battery using a vacuum flask and 20 mCi of radium (Corliss and Harvey, 1964, Moseley and Harling, 1913).

After World War II, serious interest in radioisotope power systems in the U.S. was sparked by studies of space satellites such as North American Aviation's 1947 report on nuclear space power and the RAND Corporation's 1949 report on radioisotope power. (Greenfield,

1947, Gendler and Kock, 1949). Radioisotopes were also considered in early studies of nuclear-powered aircraft (Corliss and Harvey, 1964).

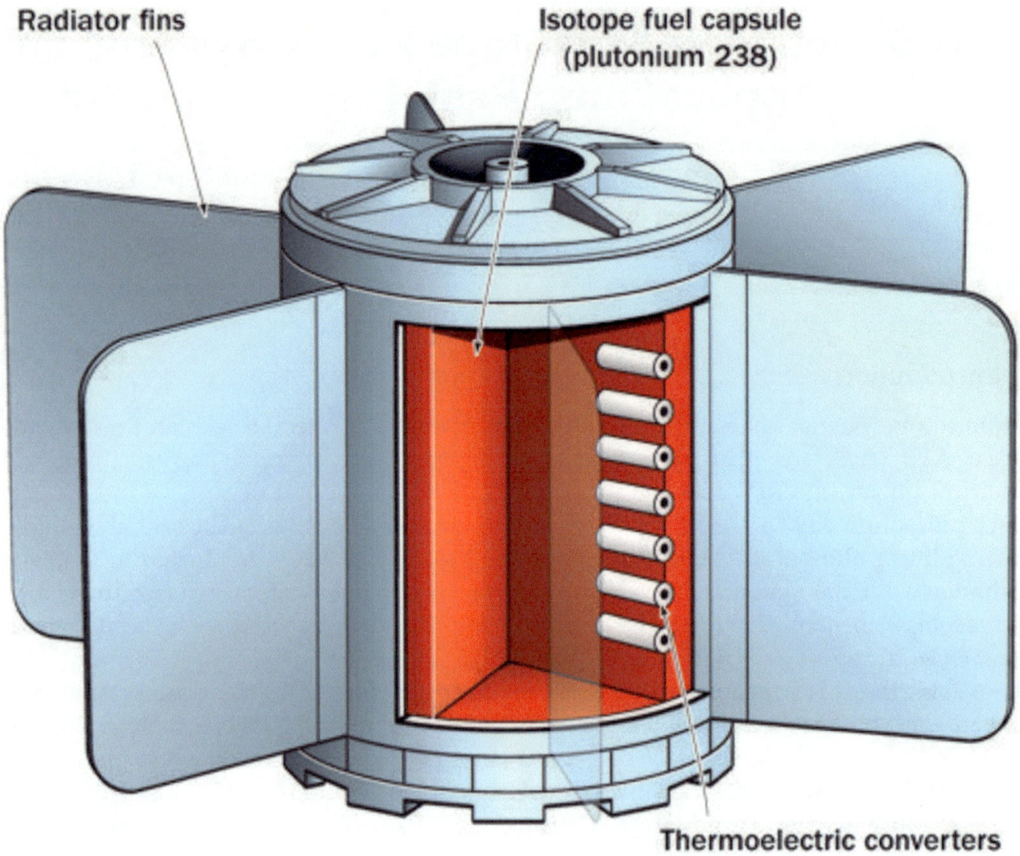

Fig. 1. Cutaway view of a radioisotope power source (RPS) (Image credit: DOE).

In 1951, the U.S. Atomic Energy Commission (AEC) signed several contracts to study a 1-kWe space power plant using reactors or radioisotopes. Several of these studies, which were completed in 1952, recommended the use of RPS (Corliss and Harvey, 1964). In 1954, the RAND Corporation issued the summary report of the Project Feedback military satellite study in which radioisotope power was considered (Lipp and Salter, 1954).

Paralleling these studies, in 1954, K. C. Jordan and J. H. Birden of the AEC's Mound Laboratory conceived and built the first RTG using chromel-constantan thermocouples and a polonium-210 (^{210}Po or Po-210) radioisotope heat source (see Figure 2). While the power produced (1.8 mWe) was low by today's standards, this first RTG showed the feasibility of RPS. A second "thermal battery" was built with more Po-210, producing 9.4 mWe. Jordan and Birden concluded that the Po-210 "thermal battery" would have about ten times the energy of ordinary dry cells of the same mass (Jordan and Birden, 1954).

Fig. 2. The first radioisotope thermoelectric generator (RTG).
Figure from the Jordan and Birden 1954 report via (Corliss and Harvey, 1964).The heat source consisted of a 1-cm-diameter sphere of 57 Ci (1.8 Wt) of ^{210}Po inside a capsule of nickel-coated cold-rolled steel all inside a container of Lucite. The thermocouples were silver-soldered chromel-constantan. The "thermal battery" produced 1.8 mWe.

2. Early SNAP program

The AEC began the Systems for Nuclear Auxiliary Power (SNAP) program in 1955 with contracts let to the Martin Company (now Teledyne) to design SNAP-1 and to the Atomics International Division of North American Aviation, Inc. to design SNAP-2. (Under the AEC nomenclature system, the odd-numbered SNAPs had radioisotope heat sources and the even-numbered SNAPs had nuclear fission reactor heat sources.) SNAP-1 was to provide 500 We using the then readily available fission product radioisotope cerium-144 (^{144}Ce) (Corliss and Harvey, 1964).

The Martin Company began with a 133-We RPS design using ^{144}Ce as the radioisotope fuel and a Rankine thermal-to-electric conversion system. From this came the 500-We SNAP-1 RPS design based on ^{144}Ce fuel and a Rankine conversion system (see Figure 3) (Corliss and Harvey, 1964). The use of a dynamic conversion system in the first RPS is a key historical fact in understanding the current focus on developing an Advanced Stirling Radioisotope

Generator (ASRG) (see Section 10). Depending on the design, dynamic conversion systems can provide double, triple and even quadruple the efficiency of state-of-practice thermoelectric conversion systems which means much less radioisotope fuel would be used to achieve the same electrical power (or, conversely, much more electrical power can be produced for the same quantity of radioisotope fuel used in an RTG).

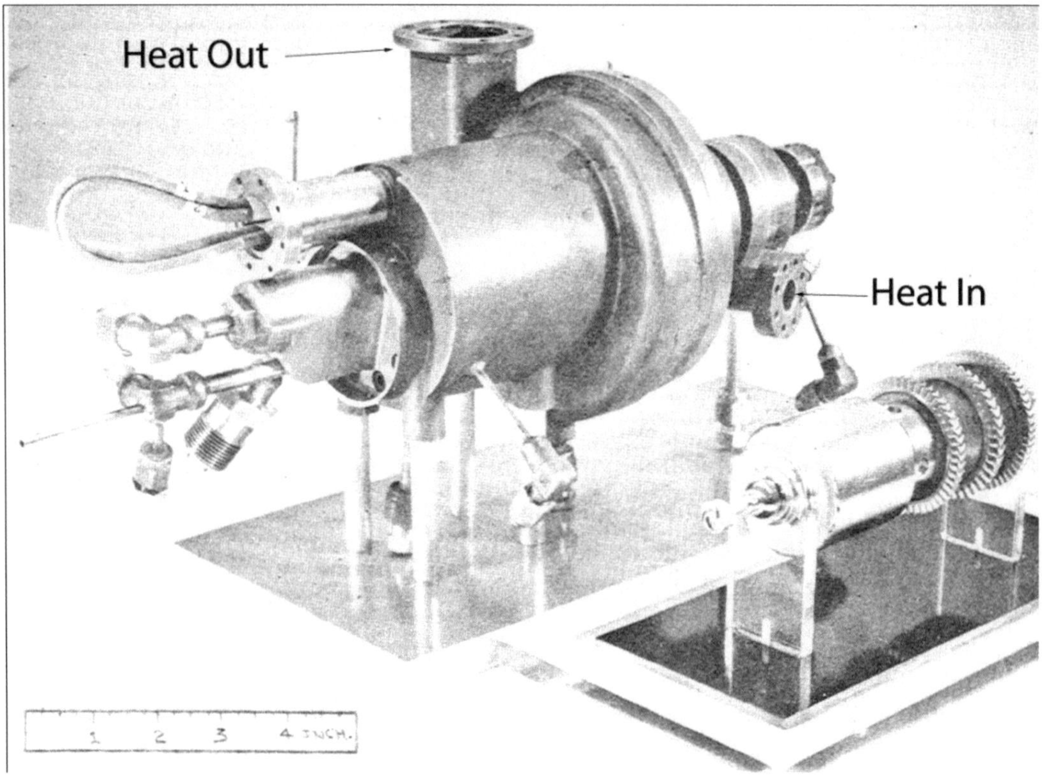

Fig. 3. SNAP-1 turbomachinery package with the shaft assembly shown separately, ruler dimensions are in inches (TRW via Corliss and Harvey, 1964).

In parallel with the SNAP-1 program a series of radioisotope power sources were studied under the umbrella of the SNAP-3 program that was based largely on using thermoelectric elements in the converter. The early SNAP-3 generators were to use ^{210}Po as the fuel but by the late 1950s it was clear that sufficient quantities of ^{238}Pu would be available to provide the fuel for small RTGs. Plutonium-238 provided a number of features that made it more attractive than ^{144}Ce or ^{210}Po, including a longer half-life (87.7 years) and a more benign radiation emission (alpha particles, which can be stopped by material as thin as a sheet of paper) (Corliss and Harvey, 1964).

Safety is the principal design requirement in the use of RPS, so the heat source is designed to contain or immobilize the fuel throughout a range of postulated accidents such as explosions and atmospheric reentries. Over the years this safety design work has led to the development of the general-purpose heat source (GPHS) module, which is the basic building block of U.S. RPS (Bennett, 1995).

All of the U.S. RPS that have flown have been either RTGs or RHUs, (see Fig. 4).

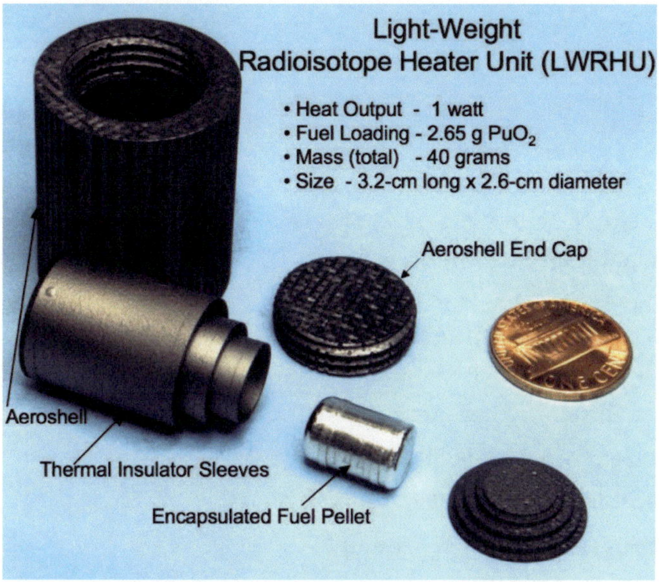

Fig. 4. Light-Weight Radioisotope Heater Unit (LWRU) (DOE)

As of 2010, as shown in Table 1, the U.S. has launched 45 RTGs, hundreds of RHUs and one space nuclear fission reactor. Of the RTGs flown, two different types of thermoelectric materials have been employed: telluride-alloy based or silicon-germanium-alloy based. The following sections will discuss these RTGS to be followed by sections discussing current efforts in radioisotope power sources.

3. The early telluride-based RTGs

The initial and current thermoelectric material of choice is based on telluride technology alloyed with lead (Pb-Te) that, to a first approximation, can be used from room temperature to about 900 K before materials properties become an issue. Above 900 K, the U.S. has had great success with a silicon-germanium alloy (Si-Ge) that has operated exceedingly well at temperatures of about 1300 K.

For the upcoming Mars Science Laboratory (MSL) mission, the U.S. will use a telluride-based thermoelectric material because it meets the requirements of being able to operate both in space on the way to Mars and on the surface of Mars with its dusty, cold, carbon dioxide atmosphere (see Section 8). The successes of the earlier (1976 era) Viking Mars Landers 1 and 2 using SNAP-19 telluride-based technology support this decision.

3.1 SNAP-3B RTGs

The SNAP-3B RTG evolved out of the overall SNAP-3 program with the goal of providing 2.7 We to the U.S. Navy's Transit 4A and Transit 4 B navigational satellites. In particular, the SNAP-3B RTGs were to provide power to the crystal oscillator that was the heart of the electronic system used for Doppler-shift tracking, a precursor of today's global positioning system (Dick and Davis, 1962, JHU/APL, 1980). Both RTGs provided power to their respective spacecraft for over 10 years (Bennett, et al., 1983). Figure 5 shows models of the SNAP-3B RTG and the successor SNAP-9A RTG.

Transit Navy Navigational Satellites

- Transits 4A and 4B (1961) SNAP-3B (2.7 We)
- Transits 5BN-1, 5BN-2 (1963) and **5BN-3 (1964)* SNAP-9A (>25 We)
- Transit TRIAD (1972) Transit-RTG (35 We)

SNAPSHOT Space Reactor Experiment

- SNAP-10A nuclear reactor (1965) (≥500 We)

Nimbus-B-1 Meteorological Satellite

- **SNAP-19B RTGs* (1968) (2 @ 28We each)

Nimbus-3 Meteorological Satellite

- SNAP-19B RTGs (1969) (2 @ 28 We each)

Apollo Lunar Surface Experiments Packages

- Apollo 12 (1969), **13 (1970)*, 14 (1971), 15 (1971), 16 (1972), 17 (1972) SNAP-27 (>70 We each)

Lincoln Experimental Satellites (Communications)

- LES 8 and LES 9 (1976) MHW-RTG (2/spacecraft @ ~154 We each)

Interplanetary Missions

- Pioneer 10 (1972) and Pioneer 11 (1973) SNAP-19 (4/spacecraft @ ~40 We each)
- Viking Mars Landers 1 and 2 (1975) SNAP-19 (2/Lander @ ~42 We each)
- Voyager 1 and Voyager 2 (1977) MHW-RTG (3/spacecraft @ >156 We each)
- Galileo (1989) GPHS-RTG (2 @ 287 We each)
- Ulysses (1990) GPHS-RTG (282 We)
- Cassini (1997) GPHS-RTG (3 @ >290 We each)
- New Horizons (2006) GPHS-RTG (1 @ 245.7 We)

(Spacecraft/Year Launched/Type of Nuclear Power Source/Beginning-of-Mission Power)
Note: SNAP is an acronym for Systems for Nuclear Auxiliary Power
MHW-RTG = Multi-Hundred Watt Radioisotope Thermoelectric Generator
GPHS-RTG = General-Purpose Heat Source Radioisotope Thermoelectric Generator
* Denotes system launched but mission unsuccessful

Table 1. Uses of Space Nuclear Power By The United States

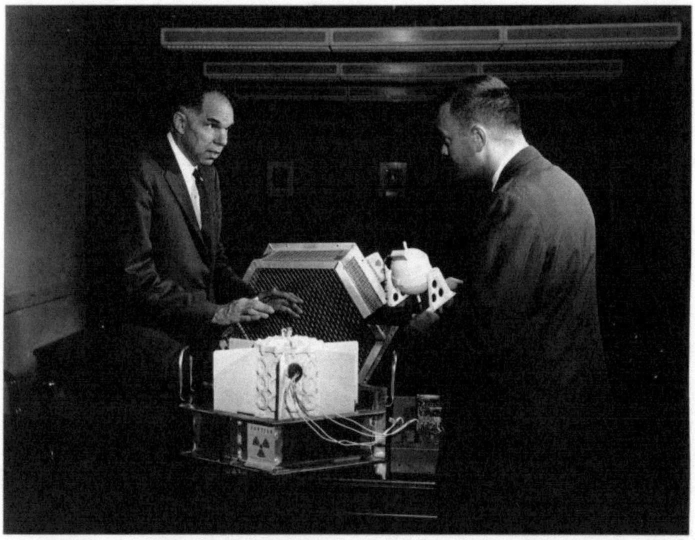

Fig. 5. Nobel Laureate Glenn T. Seaborg, Chairman of the U.S. Atomic Energy Commission, with his hands over a model of the SNAP-9A RTG and program manager Major Robert T. Carpenter holding a model of the SNAP-3B RTG (circa 1963), (AEC). The SNAP-3B RTG produced 2.7 We in a package 12.1-cm in diameter and 14-cm high with a mass of 2.1 kg. The SNAP-9A RTG produced over 25 We at beginning of mission (BOM) within a mass of 12.3 kg and a main body that was 22.9 cm in diameter and 21.3 cm high. (Image credit: AEC)

3.2 SNAP-9A RTGs

The success of the SNAP-3B RTGs on Transits 4A and 4B gave the Johns Hopkins University Applied Physics Laboratory (JHU/APL) confidence to select the next-generation RTG, known as SNAP-9A, to provide all the power for its Transit 5BN-1 and 5BN-2 navigational satellites. The objective for each SNAP-9A was to provide 25 We at beginning of mission (BOM) at a nominal 6 V for five years in space after one year of storage on Earth. The two SNAP-9As showed that RTGs could be easily integrated into a spacecraft to provide all of the electrical power (Bennett, et al., 1984, JHU/APL, 1980). Figure 5 provides a size comparison between the SNAP-9A and its predecessor the SNAP-3B.

4. SNAP-19 RTGs

The development work on the SNAP-9A RTG provided the technology that led to the SNAP-19 RTGs which were the first use of nuclear power in space by NASA.

4.1 Nimbus III

In 1969, NASA successfully launched the Nimbus III meteorological satellite powered by two SNAP-19 RTGs and solar arrays. The two SNAP-19 RTGs, which produced 56.4 We at launch, provided about 20% of the total power of the spacecraft. Had the SNAP-19 RTGs not been onboard Nimbus III, the power would have fallen below the load line about two weeks into the mission because of solar array degradation (Bennett, et al., 1984).

4.2 Pioneers 10 and 11

In 1972, NASA began its exploration of the outer Solar System with the launch to Jupiter of Pioneer 10 powered by four SNAP-19 RTGs which produced a total of 161.2 We at BOM. The next year Pioneer 10 was followed by the Pioneer 11 spacecraft which was also powered by four SNAP-19 RTGs. In the cold, dark, radiation-rich environment of the Jovian system, nuclear power was the only viable option at that time. Because the SNAP-19 RTGs performed so well, NASA was able to retarget Pioneer 11 to go to Saturn after its flyby of Jupiter. Again, the RTGs performed very well, providing steady power to the spacecraft and its scientific instruments, thus allowing scientists their first close-up measurements of the second largest planet in the Solar System (Bennett, et al., 1984).

4.3 Viking Landers 1 and 2

In anticipation of the 200th anniversary in 1976 of the signing of the U.S. Declaration of Independence, NASA launched the two Viking missions in 1975, each launch carrying an Orbiter and a Lander. Each Lander was powered by two SNAP-19 RTGs specially modified to work on the surface of Mars (see Fig. 6). The 35-We Viking SNAP-19 RTGs contained a special dome allowing an interchange of internal gases (initial fill 90:10 helium-argon; reservoir fill 95:5 argon-helium) during operation on the surface of Mars. This allowed for reduced pre-launch temperatures and maximum power output on Mars. All four SNAP-19 RTGs easily met the 90-day operating requirement of the Landers and went on to power the Landers for up to six years giving scientists their first extraordinary in-situ views of the surface of Mars (Bennett, et al., 1984).

Fig. 6. Viking Lander model showing the location of the two SNAP-19 RTGs. The average power per RTG was 42.7 We at BOM. The overall RTG diameter (across fins) was 58.7 cm and the overall length was 40.4 cm. The mass was 15.2 kg. (Image credit: NASA/JPL/Caltech/ERDA/Teledyne)

The success of the Viking SNAP-19 RTGs was a key factor in the selection of the telluride-thermoelectric-based Multi-Mission Radioisotope Thermoelectric Generator (MMRTG) for the upcoming MSL mission (see Section 8).

5. Transit RTG

The successful use of the SNAP-9A RTGs on the Transit 5BN series of Navy navigational satellites led JHU/APL to use a new telluride-based RTG called the "Transit RTG" on its TRIAD navigational satellite. The Transit RTG was based on the SNAP-19 radioisotope heat source design although in this case radiatively coupled to a telluride-based thermoelectric converter instead of being conductively coupled as in the SNAP-19 and SNAP-9A RTGs.

The Transit RTG, which was designed to be modular, produced over 35 We at BOM within a mass of about 13.6 kg. The use of a lower hot-junction temperature (~674 K for the Transit RTG versus ~790+ K for the SNAP-19 RTGs) in a vacuum environment eliminated the SNAP-19 practice of using hermetic sealing and a cover gas to inhibit the sublimation degradation that could cause a reduction in cross section and subsequent increase in electrical resistance of the thermoelectric material. (Lowering the hot junction temperature is also one of the strategies adopted for the MMRTG.) While the TRIAD spacecraft had various problems, the Transit RTG operated well beyond its five-year requirement (Bennett, et al., 1984).

6. SNAP-27 on Apollo

For the Apollo missions to the Moon, RTGs were a natural choice to power scientific instruments during the long (14-Earth-day) lunar night. To provide this power, the U.S. Atomic Energy Commission (AEC) provided NASA with SNAP-27 RTGs built by General Electric (GE, now part of Lockheed-Martin). The SNAP-27 RTGs were designed to provide at least 63.5 We at 16 V one year after lunar emplacement. (In the case of Apollo 17, the requirement was 69 We two years after emplacement). Figure 7 shows Apollo 12 astronaut Alan L. Bean removing the SNAP-27 fuel-cask assembly from the Lunar Module on 19 November 1969. This was the first use of electricity-producing nuclear power on the Moon.

All five SNAP-27 RTGs (Apollo 12, 14, 15, 16, 17) exceeded their mission requirements in both power and lifetime thereby enabling the Apollo Lunar Surface Experiment Packages (ALSEPs) to gather long-term scientific data on the internal structure and composition of the Moon, the composition of the lunar atmosphere, the state of the lunar interior, and the genesis of lunar surface features (Pitrolo, et al., 1969, Bates, et al., 1979). On Apollo 11 the experiment package deployed on the lunar surface was named Early Apollo Scientific Experiments Package (EASAP) and consisted of the laser ranging retro-reflector (LRRR, also deployed on each following Apollo mission, and are still in use today) and the passive seismic experiments package (PSEP). The PSEP utilized 2 RHUs called the Apollo Lunar Radioisotopic Heater (ALRH) for thermal control (Apollo 11 Lunar Landing Mission Press Kit, 1969) and also had a solar array power system that lasted three weeks. The ALRHs contained ~34 gm of ^{238}Pu producing 15 W_{th} each. The subsequent PSEP stations utilized power from the SNAP-27 RTGs.

Fig. 7. Apollo 12 astronaut Alan L. Bean removing the SNAP-27 fuel-cask assembly from the Lunar Module. The SNAP-27 converter is shown in front of Bean ready to receive the fuel-cask assembly. (NASA)

Fig. 8. Artist's concept of a Voyager spacecraft flying by Jupiter and Saturn. The three MHW-RTGs are shown on the boom above the spacecraft. The average power of each MHW-RTG was 158 We. The overall diameter was 39.73 cm and the length was 58.31 cm. The average flight mass for a Voyager MHW-RTG was 37.69 kg.
(Image credit: NASA/JPL/Caltech)

7. Silicon-germanium RTGs

With NASA developing the higher-powered Voyager 1 and Voyager 2 spacecraft (see Figure 8) as the next generation of outer planet explorers the bar was raised for RTG performance. To meet this demand, the AEC funded GE (now part of Lockheed-Martin) to develop the Multi-Hundred Watt Radioisotope Thermoelectric Generator (MHW-RTG), which was based on the use of a silicon-germanium alloy. Silicon-germanium, as noted earlier, can be operated at higher temperatures (~1300 K) than the telluride-based thermoelectrics (~800-900 K). Higher temperatures mean higher heat rejection temperatures, which mean smaller radiators hence lower unit masses. Combining the higher temperature with multifoil insulation (instead of bulk insulation) and vacuum operation (instead of using a cover gas) can yield a specific power that is 40% to over 70% higher than that of a telluride-based RTG (Bennett, et al., 1984). The basic layout of a silicon-germanium RTG is shown in Figure 9.

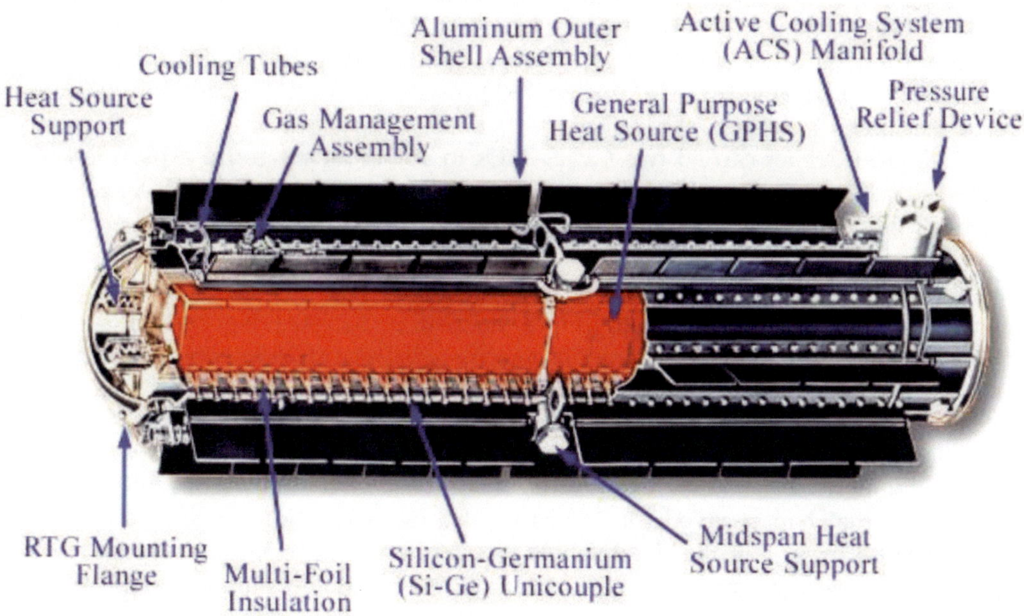

Fig. 9. Cutaway of the General-Purpose Heat Source Radioisotope Thermoelectric Generator (GPHS-RTG).

The GPHS-RTG can produce over 300 We at initial fueling. The overall diameter is 42.2 cm and the length is 114 cm. The mass is 55.9 kg (Image credit: DOE).

7.1 MHW-RTG

The MHW-RTG objective was to provide at least 125 We after five years in space. It was designed to produce at least 150 We at BOM, making it the highest-powered RTG at the time (1970s). Once the program was under way, the U.S. Air Force requested four MHW-RTGs for its communications satellites Lincoln Experimental Satellites 8 and 9 (LES-8/9) (Bennett, et al., 1984). As it turned out, LES-8/9 were launched prior to the Voyager launches (1976 versus 1977). Each LES carried two MHW-RTGs. The MHW-RTGs performed so well that the two communications satellites were used for years, including in the first Gulf War and to relay e-mail messages from stations in Antarctica.

Each Voyager spacecraft carried three MHW-RTGs (see Figure 8). The MHW-RTGs performed so well that Voyager 2 was retargeted after its flyby of Saturn (1981) to fly by Uranus and Neptune giving the human race its first close-up views of those distant worlds. Both Voyagers are still operating, almost 34 years after launch.

7.2 General-purpose Heat Source Radioisotope Thermoelectric Generator (GPHS-RTG)

For the Galileo and Ulysses missions the U.S. Department of Energy funded GE (now Lockheed Martin) to develop the General-Purpose Heat Source Radioisotope Thermoelectric Generator (GPHS-RTG), a power source essentially equivalent to two MHW-RTGs (see Figure 9). Where the MHW-RTG produced at least 150 We at BOM, the GPHS-RTG was capable of producing 300 We at BOM. Where each MHW-RTG had 312 silicon-germanium thermoelectric elements (called "unicouples"), each GPHS-RTG had 572 unicouples (Bennett, et al., 1984, Bennett, et al., 2006).

NASA's Galileo Orbiter carried two GPHS-RTGs to power its successful exploration of the Jovian system. The Ulysses spacecraft, which was built by the European Space Agency (ESA), carried one GPHS-RTG for its exploration of the polar regions of the Sun (Bennett, et al., 2006).

In 1997, NASA again used the GPHS-RTG, this time three of them to power the Cassini spacecraft that is still in orbit around Saturn. The GPHS-RTGs have performed so well that the mission has been extended several times (Bennett, et al., 2006). Figure 10 illustrates the progress that has been made in RTG performance – *in the span of a little over 30 years the power produced by a space RTG has increased over one-hundredfold!*

The most recent launch of the GPHS-RTG was in 2006 on the New Horizons spacecraft, which is traveling to Pluto. Because of the unavailability of a full complement of fresh Pu-238 fuel, the GPHS-RTG for New Horizons utilized some existing fuel that had decayed for 21 years since its production, yielding 245.7 We of power at BOM instead of the possible 300 We. Still, it is expected that the GPHS-RTG will provide sufficient power (~200 We) at the time of Pluto encounter to meet all of the mission's scientific and operational requirements. Once Pluto and its principal satellite Charon have been visited, New Horizons is designed to continue beyond to explore Kuiper Belt Objects (KBOs) (Bennett, et al., 2006).

Changes have been made in the general-purpose heat source (GPHS) that is the heart of the GPHS-RTG. For New Horizons, additional aeroshell material was added which increased the mass of the RTG. Additional material increases are planned for the GPHS modules to be used to power the MMRTG for MSL. While these changes have the effect of increasing the mass of the GPHS-RTG over the Galileo/Ulysses GPHS-RTGs there are design improvements, which could recreate the high specific power of the GPHS-RTG (Vining and Bennett, 2010).

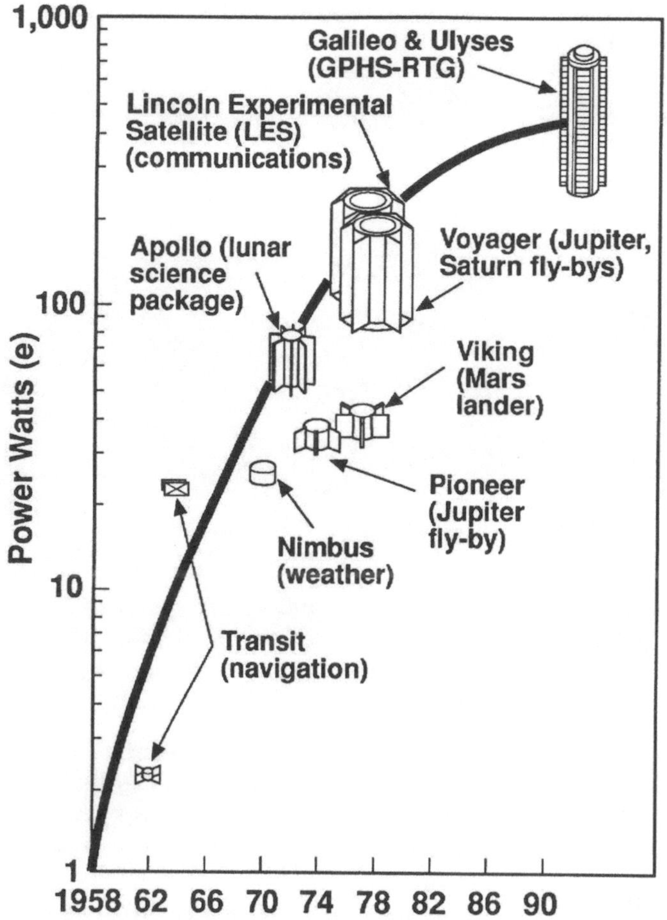

Fig. 10. Progress in RTG development. (Rockwell, 1992)

8. Multi-mission Radioisotope Thermoelectric Generator (MMRTG)

Following the successes of such flagship missions as Galileo and Cassini, NASA turned its attention to providing smaller "faster, better, cheaper" science spacecraft. In looking for an RPS which would satisfy that mandate along with being able to operate both in space and on the surface of a planetary body (e.g., Mars), a joint NASA/DOE team recommended development of the Multi-Mission Radioisotope Thermoelectric Generator (MMRTG) along with the development of the higher efficiency Advanced Stirling Radioisotope Generator (ASRG) (see Section 10) (unpublished Report of the RPS Provisioning Strategy Team, 2001).

The MMRTG, built by Rocketdyne and Teledyne, is based on the telluride thermoelectric technology used in the SNAP-19 RTG program which had shown that it could work in space (Nimbus III, Pioneers 10/11) and on a planetary surface (Viking Landers 1 and 2). The first mission to employ the MMRTG will be the Mars Science Laboratory (MSL), whose rover has been named "Curiosity" (see Figure 11). The 900-kg MSL is scheduled to be launched in the late fall of 2011 to arrive at Mars in August 2012.

Fig. 11. Artist's concept of the Mars Science Laboratory (MSL) Curiosity rover with the MMRTG shown attached to the back end (right side in the picture). MSL is ~3 m long (not including the arm), 2.7 m wide, 2.1 m tall with a mass of 900 kg. The arm can reach about 2.1 m. (Image credit: NASA/JPL/Caltech).

The overarching science goals of the MSL mission are to search for clues about whether environmental conditions (such as the existence of water for significant periods) could support microbial life today or in the past, and to assess whether the environment has favored the preservation of this evidence. MSL will be the first interplanetary mission to use a sky crane to land and the first to use guided entry to land in a precise location. MSL is designed to last for one Mars year (~687 Earth days) and to travel 20 km during its prime mission.

The MMRTG is designed to provide about 110 We on the surface of Mars at 28 to 32 V. The conversion is achieved using 16 thermoelectric modules of 48 telluride-based thermoelectric elements (Hammel, et al., 2009). The MMRTG is designed to have a minimum lifetime of 14 years. The MMRTG employs a flexible modular design approach that would allow the MMRTG concept to meet the power requirements of a wide range of missions.

Figure 12 shows a cutaway of the MMRTG. The MMRTG gets its ~2-kWt of thermal power from eight GPHS modules, the same heat source technology that was successfully used in the GPHS-RTGs and is planned for use in the ASRG. Like the GPHS-RTG, the converter housing and the eight heat rejection (radiator) fins are made of aluminum. The core assembly with 16 thermoelectric modules, each containing 48 couples (see Figure 13), are located under the eight fins with eight pairs of two modules aligned axially (Hammel, et al., 2009).

The thermoelectric modules are spring loaded to enhance conduction of heat from the GPHS modules and to enhance conduction of heat from the cold junction of the thermoelectric elements into the module bar and then into the converter housing. A bulk insulation system composed of the material Min-K reduces heat losses, in effect forcing the heat to travel through the thermoelectric elements. To enhance reliability the thermoelectric couples are electrically arranged in series and parallel. This redundant arrangement prevents loss of power should one or even several thermoelectric elements fail (Hammel, et al., 2009).

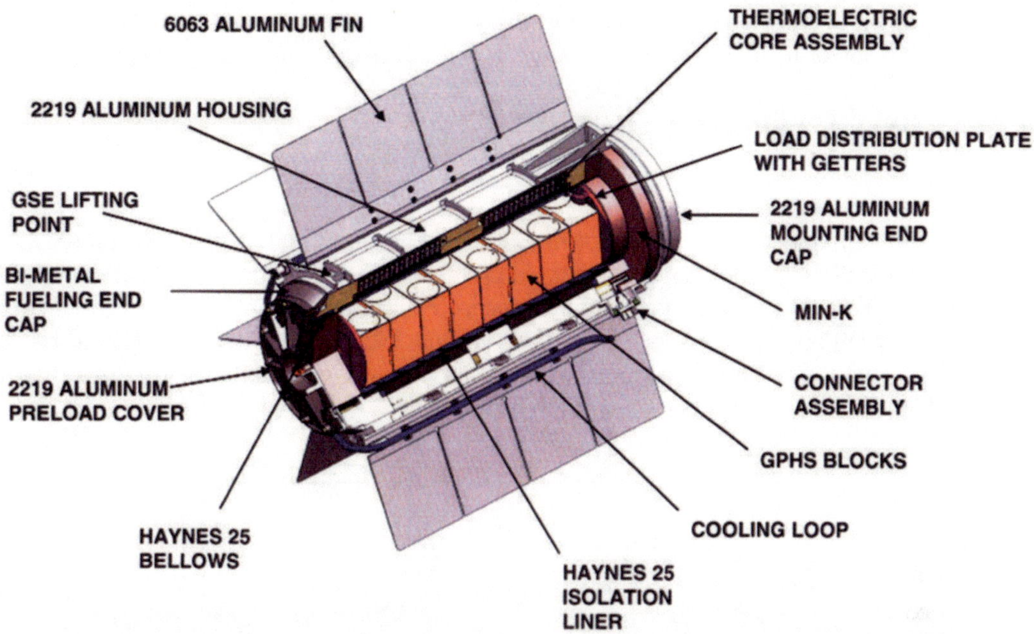

Fig. 12. Cutaway of the Multi-Mission Radioisotope Thermoelectric Generator (Hammel, et al., 2009) The MMRTG is designed to produce ~110 We at BOM with a mass of ~45 kg. The MMRTG is about 64 cm in diameter (fin-tip to fin-tip) by 66 cm long. (Image credit: DOE).

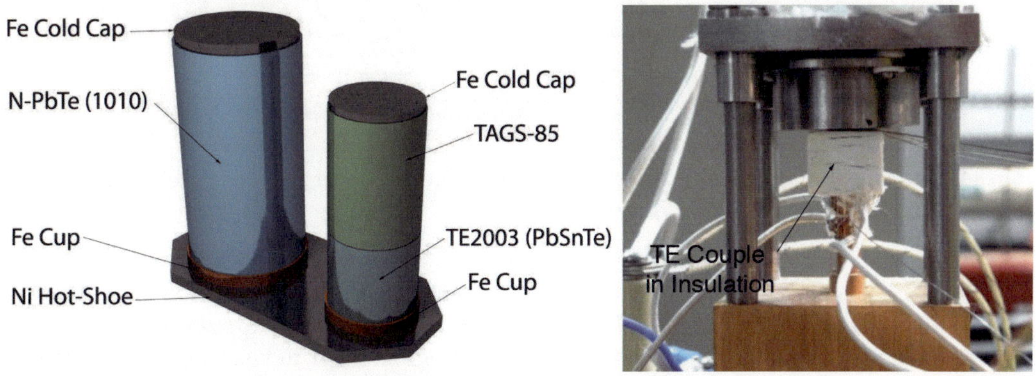

Fig. 13. The MMRTG Thermoelectric (TE) Couple in illustration and in a Test Fixture (Image credit: NASA/JPL/Caltech)

Like the SNAP-19 RTGs, the MMRTG is a sealed RTG with a cover gas. (The MHW-RTGs, GPHS-RTGs and Transit RTG were operated in a vacuum.) The heat source is sealed from the converter by a thin metal liner. Helium buildup from the natural decay of the Pu-238 fuel is prevented by venting it directly to the exterior of the MMRTG. The hermetically sealed converter contains an argon cover gas that reduces parasitic heat losses and protects the thermoelectric elements. With this venting and cover gas arrangement the MMRTG can operate in space or in an atmosphere (e.g., the surface of Mars or Titan) (Hammel, et al., 2009).

Modeling of the MMRTG performance indicates that the MMRTG will be able to provide the necessary power to enable MSL to achieve its objectives (Hammel, et al., 2009).

9. Dynamic Isotope Power System (DIPS)

The Dynamic Isotope Power System (DIPS) program was initiated in 1975 to provide increased power from radioisotope heat sources by using more efficient dynamic conversion systems (Brayton and Rankine). The precedent had been established in the 1950s with the SNAP-1 program (see Section 2) with its mercury Rankine conversion system and the SNAP-2 (3 kWe) and SNAP-8 (30 to 60 kWe) mercury Rankine space reactor programs. In terms of mass and specific power DIPS fills the gap between RTGs and nuclear reactors; in short, it could be the next logical step for increased RPS power after RTGs. Figure 14 illustrates the basic features of a representative DIPS (either Brayton or Rankine) (Bennett and Lombardo, 1989).

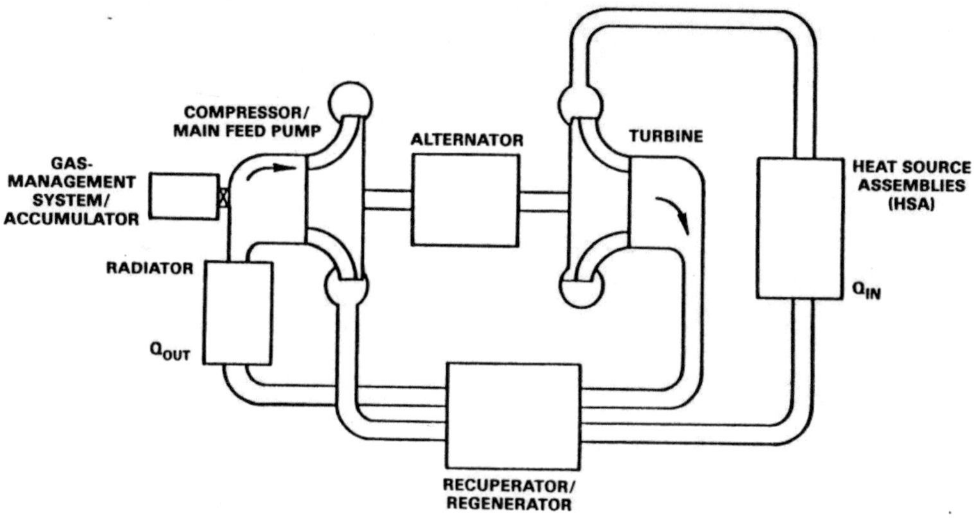

Fig. 14. Functional diagram of a generic Dynamic Isotope Power System (DIPS) (Bennett and Lombardo, 1989).

The original DIPS program was focused on producing a 1.3 kWe radioisotope power source with a mass of ≤204 kg using either a Brayton conversion system or an organic Rankine conversion system. The Brayton conversion system built upon the experience of NASA and its contractors (e.g., Garrett Corporation) dating from 1965 in developing a 2 to 10 kWe closed Brayton cycle (CBC) power system. In parallel, work on ground-based Rankine cycle systems led what was then Sundstrand Corporation to propose using Dowtherm A or toluene as a working fluid in order to avoid the corrosion issues with liquid-metal Rankine systems (Bennett and Lombardo, 1989).

Based on actual hardware tests, both the CBC and the organic Rankine cycle (ORC) were shown to be capable of meeting the DIPS goals. The organic Rankine cycle was chosen for further testing. The lack of a mission led to the termination of the program in 1980 but in 1986 the U.S. Air Force expressed interest in having a DIPS for its Boost Surveillance and Tracking System (Bennett & Lombardo, 1989). The DIPS program was restarted with Rocketdyne as the systems contractor and Sundstrand and Garrett as subcontractors.

This time the CBC was chosen and Rocketdyne developed a basic 2.5-kWe DIPS module that could be used in space or on planetary surfaces. Figure 15 illustrates the basic components of a 2.5-kWe modular DIPS power conversion unit (PCU). While again changing national priorities did not allow DIPS to be developed into a flight system, the basic technology exists to provide an RPS with powers spanning the range from 2 to ≥10 kWe (Rockwell, 1992). Section 10 describes a lower power successor to DIPS, the Advanced Stirling Radioisotope Generator (ASRG) that will provide increased efficiency and use less Pu-238 fuel than existing RTGs

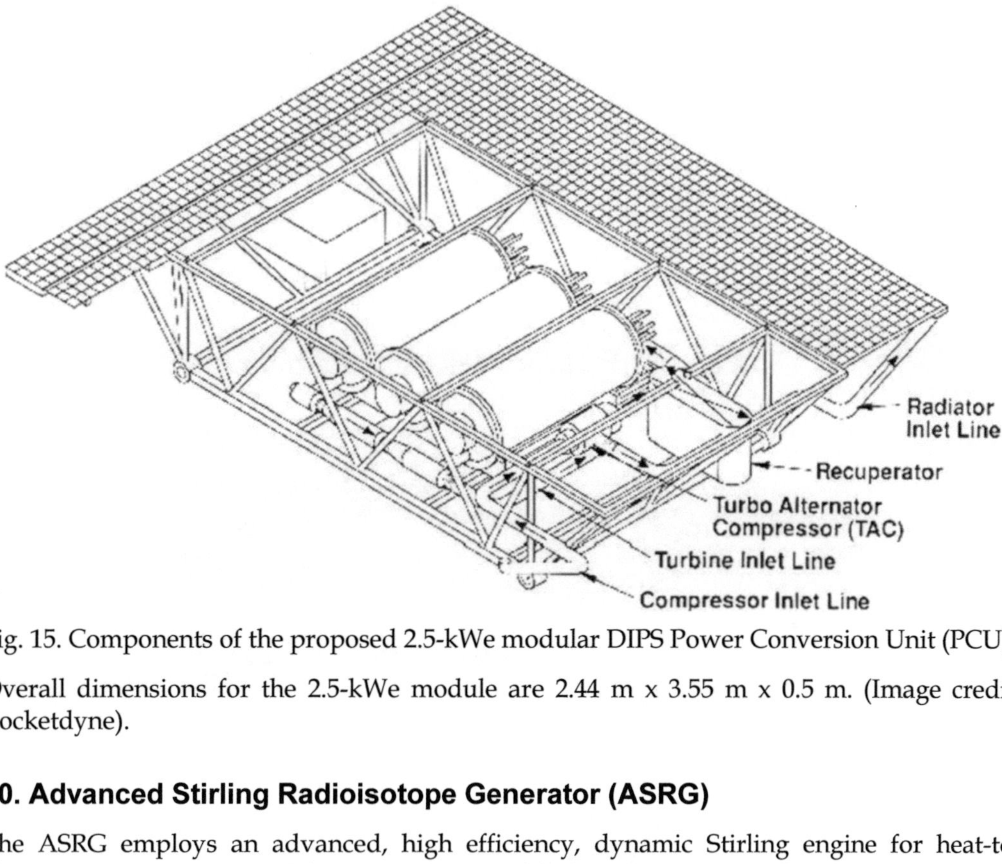

Fig. 15. Components of the proposed 2.5-kWe modular DIPS Power Conversion Unit (PCU).

Overall dimensions for the 2.5-kWe module are 2.44 m x 3.55 m x 0.5 m. (Image credit: Rocketdyne).

10. Advanced Stirling Radioisotope Generator (ASRG)

The ASRG employs an advanced, high efficiency, dynamic Stirling engine for heat-to-electric power conversion. This process is roughly four times more efficient than presently utilized thermoelectric devices. As a result, the ASRG produces comparable power to the MMRTG with only one quarter of the Pu-238, extending the supply of radioisotope fuel available for future space science missions. The higher efficiency provided by dynamic systems, such as the ASRG, could become an enabling power system option for higher power kilowatt class power systems envisioned for flagship class science spacecraft, large planetary rovers, and systems in support of human exploration activities.

The ASRG utilizes an advanced Stirling free-piston heat engine consisting of two major assemblies, the displacer and piston, that reciprocate to convert heat to electrical power as shown in Figure 16. Heat from the GPHS module is conductively coupled to the heater head (not shown). Helium is used as the working fluid and is hermetically contained within the convertor enclosure. The displacer shuttles helium between the expansion space where heat is received and compression space, where waste heat is removed. The changes in pressures and volumes of the convertor working spaces drive the power piston that reciprocates to produce AC electrical power via a permanent magnet linear alternator (Hoye, et al. 2011).

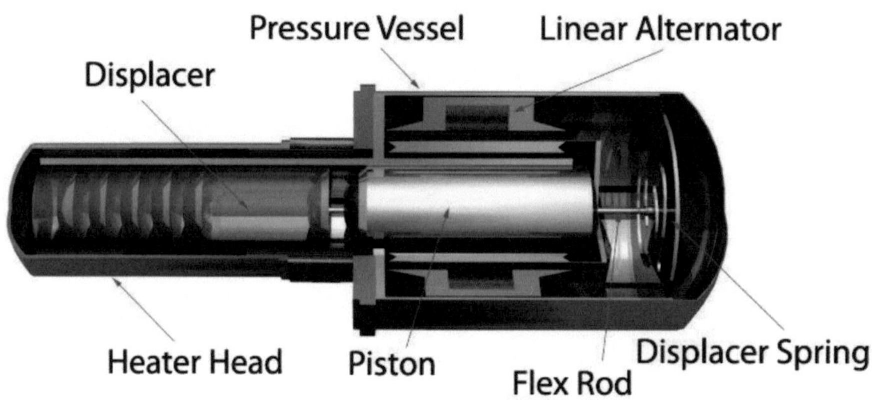

Fig. 16. Advanced Stirling Convertor (Image credit: Kristin Jansen, NASA Glenn).

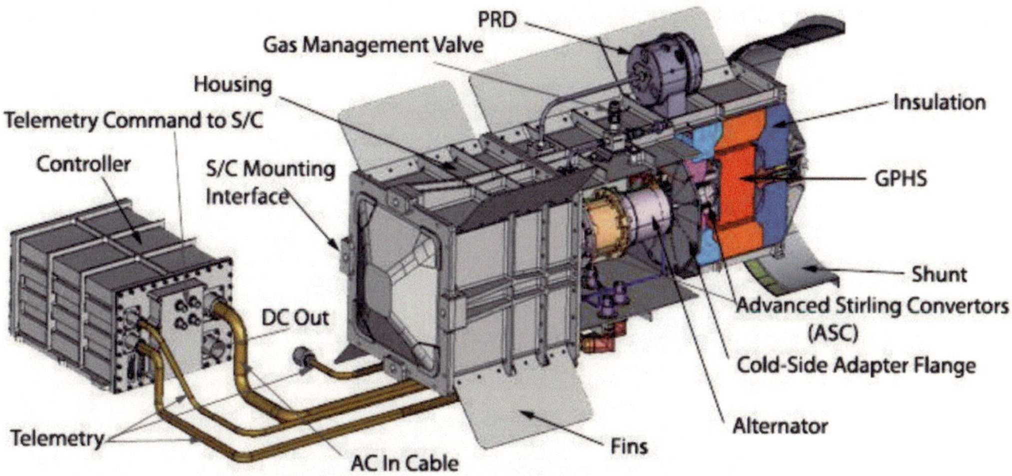

Fig. 17. Cutaway view of the ASRG (Image credit: Lockheed Martin).

Figure 17 shows one of the two GPHS heat sources, one for each advanced Stirling convertor (ASC). Each GPHS module is surrounded by insulation to minimize heat leakage thus maximizing heat input to the convertors. The displacer side of the ASC is toward the GPHS module and cold-side is attached to the housing via the cold-side adapter flange (CSAF). The housing and attached fins provide a view to the environment to maintain sufficient heat rejection. During ground storage and launch pad operations a slightly positive pressure of inert gas is maintained via the Gas Management Valve. The gas within the housing helps dissipate heat not rejected by the ASC via the CSAF. This gas is permanently vented to vacuum by the Pressure Relief Device (PRD) to achieve full operating power in space vacuum.

Operating frequency of the Stirling convertors is 102.2 Hz AC. The controller converts AC current to DC current for a typical 28-34 V spacecraft electrical bus. The shunt maintains a required load on the ASRG when it is not connected to a spacecraft, such as during storage or spacecraft integration. The controller also maintains synchronized displacer/piston movement of the two directionally opposed Stirling convertors to minimize induced disturbance to the spacecraft and its precision instrumentation. The ASRG is capable of producing 45% of total power should one ASC fail to operate. Health monitoring of the ASRG is provided by telemetry signals to the spacecraft and then transmitted back to Earth. The ASRG has an autonomous control system since space distances do not allow for direct operator control. The controller has electronics for each ASC plus a third redundant circuit to replace a failed card thus increasing overall system reliability.

The ASRG is being developed under joint sponsorship by NASA and DOE for potential flight on a future NASA mission opportunity. Projected mass of the flight unit is 32 kg or less. The flight units are anticipated to produce over 130 We in the vacuum of space and at an effective sink temperature of 4 K (deep space). Other applications on planetary bodies would either increase or decrease the power output depending on the temperature and atmosphere of the environment.

ASRG efforts leading up to its flight readiness began in 2000. An engineering unit (EU) was built in 2008 by Lockheed Martin Space Systems incorporating the advanced Stirling convertor manufactured by Sunpower. Characterization testing was performed for typical launch and space environments.

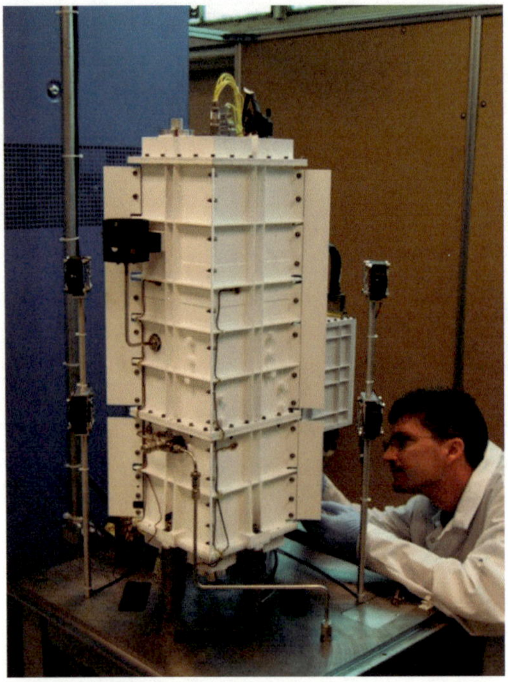

Fig. 18. ASRG engineering unit readied for extended testing at NASA Glenn Research Center (Photo Credit, NASA Glenn).

After successful characterization tests the EU was put on extended operation with electrically heated GPHS simulators as shown in figure 18. A cold gas is passed over the EU to maintain proper thermal operating conditions. The EU is planned to demonstrate 14,000 hours of operation in validating the design's viability for flight on a NASA science mission (Lewandowski and Schreiber, 2010). The EU has achieved over 11,000 hours (April, 2011) with a prototypical flight controller.

11. Future radioisotope technologies

In addition to the MMRTG and ASRG efforts are underway to develop other technologies such as advanced thermoelectric couples (ATEC), thermophotovoltaic (TPV) systems and the Stirling duplex system.

The ATEC effort is developing and demonstrating thermoelectric couples with efficiencies greater than 10% with degradation losses less than 1% per year. Part of this effort is developing high temperature complex advanced materials with twice the state of practice efficiency.

TPV uses photovoltaic cells tuned to certain spectra emitted by a radioisotope heat source. Development efforts include studies of the optical properties and optimization of the emitter, filters and collectors to achieve efficiencies of greater than 15% with low degradation rates.

The Stirling duplex concept combines a power convertor as in the ASRG with a thermodynamic Stirling cycle cooler. This concept provides both power in the range of 100-300 We and 1100 Wth cooling, allowing its potential use for missions in harsh high-temperature environments.

12. Human exploration missions

To this point, discussions have focused on lower-power science mission objectives. However, the concept of high efficiency energy conversion could find application in human exploration, particularly relating to the Moon and Mars. Multi-kilowatt power level radioisotope systems have application for the Moon due to its long 356-hour night period and on Mars due to its distance from the Sun, atmospheric dust attenuation and short winter day periods at higher latitudes.

Fig. 19. A concept for a multi-Kilowatt radioisotope power system deploying a fission reactor on the surface of Mars, (Artist concept credit: Bob Souls, John Frassanito and Associates, Courtesy of NASA).

Radioisotope power systems provide continuous power avoiding necessity of solar arrays and battery or fuel cell storage (for night time energy) and the wait to recharge them. Relative mission risk could also be reduced since solar array size for many of these applications are too large and therefore require stowage during mobility and subsequent deployment for recharging multiple times. For Mars, radioisotope systems could be used in several applications (Cataldo, 2009). Opportunities to travel to Mars occur every 26 months. Many human mission scenarios call for pre-deployment of equipment on the opportunity preceding the piloted flight to simplify launch and mission logistics. One concept for a radioisotope system is a power cart capable of several applications with mobility as a key feature. For example, supplying power to deploy a shielded fission reactor several kilometers from a habitat could be accomplished in several days without the use of large solar arrays. The power cart would have communications capability to Earth via orbiting assets to control its movements and its re-location to future sites.

Once the crew arrives on a subsequent opportunity, the power cart could be used with a pressurized rover. This scenario would allow the crew to perform long-range roving (over ~100's km) to significantly extend the scope of exploration from a single landing site. The power cart could also provide back-up power to the habitat should that become required. Should the base power system be solar instead of fission power a radioisotope-powered habitat back-up system could save significant mass during decreased array output during due to a global dust storm. A radioisotope power system could offer significant flexibility in mission planning for human missions as well as robotic science missions. In addition, since a crew would be available for repairs, maintenance or upgrades, the radioisotope fuel could be placed in a redundant cart with new conversion hardware extending the use of the fuel for many follow-on missions (Cataldo, 2009).

13.Conclusions

The U.S. has had a very successful 50 years of using RPS to power some of the most challenging and scientifically rewarding space missions in human history. These RPS have provided power at or above that required levels, and generally for longer than the original mission specification. RPS can truly be an enabling technology for both robotic probes and human exploration of the Solar System and beyond.

14. References

Apollo 11 Lunar Landing Mission Press Kit, NASA, Release NO: 69-83K, June, 1969

Bates, J. R., Lauderdale, W. W. & Kernaghan, ALSEP Termination Report, NASA, Reference Publication 1036, 1979

Bennett, G. L., Lombardo, J. J. & Rock, B. J. (1984). US Radioisotope Thermoelectric Generators in Space. *The Nuclear Engineer*, Vol. 25, No. 2, March/April 1984, pp. 49-58, ISSN 0262-5091. Reprinted from the paper "U.S. Radioisotope *Thermoelectric Generator Space Operating Experience* (June 1961 - December, 1982)", *18th Intersociety Energy Conversion Engineering Conference*, Orlando, Florida, 21-26 August 1983, American Institute of Chemical Engineers, New York, New York, 1983, ISBN 10 0816902534

Bennett, G. L. and Lombardo, J. J. (1989), The Dynamic Isotope Power System: Technology Status and Demonstration Program. Chapter 20, In: *Space Nuclear Power Systems* 1988, El-Genk, M.S. and Hoover, M. D., Orbit Book Publishing Company, ISBN-10 0894640291, Malabar, Florida, 1989

Bennett, G. L. (1995). Safety Aspects of Thermoelectrics in Space, In: *CRC Handbook of Thermoelectrics*, Rowe, D. M., CRC Press, ISBN-10 9780849301469, New York, 1995

Bennett, G. L. et al. (2006). Mission of Daring: The General-Purpose Heat Source Radioisotope Thermoelectric Generator (GPHS-RTG). AIAA 2006-4096, *4th International Energy Conversion Engineering Conference*, San Diego, California, 2006

Cataldo, R. L. (2009). Power Requirements for the NASA Mars Design Reference Architecture (DRA) 5.0. *Proceedings of Nuclear and Emerging Technologies for Space 2009*, Atlanta, GA, 2009

Corliss, W. R. & Harvey, D. G. (1964). *Radioisotopic Power Generation*, Prentice-Hall, Inc., Library of Congress Catalog Card Number 64-7543, Englewood Cliffs, New Jersey

Dick, P. J. & Davis, R. E. (1962). Radioisotope Power System Operation in the Transit Satellite. Paper No. CP 62-1173. *American Institute of Electrical Engineers Summer General Meeting*, Denver, Colorado, 17-22 June 1962

Gendler, S. L. & Kock, H. A. (1949). Auxiliary Power Plant for the Satellite Rocket: A Radioactive Cell-Mercury Vapor System to Supply 500 watts for Durations up to One Year, RAND Corporation, Santa Monica, California, 1949

Greenfield, M. A. (1947). *Studies on Nuclear Reactors, 6: Power Developed by Decay of Fission Fragments*, NAA-SR-6, North American Aviation, Inc., Los Angeles, California

Hammel, T. E., Bennett, R., Otting W. & Finale, S. (2009) Multi-Mission Radioisotope Thermoelectric Generator (MMRTG) and Performance Prediction Model. AIAA 2009-4576, *7th International Energy Conversion Engineering Conference,* Denver, Colorado, 2009

Hoye, T. J., Tantino, D. C., Chan, J. (2011), Advanced Stirling Radioisotope Generator Flight System Overview (2011). *Proceedings of Nuclear and Emerging Technologies for Space 2011*, Albuquerque, NM, 2011

Johns Hopkins University Applied Physics Laboratory (JHU/APL). (1980). *Artificial Earth Satellites Designed and Fabricated by the Johns Hopkins University Applied Physics Laboratory*, JHU/APL Report SDO-1600 (revised), 1980.

Jordan, K. C. & Birden, J. H. (1954), *Thermal Batteries Using Polonium-210*, MLM-984, Mound Laboratory, Miamisburg, Ohio, 1954

Lipp, J. E. and Salter, R. M., eds. (1954), *Project Feed-Back, Summary Report* (2 volumes) R-262, RAND Corporation, Santa Monica, California, 1954

Lewandowski, E.J. and Schreiber, J.G., (2010). Testing to Characterize the Advanced Stirling Radioisotope Generator Engineering Unit, *Proceedings of the Eighth International Energy Conversion Engineering Conference* 2010, American Institute for Aeronautics and Astronautics, Nashville, TN, 2010

Moseley, H. G. J. & Harling, J. (1913). The Attainment of High Potentials by the Use of Radium. *Proceedings of the Royal Society* (London), A, 88 (1913), p. 471

Pitrolo, A. A., Rock, B. J., Remini, W. C. & Leonard, J. A. (1969). SNAP-27 Program Review, Paper 699023 in *Proceedings of the 4th Intersociety Energy Conversion Engineering Conference*, held in Washington, D.C., 1969, American Institute of Chemical Engineers, New York

Report of the RPS Provisioning Strategy Team, 8 May 2001

Rockwell International. (1992). *Dynamic Isotope Power Systems (DIPS) for Space Exploration – Technical Information*, BC92-68, Rocketdyne Division, Canoga Park, California

Vining, C. B. & Bennett, G. L. (2010). Power for Science and Exploration: Upgrading the General-Purpose Heat Source Radioisotope Thermoelectric Generator (GPHS-RTG). AIAA-2010-6598, *8th International Energy Conversion Engineering Conference*, Nashville, Tennessee, 2010